# *IC Op-Amp Cookbook*

# IC Op-Amp Cookbook

*Third Edition*

WALTER G. JUNG

HOWARD W. SAMS & COMPANY
*A Division of Macmillan, Inc.*
*4300 West 62nd Street*
*Indianapolis, Indiana 46268 USA*

THIRD EDITION
FIFTH PRINTING—1989



International Standard Book Number: 0-672-22453-4

Library of Congress Catalog Card Number: 86-60253

Illustrated by: *Ralph E. Lund*
Designed by: *T. R. Emrick*
Cover art by: *Sherri L. VonHartman*

*Printed in the United States of America.*

# Contents

## Part III *Appendix*

# *Preface to Third Edition*

For this third edition of *IC Op-Amp Cookbook,* relatively little has changed in the book's overall general format, with regard to the second edition of 1980. After 20 years of the monolithic IC op amp and a dozen years since the book's first appearance, a high level of general maturity has been reached in the IC op-amp field. Over the years, the consistent feedback on this book has been that its success as a standard work lies largely in its practicality. So, this successful theme continues in this edition. However, this is not to say, by any means, that new features in the present edition are few in number.

In the introductory sections of the book, the main new feature is the substantial expansion of the IC op-amp-device family coverage within Chapter 2 to include a broad overview of the many significant new devices of the last few years. Both the device discussions, themselves, and the easy-to-use selection tables have been updated and expanded to reflect the most modern examples of current popular use. With the information of this chapter, the user can quickly zero in on a specific device for a given application type.

The remainder of the book, encompassing Chapters 4 through 10, focuses on applications examples as previously. In these sections are contained hundreds of circuits which have been found to be useful to the practicing circuit designer. These chapters include voltage/current regulators, references and power supplies, signal processing, log/antilog and multiplier circuits, amplifier techniques, comparator uses, integrators and differentiators, and signal-generation methods. Many new examples of useful circuits have been added within these categories, as well as updating the older circuits to reflect current devices. Throughout all sections, literature references have also been expanded and updated.

The book concludes with several appendix entries, which include manufacturers' part number systems, device source information, and a popular feature—selected data sheets for standard IC op-amp devices.

Users of the previous editions of this book will likely find this new edition of the *IC Op-Amp Cookbook* to be a useful update and expansion of their information resources. To all of these thousands of users, the author expresses thanks, along with best wishes towards enjoyment of the new edition. New users of op-amp technology should also find the book useful as a single reference, to both introduce the basic concepts as well as to give a broad perspective on typical examples of usage. Feedback from all readers is welcome towards future revisions.

WALTER G. JUNG

# *Preface to Second Edition*

The preface to the first edition is generally totally applicable to the new edition as well. However, in addition to the basic intent of the first edition, the following changes highlight the second edition.

Retaining the basic style and format of the first edition, the second edition is greatly expanded in terms of device coverage, in keeping with state-of-the-art advancements. The introductory sections of the book reflect this, outlining a wide gamut of bipolar and FET op-amp devices, of various performance capabilities, in both single and multiple formats. Although the proliferation of devices prevents widespread detail, the numerous charts of this section facilitate selection of an appropriate device.

The applications section of the book is revised somewhat in chapter format, but expanded considerably in terms of the range of applications. Including Chapters 4 through 10, topical headings are voltage and current regulators, signal processing, log and multiplier techniques, amplifier techniques, comparators, integrators and differentiators, and signal generation. The applications discussed are updated both in terms of their range and scope, and also in terms of the appropriate applicable devices, reflecting the latest technology.

First-edition sections on "Audio" and "Unique Devices" have been expanded into separate stand-alone texts so that greater emphasis might be placed on more general uses within this book.

The second edition concludes with an appendix that details part-number breakdowns for various manufacturers and source information and gives data sheets for a selected variety of devices.

Walter G. Jung

# *Preface to First Edition*

At the time this book was conceived, several texts and handbooks concerning operational amplifiers had already been published. Indeed, many of them are well accepted, and the reader will find them referenced at appropriate points within this one. No attempt is made to detract from these previous efforts; however, it can be said that none addresses itself specifically and wholly to the IC op amp, with an in-depth discussion of applications for specific standard devices. Therein lies the philosophy behind this book and why it is intended to be an appropriate complement to the already existing literature.

Depending on one's particular background, working environment, etc., we each tend to take a slightly different approach to a design problem. There are, however, certain ground rules that the greatest percentage of people will follow in arriving at solutions to these problems. In the use of operational amplifiers, the trend has been toward the IC op amp, and among the numerous types available, there are definite standards that fulfill the greatest percentage of needs. Recognizing the inherent economy and widespread appeal of the IC op amp, this book attempts to address the considerations in applying it with maximum effectiveness in a wide variety of circuits. The name chosen, *IC Op-Amp Cookbook,* is intended to convey this concept. But, as will be seen, the book is more than just a "cookbook"; it involves general applications considerations as well as a substantial amount of theory to support the applications illustrated.

Some insight into the rationale behind the format of the book is helpful in assessing its value to the individual reader. If one is to undertake the task of discussing the uses of IC op amps, there are certain general requirements that must be met. Two of the biggest considerations are style and content—decide what to say and how best to say it. Before getting into the actual contents, a word about the style chosen is appropriate. "Style," as referred to here, is not style in a grammatical sense but is the level on which the book is written. The author firmly believes that practical writing is more useful than pure theory to greater numbers of people. Thus, while there is theory in this book, the main emphasis is on what the theory means in practical applications. The real world of the IC op amp is much more concerned with the details of making a device work as intended rather than making equations balance. So the format chosen is one of "down-to-earth" discussions, as viewed from the *user's* standpoint. This is intended to appeal to users on various levels, from student to technician to design engineer, and, hopefully, to satisfy the needs of all. While a universally appealing book is perhaps somewhat unrealistic, the broadness of applicability can certainly be optimized by keeping everything as clear and understandable as possible while discussing problems meaningful to all users.

The "content" of this book also takes on the prerequisite of broad appeal. Thus, obvious inclusions are such op-amp standards as the 709, 101, and 741. But the scope of the book is not limited to general-purpose devices; it also encompasses other more-specialized devices—those that are optimized for certain specific performance parameters. The specialized types chosen here are also standard type numbers, as evidenced by broad industry acceptance.

With the complement of devices chosen, a wide range of applications is thus possible, with optimized performance available in terms of virtually any specific criterion. This factor should be of interest to the designer seeking high performance by using op amps. Throughout the book, the discussions are directed to indicate the optimum device for a specific use, and considerations for its application, such as speed, accuracy, dc stability, power consumption, and other performance parameters, are covered.

Given this general format, there is one additional consideration that is generally applicable to circuitry discussions. An applications book may be used in various degrees by different owners. A person new to a particular technique may want to study circuits and theory in some detail. For this type of use, the supportive theoretical background is intended, and appropriate references are listed for further reading in most instances. Others may desire a circuit to fulfill a certain function, and will make minimal use of the text. For this type of use, well-annotated schematics are provided, replete with values and suggestions for modifications, if desired. Wherever possible, the pertinent design equations, which summarize the function of the circuit, are also provided.

WALTER G. JUNG

# Acknowledgments

An undertaking of this nature is only successful with aid and input from various sources. The author has been most fortunate during the writing of this and the previous two editions of this book to have had the benefit of technical inputs from many sources, and this is gratefully acknowledged.

In the first edition, thanks were extended to Bob Dobkin, Carl Nelson, and Tom Frederiksen of National Semiconductor; to Bob Ricks of Fairchild Semiconductor; to Don Kesner and Don Aldridge of Motorola Semiconductor; to Don Jones of Harris Semiconductor; to Hal Wittlinger of RCA; to Werner Hoeft of Signetics; and to Bob Jones of AAI Corporation.

In the second edition, more thanks were added, to Lew Counts, Jeff Riskin, and Dan Sheingold of Analog Devices; to Dennis Haynes of Burr-Brown; to Dick Whitehead of Harris Semiconductor; to Bob Pease, Nello Sevastopoulos, and Jim Solomon of National Semiconductor; to George Erdi, Shelby Givens, and Donn Soderquist of Precision Monolithics; to Harry Gill and Jim Schmoock of Raytheon; to Guy Caputo of Signetics; and to Dale Pippenger of Texas Instruments.

For this new edition, the author would like to reiterate the thanks to all of those noted above, as much of the material which arose from their inputs still remains. In addition, new and/or renewed thanks go now also to Lew Counts, Dan Sheingold, Don Travers, and Scott Wurcer of Analog Devices; to Neil Albaugh of Burr-Brown; to Jim Williams of Linear Technology Corporation; to Bill Davis of Motorola; to Bob Pease and Hugh Wright of National Semiconductor; to Tom Cate and Andy Jenkins of Precision Monolithics; to Chuck Vinn of Raytheon; to Hal Wittlinger of RCA; to Tom DeLurio of Signetics; and to Dale Pippenger of Texas Instruments. All provided useful inputs of one form or another related to the book. The author was particularly grateful to receive countless helpful comments from Neil Albaugh and Bob Pease.

For the use of portions of their technical material, the aid of the following companies is gratefully acknowledged: Analog Devices, Burr-Brown, Linear Technology Corp., National Semiconductor Corp., Precision Monolithics, and RCA Solid State Division.

Finally, and most certainly not least, my sincere thanks to my wife, Anne, for her encouragement and support during another edition of this book.

W.G.J.

*To Anne, with my love and appreciation*

# *Introduction and Historical Background*

Before discussing modern operational amplifiers, we should first review the basics of what an operational amplifier (op amp) is. We will do this in two stages: (1) discuss what an op amp is in the *ideal* sense, and (2) discuss what an op amp *actually* is within the limitations of integrated circuits.

In many applications, the differences between ideal and actual (nonideal) operation will be so small that for all practical purposes they will be negligible. In other applications, the differences will be slight, yet still measurable. Often the overall performance can be enhanced by careful selection of the type of device and by optimizing the parameters that are under our control. This requires an understanding of both the basic concepts of op amps and the specific devices now available.

The original concept of the operational amplifier came from the field of analog computers, where operational techniques were used as early as the 1940s. The name *operational amplifier* derives from the concept of an extremely high-gain differential-input dc amplifier, the operating characteristics of which were determined by the feedback elements used with it. By changing the types and arrangement of the feedback elements, different analog *operations* could be implemented; to a large degree, the overall circuit characteristics were determined *only* by these feedback elements. Thus, the same amplifier was able to perform a variety of operations, and the gradual development of operational amplifiers grew into the beginning of a new era in circuit-design concepts.

Early operational amplifiers used the basic hardware of that era—the vacuum tube. Significantly widespread use of op amps did not really begin until the 1960s, when solid-state techniques were applied to op-amp circuit design. This was first evidenced by solid-state op amp modules, which realized the internal circuitry of the op amp in a discrete solid-state design. Then, in the mid 1960s, the first *integrated circuit* (IC) op amps were introduced. Within a few years, IC op amps became a standard design tool encompassing applications far beyond the original scope of analog computer circuits.

With the mass-production capabilities of IC manufacturing techniques, IC op amps became available in large volume, which, in turn, lowered their cost. Today, a general-purpose IC op amp having a gain of 100 dB, an input-offset voltage of 1 mV, an input current of 100 nA, and a bandwidth of 1 MHz can be purchased for less than a dollar. The amplifier, which was a system comprised of many discrete components in the early days, has evolved into a discrete component itself—a reality that has changed the entire picture of linear circuit design.

With this highly sophisticated parcel of gain available at passive-component prices, discrete active-component designs have become a waste of time and money for most dc and low-frequency applications. Clearly, the IC op amp has redefined the "ground rules" for electronic circuits by placing the emphasis of circuit design more on a systems basis. Intelligent application of these "components that are systems" is what we must now concern ourselves with in order to be ready for the challenges of the future.

# I

# *Introducing the IC Op Amp*

ONE

# Op-Amp Basics

## 1.1 THE IDEAL OP AMP

The basic fundamentals of the ideal operational amplifier are relatively straightforward. Perhaps the best approach to understanding the ideal op amp is to forget any conventional thinking about amplifier components—transistors, tubes, or whatever. Instead, think only in general terms and consider the amplifier in block form with only its input and output terminals. We will then discuss the amplifier in this *ideal* sense and disregard what is inside the block.

Such an idealized amplifier is shown in Fig. 1-1. This amplifier is a direct-coupled device with differential inputs and a single-ended output. The amplifier responds only to the difference voltage between the two input terminals, not to their common potential. A positive-going signal at the inverting (−) input produces a negative-going signal at the output, whereas the same signal at the noninverting (+) input produces a positive-going output. With a differential input voltage, $E_{in}$, the output voltage, $E_o$, will be $A_{vo}E_{in}$, where $A_{vo}$ is the gain of the amplifier. Both input terminals of the amplifier will always be used, regardless of the application. The output signal is single-ended and is referred to ground; thus, bipolar (±) power supplies are used.

With these functions of the input and the output in mind, we can now define the ideal properties of this amplifier, which are:

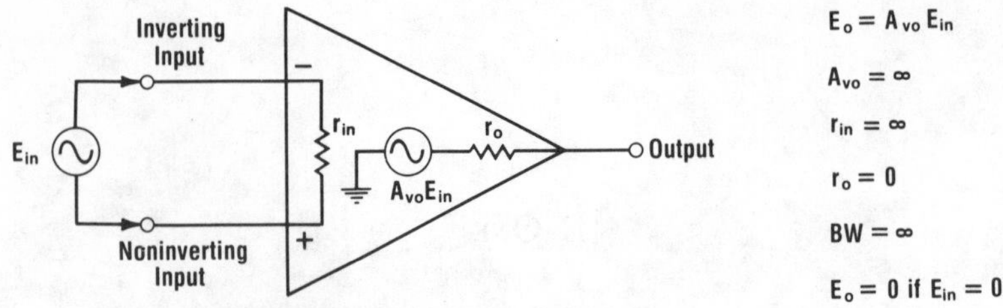


**Fig. 1-1. Equivalent circuit for an ideal operational amplifier.**

1. The voltage gain is infinite—$A_{vo} = \infty$.
2. The input resistance is infinite—$r_{in} = \infty$.
3. The output resistance is zero—$r_o = 0$.
4. The bandwidth is infinite—$BW = \infty$.
5. There is zero input offset voltage—$E_o = 0$ if $E_{in} = 0$.

From these ideal characteristics, we can deduce two very important additional properties of the operational amplifier. Since the voltage gain is infinite, any output signal developed will be the result of an infinitesimally small input signal. Thus, in essence:

1. *The differential input voltage is zero.*

Also, if the input resistance is infinite,

2. *There is no current flow into either input terminal.*

These two properties may be regarded as *axioms,* and they will be used repeatedly in op-amp circuit analysis and design. Once these properties are understood, the operation of virtually any op-amp circuit can be logically deduced. The following discussions of basic op-amp circuit configurations will serve to illustrate this point.

### 1.1.1 Basic Op-Amp Configurations

Operational amplifiers can be connected in two basic amplifying circuits: (1) the *inverting* and (2) the *noninverting* configurations. Virtually all other op-amp circuits are based in some manner on these two basic configurations. In addition, there are closely related variations on the two basic circuits, plus another basic circuit which is a combination of the first two—the *differential amplifier.*

### *The Inverting Amplifier*

Fig. 1-2 illustrates the first basic op-amp configuration—the inverting amplifier. In this circuit, the (+) input is grounded and the signal is applied to the (−) input through $R_{in}$, with feedback returned from the output through $R_f$. By applying the ideal op-amp properties previously stated, the distinguishing features of this circuit may be analyzed as follows.

Since the amplifier has infinite gain, it will develop its output voltage, $E_o$, with zero input voltage. Since the differential input to A is $E_s$, $E_s = 0$. If $E_s$ is zero, then the full input voltage, $E_{in}$, must appear across $R_{in}$, making the current in $R_{in}$

$$I_{in} = \frac{E_{in}}{R_{in}}$$

Also, since $I_s = 0$ due to infinite input impedance, the input circuit, $I_{in}$, must also flow in $R_f$; thus,

$$I_f = I_{in}$$

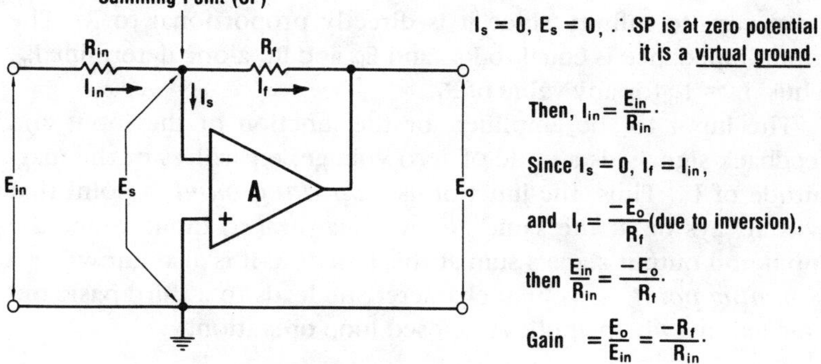


**Summary of Inverting Amplifier Characteristics:**

(1) Gain $= \frac{-R_f}{R_{in}}$, unlimited in range ($R_f$ may be 0 for 0 gain).

(2) Input Impedance $= R_{in}$.

(3) $I_f = I_{in}$, regardless of $R_f$.

(4) Summing point is a virtual ground at the same potential as the (+) input.

**Fig. 1-2. The op-amp inverting amplifier configuration.**

The output voltage, $E_o$, appears across $R_f$ and is negative due to a sign inversion in the amplifier. Stating $I_f$ in terms of $E_o$ and $R_f$, then,

$$I_f = \frac{-E_o}{R_f}$$

Since $I_{in}$ and $I_f$ are equal, we may state

$$\frac{E_{in}}{R_{in}} = \frac{-E_o}{R_f}$$

This equality may be restated in terms of gain as

$$\frac{E_o}{E_{in}} = \frac{-R_f}{R_{in}}$$

which is, in fact, the characteristic gain equation for the ideal inverting amplifier.

There are additional features of the inverting amplifier which should be noted. The gain can be varied by adjusting either $R_f$ or $R_{in}$. If $R_f$ is varied from zero to infinity, the gain will also vary from zero to infinity since it is directly proportional to $R_f$. The input impedance is equal to $R_{in}$, and $E_{in}$ and $R_{in}$ alone determine $I_{in}$. Thus, $I_f = I_{in}$ for any value of $R_f$.

The input to the amplifier, or the junction of the input and feedback signals, is a node of zero voltage, regardless of the magnitude of $I_{in}$. Thus, the junction is a ***virtual ground,*** a point that will always be at the same potential as the (+) input. Since the input and output signals sum at this junction, it is also known as a ***summing point.*** This final characteristic leads to a third basic op-amp axiom, which applies to closed-loop operation:

3. ***With the loop closed, the (−) input will be driven to the potential of the (+), or reference, input.***

This property may or may not have been obvious already from the theory of zero differential input voltage. It is, however, very helpful in understanding op-amp circuitry to regard the (+) input as a reference terminal which will command the level that ***both*** inputs assume. Thus, this voltage may be ground (as shown in Fig. 1-2) or any desired potential.

***The Noninverting Amplifier***

The second basic configuration of the ideal op amp is the noninverting amplifier, shown in Fig. 1-3. This circuit clearly illustrates the validity of Axiom 3. In this circuit, voltage $E_{in}$ is applied to the (+) input, and a fraction of the output signal, $E_o$, is applied to the (−) input from the $R_f$-$R_{in}$ voltage divider. Since no input current flows into either input terminal, and since $E_s = 0$, the voltage $E_{in}'$ is equal to $E_{in}$. Inasmuch as

$$I_{in} = \frac{E_{in}}{R_{in}}$$

then it is also true that

$$I_{in} = \frac{E_{in}'}{R_{in}}$$

Also, since

$$I_f = I_{in}$$

then

$$E_o = I_f(R_{in} + R_f)$$

and

$$E_o = I_{in}(R_{in} + R_f)$$

Substituting,

$$E_o = \frac{E_{in}}{R_{in}}(R_{in} + R_f)$$

We have, then, in terms of gain,

$$\frac{E_o}{E_{in}} = \frac{R_{in} + R_f}{R_{in}}$$

which is the characteristic gain equation for the ideal noninverting amplifier.

Additional characteristics of this configuration can also be deduced. The lower limit of gain occurs when $R_f = 0$, which yields a gain of unity. In the inverting amplifier, current $I_{in}$ always determines $I_f$, which is independent of $R_f$. (This is also true in the noninverting amplifier.) Thus, $R_f$ may be used as a linear gain control, capable of increasing gain from a minimum of unity to a max-

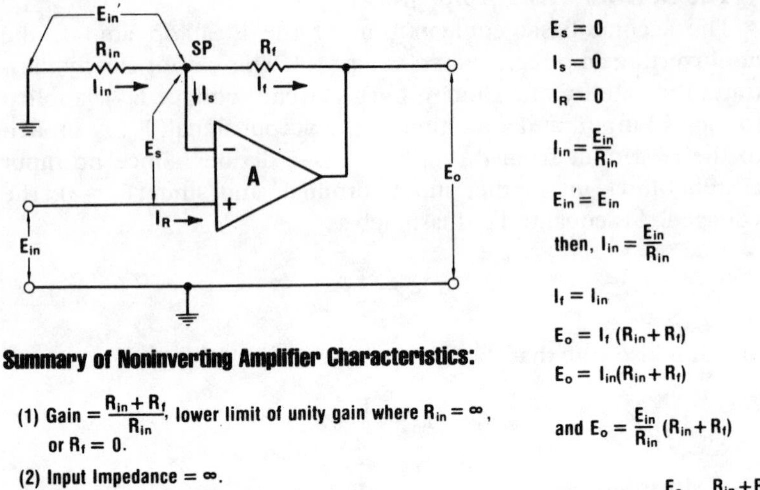


**Fig. 1-3. The op-amp noninverting amplifier configuration.**

imum of infinity. The input impedance is infinite, since an ideal amplifier is assumed.

## *Configurations Based on the Inverting and Noninverting Circuits*

*The Differential Amplifier*—A third op-amp configuration, known as the *differential amplifier,* is a combination of the two previous configurations. Although it is based on the other two circuits, the differential amplifier has unique characteristics of its own. This circuit, shown in Fig. 1-4, has signals applied to both input terminals and uses the natural differential amplification of the op amp.

The circuit may be understood by first considering the two signal inputs separately and, then, in combination. As before, $E_s = 0$, $I_s = 0$, and $I_R = 0$. In addition, $R_f' = R_f$ and $R_{in}' = R_{in}$.

The output voltage due to $E_{in_1}$ may be termed $E_{o_1}$. Using the gain equation for the noninverting circuit, and adding the effect of divider $R_{in}$-$R_f$, the output, $E_{o_1}$, will be

$$E_{o_1} = E_{in_1}\left(\frac{R_f}{R_{in} + R_f}\right)\left(\frac{R_{in}' + R_f'}{R_{in}'}\right)$$

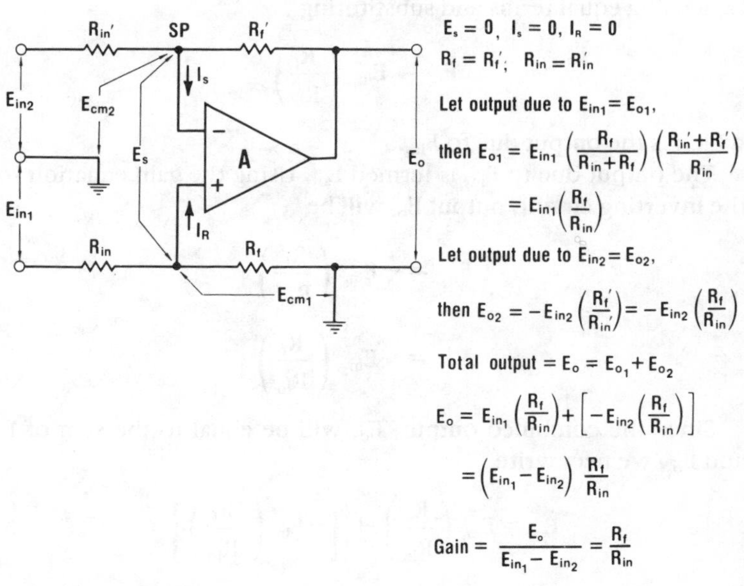


$E_s = 0, \; I_s = 0, \; I_R = 0$

$R_f = R_f'; \quad R_{in} = R_{in}'$

Let output due to $E_{in_1} = E_{o1}$,

$$\text{then } E_{o1} = E_{in_1}\left(\frac{R_f}{R_{in}+R_f}\right)\left(\frac{R_{in}'+R_f'}{R_{in}'}\right)$$

$$= E_{in_1}\left(\frac{R_f}{R_{in}}\right)$$

Let output due to $E_{in_2} = E_{o2}$,

$$\text{then } E_{o2} = -E_{in_2}\left(\frac{R_f'}{R_{in}'}\right) = -E_{in_2}\left(\frac{R_f}{R_{in}}\right)$$

$$\text{Total output} = E_o = E_{o_1} + E_{o_2}$$

$$E_o = E_{in_1}\left(\frac{R_f}{R_{in}}\right) + \left[-E_{in_2}\left(\frac{R_f}{R_{in}}\right)\right]$$

$$= \left(E_{in_1} - E_{in_2}\right)\frac{R_f}{R_{in}}$$

$$\text{Gain} = \frac{E_o}{E_{in_1} - E_{in_2}} = \frac{R_f}{R_{in}}$$

**Summary of Differential Amplifier Characteristics:**

(1) Differential-Mode Gain ($E_{in_1} \neq E_{in_2}$):

$$\frac{E_o}{E_{in_1} - E_{in_2}} = \frac{R_f}{R_{in}}.$$

(2) Common-Mode Gain ($E_{in_1} \equiv E_{in_2}$):

When

$$\frac{R_f}{R_{in}} \equiv \frac{R_f'}{R_{in}'}, = 0.$$

When

$$\frac{R_f}{R_{in}} \neq \frac{R_f'}{R_{in}'}, = \frac{R_f R_{in}' - R_f' R_{in}}{R_{in}' R_{in} + R_{in}' R_f}.$$

In terms of resistor match (worst case),

$$= 4\delta\left(\frac{R_f}{R_{in}+R_f}\right),$$

where

$\delta$ = fractional unbalance of resistors (1.0% = 0.01),

and

$R_{in}$ and $R_f$ are the nominal values.

(3) Input Impedances:

(−) input = $R_{in}'$

(+) input = $R_{in} + R_f$

differential = $R_{in} + R_{in}'$

**Fig. 1-4. The op-amp differential amplifier configuration.**

Cancelling equal terms and substituting

$$E_{o_1} = E_{in_1}\left(\frac{R_f}{R_{in}}\right)$$

which is the output due to $E_{in_1}$.

The output due to $E_{in_2}$ is termed $E_{o_2}$. Using the gain equation for the inverting circuit, output $E_{o_2}$ will be

$$E_{o_2} = -E_{in_2}\left(\frac{R_f'}{R_{in}'}\right)$$

$$= -E_{in_2}\left(\frac{R_f}{R_{in}}\right)$$

Since the combined output, $E_o$, will be equal to the sum of $E_{o_1}$ and $E_{o_2}$, we may write

$$E_o = E_{in_1}\left(\frac{R_f}{R_{in}}\right) + \left[-E_{in_2}\left(\frac{R_f}{R_{in}}\right)\right]$$

By regrouping

$$E_o = (E_{in_1} - E_{in_2})\left(\frac{R_f}{R_{in}}\right)$$

In terms of gain, this is

$$\frac{E_o}{E_{in_1} - E_{in_2}} = \frac{R_f}{R_{in}}$$

which is the gain of the stage for differential-mode signals.

This configuration is unique because it can reject a signal common to both inputs. This is due to the property of zero differential input voltage, which is explained as follows.

In the case of identical signals at $E_{in_1}$ and $E_{in_2}$, the analysis is straightforward. $E_{in_1}$ will be divided by $R_{in}$ and $R_f$ to a smaller voltage, $E_{cm_1}$, across $R_f$. Because of infinite amplifier gain and zero differential input voltage, an equal voltage ($E_{cm_2}$) must appear at SP. Since the network $R_{in}'$-$R_f'$ is identical to $R_{in}$-$R_f$, and since the same voltage is applied to both inputs, it follows that $E_o$ must be at zero potential to maintain $E_{cm_2}$ identical to $E_{cm_1}$; $E_o$ will be at the same potential as the bottom of $R_f$, which is, in fact, ground potential. This very useful property of the differential amplifier can be used to discriminate against undesirable common-mode noise components while amplifying signals that appear differentially. If the ratio

$$\frac{R_f'}{R_{in}'}$$

is identical to

$$\frac{R_f}{R_{in}}$$

the gain for common-mode signals is zero since, by definition, the amplifier has no gain for equal signals applied to both inputs.

The two input impedances of the stage are unequal. For the (+) input, the input impedance is $R_{in} + R_f$. The impedance at the (−) input is $R_{in}'$. The differential input impedance (for a floating source) is the impedance between the inputs, or $R_{in} + R_{in}'$.

*The Summing Inverter*—By utilizing the virtual-ground characteristic of the inverting amplifier's summing point, a useful modification is realized—the summing inverter (see Fig. 1-5). In this circuit, as in the inverting amplifier, $E_s = 0$, $I_s = 0$, and $I_f = I_{in}$. However, in the summing inverter, $I_{in}$ is the algebraic sum of a number of inputs, such as $I_{in_1}$, $I_{in_2}$, $I_{in_3}$, etc. Thus,

$$I_{in_1} = \frac{E_{in_1}}{R_{in_1}}$$

$$I_{in_2} = \frac{E_{in_2}}{R_{in_2}}$$

$$I_{in_3} = \frac{E_{in_3}}{R_{in_3}}$$

and

$$I_{in} = I_{in_1} + I_{in_2} + I_{in_3}$$

$$I_f = \frac{-E_o}{R_f}$$

Therefore,

$$\frac{-E_o}{R_f} = I_{in_1} + I_{in_2} + I_{in_3}$$

Finally, substituting and regrouping gives

$$E_o = -\left[E_{in_1}\left(\frac{R_f}{R_{in_1}}\right) + E_{in_2}\left(\frac{R_f}{R_{in_2}}\right) + E_{in_3}\left(\frac{R_f}{R_{in_3}}\right)\right]$$

which states that the output voltage is the inverted algebraic sum of the input voltages scaled by $R_f$. Thus, the circuit may be said to be a *scaling adder*.

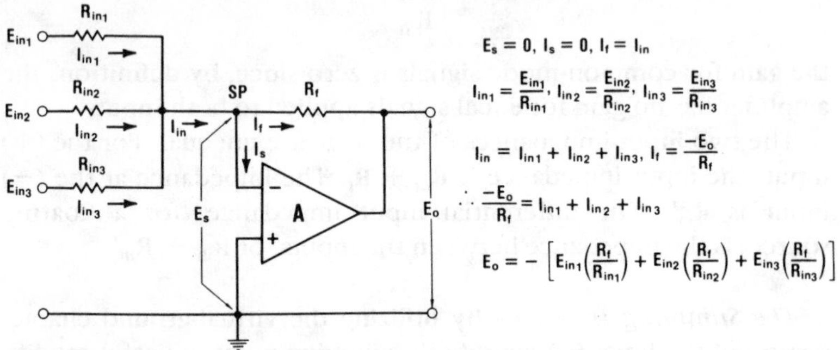


**Summary of Summing-Inverter Characteristics:**

(1) Gain (per channel) = $\frac{-R_f}{R_{in1}}, \frac{-R_f}{R_{in2}}, \frac{-R_f}{R_{in3}}$, etc.

(2) $R_f$ varies gain of all inputs; $R_{in1}$, $R_{in2}$, $R_{in3}$, etc., vary individual gains.

(3) Composite output = $E_o = -\left[E_{in1}\left(\frac{R_f}{R_{in1}}\right) + E_{in2}\left(\frac{R_f}{R_{in2}}\right) + E_{in3}\left(\frac{R_f}{R_{in3}}\right)\right]$.

(4) Input impedance = $R_{in1}$, $R_{in2}$, $R_{in3}$, etc.

**Fig. 1-5. The op-amp summing inverter.**

The overall gain of the circuit is set by $R_f$, which, in this respect, behaves as in the basic inverting amplifier. The gains of the individual channels are scaled independently by $R_{in1}$, $R_{in_2}$, $R_{in_3}$, etc. Similarly, $R_{in_1}$, $R_{in_2}$, and $R_{in_3}$ are the input impedances of the respective channels.

Another interesting characteristic of this configuration is the fact that linear signal mixing takes place at the summing point without interaction between inputs, since all signal sources feed into a virtual ground. The circuit can accommodate any number of inputs by adding additional input resistors at the summing point, such as $R_{in_4}$, $R_{in_5}$, etc.

Although the preceding circuits have been described in terms of input and feedback resistances, the resistors may be replaced by complex elements, and the basic op-amp axioms will still hold true. Two circuits that demonstrate this are further modifications of the inverting amplifier.

*The Integrator*—It has been shown that both basic op-amp configurations act to maintain the feedback current, $I_f$, equal to $I_{in}$ at all times. A modification of the inverting amplifier which takes advantage of this characteristic is the integrator, shown in Fig. 1-6. An input voltage, $E_{in}$, is applied to $R_{in}$, thus developing current $I_{in}$. As in the basic inverter, $E_s = 0$, $I_s = 0$, and $I_f = I_{in}$. The feedback element in the integrator is a capacitor, $C_f$. Therefore, the constant current, $I_f$, in $C_f$ builds a linear voltage ramp across $C_f$. The output voltage is thus an integral of the input current, which is forced to charge $C_f$ by the feedback loop. The change in voltage across $C_f$ is

$$-\Delta E_o = \frac{I_{in}\Delta t}{C_f}$$

which makes the output change per unit of time

$$\frac{\Delta E_o}{\Delta t} = \frac{-E_{in}}{R_{in}C_f}$$

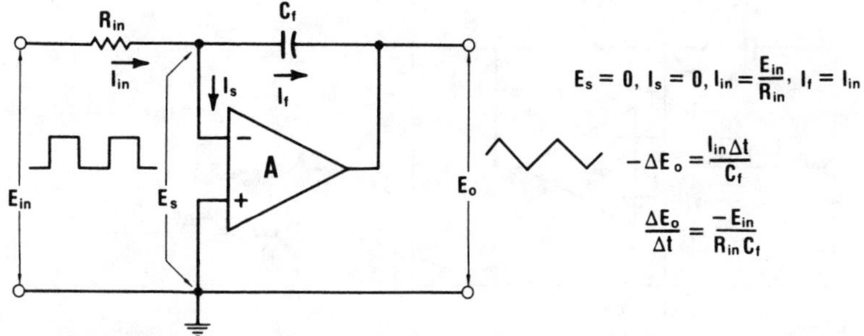


**Summary of Integrator Characteristics:**

(1) Circuit integrates input current, thus

$$\frac{\Delta E_o}{\Delta t} = \frac{-E_{in}}{R_{in}C_f}.$$

(2) Input impedance = $R_{in}$.

**Fig. 1-6. The op-amp integrator.**

As with the other inverting amplifier configurations, the input impedance is simply $R_{in}$.

*The Differentiator*—A second modification of the inverting amplifier, which also uses the current in a capacitor to advantage, is the differentiator shown in Fig. 1-7. In this circuit, the positions of R and C are reversed from those in the integrator, placing the capacitive element in the input network. Thus, the input current is made to be proportional to the rate of change of the input voltage:

$$I_{in} = \frac{\Delta E_{in} C_{in}}{\Delta t}$$

Again,

$$I_f = I_{in}$$

and since

$$-E_o = I_f R_f$$

$$E_o = -I_f R_f$$

Substitution gives

$$E_o = \frac{-\Delta E_{in} R_f C_{in}}{\Delta t}$$

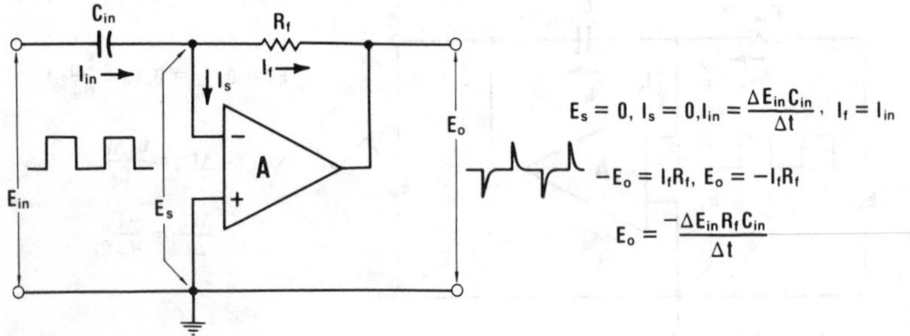


**Summary of Differentiator Characteristics:**

(1) Circuit responds to the rate of change of input voltage $E_{in}$, thus

$$E_o = \frac{-\Delta E_{in} R_f C_{in}}{\Delta t}$$

**Fig. 1-7. The op-amp differentiator.**

*The Voltage Follower*—A special modification of the noninverting amplifier is the unity-gain stage shown in Fig. 1-8. In this cir-

**Bookmark**

*HOWARD W. SAMS & COMPANY*

DEAR VALUED CUSTOMER:

Howard W. Sams & Company is dedicated to bringing you timely and authoritative books for your personal and professional library. Our goal is to provide you with excellent technical books written by the most qualified authors. You can assist us in this endeavor by checking the box next to your particular areas of interest.

We appreciate your comments and will use the information to provide you with a more comprehensive selection of titles.

Thank you,

Vice President, Book Publishing
Howard W. Sams & Company

COMPUTER TITLES:

Hardware
- ☐ Apple 140
- ☐ Macintosh 101
- ☐ Commodore 110
- ☐ IBM & Compatibles 114

Business Applications
- ☐ Word Processing J01
- ☐ Data Base J04
- ☐ Spreadsheets J02

Operating Systems
- ☐ MS-DOS K05
- ☐ OS/2 K10
- ☐ CP/M K01
- ☐ UNIX K03

Programming Languages
- ☐ C L03
- ☐ Pascal L05
- ☐ Prolog L12
- ☐ Assembly L01
- ☐ BASIC L02
- ☐ HyperTalk L14

Troubleshooting & Repair
- ☐ Computers S05
- ☐ Peripherals S10

Other
- ☐ Communications/Networking M03
- ☐ AI/Expert Systems T18

ELECTRONICS TITLES:
- ☐ Amateur Radio T01
- ☐ Audio T03
- ☐ Basic Electronics T20
- ☐ Basic Electricity T21
- ☐ Electronics Design T12
- ☐ Electronics Projects T04
- ☐ Satellites T09
- ☐ Instrumentation T05
- ☐ Digital Electronics T11

Troubleshooting & Repair
- ☐ Audio S11
- ☐ Television S04
- ☐ VCR S01
- ☐ Compact Disc S02
- ☐ Automotive S06
- ☐ Microwave Oven S03

Other interests or comments: ____________________

Name ____________________

Title ____________________

Company ____________________

Address ____________________

City ____________________

State/Zip ____________________

Daytime Telephone No. ____________________

*A Division of Macmillan, Inc.*
*4300 West 62nd Street*
*Indianapolis, Indiana 46268*

22453

cuit, $R_{in}$ has increased to infinity, $R_f$ is zero, and the feedback is 100%. $E_o$ is then exactly equal to $E_{in}$, since $E_s = 0$. The circuit is known as a "voltage follower" since the output is a unity-gain in-phase replica of the input voltage. The input impedance of this stage is also infinite.

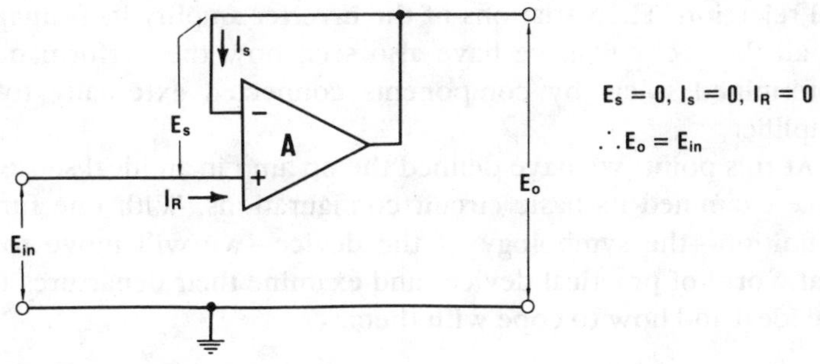


**Summary of Voltage-Follower Characteristics:**

(1) Gain = 1.

(2) Input impedance = ∞.

**Fig. 1-8. The op-amp voltage follower.**

### *Summary of the Basic Amplifier Configurations and Their Characteristics*

All of the circuit characteristics that have been described are important because they are a basis for the entire foundation of op-amp circuit technology. The five basic criteria that describe the ideal amplifier are fundamental, and from these evolve the three main axioms of op-amp theory, which bear repeating:

1. *The differential input voltage is zero.*
2. *There is no current flow into either input terminal.*
3. *With the loop closed, the (−) input will be driven to the potential of the (+), or reference, input.*

These three axioms have been illustrated in all of the basic circuits and their variations. In the inverting configuration, the concepts of zero input current and zero differential input voltage give rise to the concepts of the summing point and the virtual ground, where the inverting input is held by feedback to the same potential as the grounded noninverting input. Using the concept of the noninverting input as a reference terminal, the noninverting amplifier and voltage follower illustrate how an input voltage is indirectly scaled through negative feedback to the inverting input, which is forced to follow at an identical potential. The full differential configuration combines these concepts, illustrating the ideal of simultaneous differential amplification and common-mode signal rejection. The variations of the inverter amplify its principles. In all these circuits, we have also seen how the performance is determined solely by components connected externally to the amplifier.

At this point, we have defined the op amp in an ideal sense and have examined its basic circuit configurations. With one further definition—the symbology of the device—we will move to the real world of practical devices and examine their departures from the ideal and how to cope with them.

### 1.1.2 The Standard Op-Amp Schematic Symbol and Its Use

One additional basic tool of the op amp is its schematic symbol. This is fundamental, for a correctly drawn schematic transmits a great deal of knowledge about the workings of a circuit. The preferred op-amp symbol, shown with annotations, is illustrated in Fig. 1-9.

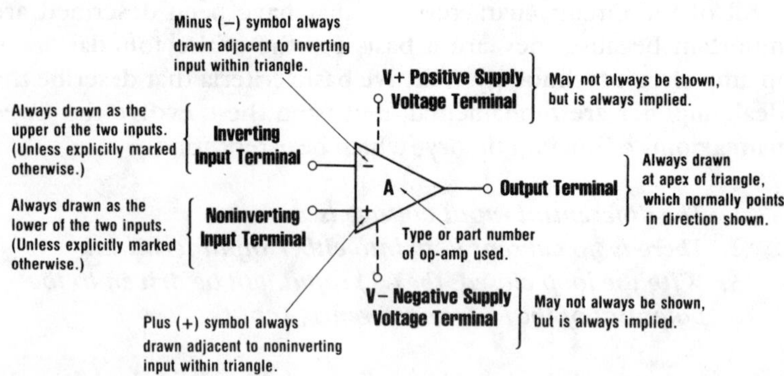


**Fig. 1-9. The standard op-amp schematic symbol.**

The basic symbol is the triangle, which is generally understood to imply amplification. The inputs are at the base of the triangle, with the output at the apex. In accordance with normal signal-flow convention, the symbol is drawn with the apex (output) to the right, but this may be altered if necessary for clarification of other circuit details.

The two inputs are usually drawn as shown with the noninverting (+) input the lower of the two. Exceptions to this rule occur in special circumstances where it would be awkward to retain the standard arrangement. Furthermore, the two inputs are always clearly identified by (+) and (−) symbols, which are adjacent to their respective leads and are drawn within the body of the triangle.

Supply-voltage leads are preferably drawn extending above and below the triangle as shown. These may not be shown in all cases (in the interest of simplicity), but they are always implied. Generally, in sketches, it is sufficient to use the 3-lead symbol to convey an intent, with power connections understood.

Finally, the type or part number of the device used is centered within the body of the triangle. If the circuit is a general one indicating any op amp, the symbol used is A (or $A_1$, $A_2$, etc.).

## 1.2 THE NONIDEAL OP AMP

So far we have examined the op-amp concept only in a general sense, assuming idealized parameters. In a real-world situation, the ideal amplifier does not exist. However, it is important to stress the idealized line of thinking because, in many applications, the differences between ideal and actual are close to negligible. Furthermore, we should always be aware of how closely we actually approach "idealized" performance. We will now examine the factors that determine how small these differences are and learn how to control performance so as to minimize them. As might be expected, these errors arise due to departures from the ideal characteristics set forth in Section 1.1.

### 1.2.1 Errors Due to Finite Open-Loop Gain

A principal source of error in op-amp applications is the factor contributed by finite open-loop gain. General-purpose IC op amps have a typical dc open-loop gain close to 100 dB, which, while large, is certainly a long way from being infinite. Also, the actual gain figure will vary from unit to unit because of the normal production tolerances associated with IC manufacturing processes. In a given device, the open-loop gain will also vary to some degree with temperature, loading, and supply voltage. In addition to this, the open-loop gain of an op amp decreases with frequency because of the necessity for correcting the overall phase response to ensure stability under closed-loop conditions. As a result, a typical open-loop gain-versus-frequency plot looks something like Fig. 1-10. Thus, in reality, the open-loop gain characteristics of practical IC op amps are nonideal in two of the basic criteria: (1) the dc gain is not infinite and (2) the bandwidth is not infinite.

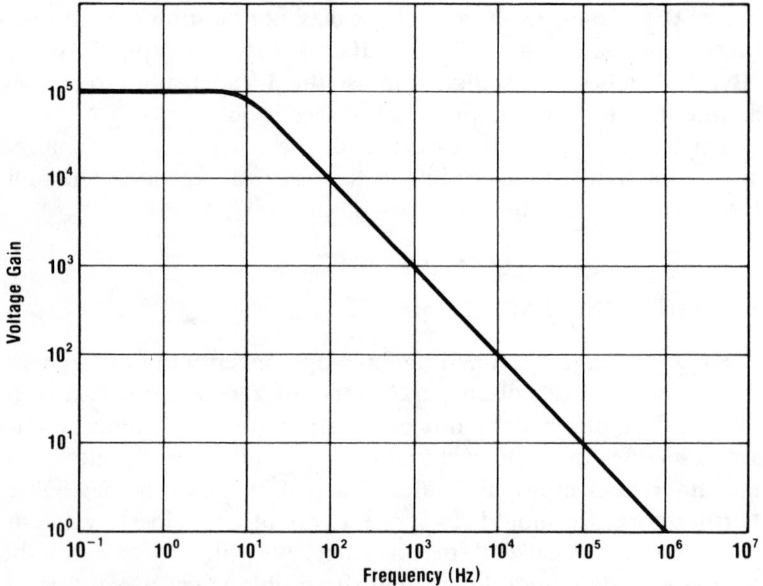


**Fig. 1-10. Typical op-amp open-loop gain response.**

To express the error factor due to a finite open-loop gain, we must again examine the closed-loop inverting and noninverting configurations and define a few more parameters associated with each. Assume all other characteristics of the amplifier are ideal (for the moment) except for the finite gain, $A_{vo}$.

Fig. 1-11A illustrates an inverting amplifier which has a finite gain, $A_{vo}$. $R_f$ and $R_{in}$ are the feedback and input resistors, and the $E_{in}$ generator is assumed to have zero source impedance. If we examine the path of the feedback signal from $E_o$, we note that it undergoes an attenuation due to the presence of $R_f$ and $R_{in}$. If we term the attenuated signal that appears at SP as $E_e$, then $E_e$ is related to $E_o$ by

$$\frac{E_e}{E_o} = \frac{R_{in}}{R_{in} + R_f}$$

This introduces a new term, $\beta$, which is defined as the *feedback attenuation factor* and is expressed as

$$\beta = \frac{R_{in}}{R_{in} + R_f}$$

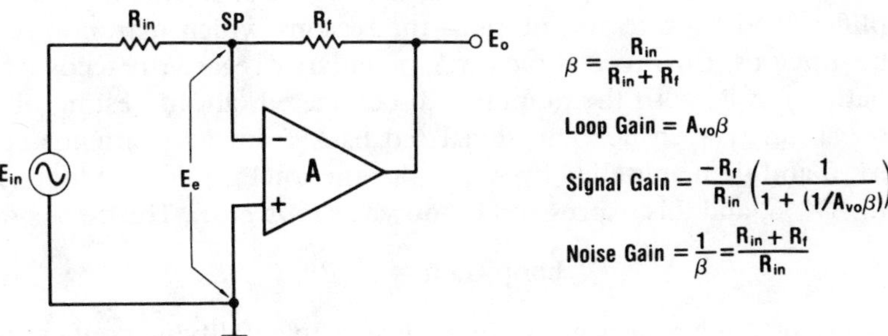


*(A) Inverting configuration.*

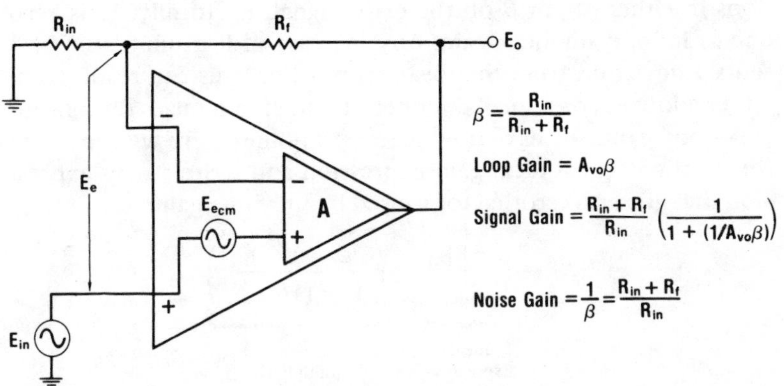


*(B) Noninverting configuration.*

**Fig. 1-11. Gain errors in inverting and noninverting configurations.**

By definition, $\beta$ is the ratio of the feedback signal to the output signal. It can reach a maximum of unity (or zero attenuation) under the condition of 100% feedback. The feedback divider, which determines $\beta$, takes into account the total attenuation between the amplifier output and the summing point. This includes frequency-dependent effects due to reactive elements (if any) and, in general, all attenuation. Therefore, to be precise, $R_f$ and $R_{in}$ of Fig. 1-11A symbolize the total equivalent feedback attenuation network. (This will be more evident as we examine other error sources.) The signal, $E_e$, is amplified by the open-loop gain of the amplifier ($A_{vo}$) and produces output voltage $E_o$.

At this point, we have traversed the complete path around the feedback loop—through the feedback network, through the amplifier, and back to the output—for reasons which may not yet be obvious. However, if we can momentarily regard the feedback path from $E_o$ with the amplifier disconnected, an interesting observation will ensue. The signal fed back from $E_o$ is attenuated by $\beta$ and then amplified by $A_{vo}$. Thus, the total gain of the loop is $A_{vo} \times \beta$, and this expression is known as *loop gain.* Thus,

$$\text{Loop Gain} = A_{vo}\beta$$

Loop gain is a measure of how closely an amplifying configuration approaches the ideal. If $A_{vo}$ were really infinite, we would not be concerned with the total gain around the feedback loop. This may be intuitively appreciated by visualizing the effects of variations in either $A_{vo}$ or $\beta$ on the error signal, $E_e$. Ideally, $E_e$ is zero due to infinite amplifier gain. Any increase of $E_e$ from zero represents a deviation from the ideal. In practice, this can result from either a low $A_{vo}$ or a small $\beta$, either of which minimizes their product—loop gain. We have now reached a point where we can relate this to the original ideal gain expression and write a new expression that gives an error factor caused by the finite gain, $A_{vo}$

$$A_v = \underbrace{\left(\frac{-R_f}{R_{in}}\right)}_{\text{Ideal Expression}} \underbrace{\left(\frac{1}{1 + (1/A_{vo}\beta)}\right)}_{\text{Error Multiplier}}$$

This equation is separated into two parts to illustrate the modification of the ideal by $A_{vo}\beta$. As the right-hand term of the denominator of the error multiplier approaches zero, the error multiplier approaches unity until $A_{vo} = \infty$, the point at which the equation reduces to the ideal. Thus, the greater $A_{vo}\beta$ is, the closer the amplifier is to ideal performance. In practice, this means that either a greater $A_{vo}$ or a greater $\beta$ (more feedback) maximizes $A_{vo}\beta$. From this, it is obvious that high-gain stages (which require a small $\beta$) require more open-loop gain to approach ideal performance.

To examine the effects of loop gain on the noninverting amplifier, we refer to Fig. 1-11B. Although the method of signal amplification is different in this configuration, the feedback path is identical to that of the inverter. Therefore, $\beta$ is determined in exactly the same manner, and the same general considerations apply concerning $A_{vo}\beta$. There is, however, an additional gain-error term in the noninverting amplifier which is due to the presence of a common-mode voltage, $E_{cm}$. In the inverting amplifier, there is no common-mode input voltage since both inputs are effectively grounded. In a noninverting stage, the inputs see a common-mode voltage equal to the input voltage, or

$$E_{cm} = E_{in}$$

Since a real amplifier responds to common-mode voltages, this error source must be taken into account. The relationship between the common-mode voltage of an op amp and its common-mode error voltage is expressed as a ratio, termed the ***common-mode rejection ratio*** (CMRR). The existence of a finite CMRR causes this input common-mode error voltage to be generated, which can be expressed as a fraction of the common-mode input voltage

$$E_{e_{cm}} = \frac{E_{cm}}{CMRR}$$

It is convenient to model this error voltage as a separate generator in series with an ideal amplifier and consider its effect on the gain. By expressing the error voltage as a fraction, we can include it with $A_{vo}$ and generate a new expression which includes the effects of common-mode gain

$$\text{Common-Mode Error} = \frac{E_{e_{cm}}}{E_{cm}}$$

Since

$$E_{e_{cm}} = \frac{E_{cm}}{CMRR}$$

we may substitute, leaving

$$\text{Common-Mode Error} = \pm \frac{1}{\text{CMRR}}$$

We must add the qualifier of (±) since common-mode error can be an error in either direction; it can either add to or subtract from the input signal. This uncertainty prevents a simple cancellation of common-mode error by adjusting $R_{in}$ or $R_f$.

We can now write the modified gain expression for the noninverting amplifier:

$$A_v = \underbrace{\left(\frac{R_{in} + R_f}{R_{in}}\right)}_{\substack{\text{Ideal} \\ \text{Expression}}} \underbrace{\left(\frac{1}{1 + (1/A_{vo}\beta)}\right)}_{\substack{\text{Error} \\ \text{Multiplier}}}$$

To summarize the gain error for both configurations due to finite or modified $A_{vo}$, we can express the error as a percentage:

$$\%\ \text{Error} = \frac{100}{A_{vo}\beta + 1}$$

Thus, a loop gain of 100 (40 dB), for example, will yield a gain error of approximately 1%.

As the reader may already have noted, there is a basic difference between the inverting and noninverting configurations in the method of signal amplification, but no difference in the method of calculating $\beta$. As a result, for equivalent *signal* gains, inverting and noninverting stages will have a different $\beta$, and thus will have different loop gains. This is also illustrated in Fig. 1-11.

If we examine a unity-gain condition for both configurations, it will be noted that for the inverting stage

$$\frac{R_f}{R_{in}} = 1$$

or

$$R_f = R_{in}$$

but for the noninverting stage

$$\frac{R_{in} + R_f}{R_{in}} = 1$$

or

$$R_{in} = \infty$$

The $\beta$ equations for the two configurations are then

$$\beta \text{ (inverting)} = \frac{R_{in}}{R_{in} + R_f}$$

$$= \frac{R_{in}}{2R_{in}}$$

$$= 0.5$$

and

$$\beta \text{ (noninverting)} = \frac{\infty}{\infty + R_f}$$

$$= 1$$

Thus, it can be seen that for equivalent signal gains, the two stages do not have the same $\beta$. This difference is most apparent at the lowest gain that both stages are capable of—unity, where the $\beta$ of the inverting stage is half that of the noninverting stage. The point of this discussion is that for the two stages, signal gain and $\beta$ do not have the same relationship. In order to compare the two stages directly in terms of $\beta$ (and, hence, loop gain), we must introduce a new term.

The parameter that is defined identically for both configurations is $\beta$. Therefore, a gain defined in terms of $\beta$ would be a useful basis for comparing the two basic stages. This gain is the reciprocal of $\beta$ and is termed *noise gain*:

$$\text{Noise Gain} = \frac{1}{\beta}$$

$$= \frac{R_{in} + R_f}{R_{in}}$$

which is the ideal equation. The term that includes error factors due to nonideal gain is the *true noise gain* or *closed loop gain*:

$$\text{Closed-Loop Gain} = A_{cl} = \underbrace{\left(\frac{1}{\beta}\right)}_{\substack{\text{Ideal}\\ \text{Expression}}} \underbrace{\left(\frac{1}{1 + (1/A_{vo}\beta)}\right)}_{\substack{\text{Error}\\ \text{Multiplier}}}$$

Noise gain is the inverse of $\beta$. Thus, like $\beta$, it takes into account all feedback resistance and includes frequency effects. Noise gain is the gain seen by amplifier input referred noise and error (offset and drift) parameters. It is often significantly different from signal gain. It is most important that this point be understood because noise gain—not signal gain—is the parameter used in calculations of closed-loop amplifier performance. To illustrate by means of the unity-gain example, the inverting stage has a noise gain of 2, whereas the noninverting stage has a noise gain of 1. In this example, then, the inverting stage would amplify input error components by a factor of 2.

At this point, it is appropriate to summarize the various gain definitions and bring them all into perspective by means of graphical illustration in Fig. 1-12. The definitions are:

1. *Open-Loop Gain ($A_{vo}$)*—The open-loop voltage gain of the basic amplifier without feedback but with loading. This includes frequency dependence.
2. *Signal Gain ($A_v$)*—The closed-loop voltage gain of the amplifying circuit for signals applied, as appropriate to the configuration.
3. *Feedback-Loop Attenuation ($\beta$)*—The voltage attenuation of the feedback network including all impedances.
4. *Noise Gain ($1/\beta$)*—The voltage-gain response given by the inverse of feedback-loop attenuation.
5. *Loop Gain ($A_{vo}\beta$)*—The net gain around the broken feedback loop as seen from the feedback-network input terminal back to the amplifier output. This includes frequency dependence.
6. *Closed-Loop Gain ($A_{cl}$)*—The noise gain including the effects of loop gain.

Fig. 1-12 illustrates these definitions on a Bode plot, the basic tool for graphing amplifier gain versus frequency. This is a log-log plot which pictures the entire amplifier response. Gain may be shown either numerically or in dB, but it is conventionally shown in dB; thus, this terminology is used in Fig. 1-12. With the various gains expressed in dB, they may be added and subtracted, which considerably simplifies the discussion.

The response, $A_{vo}$, of the basic amplifier is shown as equal to 100 dB at frequencies below 10 Hz. Above 10 Hz, $A_{vo}$ becomes

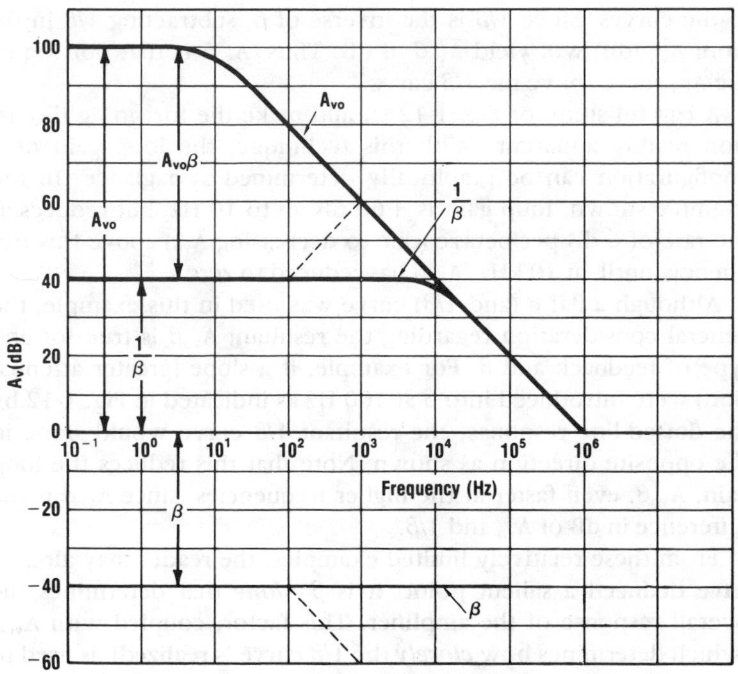


**Fig. 1-12. Bode plot illustrating the relationships of $A_{vo}$, $\beta$, $1/\beta$, and $A_{vo}\beta$.**

frequency-dependent and reduces 6 dB per octave (or 20 dB per decade) until at 1 MHz, $A_{vo}$ = 0 dB, or unity. Such a response is typical of many of the general-purpose IC op amps that will be discussed in this book.

To illustrate $\beta$ and $1/\beta$, a $\beta$ of 0.01 or −40 dB will be assumed, with no reactive effects for simplicity. Therefore, $\beta$ is drawn as a flat line (no frequency dependence) at −40 dB on the $A_v$ scale. The reciprocal, $1/\beta$, is drawn as a flat line at $1/\beta = 1/0.01 = 100$, or +40 dB on the $A_v$ scale.

Loop gain, or $A_{vo}\beta$, is the resultant of the combination of $A_{vo}$ and $\beta$, including frequency dependence. Since $A_{vo}$ is constant below 10 Hz, the product of $A_{vo}\beta$ will also be constant below this frequency. Working in dB, $A_{vo}\beta$ can be arrived at by simply adding $A_{vo}$ and $\beta$. Thus, $A_{vo}\beta$ in dB is +100, −40, or +60 dB (below 10 Hz).

Rather than draw an additional curve referred to the zero-gain axis for $A_{vo}\beta$, we can make use of the information already implicit in the curves. Since $1/\beta$ is the inverse of $\beta$, subtracting $1/\beta$ in dB from $A_{vo}$ (dB) will yield $A_{vo}\beta$ in dB. Thus, $A_{vo}\beta$ is that portion of the $A_{vo}$ curve above the $1/\beta$ curve.

A careful study of Fig. 1-12 should make the foregoing discussion readily apparent. With this technique, the loop gain of a configuration can be graphically determined at a glance. In the example shown, loop gain is +60 dB up to 10 Hz, but reduces at the rate of 6 dB per octave (due to decreasing $A_{vo}$) above this frequency, until, at 10 kHz, $A_{vo}\beta$ has reduced to zero.

Although a flat $\beta$ (and $1/\beta$) curve was used in this example, the general consideration regarding the resultant $A_{vo}\beta$ is true for any type of feedback and $\beta$. For example, if a slope (greater attenuation) were introduced into $\beta$ at 100 Hz as indicated in Fig. 1-12 by the dotted-line response, the resultant $1/\beta$ curve would slope in the opposite direction as shown. Note that this reduces the loop gain, $A_{vo}\beta$, even faster at the higher frequencies, since $A_{vo}\beta$ is the difference in dB of $A_{vo}$ and $1/\beta$.

From these relatively limited examples, the reader may already have deduced a salient point: It is $\beta$ *alone* that determines the overall response of the amplifier. This factor, coupled with $A_{vo}\beta$ (which determines how *closely* the $1/\beta$ curve is realized), is used to implement feedback configurations.

In an actual design example, the $1/\beta$ curve would be used as a starting point, since it is this curve that represents the final frequency response of the amplifier. Given $1/\beta$ as an objective, the required $\beta$ may be drawn as its inverse, which defines the characteristics of the feedback network. In practice, a certain requirement for $1/\beta$ will dictate the choice of amplifier to yield a specified loop gain for the required accuracy. This explanation is somewhat simplified, but in sections to follow, we will discuss this and other points in greater detail.

### *The Curative Properties of Feedback*

Operational feedback, as we have discussed, is a means by which the designer can exercise control over total circuit performance. Feedback has a general influence that can be called a "curative" property. This property generally improves such factors as linearity, gain stability, and input and output impedances. In all cases, the degree of improvement is related to loop gain, or the amount of working feedback the amplifier uses in any given situation. Expressions governing the extent of improvement will now be discussed.

*Gain Stability*—One reason for using a feedback amplifier is to develop a transfer gain that is relatively independent of amplifier characteristics. Indeed, the entire reasoning behind the operational amplifier is to render variations in its open-loop characteristics negligible, and make the amplifier properties dependent only on the external components.

Variations in gain for a feedback amplifier may be calculated by

$$\Delta A_v = \frac{\Delta A_{vo}}{1 + A_{vo}\beta}$$

It can be seen that gain variation is controlled almost entirely by the loop gain if $A_{vo}\beta \gg 1$. This expression is a general one, applicable to any configuration. If $A_{vo}$ and $\beta$ are modified by other nonideal terms, the corrected $A_{vo}'$ and $\beta'$ should be used in calculations. This statement also applies, of course, for the remaining expressions.

*Linearity*—Feedback reduces open-loop distortion and nonlinearity to lower closed-loop levels according to

$$THD_{cl} = \frac{THD_{ol}}{1 + A_{vo}\beta}$$

where

THD is the total harmonic distortion in percent.

*Input and Output Impedances*—Input and output impedances of op-amp configurations are also transformed (in the positive direction) by feedback. The output resistance ($r_o$) of any op amp is reduced by feedback to a new resistance, $R_o$

$$R_o = \frac{r_o}{1 + A_{vo}\beta}$$

The input impedance is transformed upward or downward, depending on the configuration. In the inverting configuration shown in Fig. 13A, the input impedance, $R_{in}$, seen at the summing point is

$$R_{in} = \frac{r_{in} \parallel R_f}{1 + A_{vo}\beta}$$

If $R_f \ll r_{in}$ (usually the case) and $A_{vo}\beta \gg 1$, a useful approximation is

$$R_{in} \cong \frac{R_f}{A_{vo}\beta}$$

This clearly demonstrates the virtual-ground properties of the inverting configuration, as $R_f$ is "transformed" by feedback to a much lower value. This makes the inverting stage a natural one for processing the output from current generators.

The noninverting configuration is shown in Fig. 1-13B, and its property is to transform the input resistance to a higher value.

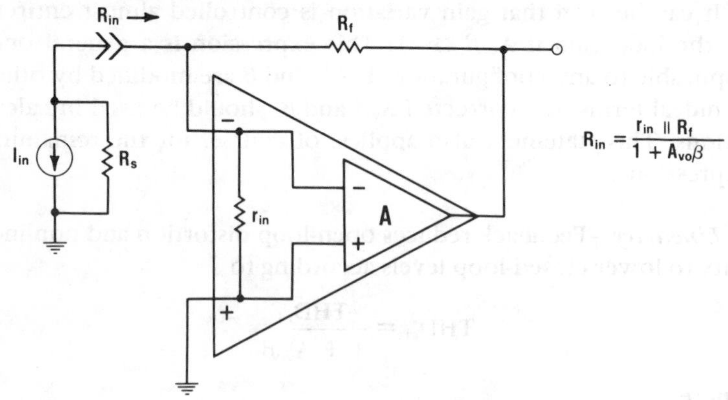


*(A) Inverting configuration.*

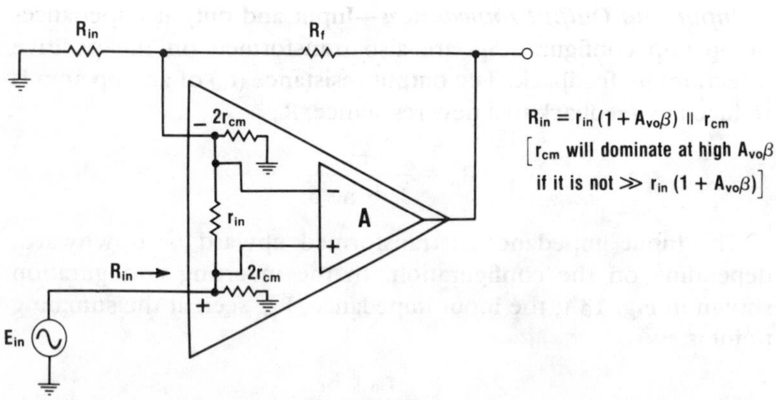


*(B) Noninverting configuration.*

**Fig. 1-13. Effects of feedback on input resistance.**

Since the voltages across both ends of $r_{in}$ are the same (due to zero input-voltage differential), $r_{in}$ is effectively "bootstrapped" by the feedback. The two common-mode resistances, each of value $2r_{cm}$, are not bootstrapped, however, and they appear in parallel with the input terminal. Thus,

$$R_{in} = r_{in}(1 + A_{vo}\beta) \parallel r_{cm}$$

At high loop gains, such as with a voltage follower, the common-mode input resistance will usually be the upper limit on the input impedance of the noninverting stage.

From these few observations, it can be seen that loop gain is the primary determinant of feedback performance, and its effect is always positive. Thus, it should be kept in mind that in optimizing feedback, the parameter $A_{vo}\beta$ is the key performance parameter.

### 1.2.2 Errors Due to Finite Input Resistance

Another factor having definite effects on op-amp circuit performance, in both the inverting and noninverting configurations, is finite input resistance. In this section, we will examine these effects, with all other parameters assumed to be ideal.

#### *Inverting Configuration*

Fig. 1-14A illustrates the inverting amplifier connection with the addition of the input resistance, $r_{in}$. The effect of the input resistance, which appears in parallel with the summing point, is to modify $\beta$. $R_{in}$ and $r_{in}$ in parallel form a new $\beta$, $\beta'$, which is lower than in the ideal (unloaded) network. Thus,

$$\beta' = \frac{R_{in} \parallel r_{in}}{(R_{in} \parallel r_{in}) + R_f}$$

where

$\beta'$ is the $\beta$ with finite input resistance $r_{in}$.

The net effect of finite input impedance is therefore a reduction in $\beta$ and, thus, a reduction in loop gain.

In practice, the actual loading effects of $r_{in}$ will depend on the feedback impedances and the input impedance of the amplifier used. Generally, as long as the amplifier input resistance is 10 or more times greater than the output impedance of the feedback network, the error due to input resistance loading will be negligi-

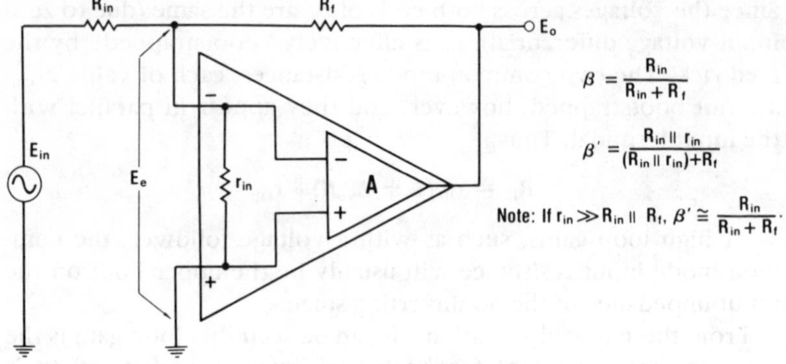


*(A) Inverting configuration.*

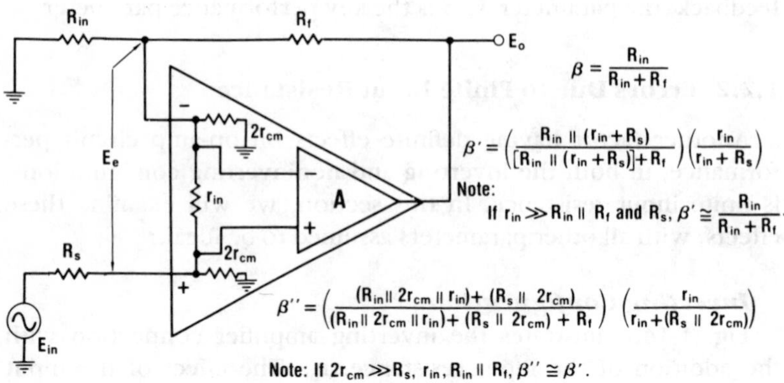


*(B) Noninverting configuration.*

**Fig. 1-14. Effects of finite input resistance.**

ble. General-purpose IC op amps provide input resistances of a few megohms, while FET-input devices achieve input resistances as high as $10^{12}$ ohms.

### *Noninverting Configuration*

Similar to the inverting configuration, the noninverting configuration can also introduce a reduction in $\beta$. In this case, however, there are two potential contributions. This is illustrated in Fig. 1-14B. If the source resistance seen by the (+) input of A is low, then the error due to amplifier input resistance reduces to the same error as in the inverting stage. If the source resistance, $R_s$, is such that it subtracts significantly from the error voltage, $E_e$, then the modified equation for $\beta$ is

$$\beta' = \left(\frac{R_{in} \parallel (r_{in} + R_s)}{[R_{in} \parallel (r_{in} + R_s)] + R_f}\right)\left(\frac{r_{in}}{r_{in} + R_s}\right)$$

Also, like the inverting configuration, a sufficiently high input impedance in the noninverting configuration can render these error sources negligible and can reduce the equation for $\beta$ to that of the classic form. This is also the case when $r_{in}$ is 10 or more times greater than the feedback source resistance, $R_s$.

Unlike the inverting configuration, the noninverting configuration is subject to errors due to common-mode resistances. The common-mode input resistance is modeled in Fig. 1-14B as two resistances of $2r_{cm}$, one at each input terminal. Taking these resistances into account, the modified equation for $\beta$ becomes extremely unwieldy:

$$\beta'' = \left(\frac{(R_{in} \parallel 2r_{cm} \parallel r_{in}) + (R_s \parallel 2r_{cm})}{(R_{in} \parallel 2r_{cm} \parallel r_{in}) + (R_s \parallel 2r_{cm}) + R_f}\right)\left(\frac{r_{in}}{r_{in} + (R_s \parallel 2r_{cm})}\right)$$

In practice, this equation is of little more than academic interest, as typical common-mode input impedances of IC op amps are in the hundreds of megohms. Therefore, errors due to common-mode input resistance loading may be neglected in almost all cases; other effects are far more dominant.

### 1.2.3 Errors Due to Nonzero Output Resistance

The nonzero output resistance of practical op amps also contributes to gain error, having a common effect on either the inverting or noninverting configurations, as illustrated in Fig. 1-15. A real amplifier, with output resistance $r_o$, can be modeled as an ideal amplifier with gain $A_{vo}$ in series with output resistance $r_o$. From this, it can be seen that the loading effects of $R_L$ and $R_f$, combined with $r_o$, form a voltage divider that reduces the ideal gain to a new gain, $A_{vo}'$. Including the effects of the $r_o$-$R_f \parallel R_L$ divider yields a new open-loop gain equation:

$$A_{vo}' = \frac{A_{vo}}{1 + [(R_L + R_f)/R_L R_f]\, r_o}$$

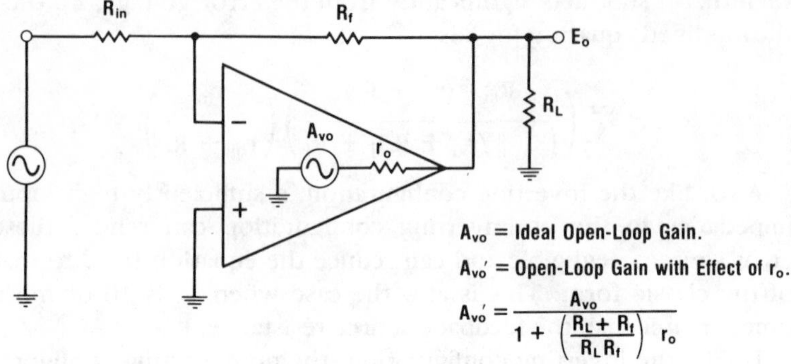


$$A_{vo}' = \frac{A_{vo}}{1 + \left(\frac{R_L + R_f}{R_L R_f}\right) r_o}$$

**Fig. 1-15. Effects of nonzero output resistance.**

From this equation, the greater the open-loop output resistance, $r_o$, the greater the attenuation of $A_{vo}$. As a result, loop gain is also attenuated as $r_o$ increases. The attenuation of open-loop gain due to loading can be a significant error in instances where $r_o$ is high in relation to the load resistance. Since there are some IC op amps with output resistances as high as several thousand ohms, this is a factor to be considered when maximizing loop gain. Generally, as long as $r_o$ is one tenth or less of the total load resistance, it will not contribute excessive error.

### *Summary of Gain Errors*

The nonideal properties of practical amplifiers discussed thus far all result in deviations from the ideal gain characteristic. This is due to a reduction in loop gain—either from a limited or reduced $A_{vo}$, or from a reduced $\beta$. Loop gain is the key parameter of feedback-amplifier configurations, since it determines the ultimate accuracy.

A complete gain equation that would include all sources of gain error is not given because it would not only be an extremely unwieldy expression but, in practice, the dominant sources of error may often be pinpointed by inspection. As a result, those error factors that warrant exact expression will usually be examined individually to generate modified figures for $A_{vo}$ and $\beta$. Quite often the "rule-of-thumb" judgments previously mentioned for such parameters as input and output resistance will give values for $A_{vo}\beta$ to within a few percent, which in most cases is sufficient accuracy.

### 1.2.4 Effects of Input-Referred Errors

In addition to the static gain errors discussed in the previous sections, there are a number of error sources in practical op amps that are referred to the input terminals. These errors include input voltage and current offsets (with their associated drifts due to temperature changes), and self-generated and power-supply-induced input noise. These nonideal parameters, together with techniques for their minimization, will be covered in this section.

***Input Offset Voltage***

While the ideal op amp has the property of zero output voltage at zero input voltage, a real amplifier does not. Practical op amps have an input offset voltage, as illustrated in Fig. 1-16. This practical amplifier is modeled as an ideal amplifier, A, with an offset voltage generator, $V_{io}$, in series with one of its inputs. It does not matter which input is used, since it can be shown that the net effect will be the same for either one. Also, the actual polarity of the offset voltage may be either positive or negative. The input offset voltage, $V_{io}$, is defined as the voltage applied between the input leads to obtain zero output voltage. Offset voltage is a characteristic independent of operating gain, and it is amplified by the noise gain of the stage. Thus, the output voltage due to $V_{io}$ is

$$V_o = V_{io}\left(\frac{R_{in} + R_f}{R_{in}}\right)$$

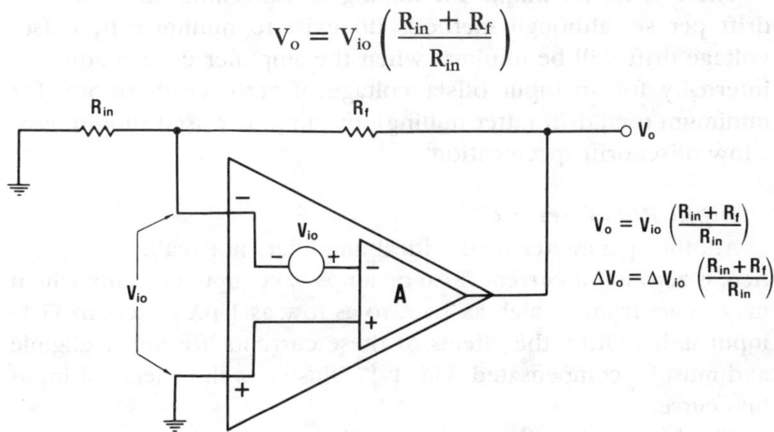


**Fig. 1-16. Input offset-voltage errors.**

Input offset voltage is due to imperfect matching in the input stage of the op amp. Different op-amp types will exhibit different degrees of matching and so will have better (or worse) input offset voltages. Offset voltage may be nulled by introducing an opposing voltage at one of the input terminals to cancel its effect, or by introducing an imbalance within the amplifier, resulting in an actual input offset of zero. In practice, control of offset voltage can be exercised either by selecting a device which yields an input offset below that required, or by nulling a device to achieve the required offset. Various methods of offset nulling will be discussed in detail in later sections.

### *Offset Voltage Drift*

With input offset voltage adjusted to zero, the remaining voltage offset error is input offset voltage drift. Although input offset may be nulled at room temperature, temperature changes will alter the state of balance in the input stage, inducing an input offset voltage drift. Offset voltage drift, being a component of input offset voltage, is subject to the same considerations; i.e., it is independent of stage gain and is amplified by the noise gain. Thus, output drift is

$$\Delta V_o = \Delta V_{io}\left(\frac{R_{in} + R_f}{R_{in}}\right)$$

There is no technique for nulling or cancelling offset voltage drift per se, although methods do exist to minimize it. Offset voltage drift will be minimal when the amplifier used is adjusted internally for an input offset voltage of zero. Furthermore, for minimum total drift (after nulling), the amplifier used should have a low offset drift specification.

### *Input Bias Current*

Another parameter of the ideal amplifier, not realized in practice, is zero input current. Real op amps have input currents which may range from as high as 1 $\mu$A to as low as 1 pA or less in FET-input units. Often the effects of these currents are not negligible and must be compensated. Fig. 1-17 illustrates the effects of input bias current.

The bias current flowing in the input terminals of an op amp can be modeled as a current source in parallel with each lead of an ideal op amp. In Fig. 1-17, these current sources are labeled $I_{ib-}$ and $I_{ib+}$. Although shown as being returned to ground, in actuality these current sources (or sinks) may be referred to the positive or

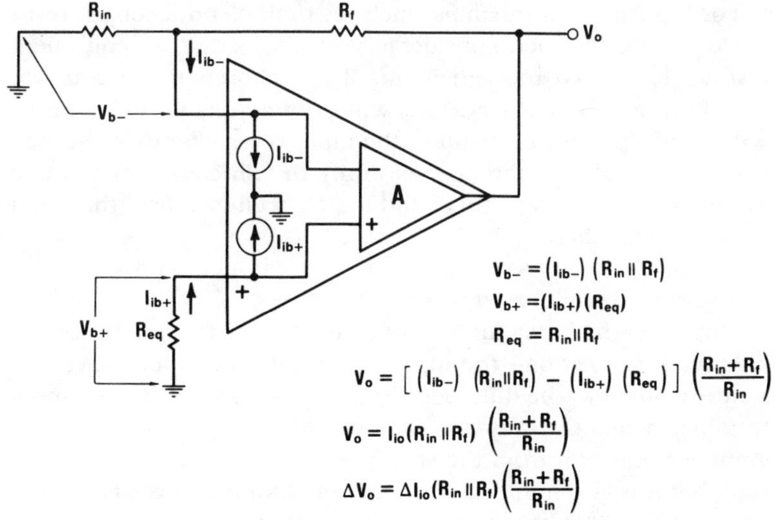


**Fig. 1-17. Input-bias and input-offset-current errors.**

negative supply, depending on the design of the individual amplifier. The currents $I_{ib-}$ and $I_{ib+}$ are independent of the common-mode voltage and flow in a return path (provided externally) at each input.

External to the amplifier, the input currents flow in the source resistances seen at each input and thus develop voltage drops. Current $I_{ib-}$ flows through $R_{in}$ and $R_f$, thus dropping a voltage:

$$V_{b-} = (I_{ib-})(R_{in} \parallel R_f)$$

For the moment, we will not consider $R_{eq}$, but we will assume the (+) input to be grounded directly. Under these conditions, the voltage drop, $V_{b-}$, due to the (−) input bias current is, in actuality, an input offset voltage (not to be confused with the input offset voltage of the amplifier, $V_{io}$, which is assumed to be ideal in this discussion).

As with $V_{io}$, $V_{b-}$ is amplified by the noise gain of the stage. Whether or not $V_{b-}$ is a significant error depends on two factors: (1) the magnitude of $V_{ib-}$ and/or (2) the magnitude of $R_{in} \parallel R_f$. In general, to minimize $V_{b-}$, either $R_{in} \parallel R_f$ must be small or a very low (<1 nA) input current must be used. This latter condition is generally true for modern FET-input units, which allow high resistances to be used.

For bipolar-input op amps, such restrictions on feedback resistance can be lessened considerably by the use of an equivalent resistance, $R_{eq}$, to the (+) input. $R_{eq}$ is chosen to be equal to $R_{in} \| R_f$. Then, if $I_{ib+} = I_{ib-}$, $R_{eq}$ will drop a voltage, $V_{b+}$, which will equal $V_{b-}$. In this manner, the input offset due to $V_{b-}$ is balanced out, and the amplifier sees only the difference in $V_{b-}$ and $V_{b+}$ as an input voltage. If $I_{ib-}$ and $I_{ib+}$ are well matched, this input offset voltage can be quite low.

### *Input Offset Current*

How well the input currents of an op amp match is described in another specification—the input offset current. Input offset current is defined as the difference in the currents into the two input terminals when the output is at zero. Thus, not only is a practical amplifier nonideal from the standpoint that it requires input current, but it is also nonideal from the standpoint that the two input currents it does require are unequal, or offset.

In practice, the offset current is typically 10% or less of the bias current in a well-matched op-amp input, although this figure can vary with different devices. In terms of circuit considerations, the only compensation for bias-current-induced offset voltage that can be provided is a nominally matched set of resistances, such as $R_{eq} = R_{in} \| R_f$. Beyond this, an amplifier must be selected for a low and stable characteristic offset current. In general, to minimize bias and offset-current effects, the source resistances seen at both inputs should be matched and maintained as low as is practicable. If the use of high resistances is required, then a low-input-current amplifier should be used. FET-input amplifiers feature initial (room temperature) bias currents in the picoampere region; however, they also have large temperature coefficients.

### *Bias- and Offset-Current Drift*

Bias current has a temperature dependence which is variable according to the type of input stage design used in the op amp. Bipolar-input op amps tend to have relatively stable input bias currents, which also makes their input offset currents relatively stable. Input offset-current drift is just as important a specification as input offset-voltage drift, because it reduces to an equivalent-offset-voltage drift in a circuit. Also, like voltage drift, this source of error must be minimized by careful device selection as the final measure of control.

FET-input devices have the lowest bias currents, although on a percentage basis, their *offset* currents are not as low as those of bipolar devices. Their input and offset currents are, however, low enough in absolute level that in many circuits they do not need offset-current compensation. A disadvantage of FET-input devices is the temperature dependence of bias (and offset) current, which roughly doubles for every 10°C of temperature rise. If very high operating temperatures are required, this characteristic tends to defeat their low-input-current superiority somewhat, although the degree to which this is true depends, of course, on the device and at how high a temperature one must operate it.

Summarizing the discussion on input offset voltage and currents, a combined model illustrating both is presented in Fig. 1-18.

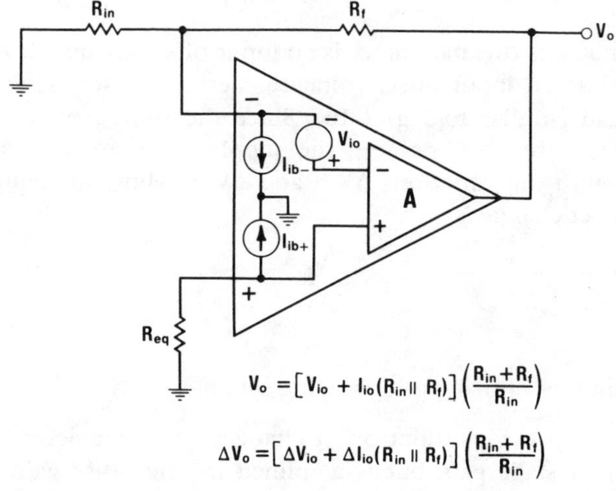


$$V_o = [V_{io} + I_{io}(R_{in} \parallel R_f)]\left(\frac{R_{in}+R_f}{R_{in}}\right)$$

$$\Delta V_o = [\Delta V_{io} + \Delta I_{io}(R_{in} \parallel R_f)]\left(\frac{R_{in}+R_f}{R_{in}}\right)$$

**Fig. 1-18. Composite input offset-voltage and -current errors.**

### *Power-Supply-Induced Input Offset Voltage*

Due to an incomplete ability to reject power-supply voltage variations, practical op amps have another source of input offset-voltage change. This source is termed power-supply rejection or, sometimes, supply-voltage sensitivity, and is a measure of how much variation appears as an equivalent input offset change for a given amount of variation in the power supplies. This parameter is illustrated in Fig. 1-19.

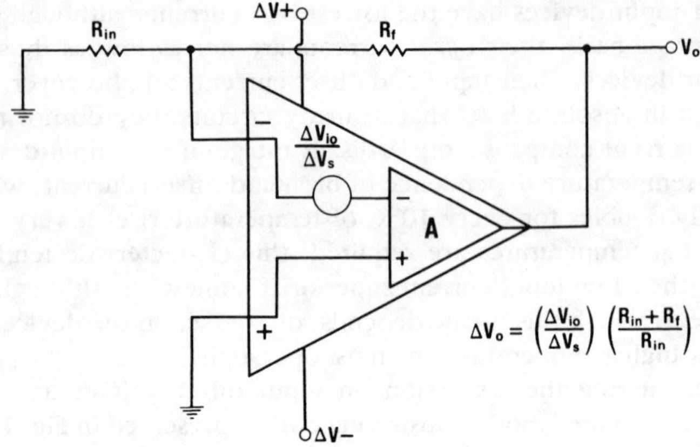


**Fig. 1-19. Power-supply-induced errors.**

Inasmuch as this parameter is an input offset change, it may be modeled as an input offset-voltage generator in series with one input lead (similar to Fig. 1-16). Since the source of this offset change is either V+ or V−, the supply leads are also shown, labeled with the variations ΔV+ and ΔV−. Thus, the equivalent input offset change is

$$\frac{V_{io}}{\Delta V_s}$$

where
$\Delta V_s$ implies both supplies varying symmetrically.

Like the previous input offset changes, this parameter is independent of stage gain but is amplified by the noise gain of the stage. Thus, the equivalent output change is

$$\Delta V_o = \left(\frac{\Delta V_{io}}{\Delta V_s}\right)\left(\frac{R_{in} + R_f}{R_{in}}\right)$$

Unlike the previous sources of input offset-voltage change, power-supply-induced offset can contain ac components as well as dc. Thus, a complete evaluation of the effects of this error component, as it appears at the output, must include the frequency response of the coupling to the input as well as the frequency response of the amplifier configuration. In general, the power-supply rejection of an op amp will deteriorate at higher frequencies. This will vary according to the op-amp design and is usually given in the form of a graph. Also, the rejection for V+ and V− is not necessarily symmetrical, particularly with respect to frequency. This can be an important factor when operating from supplies having appreciable high-frequency noise components.

The minimization of power-supply noise sensitivity can take one of two general approaches: (1) selection of an op amp for inherently low noise sensitivity, or (2) selection of a standard (general-purpose) op amp and employment of local decoupling where necessary. The evaluation of a specific application to determine which approach is the most appropriate or economical is usually left up to the designer.

### *Input Noise*

Although not specifically stated as one of the primary characteristics of an ideal op amp, it is desirable that noise-free operation be one of its virtues. However, practical amplifiers degrade the input signal by adding noise components. This noise is nearly random in nature and determines the ultimate lower limit of signal-handling capability. It is specified as equivalent input noise and, like the other input error factors, is increased by the noise gain of the stage. Unlike the other input parameters, however, input noise specification and control is by no means a straightforward process. It involves the interpretation of a number of involved specifications, as well as an understanding of the basic physics of its mechanism, to yield maximum performance. Disregarding the basic rules of noise-performance optimization, however, can result in signal-to-noise ratios that may be short of maximum by an order of magnitude or more.

A circuit model for the discussion of input-referred noise components is shown in Fig. 1-20. Similar to bias current and offset voltage, the noise generators of an amplifier are modeled as a series voltage-noise generator, $e_n$, and shunt current-noise generators, $i_{n+}$ and $i_{n-}$. These generators represent the mean values of voltage and current noise referred to the input of the amplifier. They are specified in terms of noise density, in volts-squared or amperes-squared per hertz of bandwidth. To these generators must be added two other noise sources—the thermal noise of the source resistances seen by the amplifier, which are $R_{eq}$ and $R_{in} \| R_f$.

Thus, a real amplifier has five potential sources of noise to be considered for minimization. First, there is the thermal noise of

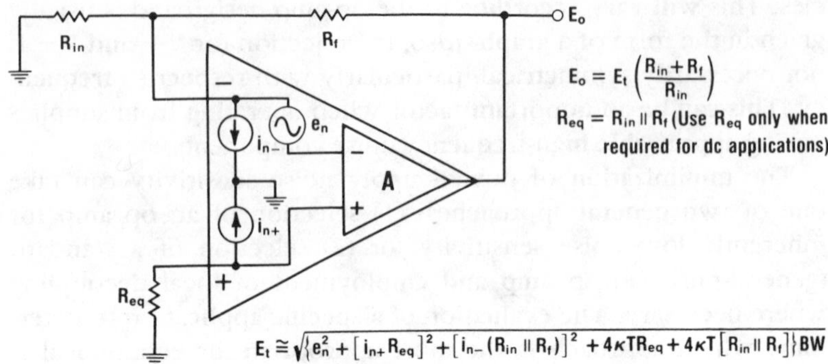


**Fig. 1-20. Input current and voltage noise errors.**

(possibly) two source resistances seen by the inputs, which is an irreducible minimum, existing even with an ideal (noiseless) amplifier. For audio applications, where the lowest noise is of greater concern than the lowest dc offset, a bias-compensation resistor, such as $R_{eq}$, should *not* be used. This will reduce noise by the simple elimination of an unnecessary noise source. Next, there are the noise-current and noise-voltage generators. For low values of source resistance, the effect of $i_n$ is often at a minimum. Under such conditions, which are usually typical for audio, $e_n$ will dominate as the source of amplifier noise. As the source resistance is increased, the effect of $i_n$ becomes larger until, at high source resistances, $i_{n+} R_{eq}$ and $i_{n-} (R_{in} \| R_f)$ are the dominant components of amplifier input noise. Thus, in specifications, these two parameters are detailed separately, with $e_n$ specified at a low source resistance and $i_n$ specified at a high source resistance. Both $e_n$ and $i_n$ are given in terms of spectral density, measured with a narrow bandwidth filter at a series of points across the useful spectrum of the amplifier. Data are given either in terms of $e_n$ (or $e_n^2$) or $i_n$ (or $i_n^2$) versus frequency, with typical graphs such as shown in Figs. 1-21A and B. These particular curves are given in $e_n/\sqrt{Hz}$ and $i_n/\sqrt{Hz}$, but other manufacturers sometimes choose to display the same information in terms of $e_n^2/Hz$ and $i_n^2/Hz$. Conversion between the two may be accomplished by simply squaring or extracting the square root as appropriate. When comparing such data in graphical form, however, it should be noted that variations in terms of $e_n^2$ will appear exaggerated over the same data expressed in terms of $e_n$. This clearly emphasizes the need for comparison of data on the same basis.

Although the spectral densities of both $e_n$ and $i_n$ are obviously not flat in the examples given (nor are they in general), it is not unreasonable to select a mean value for $e_n$ and $i_n$ if their total variation across the frequency range of the amplifier is within the same order of magnitude. An accurate calculation of total rms noise would involve integration across the bandwidth of the amplifier. The simplified approach given here will yield order-of-magnitude estimates of noise performance that are adequate for comparison purposes. Thus, given values of $e_n$, $i_n$, and bandwidth, the total noise of the circuit can be approximated as follows:

$$E_t \cong \sqrt{\{e_n^2 + [i_{n+} R_{eq}]^2 + [i_{n-} (R_{in} \parallel R_f)]^2 + 4\kappa T R_{eq} + 4\kappa T[ R_{in} \parallel R_f ]\}BW}$$

where

$E_t$ is the total circuit noise,
$e_n$ is the amplifier noise voltage in $V/\sqrt{Hz}$,
$i_n$ is the amplifier noise current in $A/\sqrt{Hz}$ ($i_{n+}$ or $i_{n-}$),
$R_{eq}$ and $R_{in} \parallel R_f$ are the source resistances ($R_{eq} = R_{in} \parallel R_f$),
$\kappa$ is Boltzmann's constant ($1.38 \times 10^{-23}$ J/K),
T is the absolute temperature (Kelvin),
BW is the noise bandwidth in Hz.

It is obvious from this expression that as $e_n$ and $i_n$ are reduced, the total noise approaches the thermal noise of $R_s$. A further point is that total noise is proportional to $R_s$ and bandwidth. An illustration of this is given in Fig. 1-21C, with the individual noise components shown in Figs. 1-21A and B. Such performance curves are typically given for IC amplifiers and may, in fact, be adequate information for comparative evaluation.

In choosing an amplifier, the requirements will often dictate a certain source resistance from which the amplifier must work. This will dictate which noise generator is dominant and, therefore, which specification must be minimized—$e_n$ or $i_n$. In general, low-input-current amplifiers, such as FET types or low-bias-current bipolar types, will have lower current noise, thus tending to be quieter with source impedances above 10,000 ohms. Below 10,000 ohms, the advantage swings to bipolar types, which have lower voltage noise. In any instance, absolute minimum values of feedback resistance should be used so that $R_s$ is composed largely of generator resistance. This means that the feedback resistances used should be low in relation to the generator resistance. Another consideration is that the noninverting configuration has less

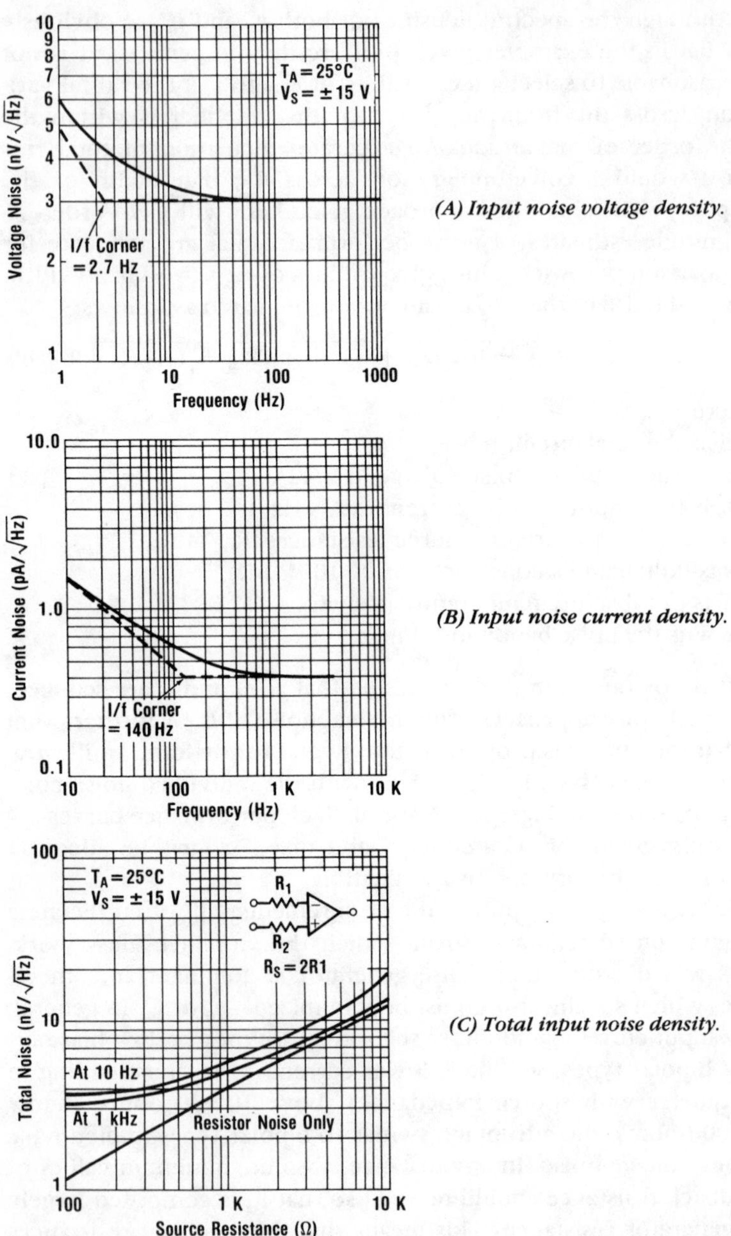


**Fig. 1-21. Typical op-amp noise performance specifications—OP-27 amplifier.** *(Printed with permission from Precision Monolithics Inc.)*

noise gain than the inverting configuration for low signal gains; therefore, it offers a distinct advantage in signal-to-noise ratio (at high gains, this advantage diminishes, however).

The equation for $E_t$ also illustrates (indirectly) another property of noise behavior and calculations. This is the rms addition property of unrelated noise voltages used to arrive at a total noise voltage.

For instance, if there are two noise voltages whose sum is to be calculated, voltages $E_1$ and $E_2$, it is done as follows:

$$E_t^2 = E_1^2 + E_2^2 + \ldots E_n^2$$

or

$$E_t = \sqrt{E_1^2 + E_2^2 + \ldots E_n^2}$$

As an example, if $E_1 = 1\ \mu V$ and $E_2 = 1\ \mu V$, $E_t$ will be $1.414\ \mu V$ (not $2\ \mu V$).

Likewise, taking a more practical example, suppose we had an amplifier circuit where there were three noise sources (expressed in $nV/\sqrt{Hz}$), which were $3\ nV/\sqrt{Hz}$, $1.3\ nV/\sqrt{Hz}$, and $4\ nV/\sqrt{Hz}$, respectively. Their sum would be

$$\begin{aligned} E_t &= \sqrt{(3 \times 10^{-9})^2 + (1.3 \times 10^{-9})^2 + (4 \times 10^{-9})^2} \\ &= 5.2\ nV/\sqrt{Hz} \end{aligned}$$

This is a realistic example, as it could represent an OP-27 op amp ($3\ nV/\sqrt{Hz}$), with a 100-ohm feedback resistance, fed from a 1000-ohm source. The net noise of such a configuration would be $5.2\ nV/\sqrt{Hz}$.

It should always be borne in mind that the total noise of an amplifier working from a given source resistance will, therefore, be dictated by a number of factors, many of which are under the designer's control. Source resistance or feedback-resistor noise can be readily determined and, obviously, such resistances should be minimized where possible.

To minimize total noise, it should be obvious from the equation for $E_t$ that as many components within the radical as possible are to be either minimized or eliminated. In a case where it is desirable to characterize the noise performance of a given op amp, the source resistance should be held to an absolute minimum for measurement of $e_n$, or at a defined value for measurement of $i_n$.

In audio work, it is usually $e_n$ that is the more important of the two, inasmuch as source resistances are always held to low values for low-noise amplifiers, and under such conditions only $e_n$ will be significant. The noise characterization procedure described in reference 7 measures op-amp $e_n$ basically, and $i_n$ optionally.

Noise of an op amp can be measured simply by placing the device in a high-gain configuration, with a defined source resistance, and measuring the output rms noise in a defined bandwidth. Of course, appropriate shielding, screening, and power-supply noise-isolation techniques must be used so as not to contaminate the reading with noise components external to the unit under test.

Two fundamental properties of noise behavior should be appreciated before noise measurements are attempted. First, bandwidth as related to the measurement of noise is not the same as 3-dB bandwidth, for fundamental reasons. Noise bandwidth is measured ideally by an infinitely fast rolloff filter, which rejects all noise components above the frequency of cutoff. A simple 6-dB/octave rolloff with a corner frequency will, of course, pass significant information above its cutoff frequency, as shown in Fig. 1-22. Therefore, if noise is to be measured by anything other than a "brick wall," or an ideal noise filter, correction factors must be used.

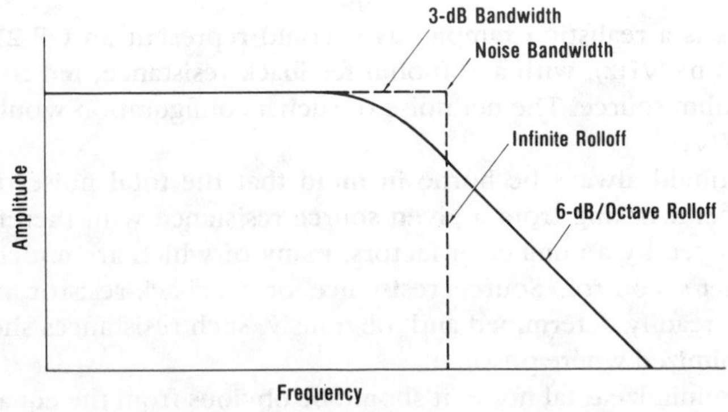


**Fig. 1-22. Relationships between noise bandwidth and 3-db bandwidth.**

For a single-pole 6-dB/octave rolloff, the relationship is

$$\text{Noise Bandwidth} = 1.57\ f_c$$

where

$f_c$ is the 3-dB small-signal bandwidth.

Faster slope filters afford greater accuracy, as they approach the ideal noise-bandwidth filter function.*

Another point is that noise should be measured by an *rms responding* voltmeter for best accuracy, because of the crest factor of noise voltages. Typical sine-wave-calibrated average-responding meters will read 1 dB low on noise, compared to a true rms meter. They can still be used, however, if they have adequate bandwidth (10 times measurement or better) and a correction factor is used. The correction factor is +1 dB (or 113%). Mathematically, this is expressed as

$$E_{rms\ (corrected)} = 1.13\ E_{reading\ (average)}$$

Crest factor errors can be minimized by avoiding near full-scale readings, which can overload on noise peaks and cause errors.†

### 1.2.5 Errors Due to Dynamic Effects

As pointed out in Section 1.2.1, the predominant error factor in most op-amp applications is limited loop gain, due most often to finite open-loop gain. In this section, the Bode plot was introduced with a graphical example of the interrelationship of closed- and open-loop performance parameters. We will now examine some frequency response considerations of op amps to determine the closed-loop performance characteristics for wide bandwidths.

#### *Gain and Bandwidth Relationships*

As discussed previously, a real op amp does not have infinite gain; it also does not have infinite bandwidth. Open-loop bandwidth begins to roll off from the full dc value at some low frequency, generally between 10 Hz and 1000 Hz, but most often between 10 Hz and 100 Hz in general-purpose op amps. As noted in Fig. 1-23, from the open-loop −3-dB point (the point at which the gain is down 3 dB from the full dc value), the rolloff is 6 dB/

---

* If a great number of high-accuracy, highly repeatable measurements are to be performed, a committed 20-kHz noise-bandwidth filter should be considered.

† For a discussion of noise-measurement techniques applicable to IC op amps, see reference 7 at the end of the chapter. Also, reference 10 is an excellent general reference work on noise, its analysis, and measurement.

octave (20 dB/decade) with increasing frequency until the *unity-gain* frequency, $f_t$, is reached. In a curve such as this, the product of gain and frequency is constant at any point on the curve. Thus, the unity-gain frequency also defines the available gain at any point along the curve.

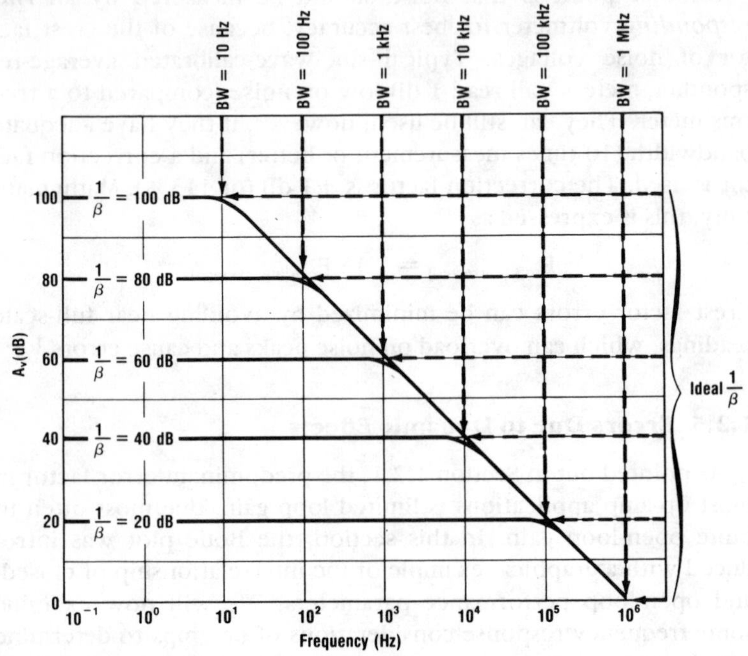


| $\frac{1}{\beta}$ | BW, @ −3-dB |
|---|---|
| 0 dB (1) | 1 MHz ($10^6$) |
| 20 dB ($10^1$) | 100 kHz ($10^5$) |
| 40 dB ($10^2$) | 10 kHz ($10^4$) |
| 60 dB ($10^3$) | 1 kHz ($10^3$) |
| 80 dB ($10^4$) | 100 Hz ($10^2$) |
| 100 dB ($10^5$) | 10 Hz ($10^1$) |

**Fig. 1-23. Illustration of gain and bandwidth interrelationship with constant GBP amplifier open-loop response.**

In Fig. 1-23, the example chosen is an amplifier with a unity-gain frequency of 1 MHz, typical of a large number of IC op amps. The product of gain and bandwidth (GBP) in an amplifier response such as this is constant; thus,

$$GBP = A_v BW$$

Graphically, the closed-loop −3-dB bandwidth point is defined as the point at which the ideal closed-loop gain curve ($1/\beta$) intersects the open-loop gain curve. An entire family of $1/\beta$ curves is drawn on this plot to illustrate the point, and this relationship is also shown by the table of gains and bandwidths. Thus, for a gain of 10 (20 dB), the bandwidth is 100 kHz and the GBP is 1 MHz. Similarly, a gain of 1000 (60 dB) yields a bandwidth of 1 kHz, and the GBP is again 1 MHz. Although the curves shown are drawn in decade increments of gain and bandwidth for clarity, intermediate values of gain (or bandwidth) will likewise yield a constant gain-bandwidth product.

The conclusion to be drawn from this is a basic mechanism of the feedback process. More feedback (exemplified by the lower gains) allows greater bandwidth, up to the limit imposed by the open-loop response of the amplifier. As far as the amplifier open-loop response is concerned, it remains unchanged regardless of the amount of feedback. Thus, the rationing of gain and bandwidth is an exchange process—if more gain is necessary, reduced bandwidth is the by-product.

One final conclusion to be drawn from this is perhaps already obvious. With regard to loop gain, there is also an exchange process for various closed-loop gains. Loop gain, as stated previously, may be evaluated graphically as that portion of the response above a given closed-loop curve. For example, loop gain for $1/\beta = 20$ dB is 100 dB, −20 dB, or 80 dB at 1 Hz. The loop gain falls to zero at the $1/\beta$ −3-dB point, or 100 kHz. For the $1/\beta = 40$ dB curve, loop gain is 100 dB, −40 dB, or 60 dB at 1 Hz. The loop gain falls to zero at 10 kHz, the $1/\beta = 40$ dB, −3-dB point. Thus, the higher the closed-loop gain is, the lower the loop gain is and, also (but just as important), the sooner the loop gain diminishes to zero, again due to the fixed gain-bandwidth product. It is sobering to consider these effects at high closed-loop gains—not only does the bandwidth shrink, but even before the −3-dB point is reached, loop gain is falling, diminishing the curative aspects of the feedback. We will now examine methods of optimizing loop gain with frequency, which at this point is obviously a nonideal property.

### *High GBP Amplifiers*

Since the limitation on the available loop gain demonstrated in Fig. 1-23 is due to a limited gain-bandwidth product, it naturally follows that a greater gain-bandwidth product allows proportionally greater loop gain, or greater bandwidth. These factors are demonstrated in Fig. 1-24. For reference, this illustration includes the original 1-MHz GBP curve (curve A), as well as a 10-MHz GBP curve (curve B), and a 100-MHz GBP curve (curve C).

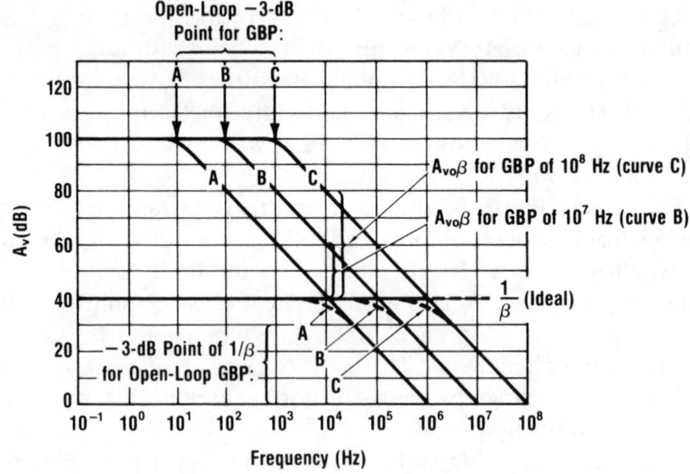


**Fig. 1-24. Gain and bandwidth interrelationships with extended gain-bandwidth products.**

The evaluation of this new parameter can be approached from a number of viewpoints. For simplicity, a single $1/\beta$ curve is shown for a gain of 40 dB, but the same general considerations apply for any closed-loop gain. As an example, curve B extends the −3-dB point for the $1/\beta$ curve to 100 kHz, and curve C extends it to 1 MHz. This is also apparent from the 10-MHz and 100-MHz gain-bandwidth products which, when divided by a gain of 100 (40 dB), do in fact yield 100 kHz and 1 MHz. Moving the frame of reference lower in frequency to observe loop-gain differences, at 10 kHz the 10-MHz open-loop response (curve B) allows 20 dB of loop gain, while the 100-MHz response (curve C) allows 40 dB of loop gain. Moving even lower in frequency, it will also be noted that the higher open-loop −3-dB points allow the full low-frequency loop gain of 60 dB to be extended to 100 Hz and 1 kHz for curves B and C, respectively. Finally, if the new curves are viewed from the standpoint of a given bandwidth (or loop gain), the higher GBP curves allow more gain to be realized in a given bandwidth. The 10-MHz curve, for example, allows 60 dB of gain in a 10-kHz bandwidth, while the 100-MHz curve allows 80 dB of gain in a 10-kHz bandwidth.

Figs. 1-23 and 1-24 should be instructive in the interrelationship of gain and bandwidth, and the necessity for a high GBP in achieving high loop gains and subsequent accuracy at high frequencies. The numerical evaluation of gain-bandwidth relationships may be done either mathematically or by using the Bode plot. It is suggested, however, that the reader become familiar with the Bode plot at the earliest opportunity as it yields a large amount of information very quickly, and it is a tool as fundamental to op-amp technology as the op-amp symbol itself.

So far, nothing has been said concerning the methods of achieving the higher gain-bandwidth products represented by curves B and C of Fig. 1-24. High GBP with stability in an op-amp circuit is one of the more "meaty" challenges with which the reader must become familiar, and this subject is discussed in the next section.

### *Frequency/Phase Compensation*

The previous section stressed the characteristics of the 6-dB/octave open-loop rolloff rate which is typical of the greatest percentage of op amps. At this point, the reader might justifiably ask: "If more gain is needed at higher frequencies, why not just extend the open-loop bandwidth?" The answer to this question is basic and fundamental to stability in closed-loop feedback systems.

*Stability Requirements*—The 6-dB/octave rolloff rate is characteristic of a simple RC low-pass lag network such as that shown in Fig. 1-25A. In such a network, the maximum phase shift is −90°, as illustrated in Fig. 1-25B. At the −3-dB point of

$$f_1 = \frac{1}{2\pi RC}$$

the phase shift is −45°. As the frequency is increased, the phase shift reaches a maximum of −90° at approximately

$$f_2 = \frac{10}{2\pi RC}$$

and then remains at −90° for any further increase in frequency.

This simple RC network has an important relationship to op-amp frequency response, as we shall see now in Fig. 1-26.

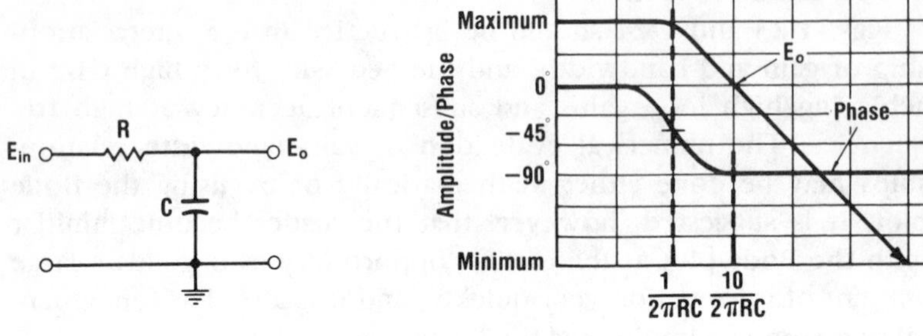


(A) *Schematic.* (B) *Amplitude and phase response.*

**Fig. 1-25. Response of a simple RC lag filter.**

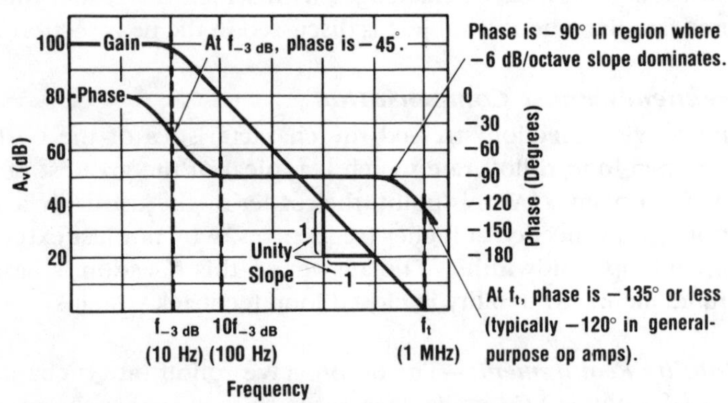


**Fig. 1-26. Open-loop gain and phase response for a general-purpose op amp.**

In all the discussions about feedback thus far, no mention of phase shift or its effect on closed-loop performance has been made. The fact is that phase shift and its control are of paramount importance in achieving closed-loop stability in a feedback system. We have been assuming a feedback loop with an inverting sign, which implies a phase shift of −180°. Consider now the effects of an additional phase shift. If, in addition to the −180° shift, we were to add more phase shift until we reached an additional −180°, we would have achieved 360° of phase shift, or an in-phase condition. If the loop gain is greater than unity at the frequency where this additional phase lag is −180°, we would also have succeeded in making an oscillator, for these are the two conditions which satisfy the criteria for oscillation. Thus, the control of phase shift and gain in feedback amplifiers is basic to achieving closed-loop stability. However, as there is a criterion for oscillation, there is also a corresponding stability criterion for feedback amplifiers. This fundamental rule can be applied to any feedback amplifier, and a stable system can be graphically determined with the aid of the Bode plot.

As may be deduced from the oscillation conditions just mentioned, the requirements for stability must reduce the total gain to less than unity before the additional 180° of phase shift is reached, in order to prevent oscillation. This means that the open-loop gain rolloff of the op amp must be done in such a manner as to prevent its phase from reaching −180° before the gain has dropped below unity. This characteristic is illustrated by the gain and phase plots of Fig. 1-26.

If the entire rolloff range of the amplifier can be made to act like the single RC lag element, then the maximum phase shift it can accumulate is −90°, which gives an additional 90° of phase margin before the −180° point of potential instability is reached. It is common practice to provide a phase margin in feedback-loop design to allow for component tolerances, capacitive loading, and other effects that can further increase the total phase shift.

In Fig. 1-26, the amplifier open-loop rolloff appears as a single lag over most of the range. Such a response is typical of general-purpose op amps with a GBP of 1 MHz. At the −3-dB point of 10 Hz, the phase shift is −45°, increasing to −90° at 100 Hz. It then remains at −90° until the frequency nears the unity-gain crossover region, where additional internal phase shifts begin to accumulate, and the phase shifts further toward −180°. At the unity-gain frequency, however, the phase shift is appreciably less than −180°, usually by a phase margin of 60° or more. A phase margin of 45° is considered the minimum acceptable figure for conservative designs.

Such an amplifier as we have just described is said to be *unconditionally stable,* or *unity-gain stable,* because it can be used with any amount of feedback (including 100%) without oscillation or instability problems. The unity-gain condition is the "worst-case" feedback state for stability because under this condition, $\beta = 1$ and $A_{vo}\beta$ is a maximum. With no feedback attenuation, there is a maximum of available gain (the full open-loop gain of the op amp) attempting to satisfy the oscillation criterion. Therefore, control of phase shift is crucial to unity-gain feedback configurations. This is the impetus behind the use of a single, lag-element-type, open-loop response; it affords stable control of open-loop phase characteristics over many decades of gain rolloff, as seen in Fig. 1-26. It was the synthesis of just such a response as appeared in Fig. 1-26 that led to the general-purpose IC op amps which we find so useful today.

*General Stability Criterion*—A more general examination of stability in IC op amps must include conditions other than the specific response shown in Fig. 1-26. These other conditions include multiple-slope rolloffs where the open-loop phase can become −180° and greater, plus special cases of feedback where even the −6-dB/octave slope of Fig. 1-26 can be unstable in special cases.

For any feedback amplifier, there exists a stability criterion which can be stated as follows:

> ***For unconditional stability, the ideal 1/β curve must intersect the open-loop response with a net slope of less than −12 dB/octave.***

Fig. 1-27 illustrates the open-loop response of a hypothetical op amp with a number of $1/\beta$ curves. The first significant difference to be noted about this open-loop curve is the fact that it has two slopes in its rolloff region. The first section rolls off at a rate of −6 dB/octave, while the second section rolls off at −12 dB/octave, or twice the rate of the first section. A −12-dB/octave rolloff will ultimately reach −180° of phase shift; thus, it is potentially unstable if the loop is closed at or near the point where the phase is −180°. We will now examine the effects of different examples of loop closure (points of intersection shown in Fig. 1-27) on stability.

In the first example, the ideal gain curve, $1/\beta_1$, is a flat gain of 60 dB which clearly intersects the open-loop curve in a −6-dB/octave region. Thus, the net slope (slope between the two curves) is −6 dB/octave (since the $1/\beta_1$ curve has a slope of zero), and the loop is stable. An additional point which indicates stability of this loop closure is that the change in slope to −12 dB/octave is a full decade higher than the intersection of the $1/\beta_1$ curve with the open-loop curve. Thus, the phase changes that will be associated

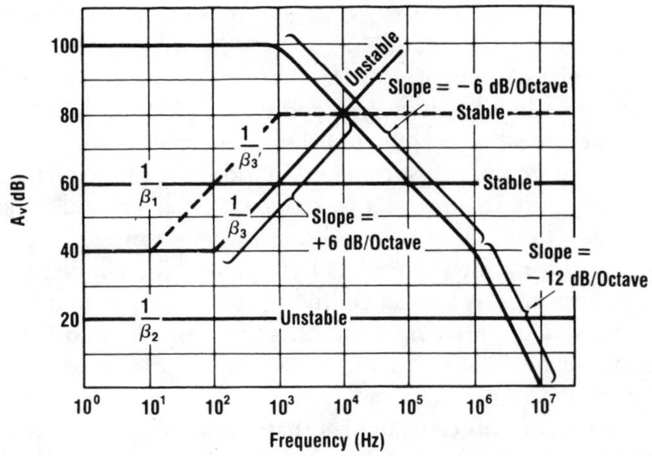


**Fig. 1-27. Stability considerations with various loop closures.**

with this slope change have not yet begun, so the open-loop phase at this point will be −90°.

In the second example, the ideal gain curve, $1/\beta_2$, is also flat, but it is at a gain of 20 dB. This curve intersects the open-loop curve in the −12-dB/octave region, and since $1/\beta_2$ has zero slope, the net slope at the intersection is −12 dB/octave. Also, since this intersection point is beyond the point of slope change from −6 dB/octave to −12 dB/octave, the open-loop phase shift will have reached −180°. Thus, this loop closure will definitely be unstable and will result in oscillation, because there is still a 20-dB gain at the point where the curves intersect.

So far the examples considered have assumed a flat frequency response for the feedback characteristic, $1/\beta$. However, this is not always the case; $1/\beta$ may include reactive elements. Reactance in the feedback path changes the phase characteristic of $1/\beta$, just as reactance also changes the phase of the open-loop characteristic. For stable closed-loop operation, both reactance effects must be considered.

A reactive feedback characteristic can affect stability even when using the so-called "ideal" 6-dB/octave open-loop rolloff. Such a case is the $1/\beta_3$ curve shown in Fig. 1-27, which is a "peaked" open-loop response characteristic, typical of a differentiator. This curve intersects the open-loop characteristic with a +6-dB/octave slope, while the open-loop response is −6-dB/octave at the intersection. Therefore, the net slope is −12 dB/octave. The phase of the $1/\beta_3$ characteristic will be 90°, since it is two decades beyond the initial slope change at 100 Hz. The open-loop phase will be −90°, since it is one decade beyond the initial slope change. Thus, the net phase is −180° and the loop closure will be unstable. To restore stability to such a closed loop, the phase characteristic must be modified in order to reduce the net slope difference at the intersection. This may be done by modifying $1/\beta_3$ to the new curve, $1/\beta_3{}'$, which intersects the open-loop characteristic in a region where $1/\beta_3{}'$ is flat. Since this region of $1/\beta_3{}'$ is one decade beyond the last slope change, the phase will be close to zero, so the net phase is reduced to the −90° of the open-loop curve; thus, the loop closure will be stable.

There are many other examples that could be used to demonstrate the working of the stability criterion, but they all reduce to control of the net phase between the ideal $1/\beta$ curve and the open-loop characteristic. With a given open-loop gain/phase characteristic, the feedback loop must be controlled in gain and phase so that the stability criterion is satisfied. Generally speaking, the −6-dB/octave slope characteristic of the general-purpose op amp allows freedom of feedback networks without much regard for stability considerations, except in certain special cases, such as the $1/\beta_3$ curve shown in Fig. 1-27.

When gain-bandwidth is to be optimized, it may be more advantageous to exercise control over the open-loop characteristic of the op amp, in order to satisfy loop gain or stability requirements. The reason is that high GBP amplifiers with uniform rolloff rates of 6 dB/octave are rare. In fact, the 100-MHz −6-dB/octave characteristic alluded to in Fig. 1-24 is not yet a reality in any IC op amp. Such a characteristic is quite difficult to realize because of inevitable high-frequency phase shifts and parasitic effects within the op-amp circuitry, as well as the external circuitry. However, over *portions* of the total response, a high gain-bandwidth product can be realized with controlled rolloff rates and controlled phase characteristics. Thus, high-gain wideband amplifiers can be built using IC op amps with externally adjustable phase/frequency compensation, which allows some shaping of open-loop characteristics.

### *Effects of Phase/Frequency Compensation*

A basic requirement of any op-amp frequency-compensation network is the control of open-loop rolloff so as to guarantee stability under closed-loop conditions. RC networks, placed at appropriate points in the op-amp circuit, cause the open-loop rolloff to assume the characteristic desired.

In general-purpose op amps, rolloff networks cause the open-loop response to assume a single rolloff slope of −6 dB/octave with an associated phase characteristic of −90° lag. This network can be self-contained within the amplifier; then, the amplifier is said to be *internally compensated.* Other op amps require external compensation networks to shape the open-loop response—either to a response similar to a general-purpose type or, with some amplifiers, to a response that may exhibit multiple slopes, such as the open-loop characteristic shown in Fig. 1-27.

It is not the purpose of this discussion to describe the design of these networks, but to impart a useful appreciation of the value of their differences, the basics of what they accomplish, and, most importantly, how to interpret different open-loop responses so as to obtain maximum use of various IC amplifiers.

Basically, interpreting an open-loop response is evaluating the gain/phase characteristic in view of a particular application. With the previous background on gain-bandwidth, loop gain, stability requirements, and the general use of the Bode plot as a framework, the reader should now be developing an appreciation for the differences between the dc and ac performance of op amps. We will now examine what frequency-compensation networks accomplish in modifying open-loop response.

In Fig. 1-28, the uncompensated open-loop response of a hypothetical op amp is shown as curve A. This op amp has a number of internal phase shifts associated with it, as is evident from its three different rates of rolloff. The first rolloff begins at 10 kHz, where the open-loop response is 3 dB down and is at a rate of −6 dB/octave. The second rolloff occurs at 1 MHz and changes to a rate of −12 dB/octave. The third rolloff occurs at 10 MHz and changes to a final slope of −18 dB/octave. As it stands, this amplifier would be usable only with loop closures of 1 MHz or less (from point A2 upward on the curve), which limits it to gains of 60 dB or more. To make this amplifier usable at lower gains (from 60 dB down to unity), compensation must be introduced to synthesize a new rolloff slope of −6 dB/octave.

The most simple and direct method of doing this is to connect a capacitor at an appropriate point within the amplifier to introduce a lag. If the frequency at which this capacitance value reacts is low

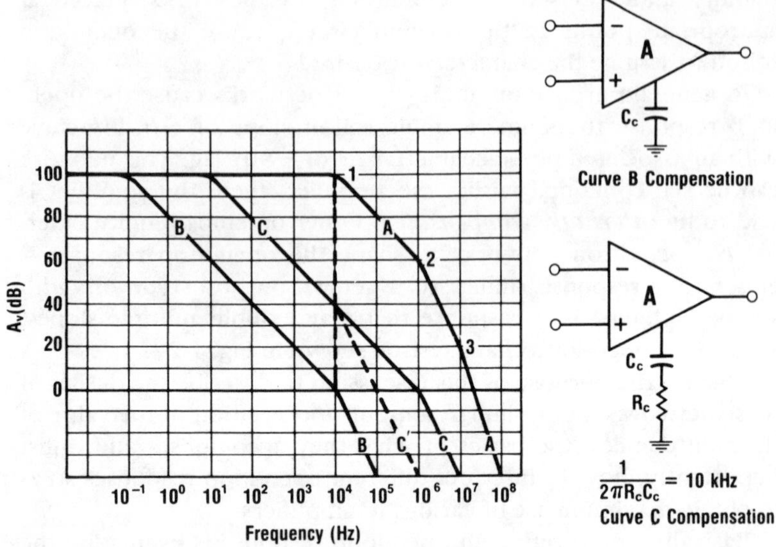


**Fig. 1-28. Phase compensation techniques.**

enough, its rolloff rate will become dominant, thus providing a single rolloff rate. To do this, the natural rolloff effects of the amplifier must be suppressed to a point where they can occur at gains less than unity. Curve B demonstrates this effect, which is aptly called a "brute force" technique. While this type of compensation will work, it ruins practically all of the useful high-frequency response of the amplifier.

A much more desirable technique is to introduce the rolloff at a higher frequency, as in curve C. A single capacitor lag introduced at this frequency will also result in a second rolloff due to the unsuppressed amplifier rolloff at 10 kHz. If left as such, curve C′ will result. This makes the amplifier usable with feedback down to 40 dB, but it is still not stable at unity gain because of the phase shift above 10 kHz. If, however, the rolloff due to $C_c$ is removed at the first rolloff frequency (10 kHz), the composite curve can be made to approximate a single rolloff. It will then appear as the complete C curve, which extends to unity gain at 1 MHz. Now the amplifier will be stable at unity gain and will have an overall characteristic which resembles a 6-dB/octave rolloff.

Compensation is by nature a rolloff of gain, done in such a manner as to control the overall phase characteristic. Control of phase is the desired objective, and gain rolloff with frequency is the means to this end. As seen in Fig. 1-28, this is costly in terms of high-frequency gain. For example, if the unmodified response (curve A) and the compensated response (curve C) are compared at 10 kHz, the open-loop characteristic (curve A) has 100 dB of gain and can be stabilized at a gain of 60 dB, which will yield a gain accuracy of 1%. By contrast, the compensated response (curve C) can only achieve a gain of 40 dB at 10 kHz, the frequency where its loop gain is zero.

From this discussion, it is hoped that the reader will be made aware of the trade-off involved in frequency compensation. While the example is not intended to portray an actual IC amplifier, it is close enough to be representative. Further study of manufacturers' data sheets will be more informative in terms of specific examples.

### *Slew Rate*

The final source of error from dynamic effects is called the *slew rate.* Simply stated, slew rate is the maximum rate of change of the amplifier output voltage. In addition to the small-signal bandwidth discussed previously, it is another aspect of op-amp performance with frequency.

Slew-rate limiting is brought about by a limitation in the internal circuitry of an op amp to drive capacitive loads, either internal or external. The capacitance that limits this slewing ability with a given amplifier is most often the compensation capacitance, either internal or external, although in some instances it is the load capacitance. At high frequencies or high rates of signal change, the current available to charge and discharge the capacitance becomes exhausted, and slew-rate limiting occurs. This is evidenced by distortion in the output, which will appear similar to the waveforms shown in Fig. 1-29. Under these conditions, the amplifier output will change at a maximum slew rate of

$$SR = \frac{\Delta E}{\Delta t}$$

Since slew rate is a large-signal performance parameter, a given rate also defines what is called the *power bandwidth,* which is

$$f_p = \frac{SR}{2\pi E_{op}}$$

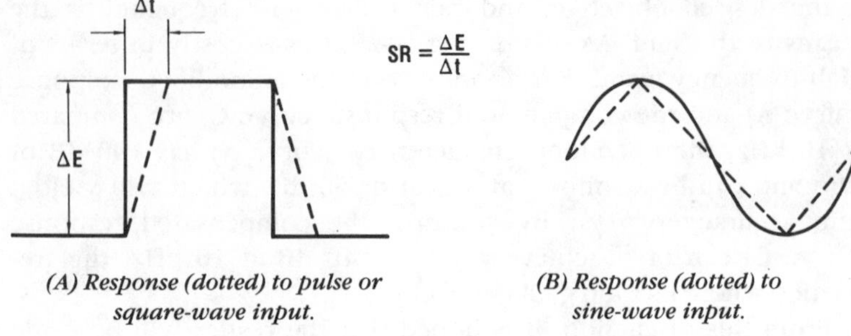


*(A) Response (dotted) to pulse or square-wave input.*

*(B) Response (dotted) to sine-wave input.*

**Fig. 1-29. Examples of amplifier output waveforms under conditions of slew-rate limiting.**

where

$f_p$ is the full power frequency,

$E_{op}$ is the peak amplifier output voltage.

Because slew rate is usually related to compensation capacitance, it follows that it is usually a variable for externally compensated amplifiers. One method used to minimize slew-rate problems is the selection of a device that can be optimized externally by adjusting the compensation (where stability requirements allow). If an amplifier is to be used at unity gain, however, a device must be chosen that will meet the slew-rate requirements under the conditions needed for stability.

### *Summary of Dynamic Effects*

Bandwidth, compensation, and slew-rate considerations are some of the most important sources of error that will be encountered in op-amp circuitry. Where appropriate in the remainder of this book, special consideration will be given to optimum ac performance in the circuits discussed.

## 1.3 IC OP-AMP SPECIFICATIONS

IC op-amp specifications can be divided into two general categories: (1) maximum ratings and (2) electrical characteristics. These specifications provide two different kinds of information about an IC op amp, and both are important. Maximum ratings give the maximum operating conditions under which the manufacturer recommends that you operate his device. Exceeding the maximum ratings can lead to immediate destruction of the device at worst, and will almost always result in degradation of the general operating characteristics. Electrical characteristics tell a more detailed story of device performance under typical operating conditions of temperature, supply voltage, loading, etc. These specifications may be defined in terms of certain limits, either minimum or maximum, depending on the particular parameter. A typical value is also given in most cases. This serves as a nominal center for design guidance. In addition to these specification limits, supplementary data are provided in the form of graphs, curves, and test circuits under which specification measurements are made. Data given in the form of curves are not normally guaranteed, whereas "performance limits" are.

This section discusses IC op-amp specifications by first designing the term and identifying its symbol, and then clarifying it in the light of practical examples wherever possible.

### 1.3.1 Maximum Ratings

Supply Voltage (V + and V−)
*The maximum allowable supply voltage that can safely be applied to the amplifier.*

Although designed for ±15-volt supplies as standard, most IC op amps will operate over a wide range of potentials, some from as low as ±1 volt and some up to ±40 volts.

Power Dissipation ($P_D$)
*The power that a particular device is capable of dissipating safely on a continuous basis while operating within a specified temperature range.*

This rating will vary slightly according to the type of package used. Ceramic packages, for example, allow the highest power dissipation. Metal packages allow the next highest dissipation, while plastic-encapsulated packages generally allow the lowest. A typical power-dissipation capability for an IC op amp is about 500 milliwatts (up to a specified temperature); the exact figure will vary according to the manufacturer and the device. Note that only the standard (TO-99) metal-can package usually allows ready attachment of a heat sink, to augment dissipation characteristics.

Operating Temperature Range ($T_{or}$)
***The range of temperature over which the device will perform within the rated specification(s).***

Military-grade devices operate from −55°C to +125°C, industrial-grade devices operate from −25°C to +85°C, and commercial-grade devices operate from 0°C to +70°C.

Differential Input Voltage ($V_{id(max)}$)
***The maximum voltage that can safely be applied between the differential input terminals without excessive current flow.***

Some amplifiers have clamping diodes connected back to back across the input terminals. These types are limited to a differential input voltage of ±0.5 volt or less. Other types are limited to ±5.0 volts or less to prevent emitter-base breakdown in the input stage. The highest differential input-voltage ratings are provided by devices using a combination pnp/npn cascode input, which provides a ±30-volt rating, or by FET-input devices with similar ratings.

Common-Mode Input Voltage ($V_{cm(max)}$)
***The maximum voltage that can safely be applied between both input terminals together and circuit common.***

There is both a positive and a negative limit to this rating, although, typically, they are equal.

Output Short-Circuit Duration
***The length of time that the amplifier can withstand a direct short circuit from the output to ground or to either supply terminal.***

Note that rating(s) can be different, dependent upon the specific case (a ground or a supply short).

Lead Temperature
***The lead temperature that the device will withstand during the soldering process for a specified time period.***

This rating is typically 300°C and 60 seconds for most devices, with some manufacturers specifying 10 seconds. Plastic packaged devices will typically be lower in terms of both temperature and time, i.e., 260°C and 4 seconds.

### 1.3.2 Electrical Characteristics

The electrical characteristics of an IC op amp provide a more detailed story of performance by defining the means and limits of behavior, usually in open-loop terms. A device is generally characterized by a range of performance limits, obtained from testing a large number of units on a production run. These limits are defined under standard conditions of room temperature and supply voltage (±15 volts). Any further conditions are specified in detail with the individual measurements—for example, source resistance, frequency, load, temperature range, etc. Typical performance values are often accompanied by either a minimum or a maximum limit, although sometimes the typical values alone are given.

In the following discussion, the various parameters are grouped into categories such as input, output, dynamic, and general characteristics.

#### *Input Characteristics*

INPUT OFFSET VOLTAGE ($V_{io}$)
*The voltage that must be applied between the input terminals through two equal resistances to obtain zero output voltage.*

Ideally, an op amp will have zero input offset voltage. In practice, however, IC amplifiers have low but definite offset voltages. Offset voltage is lowest in bipolar-input op amps because bipolar transistors tend to match extremely well. It is unusual for a bipolar-input IC op amp to have an input offset greater than a few millivolts, and general-purpose types typically have offsets of ±1 millivolt. Special selections of bipolar-input op amps have input offsets lower than ±0.5 millivolt, with a number of actively trimmed devices offering offset voltages of ±0.1 millivolt or less, some as low as 25 microvolts!

FET-input op amps are (comparatively speaking) poor in terms of input offset voltage because FETs do not tend to match nearly as well as do bipolar transistors; and the voltage being matched ($V_P$) is four to five times larger in a FET than the analogous voltage in a bipolar transistor ($V_{BE}$). Typical general-purpose FET-input amplifier offset voltages are in the 5- to 10-millivolt region, with some types having worst-case offsets that are greater. Actively trimmed FET units are available, however, with offsets to below ±1 millivolt.

Most IC op amps have provisions for internal offset null adjustment, which allows the input offset to be adjusted to zero.

Depending on the input bias-current characteristics of the particular device, the specified input offset voltage is guaranteed only up to a certain value of input resistance. In the case of general-purpose bipolar-input types, for example, this value is 10,000 ohms. Such a limitation is generally due to the offset-current characteristics (the difference in the two bias currents) of the device. With lower bias currents, offset current is also reduced; thus, higher source resistances may be used without offset-voltage degradation.

INPUT OFFSET-VOLTAGE TEMPERATURE COEFFICIENT ($\Delta V_{io}/\Delta T$)
***The ratio of change in input offset voltage to the change of circuit temperature for a constant output voltage.***

Input offset-voltage temperature coefficient, or offset-voltage drift, is specified in $\mu V/^{\circ}C$ of offset-voltage change and is an average value over the temperature range of the device. Again, due to the excellent matching characteristics of bipolar-input op amps, these types are best in terms of offset-voltage drift. This is one of the more stringent operating parameters of IC op amps and is not specified for all devices. Premium IC op amps have offset-voltage drifts of 1 $\mu V/^{\circ}C$ or less.

INPUT BIAS CURRENT ($I_{ib}$)
***The average of the two input currents.***

Input bias current is simply the current that is required at the input terminals of an op amp. It may flow either into or out of the input terminals, depending on the internal design used. In the balanced state, the two input currents are very nearly equal, but the parameter is usually specified as the average of the two input currents. In general, FET-input amplifiers will have lower bias currents, but bipolar amplifiers will have better-matched input currents (in terms of percent).

INPUT BIAS-CURRENT TEMPERATURE COEFFICIENT ($\Delta I_{ib}/\Delta T$)
***The ratio of the change in the input bias current to the change in circuit temperature.*** (This is an average value for a specified temperature range.)

Input bias-current temperature coefficient is a measure of how stable the input bias currents remain over the operating temperature range. It is most stable in bipolar-input op amps, usually rising at lower temperature extremes. On the other hand, FET-input op amps have a bias current that doubles (approximately) with every 10°C rise in temperature, because it is essentially a diode leakage current.

INPUT OFFSET CURRENT ($I_{io}$)
***The difference in the currents into the two input terminals when the output is at zero.***

Input offset current is a measure of the mismatch between the two bias currents. In general-purpose bipolar-input op amps, the input offset current is typically 10% of the bias current, but it can be much lower in premium types. FET-input devices do not match as well because there is less control over the bias currents, but this may not be relevant due to their lower (absolute) bias currents.

INPUT OFFSET-CURRENT TEMPERATURE COEFFICIENT ($\Delta I_{io}/\Delta T$)
***The ratio of the change in input offset current to the change in circuit temperature for a constant output voltage.*** (This is an average value for a specified temperature range.)

Input offset-current temperature coefficient is a measure of the change in offset voltage over the temperature range of the op amp. Like input offset-voltage temperature coefficient, it is an average value. It is one of the more stringent parameters and is not specified for all IC op amps. Also, like input offset-voltage temperature coefficient, it is better in bipolar-input devices due to better matching.

INPUT RESISTANCE ($r_{in}$)
***The resistance seen looking into either input terminal with the other input grounded.***

Input resistance in an IC op amp may be very high, as in the case of FET-input amplifiers; or it may be medium, as in the case of a bipolar-input amplifier. Input resistances for FET-input amplifiers can be as high as $10^{12}$ ohms at room temperature. General-purpose bipolar-input op amps have input resistances of 1 megohm or more, but specialized types (low input-current bipolar types) can be much higher. Ideally, of course, the input resistance is infinite, and this is very closely approximated by FET-input devices.

### Input Capacitance ($c_{in}$)

***The capacitance seen looking into either input terminal with the other input grounded.***

Input capacitance is important because of the reduction of input impedance that it represents with increasing frequency. At higher frequencies, the input impedance can become limited due to the shunting effect of capacitive reactance. This must be considered when the source resistances seen by the input(s) of the amplifier are large enough to cause appreciable phase shift at the desired operating frequency.

### Common-Mode Rejection Ratio (CMRR)

***The ratio of the input voltage range to the peak-to-peak change in input offset voltage over this range.***

An ideal op amp responds only to differential input signals and ignores signals common to both inputs. In practice, this is only an approximation; real amplifiers have a small but definite common-mode error. The ratio of the common-mode voltage to the common-mode error voltage is the common-mode rejection ratio, CMRR. It is commonly expressed in dB.

Common-mode rejection is important to noninverting or differential amplifiers because these configurations see a common-mode voltage. It is specified at dc or a very low frequency, and it rolls off with a 6-dB/octave slope starting at the corner frequency. The corner frequency is about 100 Hz for general-purpose op amps, but it is greater in high-speed types. Dc rejection ratios can range from 90 dB to 120 dB, depending on the device. They are generally higher in bipolar-input amplifiers than in FET-input amplifiers, since FETs tend to have a poor common-mode rejection compared to bipolars.

### Power-Supply Rejection Ratio (PSRR)

***The ratio of change in input offset voltage to the change in supply voltages, with the supplies varied symmetrically.***

The ability of an IC op amp to reject power-supply-induced noise and drift is an important parameter. Voltage changes on the supply lines can be coupled into the amplifier, and they appear as an equivalent input signal, regardless of the configuration used. For low-level high-gain applications, this noise source must be considered. During measurement, the supply lines are varied equal amounts, and the measured changes are referred to the input. The measured value is noted as

$$\text{PSRR} = \frac{\Delta V_{io}}{\Delta V_S}$$

where

$\Delta V_S$ is the increment of supply-voltage change.

Typical figures for power-supply rejection may be as low as 2 $\mu$V/V or as high as 200 $\mu$V/V, with the performance varying with different designs. In a given type of amplifier, there may also be different limits on the positive- and negative-supply rejection ratios; or, the manufacturer may specify the two measurements together, giving a typical value and a maximum limit. This parameter is also frequency-dependent, with the exact degree varying with the type of amplifier.

INPUT VOLTAGE RANGE ($V_{icr}$)

***The range of voltage on the input terminals for which the amplifier operates within specifications.***

Input voltage range (also called the *operating common-mode range*) is the maximum range of voltage swing that the input stage can tolerate and still operate within the specification limits of the amplifier. It is applicable to either input since, in an actual closed-loop situation, the two inputs will be at or near the same potential. In the case of a unity-gain voltage follower, the inputs are required to swing the full output range of the amplifier because the feedback is 100%. Therefore, it is necessary that the input stage be capable of operating with full specifications over the dynamic range of output swing. If it cannot, the possibility exists that the amplifier may saturate or "latch up" when its input range is exceeded.

Latch-up was a particular problem with early IC amplifiers because the maximum input range could be exceeded with normal output swings. This problem will be discussed in detail in a later section.

The 301, 741, and later amplifiers solved the input voltage-range problem with typical specifications of $\pm 13$ volts or more. The input voltage range is directly related to the supply voltages used since it is the internal biasing potentials which set the input voltage range. Thus, it is reduced for lower supply potentials and increased for higher supply potentials. Some op amps are specifically designed to accommodate input common-mode voltages which include one supply rail (i.e., 324 for the minus rail, 301 for the plus rail).

### *Output Characteristics*

OUTPUT VOLTAGE SWING ($V_{op}$)
*The peak output swing, referred to zero, that can be obtained without clipping.*

Output voltage swing is the ability of an op amp to deliver its rated voltage across a specified value of load resistance. A symmetrical output swing is usually implied, although this is not necessarily true in all cases. The output swing can become limited due to loading effects, power supplies, frequency effects, and source resistance of the amplifier. Generally, IC op amps can supply peak voltage swings to within a few volts of the supply voltages used. There are normally two different load-resistance values into which the rated voltage swing is specified; these are typically 10,000 ohms and 2000 ohms.

With ±15-volt supplies, the 301 and 741 amplifiers are rated for ±14 volts typical swing into a 10,000-ohm load, ±12 volts minimum. Into a 2000-ohm load, the figures are ±13 volts typical and ±10 volts minimum. For supply voltages other than ±15 volts, the output-swing capability will change in proportion.

For load resistances below the rated value of 2000 ohms, the output begins to drop due to the effect of current limiting. This will not damage the amplifier as long as its power-dissipation limits are observed, but it will reduce the open-loop gain due to loading.

OUTPUT SHORT-CIRCUIT CURRENT ($I_{osc}$)
*The maximum output current available from the amplifier with the output shorted to ground or to either supply.*

As a means of self-protection, IC op amps are usually provided with an internal current-limiting feature which prevents output current from rising above a level that would be destructive. This makes operation virtually foolproof from an accident standpoint, since situations will inevitably arise where an output will be short-circuited. Early op-amp types (such as the 709) were short-circuit proof only for brief durations and were subject to catastrophic failure if short-circuited for an appreciable length of time. Later op-amp types have incorporated full-time current limiting in the form of active devices which sense load-current flow and limit it to a maximum of 20 to 25 milliamperes in either direction. Since this is not a critical parameter, it is specified only as a typical value. The current is often sensed by a transistor junction voltage; thus, the current-limiting level is inversely proportional to temperature.

OUTPUT RESISTANCE ($r_o$)
***The resistance seen looking into the output terminal with the output nulled.*** (This parameter is defined only under small-signal conditions at frequencies above a few hundred hertz, in order to eliminate the influence of drift and thermal feedback.)

IC amplifiers vary in output resistance, with resistances from below 100 ohms, in general-purpose types, up to some designs with output resistances approaching 10,000 ohms. Higher output resistances will result in a loss of gain when the load impedance becomes low in relation to the output resistance because part of the output signal is lost across the output resistance. This must be considered in cases where maximum gain is important. General-purpose op amps with output resistances less than 100 ohms are specified to operate into loads of 2000 ohms or higher, and they suffer little gain loss due to loading. It is most useful, therefore, that the actual intended loading conditions are those for which the gain is specified.

In all cases, the effective output resistance is reduced by feedback. The degree of reduction is in proportion to loop gain as discussed in the section on nonideal amplifier parameters.

### *Dynamic Characteristics*

OPEN-LOOP VOLTAGE GAIN ($A_{VO}$)
***The ratio of the maximum output-voltage swing with load to the change in input voltage required to drive the output from zero to this voltage.***

The open-loop voltage gain of an op amp is one of the most important specifications because it has the greatest effect on overall accuracy. General-purpose IC op amps have typical dc open-loop gains of 100 dB when operated with rated loads of 2000 ohms. This gain does not change excessively with operating temperature or variations in the supply voltage, although individual devices will vary in these regards. Special-purpose premium devices can achieve voltage gains of 130 dB.

Although the high gain available at dc is adequate to provide sufficient accuracy for most applications, the same is not true for ac. Because of the need for a 6-dB/octave rolloff to achieve closed-loop stability, internally compensated general-purpose op amps typically have an open-loop −3-dB frequency of 10 Hz. Beyond this frequency, gain falls off at a 6-dB/octave (20-dB/decade) rate. An example of such an open-loop gain curve is the 741 response shown in Fig. 1-30A.

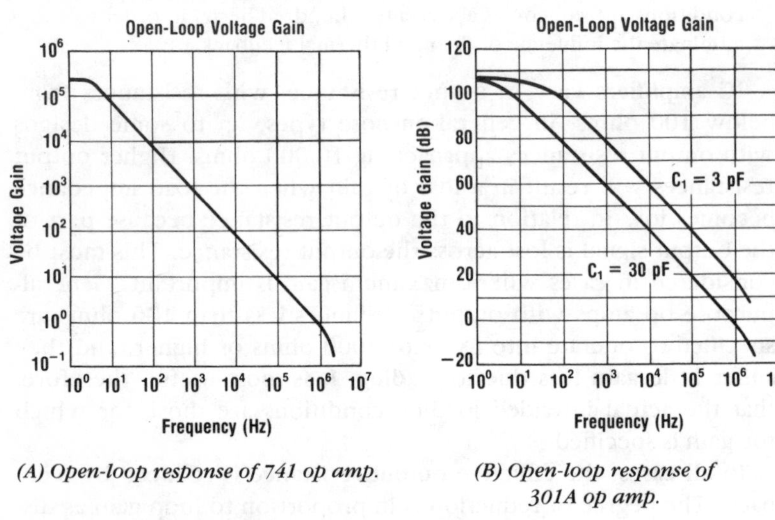


*(A) Open-loop response of 741 op amp.*

*(B) Open-loop response of 301A op amp.*

**Fig. 1-30. Typical open-loop voltage-gain characteristics.**

SMALL-SIGNAL UNITY-GAIN FREQUENCY ($f_t$)

***The frequency at which the open-loop small-signal voltage gain is unity with the amplifier compensated for unity-gain stability.***

The small-signal unity-gain frequency is a very useful amplifier specification because it yields much information about the amplifier frequency response. In many applications, unity-gain stability is required of an op amp. Thus, the most meaningful frequency-response specification for such an amplifier is the unity-gain frequency, as this defines its gain-versus-frequency capability.

For example, with a 6-dB/octave rolloff such as shown in Fig. 1-30A, the frequency where this curve crosses the unity-gain axis defines the entire gain characteristic (since the slope is fixed). Once the $f_t$ of a 6-dB/octave rolloff amplifier is known, its open-loop response can be drawn just by sketching a straight line with a $-1$ slope through the unity-gain frequency.

For the 741 example shown in Fig. 1-30A, the $f_t$ is 1 MHz, which is typical of general-purpose amplifiers. High-speed IC op amps can achieve unity-gain frequencies of 10 to 15 MHz or more, which extends their high-frequency usefulness considerably.

### Gain-Bandwidth Product (GBP)

***The product of available open-loop gain and bandwidth at a specific frequency.***

Gain-bandwidth product and unity-gain frequency are closely related, and in many instances they are actually the same thing. For example, the 741 gain-bandwidth product may also be stated as 1 MHz, since a gain of unity at 1 MHz yields a gain-bandwidth product of 1 MHz. It should also be noted that this product is constant regardless of frequency (or gain). Thus, a gain of $10^5$ at 10 Hz yields a GBP of $10^6$, or 1 MHz. In a 6-dB/octave rolloff amplifier, the gain at any frequency can be found by dividing the GBP by the bandwidth, since

$$GBP = A_v BW$$

The difference between GBP and $f_t$ becomes apparent with externally compensated amplifiers, which can achieve a higher GBP. This is illustrated in Fig. 1-30B, which shows the open-loop frequency response of a 301A amplifier compensated for two different closed-loop gains. The lower curve is identical to the 741 characteristic in Fig. 1-30A; it has an $f_t$ of 1 MHz, which is also its GBP. However, the upper curve has a GBP ten times greater, or 10 MHz, as it yields a gain of 10 at 1 MHz. Note that this curve does not allow unity-gain stability as does the lower $f_t$ curve. Therein lies the difference between $f_t$ and GBP—$f_t$ is actually a specific case of GBP, while GBP is a more general type of gain-frequency specification.

Unfortunately, GBP is not always implicit in numerical specifications, whereas $f_t$ usually is. GBP is evident in the performance curves with various compensations, and the higher these curves extend in gain and frequency, the greater is the GBP for a given amplifier. It is true that with externally compensated amplifiers, extremely high GBPs can be achieved only at high gains, but it is also obvious that these are the conditions which require the high GBP for accuracy. Several IC op amps can yield GBPs or 100 MHz or more, in some instances.

SLEW RATE (SR)
***The maximum rate of change of output voltage under large-signal conditions.***

Slew rate is the maximum rate of change of the output voltage of an amplifier under large-signal conditions, as opposed to the small-signal rise time. Unless otherwise specified, slew rate is measured and specified for closed-loop conditions, usually for a unity-gain voltage follower, which is the worst case for slew rate. The slewing rate of an amplifier will define a full-power bandwidth, the maximum frequency at which the amplifier will deliver full-rated output-voltage sine wave with less than 1% distortion. The two are interrelated by

$$SR = 2\pi E_{op} f_p$$

where

$E_{op}$ is the peak output voltage,
$f_p$ is the full-power response frequency.

Slew rate is the parameter normally specified, although full-power response may also be specified. Slew rate is measured in a standard test circuit using a full-scale input pulse with negligible rise time. The output will then change at a rate limited by the slewing ability of the amplifier.

The slewing rate of an externally compensated amplifier can be proportional to the closed-loop gain if the compensation is adjusted to optimize GBP. This is because more compensation is used at lower gains, and it is the compensation capacitance which determines slew rate. Slew-rate limitations are caused by the limited ability of the internal circuitry of the amplifier to charge and discharge this capacitance. When an operating frequency and output-voltage swing are reached which exceed the ability of the amplifier to charge and discharge the capacitance, the output no longer follows the input linearly but becomes "slew-rate-limited"; the output can then follow only at its rate limit.

In a given type of amplifier, reducing the capacitance that must be charged and discharged (the compensation capacitance) will raise the slewing rate in proportion to the capacitance reduction. Thus, higher-gain stages (which are compensated with less capacitance) have higher slew rates which, as we stated originally, makes slew rate proportional to closed-loop gain. This, of course, does not hold true for internally compensated amplifiers.

If a high slew rate is necessary concurrent with unity-gain stability, an FET-input device will most likely be suitable, as these devices are characterized by high slew rates.

TRANSIENT RESPONSE (TR)
*The 10% to 90% closed-loop step-function response of the amplifier under small-signal conditions.*

Like slew rate, the transient response of IC op amps is specified under closed-loop conditions. With a small-signal pulse input, the output will be an exponential function, with a time constant inversely proportional to the bandwidth, according to

$$\text{bandwidth} = \frac{0.35}{\text{rise time}}$$

with bandwidth and rise time normally given in MHz and $\mu$s, respectively.

The test circuit used is typically a unity-gain voltage follower, with the amplifier driving its rated values of load resistance and capacitance. The signal level used is generally 50 mV or less to ensure operation well below the slew-rate limits of the device, which is well within the small-signal region.

Since bandwidth and rise time are interrelated, it follows that devices with high unity-gain frequencies will exhibit proportionally reduced rise times.

SETTLING TIME ($t_s$)
*The time it takes for the amplifier to reach (and to remain within) a specified accuracy band about some final voltage, after the application of a step input.*

Settling time is a somewhat elusive op-amp performance specification inasmuch as it is not universally specified, and, when it is, it will be for specific feedback and load conditions. Settling time is most important with regard to d/a and a/d conversion, where it is necessary to know how fast an amplifier can acquire and settle to a given accuracy after a pulse input. Typically, settling times are in terms of 0.01% to 0.1% for a 5- or 10-volt step, for either a follower or inverter configuration.

### *General Characteristics*

POWER CONSUMPTION ($P_c$)
***The dc power required to operate the amplifier with the output at zero and with no load current.***

The power consumption of an IC op amp is the power consumed for biasing purposes. It is not usable output power because none of it is delivered to the load. It should not be confused with power dissipation, which is a measure of the ability of the package to dissipate power.

Power consumption is normally specified numerically at supply voltages of ±15 volts. It may be given in the form of milliwatts of power, or milliamperes of supply current, at the voltages from which power may be calculated. In addition, power consumption as a function of temperature and supply potential may be given.

With some IC op amps, power consumption may be controlled externally over an extremely wide range with a master bias terminal. This terminal may be used to switch the IC from completely "off" to "on" by removal or application of the bias current. In addition, the level of the current when "on" may be varied, which, in turn, varies the power consumption of the IC. These types are referred to as *programmable* types.

## 1.4 REFERENCES


1. *Applications Manual for Computing Amplifiers,* 2nd Edition. Philbrick/Nexus Research, Dedham, MA, 1966, 1984.
2. Barna, A. *Operational Amplifiers.* John Wiley & Sons, Inc., New York, 1971.
3. Borlase, W. "An Introduction to Operational Amplifiers." *Analog Devices Application Note,* September 1971. Analog Devices, Inc., Norwood, MA.
4. Giles, J. N. *Fairchild Semiconductor LIC Handbook.* Fairchild Semiconductor, Mountain View, CA, 1967.

5. Huehne, K. "Getting More Value Out of an Integrated Operational Amplifier Data Sheet." *Motorola Application Note AN-273A,* February 1972. Motorola Semiconductor Products, Inc., Phoenix, AZ.

6. "Glossary of Microelectronic Terms, Definitions, and Symbols." *JEDEC Engineering Bulletin No. 1-B,* March 1971. EIA Engineering Dept., Washington, DC.

7. Jung, W. G. *Audio IC Op-Amp Applications,* 3rd ed. Howard W. Sams & Co., Inc., Indianapolis, IN, 1986.

8. "Basic Principles and Applications of Operational Amplifiers." *Microsystems International Applications Bulletin 40001A,* 1971. Microsystems International Ltd., Ottawa, Ontario, Canada.

9. Miler, G. G. "Equivalent Input Noise Measurements on High Gain Monolithic Operational Amplifiers." *Harris Semiconductor Application Note 506,* October 1970. Harris Semiconductor, Melbourne, FL.

10. Motchenbacker, C. D., and Fitchen, F. C. *Low Noise Electronic Design.* John Wiley & Sons, Inc., New York, 1973.

11. "Integrated Circuit Operational Amplifiers." *RCA Application Note ICAN-5290,* March 1970. RCA Solid State Div., Somerville, NJ.

12. Roberge, J. K. *Operational Amplifiers—Theory and Practice.* John Wiley & Sons, Inc., New York, 1975.

13. Smith, J. I. *Modern Operational Circuit Design.* John Wiley & Sons, Inc., New York, 1971.

14. Smith, L., and Sheingold, D. "Noise and Operational Amplifier Circuits." *Analog Dialogue,* Vol. 3, No. 1, March 1969.

15. Solomon, J. E. "The Monolithic Op Amp: A Tutorial Study." *IEEE Journal of Solid State Circuits,* Vol. SC-9, No. 6, December 1974.

16. Stata, R. "Operational Amplifiers—Parts I and II." *Electromechanical Design,* September, November 1965. (Available as reprint from Analog Devices, Inc., Norwood, MA).

17. Tobey, G. E., Graeme, J. G., and Huelsman, L. P. (eds.). *Operational Amplifiers—Design and Applications.* McGraw-Hill Book Co., New York, 1971.

18. "Operational Amplifier Static Gain Errors Analysis and Nomographs." *Union Carbide Electronics AN-7,* December 1966.

19. Vander Kooi, M. K. "Predicting Op Amp Slew Rate Limited Response." *National Semiconductor Linear Brief LB-19,* August 1972. National Semiconductor Corp., Santa Clara, CA.

20. Widlar, R. J. "Drift Compensation Techniques for Integrated DC Amplifiers." *National Semiconductor Application Note AN-3,* November 1967. National Semiconductor Corp., Santa Clara, CA.

Two

# *IC Op Amps: The Evolution of General-Purpose and Specialized Types*

## 2.1 HISTORY AND DEVELOPMENT OF GENERAL-PURPOSE OP AMPS

No discussion of IC op amps would be complete without some historical background to place the different types in perspective. Due to the overwhelming proliferation of op-amp types, this is a challenging task; however, there are distinct evolutionary steps in the development of these devices. From this framework, we can at least lead up to the period of the late 1960s and middle-to-late 1970s when the diversity of types exploded, so to speak, into the broad spectrum of devices available today.

The foundation for IC op-amp design really began with the $\mu$702, the first generally accepted IC op amp, which was introduced by Fairchild Semiconductor in 1963. For a number of reasons, the 702 was not a universally applicable op amp. It had a very limited common-mode input range, relatively low voltage gain (approximately 70 dB), used odd supply voltages (such as +12 and −6 volts), and was susceptible to burnout when the output was temporarily shorted. However, the 702 did set the stage for future trends in IC op-amp design by establishing the practice of using matched-component techniques. Although the 702 design is not covered in detail here, information on it is available through the references cited at the end of the chapter.

### 2.1.1 IC Design Philosophy

The design of monolithic integrated circuitry is entirely different from that of discrete-component circuitry, because of inherent limitations in the IC fabrication process. In fabricating ICs, resistors are limited to low values, have very poor tolerances, and exhibit poor performance with changes in temperature. Complementary transistors of equal performance are not feasible without additional processing steps; thus, designs must be either npn-only or worked around the poor-quality pnp transistors that are compatible with the npn process. Capacitors of more than a few picofarads in value are not desirable because they occupy excessive chip area and, hence, increase the cost per device.

Because conventional discrete-component design techniques were not compatible with monolithic IC processing, new methods had to be devised. The old "ground rules" were discarded in favor of ones which optimized designs based on inherent properties of the monolithic IC process. A brief review of these features is appropriate before we begin to discuss them in the context of actual IC op-amp circuits.

Because of the small size of the transistors used in integrated circuitry, it is possible to lay out and interconnect a large number of them in the space formerly occupied by a single conventional transistor. This is a big advantage, since for a given area (or "chip real estate" as it is termed in the IC world), it is just as economical to create a large number of the smaller IC transistors as it is to create only one discrete transistor. Active components are inexpensive in IC designs; therefore, a larger number of them may be used without materially affecting cost. This is in direct contrast to the reasoning behind discrete-component designs, where the object is to use the minimum number of components that will satisfy the circuit requirements. As a result, an IC schematic will invariably appear to have more transistors and diodes in it than are needed. From the standpoint of cost (which is directly proportional to chip area), it is advantageous to use as many of the smaller active components as possible to accomplish a given function. Since capacitors, resistors, and transistors occupy diminishing amounts of space in that order, designs will lean toward transistors first, resistors next, and capacitors last. Often a transistor may be biased to simulate a large-value resistor because, if a resistor were used for the same function, it would occupy a much greater area. Transistors are used in many ways, a few of which are illustrated in Fig. 2-1.

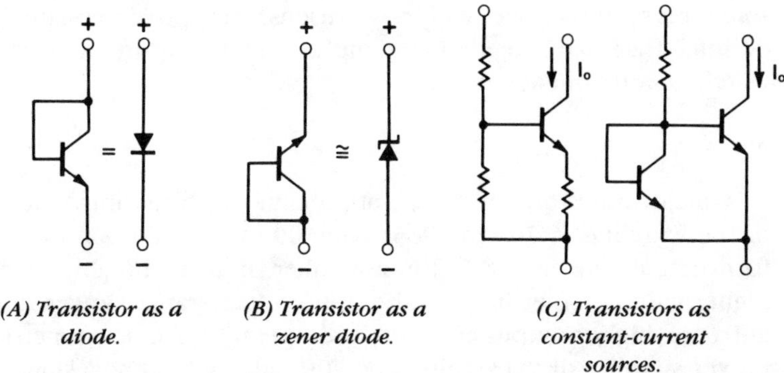


*(A) Transistor as a diode.*

*(B) Transistor as a zener diode.*

*(C) Transistors as constant-current sources.*

**Fig. 2-1. Transistors are used to simulate other IC circuit elements.**

In addition to the substantial use of transistors in various forms, another inherent advantage of the monolithic process is extensively utilized. Due to the very close physical proximity of IC components, and the fact that they are all made simultaneously on the same silicon chip (and thus are subject to identical conditions), all elements tend to match very well. Although the resistors are subject to tolerance variations of up to ±30%, the match between them is much better—within 3%. This matching characteristic not only holds true for resistors but also for transistors. Current gains and base-emitter potentials match up within a few percent, as do breakdown voltages. A pair of IC transistors laid down side by side can be matched to have base-emitter voltages ($V_{BE}$) within a millivolt of each other. This makes IC transistors naturally suited for balanced-gain stages, such as differential amplifiers, which work on the symmetry principle. In addition, diode-connected transistors can be used readily as matched pairs. Since the components on an IC chip are all within a few thousandths of an inch of each other and are an integral part of the chip, these matching characteristics also track very well with temperature changes.

More recent technological advances are ion implantation and active trimming. Ion implantation first allowed high-quality JFET transistors to be fabricated along with bipolar transistors in the 356 device series. Another FET-type amplifier family includes the 3130 and 3140 units. Active trimming, either by laser or zener shorting techniques, allowed very low (<100 $\mu$V) input-voltage offsets, as well as low drift, to be achieved.

To illustrate the preceding points and to give the discussion some perspective, we will now discuss several representative designs that are excellent examples of the optimum use of matched-component design.

### 2.1.2 The μA709

A major milestone in IC evolution came with the introduction of the Fairchild μA709 in 1965. The 709 improved upon the performance of the earlier 702 in a number of areas; notably, it had higher gain, a larger input and output voltage range, lower input currents, higher output currents, and operation from symmetrical power supplies of ±15 volts. The 709 quickly became a standard type number, and is still being produced. So universal was the acceptance of the 709 that it must be regarded as a classic among IC op amps. Although many of its individual performance parameters have long since been eclipsed, the 709 remains a unique blend of performance, versatility, and economy—three ingredients that are a hard combination to beat.

#### *μA709 Circuit Description*

The design principles established with the 702 were exercised again in the 709, with some new advances also being introduced. In addition, many of the problems encountered in applying the 702 were eliminated with the 709 design.

Fig. 2-2 is the schematic diagram of a 709 amplifier. You can see immediately that a large number of transistors are used (15) and that the resistance values chosen are fairly low—$R_{15}$ is the highest at 30,000 ohms. This is in keeping with the basic design philosophy set forth in the 702. Still, this entire circuit does not occupy an excessive chip area—the 709 is a 0.055-inch (1.4-millimeter) square chip of silicon.

*Input Stage*—In the 709, $Q_1$ and $Q_2$ form the input differential-amplifier pair. Each is operated at a collector current of 20 microamperes by the constant-current transistor, $Q_{11}$. This low current level establishes one important advantage—a big reduction in the input bias current to $Q_1$ and $Q_2$. With a typical $\beta$ of 100 for $Q_1$-$Q_2$, the input current is 200 nanoamperes, or about 1/20 that of the 702. To obtain a 40-μA current by conventional means would require a very high resistance in the emitter of $Q_{11}$. In addition, to increase the negative common-mode input range of $Q_1$ and $Q_2$, $Q_{11}$ must be biased close to the V− line; otherwise it would saturate as $Q_1$ and $Q_2$ swing

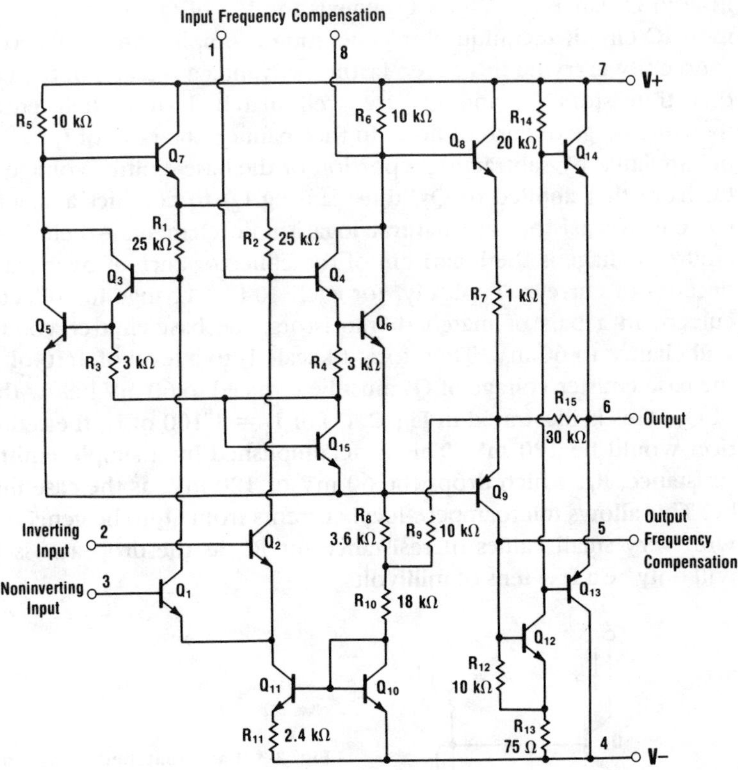


**Fig. 2-2. Schematic of μA709 op amp.**

negative. From this standpoint, $Q_{11}$ should swing to at least −10 volts peak. The solution to the biasing of $Q_{11}$ at 40 μA was a new type of constant-current source made up of $Q_{10}$ and $Q_{11}$.

In a pair of matched transistors, such as those that are readily available in a monolithic IC, the base-emitter voltages that are required for given collector currents form an extremely predictable parameter. Not only will these voltages be within 1 or 2 mV for equal collector currents, the *difference* in the two base-emitter voltages for a *ratio* of currents is also very predictable. This is illustrated by Fig. 2-3. With a current, $I_1$, flowing in diode-connected $Q_1$, $Q_1$ will develop a base-emitter voltage, $V_{BE1}$, which is proportional to this current. For an identical current to flow in $Q_2$, $Q_2$ is connected directly in parallel with the base-emitter junction of $Q_1$. Since the two transistors are matched, an identical current, $I_2$, will flow in $Q_2$ since the $V_{BE}$ of $Q_2$ equals the $V_{BE}$ of $Q_1$. This is a common IC circuit technique for generating a simple constant-current source. By carrying this a step farther, advantage is taken of the fact that transistors $Q_1$ and $Q_2$ are well matched over their entire dynamic range of current flow. In this manner, the ratio of $I_1$ to $I_2$ is manipulated by subtracting a portion of the base-emitter voltage of $Q_1$ from that applied to $Q_2$, thus causing $Q_2$ to conduct a smaller current. Transistors are natural logarithmic elements—their base-emitter voltage is the logarithm of the collector current over many decades of current. Typically, for each 10-to-1 change in collector current in a pair of matched transistors, the base-emitter voltage will change to 60 mV. Therefore, to scale $I_2$ to a level of 1/10 of $I_1$, the base-emitter voltage of $Q_2$ must be reduced to 60 mV below that of $Q_1$. This is illustrated in Fig. 2-4. For $I_2 = 1/100$ of $I_1$, the reduction would be 120 mV. This is accomplished by a simple emitter resistance, $R_E$, which drops the 60 mV or 120 mV, as the case may be. This allows microampere-level currents from $Q_2$ to be generated with very small values of resistance for $R_E$, as the drop across $R_E$ will only be a few tens of millivolts.

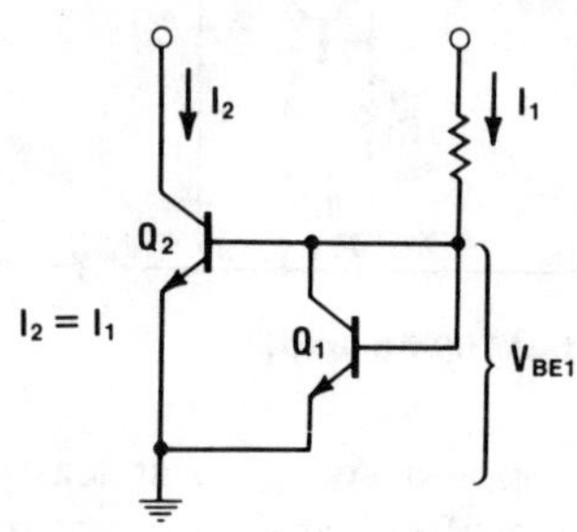


**Fig. 2-3. Basic matched-transistor current source.**

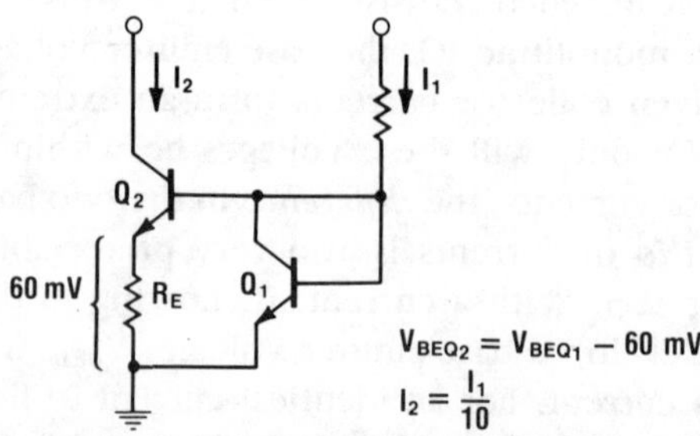


**Fig. 2-4. Logarithmic matched-transistor current source.**

Referring to Fig. 2-2, we can see how this is done in the 709 with $Q_{10}$ and $Q_{11}$. $Q_{10}$ is the reference transistor and $R_{11}$ represents the $V_{BE}$ scaling resistor, $R_E$ (Fig. 2-4). $R_E$ subtracts a portion of the $V_{BE}$ of $Q_{10}$ from that of $Q_{11}$, forcing $Q_{11}$ to conduct at a much lower current. An additional advantage of this technique is the dynamic range of voltage it allows at the collector of $Q_{11}$. Since the base of $Q_{11}$ is only 1 $V_{BE}$ above the V− line, the emitters of $Q_1$ and $Q_2$ can drop very nearly to the $V_{BE}$ of $Q_{11}$ (or very close to −15 V) before $Q_{11}$ saturates. This maximizes the negative-input common-mode range of $Q_1$-$Q_2$.

Another interesting feature of the $Q_1$-$Q_2$ input stage is its performance with temperature changes. The output current from $Q_{11}$ varies with temperature, but it does so in such a manner that it complements the transconductance variations of $Q_1$ and $Q_2$ with temperature changes. As a consequence, the resultant combination of these two opposite effects is a voltage-gain characteristic of the input stage, which stays constant within a few percent over the operating temperature range. In addition, the variations in the operating current of the $Q_{11}$ current source with the supply voltage are reduced to the relationship of the current in $Q_{11}$ to the current in $Q_{10}$. The current in $Q_{11}$ is related logarithmically rather than directly to the current in $Q_{10}$. Although the current in $Q_{10}$ will vary with changes in the supply voltage, the resultant effect on the current in $Q_{11}$ is very much reduced due to this logarithmic relation, so it is practically unaffected by supply-voltage changes.

*Second Stage*—The second stage of the 709 op amp is diagrammed in simplified fashion in Fig. 2-5. $R_1$ and $R_2$, the load resistors for the first stage, are high in value for high gain. The balanced-biasing feature possesses a property of immunity to supply-voltage changes. The full circuit is shown in the schematic of Fig. 2-2. Darlington connections are used for $Q_5$ and $Q_6$ to minimize loading on $R_1$ and $R_2$. An emitter follower, $Q_7$, is used to feed $R_1$ and $R_2$. This removes the input-stage currents flowing in $R_1$ and $R_2$ from $Q_5$ and improves the overall balance. The remaining element of the second stage is diode-connected $Q_{15}$, which stabilizes the Darlington pairs, $Q_3$-$Q_5$ and $Q_4$-$Q_6$, by removing excess leakage currents that occur at high temperatures.

In the 709, a high common-mode range is provided by second-stage transistors $Q_5$ and $Q_6$, which return to V− through bias resistors $R_8$ and $R_{10}$. In addition, the static voltage on $R_1$ and $R_2$ is made

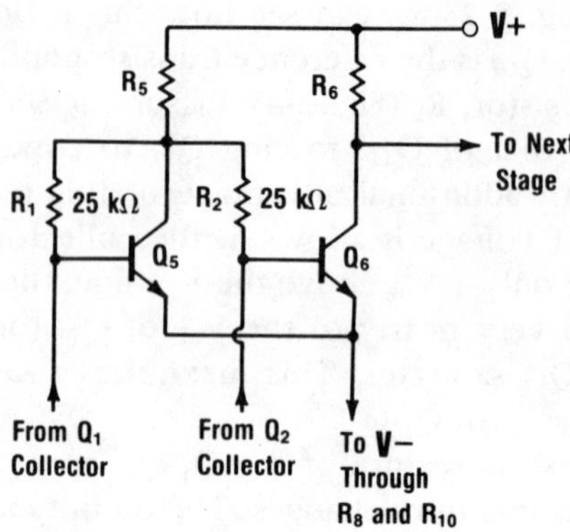


**Fig. 2-5. Simplified schematic of μA709 second stage.**

high enough so that $Q_1$ and $Q_2$ can swing typically to +10 volts with +15-volt supplies. The second-stage output is developed across $R_6$ and is buffered by emitter follower $Q_8$.

*Output Stage*—The output stage is composed of $Q_9$, $Q_{12}$, $Q_{13}$, and $Q_{14}$. For a wide range of voltage swing with moderate current output capability, a class-B output stage is highly desirable, as it provides the requisite power with low standby current. The 709 uses a class-B output stage consisting of complementary emitter followers $Q_{13}$ and $Q_{14}$. This yields an inherently low output impedance with the advantage that both transistors cannot conduct simultaneously due to their direct base connection. A high-voltage gain to drive the $Q_{13}$-$Q_{14}$ pair is provided by common-emitter amplifier $Q_{12}$ and is driven by common-base stage $Q_9$, which has no current gain. Its primary purpose is to shift the high dc level at $Q_8$ to a lower dc level that is compatible with the base-drive requirements of $Q_{12}$. The entire output stage is, in essence, a single-stage voltage amplifier ($Q_{12}$) with a buffered output. The voltage gain of the circuit would be essentially the ratio of $R_{14}/R_{13}$, or 20,000/75 = 267, were it not for $R_{15}$ and $R_7$. These resistors provide local feedback around the output stage. $R_{15}$ is the feedback resistor, with $R_7$ serving as the input resistor. The gain of this local feedback loop from $Q_8$ to the output is $R_{15}/R_7$, or 30,000/1000 = 30. The feedback is used mainly to correct for the "dead zone" of transistors $Q_{13}$ and $Q_{14}$ as the output circuit swings above and below ground level. In addition, it further lowers the output impedance of $Q_{13}$-$Q_{14}$ by virtue of the negative feedback through $R_{15}$.

Two new components are introduced into the design in the output stage: $Q_9$ and $Q_{13}$, which are pnp transistors. Both of these transistors are made with the same process that is optimized for the remaining transistors, which are all npn. As pnp transistors go, $Q_9$ and $Q_{13}$ are of poor quality, having current gains of 10 or less. However, this represents no great difficulty since the circuit is designed to allow for these compromises. $Q_9$ is used only as a level shifter and, thus, low-current gain does not present a problem since an excess of drive is available from emitter follower $Q_8$. $Q_{13}$ also has relaxed performance requirements—in reality, it need function only as a diode, since an excess of drive is available from $Q_{12}$. The gain it may provide, however, can be regarded as a bonus.

A point not obvious from an examination of the circuit is the relative size of $Q_{13}$ and $Q_{14}$. These transistors, like most of those used in linear ICs, are of small geometry and have a rapid falloff in gain when driven into high-current operation. As a result, the output stage protects itself against short circuits, at least for short durations, by the self-limiting current of the output transistors.

*Frequency Compensation*—As discussed in the section on basic theory, a prerequisite for stability in a feedback amplifier is the control of phase shift in the amplifier before feedback is applied. Amplifier phase shift must be less than 180° at the point where the open-loop and closed-loop gain curves intersect. In the 709 op amp, the control of open-loop gain/phase characteristics is accomplished by three frequency compensation components, as shown in Fig. 2-6A. Components $R_1$ and $C_1$ are called the input compensation, and $C_2$ is called the output compensation. These three components shape the open-loop response and, with the value noted, yield the family of open-loop responses shown in Fig. 2-6B. These are the networks necessary for stability in configurations of 0-, 20-, 40-, and 60-dB closed-loop gains. The impedance at the 709 compensation terminals is high enough that full compensation can be realized with reasonably small capacitors (5000 pF maximum). These networks provide adequate stability with a 709 device under worst-case conditions of temperature, supply voltage, etc. A summary of the μA709 electrical characteristics is given in Table 2-1.

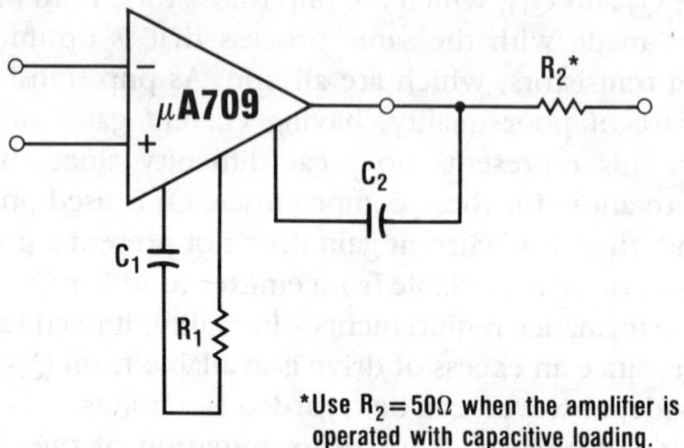


*(A) Frequency compensation circuit.*

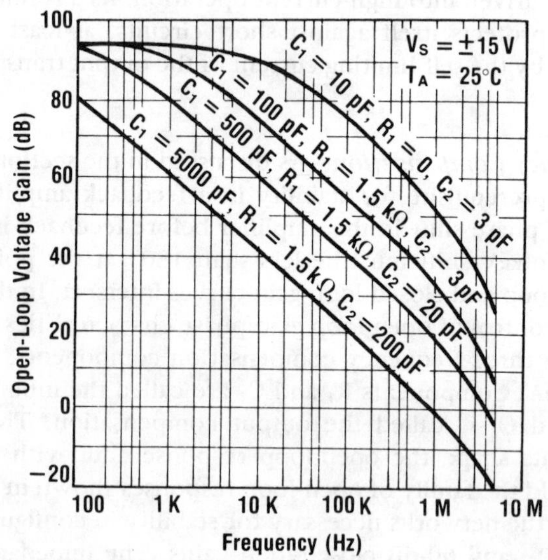


*(B) Open-loop frequency response for various compensation values.*

**Fig. 2-6. Frequency compensation methods and responses for the 709 op amp.**

**Table 2-1. Typical Electrical Characteristics of the $\mu$A709 Operational Amplifier ($T_A = 25°C$, $V_S = \pm 15$ V)**

| Parameter | Value | Unit |
|---|---|---|
| Input Offset Voltage | 1 | mV |
| Input Offset Current | 50 | nA |
| Input Bias Current | 200 | nA |
| Input Impedance | 400 | kΩ |
| Input Common-Mode Range | ±10 | V |
| Common-Mode Rejection Ratio | 90 | dB |
| Voltage Gain | 45,000 | |
| Output Voltage Swing | | |
| $R_L = 10$ kΩ | ±14 | V |
| $R_L = 2$ kΩ | ±13 | V |
| Output Impedance | 150 | Ω |
| Power Consumption | 80 | mW |
| Power-Supply Rejection Ratio | 25 | μV/V |

### 2.1.3 The 101 and 741 Families

***The LM101****

Introduced in 1967 by National Semiconductor Corporation, the LM101 was the next evolutionary step in the history of IC op-amp technology. The 709 devices are regarded as the "first generation" of IC op amps; the arrival of the LM101 began the second generation of IC op amps.

For the same reasons that the 709 has historical and applicational importance, so also does the LM101. It is historically important because it represents the next level of technology. And it is also important conceptually, for it is still the basis for an entire range of general-purpose devices which are variations on its basic design. Like the 709, we will examine the 101 closely because once it is understood, the entire range of similar devices within the second generation can also be understood. This encompasses a large percentage of the op amps still in general-purpose usage.

***LM101 Circuit Description***

The objectives of the 101 design were to eliminate the major problem areas associated with the 709. These were:

* The generic family is described here in terms of the "101" part number, which is the full (military) temperature-range device. The applications (and later) sections of the book are in terms of commercial devices, such as the LM301A (vis-a-vis LM101A).

1. Inadequate short-circuit protection.
2. Complicated frequency-compensation networks.
3. Latch-up when common-mode range was exceeded.
4. Too low differential input voltage.
5. Excessive power dissipation.
6. Sensitivity to capacitive loading and susceptibility to oscillations.
7. Limited supply-voltage range.

The 101 design solved all of these problems and also added some additional refinements. Gain was increased to 160,000 over the 45,000 of the 709, and the useful range of supply operation was increased from ±5 volts to ±20 volts. The device retained the 709 pin configuration for input, output, and power-supply leads. A simplified circuit of the 101 is shown in Fig. 2-7. Although this is a simplified schematic, one thing that is immediately obvious is the absolute minimum use of resistors and the use of a large number of active devices.

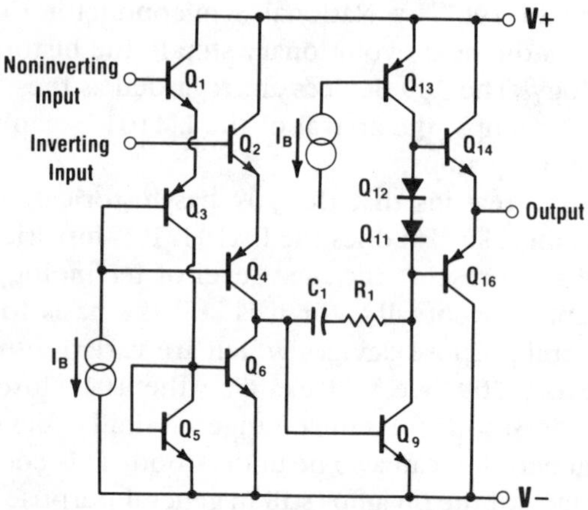


**Fig. 2-7. Simplified schematic of LM101 op amp.**
*(Courtesy National Semiconductor Corp.)*

To accomplish the primary objective of simple frequency compensation, the 101 uses a two-stage amplifier design; the fewer the number of stages in a device, the fewer the phase shifts with which to contend. However, to meet the high gain required from the device, these two stages must provide the gain previously supplied by three stages in the 709. This means that the gain per stage must be highly optimized to achieve a total gain of 100 dB for the entire amplifier. In the 101, this is done by using active loads, which increase the available gain per stage to a maximum without resorting to undesirable high resistances.

An illustration of this is shown in Fig. 2-8A. A basic common-emitter amplifier will have a gain equal to the ratio of the load resistance ($R_L$) to the dynamic emitter impedance ($R_e$). Gain may be optimized by raising $R_L$ to maximum, but a limitation is soon reached. Since high values of resistance are not desirable in an IC, operation with resistive loads at collector currents of 20 $\mu$A is prevented because the load resistance required for high gain would be over one megohm. However, with a transistor as a load as in Fig. 2-8B, the situation is different. The output impedance of a transistor biased for a constant collector current is extremely high and can easily achieve the required one-megohm level. An additional bonus of this technique is the fact that it requires a minimum of voltage drop across the transistor for operation (as opposed to a resistor), and this extends the dynamic range over which the stage can swing before reaching saturation.

*First Stage*—Referring to Fig. 2-7, we can see how active loading is utilized in the 101 op amp. $Q_3$ and $Q_4$ in this stage form an equivalent pnp differential pair. Since monolithic pnp transistors have low gains, their inputs in the 101 are buffered by high-gain emitter followers $Q_1$ and $Q_2$. The net equivalent is a composite, high-gain, pnp differential pair. The outputs of $Q_3$ and $Q_4$ are loaded by active loads $Q_5$ and $Q_6$. This stage also serves as a differential to single-ended converter, transforming the push-pull output from $Q_3$-$Q_4$ into an equivalent single-ended drive at the collector of $Q_6$. One may also note that the common-mode input range of this stage is quite high; $Q_1$ and $Q_2$ may swing positive to a limit equal to the V+ line; the negative limit is a level $3V_{BE}$ above the V− line. This feature removes any tendencies toward latch-up due to input stage saturation.

The gain of this input stage is 60 dB. An additional feature is a very high differential input-voltage rating, due to the high base-emitter breakdowns of the pnp transistors. The fact that the input can tolerate differential inputs of ±30 volts removes restrictions on differential input swings.

*Second Stage*—The second stage of the 101 is composed of a

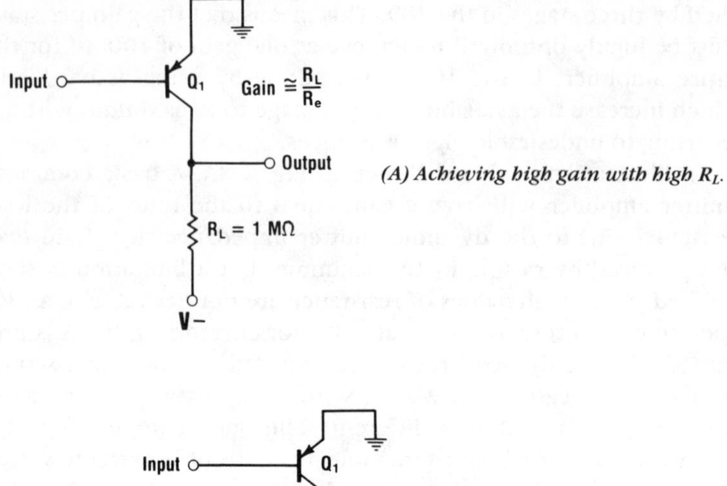


*(A) Achieving high gain with high $R_L$.*

*(B) Achieving high gain with active loading.*

**Fig. 2-8. Methods of achieving high single-stage gain in integrated circuitry.**

common-emitter amplifier, $Q_9$, loaded by an active transistor current source, $Q_{13}$. Again, this is an example of active loading to achieve high gain in a single stage. This high-gain voltage amplifier is buffered to a low impedance level by a class-AB emitter follower composed of $Q_{11}$ and $Q_{12}$, $Q_{14}$ and $Q_{16}$. At this point we are ready to consider the composite circuit as a whole, which is shown in Fig. 2-9. Fig. 2-10 is a two-stage model of the 101 (and related devices).

Frequency compensation of the 101 is accomplished by a single external 30-pF capacitor connected between pins 1 and 8 as shown in Fig. 2-10. This capacitor makes the second stage a Miller integrator and forces the gain to roll off from its open-loop value of 104 dB, beginning at 10 Hz and continuing at a rate of 6 dB per octave down through the unity-gain frequency of 1 MHz. This single-

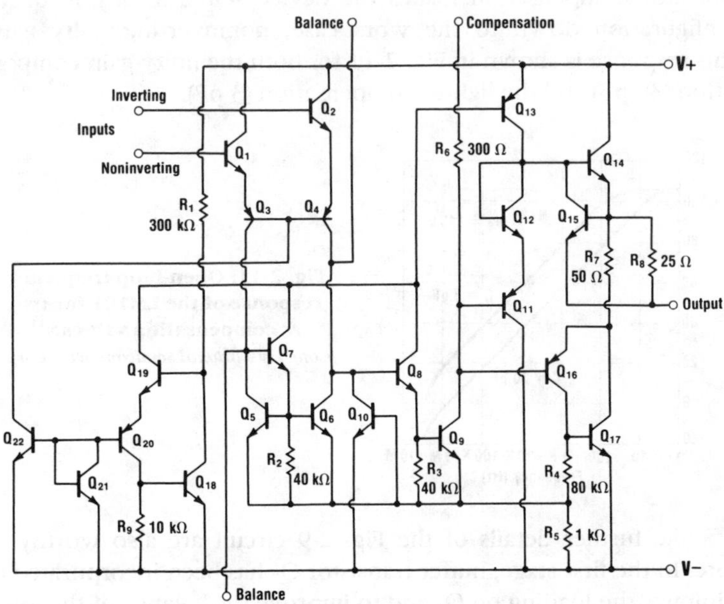


**Fig. 2-9. Complete schematic of LM101 (LM101A, etc.).**
*(Courtesy National Semiconductor Corp.)*

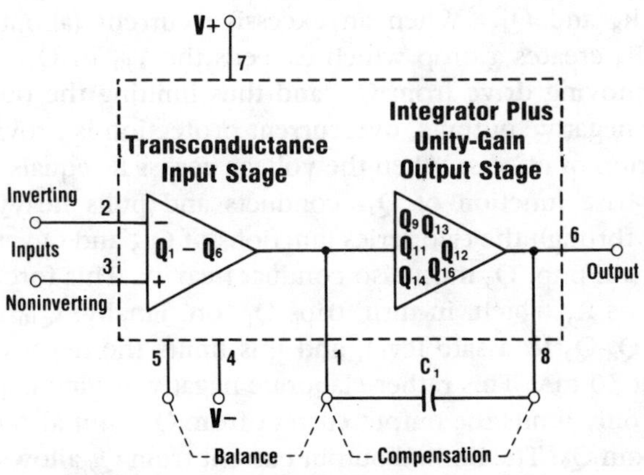


**Fig. 2-10. Two-stage model of LM101 (LM101A, etc.).**

capacitor compensation makes the device stable in any feedback configuration down to the worst-case, noninverting unity gain. This response is shown in Fig. 2-11 for both the unity-gain compensation (30 pF) and the lighter compensation (3 pF).

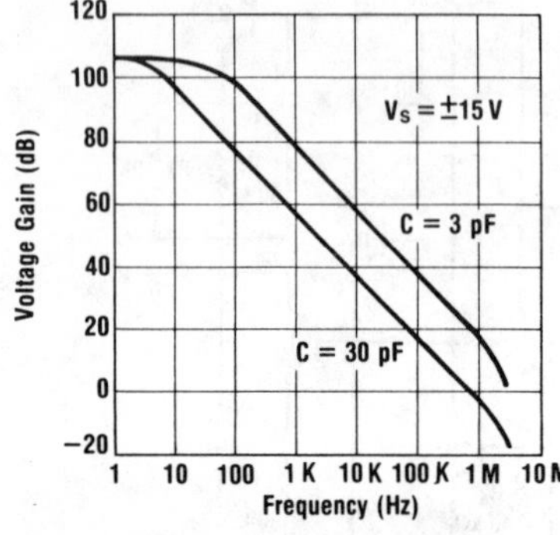


**Fig. 2-11. Open-loop frequency response of the LM101 for two compensation values.**
*(Courtesy National Semiconductor Corp.)*

Some further details of the Fig. 2-9 circuit are also worthy of note. In the first stage, buffer transistor $Q_7$ has been incorporated to minimize the loading on $Q_3$ and to improve the balance of the input stage differential to single-ended connector. In a similar manner, $Q_8$ drives $Q_9$ in Darlington fashion to prevent loading of $Q_4$-$Q_6$, thus maximizing the gain of the input stage. In the output stage, $Q_{11}$ and $Q_{12}$ provide a slight forward bias to $Q_{14}$ and $Q_{16}$ to minimize crossover distortion. Current limiting is also added for both directions of output swing. $Q_{14}$ is protected against overcurrent for positive outputs by $R_8$ and $Q_{15}$. When an excessive current (about 30 mA) through $R_8$ creates a drop which exceeds the $V_{BE}$ of $Q_{15}$, $Q_{15}$ conducts, removing drive from $Q_{14}$ and thus limiting the output current. For negative outputs, overcurrent protection is provided by a combination of effects. When the voltage across $R_7$ equals $2V_{BE}$, the collector-base junction of $Q_{15}$ conducts and pulls heavy current from $Q_{11}$ through the two series junctions of $Q_{15}$ and $Q_{12}$. Since $Q_{11}$ is a low-gain pnp, $Q_9$ must also conduct heavily. This forces a large drop across $R_5$ which, in turn, trips $Q_{10}$ on. Finally, $Q_{10}$ limits the drive to $Q_8$-$Q_9$ to a safe level, and this limits the negative output current at 30 mA. This rather elaborate negative-swing current limiting not only limits the output current from $Q_{16}$, but also limits the output from $Q_9$. The limited output current from $Q_9$ allows external diodes to be connected to the compensation terminal (pin 8) to clamp the voltage swing of the output stage to defined voltage limits, a valuable feature for comparator applications. The remaining transistors in the circuit ($Q_{18}$-$Q_{22}$) are used in providing a self-regulating input bias to the input stage and $Q_{13}$; they are not part of the signal-flow path.

The 101 again demonstrates the use of a large ratio of active to passive components, containing 22 transistors but only 9 resistors. Performance and applicational ease are both improved from the 709. A summary of LM101 performance is given in Table 2-2. All the intended objectives were accomplished. And, in addition, more open-loop gain, a wider supply-voltage range, and a facility for use as a voltage comparator were realized.

**Table 2-2. Typical Electrical Characteristics of the LM101 Operational Amplifier ($T_A = 25°C$, $V_S = \pm 15$ V)**

| Parameter | Value | Unit |
|---|---|---|
| Input Offset Voltage | 1 | mV |
| Input Offset Current | 40 | nA |
| Input Bias Current | 120 | nA |
| Input Resistance | 0.8 | MΩ |
| Voltage Gain | 160,000 | |
| Output Voltage Swing | ±14 | V |
| Input Voltage Range | ±13 | V |
| Common-Mode Rejection Ratio | 90 | dB |
| Power-Supply Rejection Ratio | 90 | dB |
| Temperature Range | −55 to +125 | °C |
| Temperature Drift | 3 | μV/°C |
| Supply-Voltage Range | ±5 to ±20 | V |
| Power Consumption | 50 | mW |

### *The LH101*

The first self-contained, internally compensated op amp—the LH101—was introduced by National Semiconductor in January of 1968. The LH101 is a hybrid combination of an LM101 chip and a 30-pF capacitor in a single package. It has exactly the same specifications as an LM101 that is compensated externally with 30 pF of capacitance. Pinout is the same except for the lack of internal offset null and the inaccessibility of the clamp point.

### *The μA741*

In May of 1968, Fairchild Semiconductor introduced the μA741, an internally compensated IC op amp with ac and dc performance characteristics very similar to those of an LM101 that is compensated for unity gain. A schematic of the μA741 is shown in Fig. 2-12.

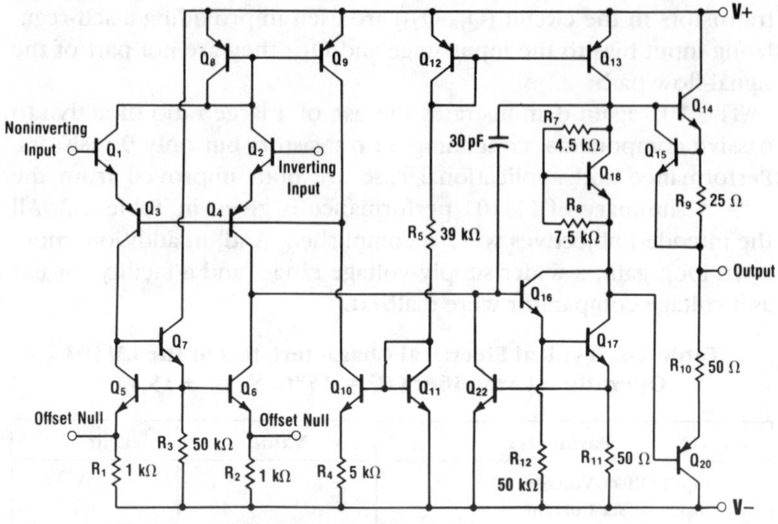


**Fig. 2-12. Schematic of μA741 op amp.**

Although there are substantial differences in the internal circuitry of the 101 and the 741, these are associated mainly with biasing and the method of short-circuit protection. The 741 may also be modeled as a two-stage amplifier similar to the 101, as shown in Fig. 2-10 and as discussed in the previous section.

The main difference between the LM101 and the μA741 is that the 741 accomplishes frequency compensation by means of an on-chip capacitor. This marked the first time that frequency compensation had been accomplished by means of an on-chip capacitor in a monolithic IC op amp. An additional difference in the 741 is the method of offset null. Offset null is accomplished in the 741 by adjusting the differential currents in $Q_5$ and $Q_6$ at their emitters, as shown in Fig. 2-12 (pins 1 and 5 in Fig. 2-13). In the 101 op amp, offset null is accomplished by adjusting the currents in $Q_5$ and $Q_6$ at their collectors (pins 1 and 5 of Figs. 2-10 and 2-14). Both methods perform the same function with regard to dc—a slight imbalance of the collector currents of $Q_3$ and $Q_4$ to null the input offset voltage. The differences in the actual external connections are shown in Figs. 2-13 and 2-14.

### *The LM101A*

In December, 1968, National Semiconductor introduced the LM101A, an improved version of the LM101. The schematic of the

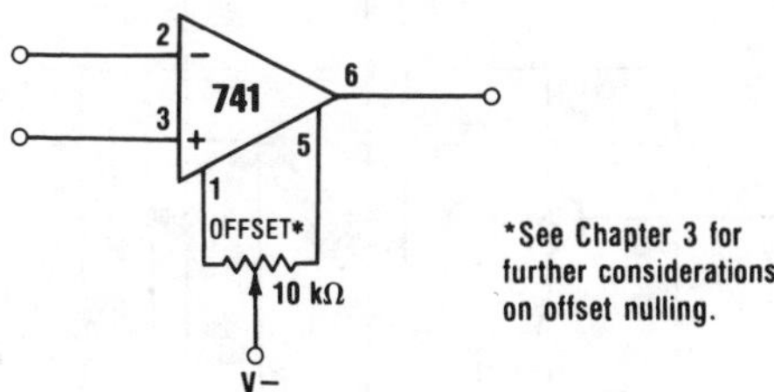


**Fig. 2-13. Offset null circuit for 741 op amp.**

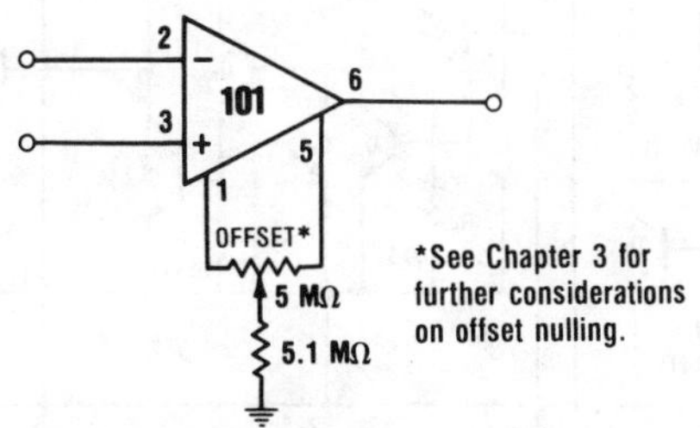


**Fig. 2-14. Offset null circuit for 101 op amp.**

LM101A is shown in Fig. 2-15. As will be noted, there is little difference in the appearance of the circuit from that of the LM101. The portion changed is the biasing circuit for the input stage ($Q_{18}$-$Q_{22}$), which allows better control of input-current characteristics over the operating temperature range. In addition, certain processing refinements for input transistors $Q_1$ and $Q_2$ resulted in much lower input and offset currents (typically 30 nA and 1.5 nA, respectively), or about a factor of 10 better than the 709. No changes were made in the other characteristics, and the LM101A is a pin-for-pin replacement for the LM101.

### *The LM107*

National Semiconductor introduced the LM107 and the LM101A at approximately the same time. The LM107 featured the same improved dc performance as the LM101A, except that the 30-pF frequency-compensation capacitor was built into the silicon chip, similar to the 741.

### *The μA748*

In 1969, Fairchild Semiconductor introduced the μA748, an op amp with the dc performance of a 741 but with external frequency compensation, similar to the LM101. The pinout for the

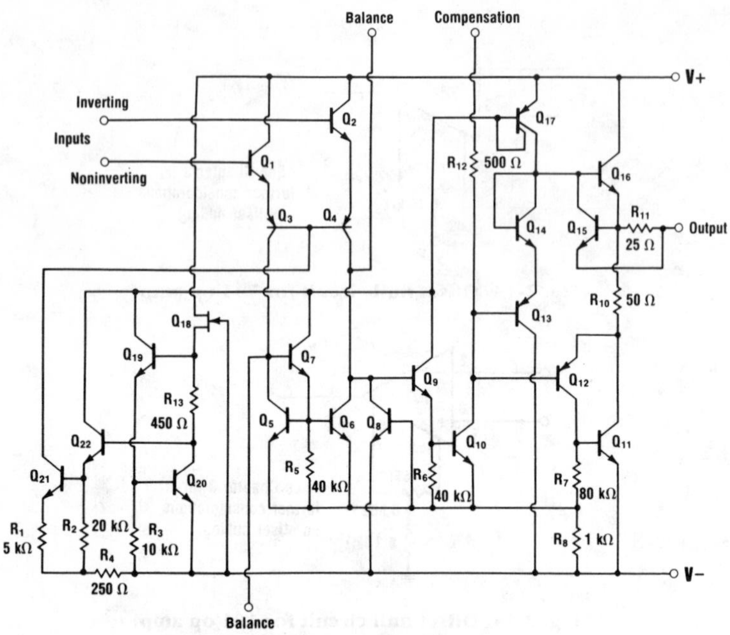


**Fig. 2-15. Schematic of LM101A op amp.**
*(Courtesy National Semiconductor Corp.)*

748 is functionally identical to the 101, although the circuit designs are different.

### *Comparison of the 101 and 741 Families*

At this point, it may already be obvious that there are many similarities between the basic 101 and 741 designs. In fact, if we were to redraw the simplified schematic of the 101, we would find that it applies equally well to all of the general-purpose types. By assigning multiple reference designations to the transistors, we can further illustrate this basic similarity, as shown in Fig. 2-16. Note that even though there are considerable differences in the 101, 101A, and 741/748 designs, any one of them can be reduced to a functionally equivalent circuit, such as that shown in Fig. 2-16. And, as stated previously, all of these amplifiers can be modeled as a basic 2-stage design, with frequency compensation accomplished by an integrating capacitor connected around the second stage. This basic design concept is the principle behind the greatest percentage of op amps used today and virtually all of the modern bipolar general-purpose types. There are only a few subtleties that really separate the 101 and 741 families. Some important characteristics of the 101 and the 741 families should now be defined.

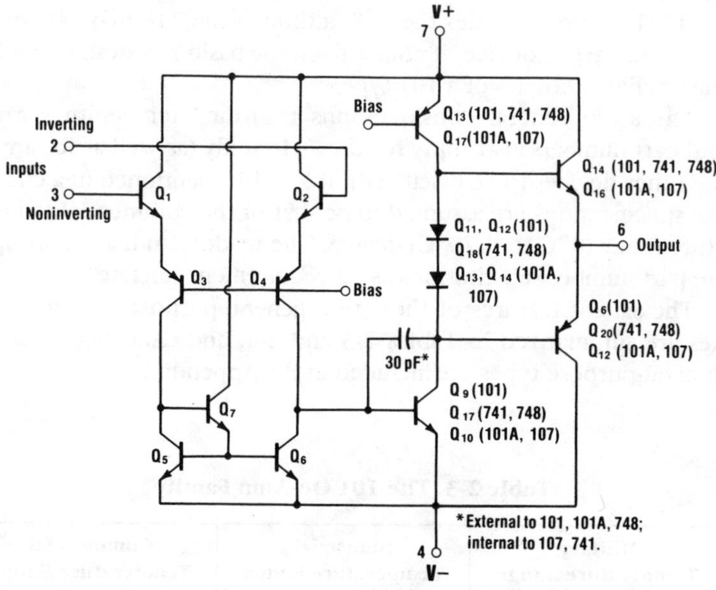


**Fig. 2-16. Basic functional circuit diagram of the 101 and 741 op-amp families.**

*101 Family*—Any "101-type" amplifier can be said to belong to the 101 family. This category includes the basic LM101, the LH101 (both now obsolete), the LM101A, and the LM107. It also includes corresponding devices of all temperature ranges. For example, LM101-type op amps are supplied in three basic temperature ranges: (1) −55°C to +125°C (the LM1— series), (2) −25°C to +85°C (the LM2— series), and (3) 0°C to +70°C (the LM3— series). Thus, an LM301A is the 0°C to +70°C temperature-range equivalent of an LM101A. Both are "101-type" devices, since they belong to the same family. Throughout the remainder of the book, the numbers used in referring to a device will be the commercial part number; e.g., a "301A" amplifier for 0°C to +70°C operation, such as an LM301A. The reader should be cautioned, however, that there are often specification differences between military- and commercial-grade devices, such as supply voltages, gain, offset, etc. (The military-grade devices will almost always have superior ratings and, consequently, higher cost.)

*741 Family*—As with the 101 family, any "741-type" amplifier can be said to belong to the 741 family. This includes the basic 741 device and, also, multiple versions, such as the 747, 1558, 348, and MC4741. It also includes the 748, although the 748 may be said to be a "gray-area" device; it comes from the basic 741 design, but has the ac characteristics of a 101 type.

The same general considerations regarding temperature range and part numbers also apply for the 741 family (as well as for any of the other devices to be discussed). If a 741 is mentioned in a circuit, the specifications are assumed to be that of the commercial temperature range (0°C to +70°C) device. The reader can make appropriate part number substitutions as his requirements dictate.

The salient features of these two general-purpose op-amp families are summarized in Tables 2-3 and 2-4, and data sheets for the general-purpose types are included in the Appendix.

**Table 2-3. The 101 Op-Amp Family***

| Military Temperature Range −55 °C to +125 °C | Industrial Temperature Range −25 °C to +85 °C | Commercial Temperature Range 0 °C to +70 °C |
|---|---|---|
| LM101 | LM201 | |
| LM101A | LM201A | LM301A |
| LM107 | LM207 | LM307 |
| LH2101A | LH2201A | LH2301A |

***Comments***

Basic device of family is the (original) LM101, a general-purpose, externally compensated, IC op amp. It is frequency-compensated with a single 30-pF capacitor. It features a wide input and output voltage range with freedom from latch-up, output short-circuit protection, controlled power-dissipation characteristics, operation from a wide range of supply voltages with little change, and relative immunity to oscillation tendencies. It has been supplanted by the LM101A (listed below).

The LH101 is an LM101 chip with a 30-pF capacitor added to form an internally compensated op amp in hybrid (2-chip) form (but no longer available).

The LM101A is similar to the LM101 but with better dc input characteristics.

The LM107 is an LM101A with internal compensation.

The LH2101A is a dual LM101A.

*The designations shown are not intended for ordering purposes. For complete ordering information (package styles, etc.), consult the manufacturer's catalog or individual data sheets.

**Table 2-4. The 741 Op-Amp Family***

| | Military Temperature Range −55 °C to +125 °C | Commercial Temperature Range 0 °C to +70 °C |
|---|---|---|
| Single | μA741 | μA741C |
| | μA741A | μA741E |
| | μA748 | μA748C |
| Dual | μA747 | μA747C |
| | μA747A | μA747E |
| | MC1558 | MC1458 |
| Quad | LM148 | LM348 |
| | MC4741 | MC4741C |

***Comments***

Basic device of family is the μA741. The μA741 is a general-purpose, internally compensated, IC op amp. It features a wide input and output voltage range with freedom from latch-up, output short-circuit protection, operation from a wide range of supply voltages, and relative immunity to oscillation tendencies.

The μA741A is similar to the μA741, but with tighter characteristics (to military specifications MIL-M-38510/10101).

The μA748 is a μA741 without internal compensation. Compensation and offset null are similar to the LM101. The dc specifications are identical to the μA741.

The μA747 is a pair of 741 op amps in a single 14-pin or 10-pin package.

The MC1558 is a pair of 741 op amps in a single 14-pin or 8-pin package.

The LM148 and MC4741 are quad 741 op amps in a single 14-pin package.

*The designations shown are not intended for ordering purposes. For complete ordering information (package styles, etc.), consult the manufacturer's catalogs or individual data sheets.

There are many other considerations involved in selecting one particular type over another. Ease of application and amount of necessary peripheral hardware are often considerations, not only from a convenience standpoint but also from the standpoint of board space, the number of parts to be stocked, etc. The 741 and its many derivatives will rate favorably in these regards if optimum compensation is not required. If the application requires optimizing speed or overcompensation, then an externally compensated unit, such as a 748 or 101A, may be favored. In other situations, dc requirements may dictate a 101A. Although these three types will fulfill most general-purpose applications, no one type is overwhelmingly superior in all situations. Each situation should be evaluated on its own requirements. Where a multiplicity of devices is necessary, the multiple units will be found attractive (see tables).

For applications that are obviously out of the general-purpose area (high voltage, high speed, low input current, etc.), there are more specialized types from which to choose. These types cover a broad range of remaining applications and will round out the devices to be discussed in this book.

A great many 741- and 101-type devices are still actively manufactured, but many of the 101-type devices today are largely for replacement purposes. The generic "741" spec has been eclipsed by many newer devices, but it will be available for some time.

### 2.1.4 Multiple Units

#### *The RC4558*

The RC4558 (Raytheon, 1974) is also a popular device. Although the RC4558 is not a "true" 741 in the sense that it uses nonidentical circuit topology, it nevertheless is considered *functionally* so by many. It is a dual unit, offered in the same pinout as the MC1458 (this 8-pin dual-amplifier pinout has become standard).

The schematic diagram representative of one of the RC4558 circuits is shown in Fig. 2-17. As can be noted, the major difference from a 741 circuit is use of pnp input transistors, as opposed to the npn's in the 741. This, of course, reverses the direction of the bias current in the source resistance(s), which may affect the performance of some circuits. It is for this reason that the 4558 (and related units) cannot be termed a "true" 741, but it equals or betters the performance of the 741 in other regards.

The RC4558 features improved performance over a standard 741 in terms of noise, slew rate, and bandwidth. A great many other op amps are patterned after the pnp-type input topology, and it appears in quad form as the RC4136. The RC4136 was the first quad op amp to appear on the market, and it enjoys wide popularity and multiple sourcing, as does the RC4558.

#### *The LM324*

The LM324 quad op amp (National Semiconductor, 1974) has become an extremely popular device. The 324 circuit is also not a "true" 741, as it uses pnp transistors; it also has a class-AB output

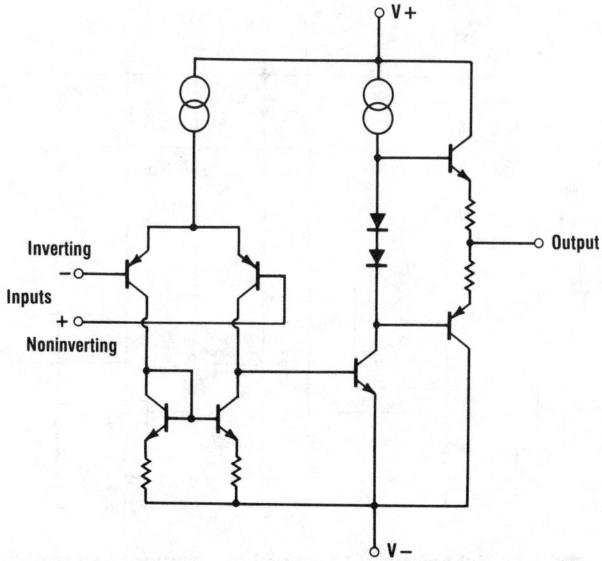

**Fig. 2-17. Simplified schematic of 4558-/4136-type op amp.**

stage. The LM324 is similar to a 741 in terms of speed and input currents, but there the resemblance ends. Designed with different objectives in mind, the 324 is useful for multidevice applications which require very low power operation, operation from low-voltage (5 V) single (as opposed to split) power supplies, and higher output currents.

A schematic of the LM324 circuit is shown in Fig. 2-18. A Darlington-connected pnp differential input allows linear operation of the input terminals at the V− supply level, or ground. Likewise, the output can swing to ground, and these two features optimize the utility of the device for single-supply use. To minimize quiescent power, the output stage of the device is biased in a class-B mode. This results in a total quiescent current for the four amplifier circuits of less than 1 mA. The output stage can also source or sink substantially large currents, up to 40 mA.

The LM324 is available in a version with tightened dc parameters—the LM324A. A dual-amplifier version of the device is the LM358. Both of these units are widely second-sourced. An LM324 data sheet is included in the Appendix.

The LM324 established some basic performance features that have become standard on a number of devices, namely the input/

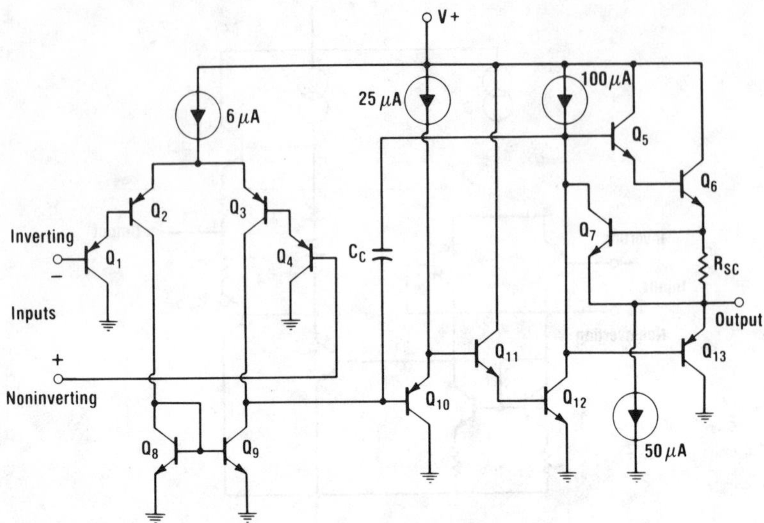


**Fig. 2-18. Schematic of LM324/324A op amps (¼ shown).**

output range, which includes V− (or ground). A related device that has these same features is the Motorola MC3403; however, the 3403 uses a class-AB output stage to eliminate the attendant crossover distortion of the 324's class-B design. Consequently, the 3403 consumes a greater quiescent current. The 3403-type circuit is available in dual form as the MC3458 or μA798, and in single form as the μA799.

### *The LM348 (LM349)*

The LM348 and LM349 quad op amps (National Semiconductor) are true 741 quad devices, with comparable input/output ranges and npn inputs. The 348 specifications are very similar to the 741 in ac terms, and meet or exceed the 741 in dc terms. The LM349 is a decompensated version of the LM348, stable at a minimum closed-loop gain of 5.

The LM348 (LM349) circuit is generally similar to the 741 circuit of Fig. 2-12 (for each ¼ section, with no nulling). The input stage employs a transconductance reduction technique to reduce the compensation capacitance and, thus, the overall chip area. This, however, has no external effect on performance, which is (as stated) equal to or better than a 741. The LM348 and LM349 are second-sourced by a number of manufacturers. A similarly performing device to the LM348 is the Motorola MC4741, which, in turn, is also multiple-sourced.

### 2.1.5 FET-Input Devices

Perhaps the greatest technological advance since the earlier editions of this book were published has been the maturation of the FET-input IC op amp. Prior to 1974, there were a number of monolithic FET op amps, such as the ICL8007, $\mu$A740, and NE536. However, these first-generation devices tended to be much less than ideal in dc characteristics, with typically high offset voltages and drift, and poor CMRR. The second generation devices of the late 1970s conquered many problems, and FET designs have since then continued to evolve.

Today (circa 1985), there is literally a plethora of FET op amps, some with outstanding characteristics at very reasonable cost, in both JFET and MOSFET types. FET op amps do, of course, greatly lessen (or remove) constraints on bias currents, and general-purpose FET amplifiers have ac characteristics that are, in most cases, superior to the 741 bipolar prototype.

The following section begins with some introductory design discussions and relates FET input general-purpose units to their "741" bipolar counterpart.

#### *The CA3130*

The CA3130 (RCA, 1974) represented the first of a new wave of FET-input op amps. This group of devices featured many significant performance improvements, vis-a-vis previous FETs.

A schematic of the CA3130 is shown in Fig. 2-19. As can be noted, the input stage uses a PMOS FET transistor pair, $Q_6$ and $Q_7$. These devices provide an extremely low input current, typically 5 pA. This factor is perhaps the single most outstanding performance attraction of the MOS devices. Input differential overvoltage protection is provided by diodes $D_5$ through $D_8$. $Q_6$ and $Q_7$ feed a differential-to-single-ended current mirror, $Q_9$-$Q_{10}$. The second-stage gain is provided by $Q_{11}$, with phase/frequency compensation provided by an external capacitor (pins 1 and 8). $Q_{11}$ drives a CMOS drain-loaded complementary inverter pair, $Q_8$-$Q_{12}$. This stage provides 15–20 mA of output drive, and the unique property of a rail-to-rail output swing (such as that found in CMOS logic devices).

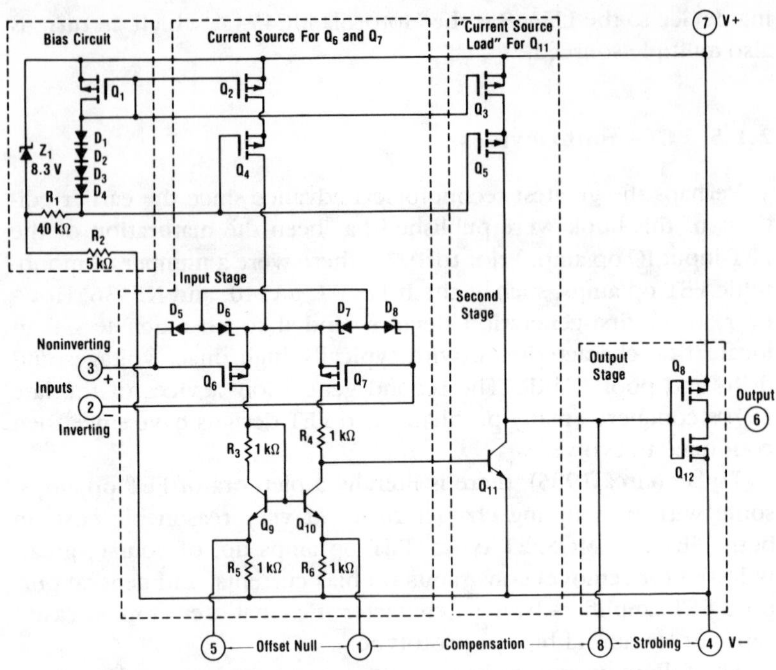


**Fig. 2-19. Schematic of the CA3130 op amp.**
*(Courtesy RCA Solid State Division)*

The CA3130 is rated for a 5- to 15-volt supply voltage (16 volts maximum). The input stage can respond to common-mode voltages to V− (or ground level), and, in fact, to as low as 0.5 volt *below* the pin 4 potential. This factor, coupled with the rail-to-rail output swing makes the 3130 useful in single-supply systems.

The CA3130 is a popular device and its characteristic performance is available in an internally compensated version, the CA3160. In addition, a dual version of the device is also available, the CA3260.

### *The LF355/356/357 Series*

Another early FET-device development was the LF356 series (National Semiconductor, 1975) which included the LF355, LF356, and LF357 devices. This device family has had a major impact on IC op-amp technology for a number of reasons, but a basic reason for the importance of the LF355/356/357 series is that it marked the first use of ion implantation in an IC op amp.

Ion implantation is a processing step which (in general) allows a greater degree of control over the *uniformity* of device characteristics. With regard to op-amp circuitry, the important thing it provides is the ability to simultaneously fabricate well-matched p-channel JFET transistors, as well as bipolar transistors. Not only is the matching improved, which results in better dc characteristics, but the JFET devices also are capable of much wider bandwidth than their counterpart device in all bipolar circuits, the lateral pnp. Therefore, in the LF355/356/357 series, a *combination* of such very desirable op-amp parameters was achieved. Ion implantation is now widely used in the industry in a number of op amps, as well as other ICs. (For further information, see the references at the end of the chapter.)

The LF355/356/357 circuit represents a substantial departure from the classical two-stage topology of the 741 designs and its many variations. The circuit is shown in Fig. 2-20.

As can be noted, this basic circuit is used for all three devices of the series: the LF355 and LF356 (just as shown), and the LF357 with a 2-pF compensation capacitor. The 355 and 356 are unity-gain stable, while the 357 is stable at a minimum gain of 5.

The adjustment of the input offset voltage of $J_1$-$J_2$ is accomplished by current sources $J_7$-$J_8$ (via an external potentiometer), with no disturbance of the operating current in $J_1$-$J_2$. This offset trim technique results in little or no increase in drift.

A summary of the specifications of this device series is given in Table 2-5. The dc specifications of all devices in the series are essentially similar; the major difference is in terms of speed; i.e., SR and $f_t$. The LF356 and LF357 devices are also capable of low-noise operation, particularly in terms of current noise, while the LF355 is a lower-power-consumption device.

The LF355/356/357 device series is well established as a standard FET-amplifier type and has multiple sources. In addition, a number of related devices have been introduced which are functionally similar, but offer improved specifications (see selection table).

### *The CA3140*

The CA3140 (RCA, 1976) like its predecessor, the CA3130, features a low-bias-current (10 pA) MOSFET-input stage. The CA3140

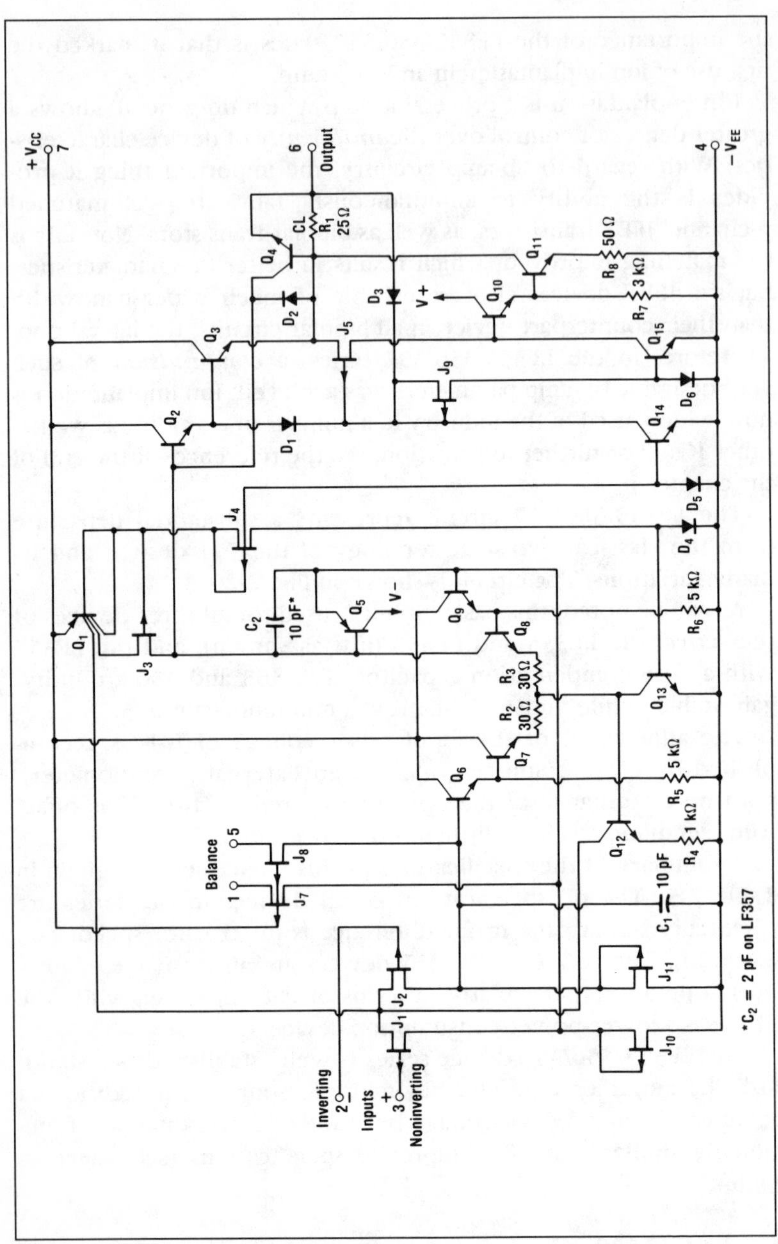


**Fig. 2-20. Schematic of LF355/356/357 op amps.**

*(Courtesy National Semiconductor Corp.)*

**Table 2-5. LF355/356/357 Specifications**

| Unit | Value |
|---|---|
| $V_{io}$ | 2–10 mV (max) |
| $I_{ib}$ | 50–200 pA (max) |
| $\Delta V_{io}/\Delta t$ | to 5 μV/°C (max) |
| $f_t$ | 4 MHz } LF356 |
| SR | 10 V/μs } LF356 |
| $t_s$ (0.01%) | 1.5 μs } LF356 |
| $e_n$ | 12 nV/$\sqrt{Hz}$ } LF356, LF357 |
| $i_n$ | 0.01 pA/$\sqrt{Hz}$ } LF356, LF357 |
| CMRR, PSRR | 80 dB (min) |

has features more like a general-purpose op amp, however, in that it is capable of full-rated supply-voltage ranges of ±15 volts, and it also uses a bipolar output stage.

The schematic of Fig. 2-21 shows the circuit in detail. $Q_9$ and $Q_{10}$ are PMOS FET-input devices, like the analogous units in the CA3130 (Fig. 2-19). Similar also to the CA3130 is the current mirror, $Q_{11}$-$Q_{12}$, and the second stage, $Q_{13}$. Unlike the CA3130, however, the output stage of the CA3140 has no voltage gain and consists of emitter followers $Q_{17}$-$Q_{18}$ for positive swings, and $Q_{21}$-$Q_{16}$ for negative swings. The output stage swings very close to V− (or ground) potential for low currents, by virtue of current source $Q_{15}$.

Like the CA3130, the input of the CA3140 can respond to common-mode voltages which include V− (or ground). The device is characterized for operation over a range of supply voltages from +5 to +36 volts (or ±18 volts). The 3140 is second-sourced and a dual version, the CA3240, is available. In addition, related low-voltage and low-bias-current devices are available, the CA3440 and the CA3420.

### *The TL081, TL071 Series*

The TL084 (Texas Instruments, 1976), a quad JFET-input op amp, was the first of a series of single and dual, as well as quad, devices. The circuit, shown in simplified form in Fig. 2-22, is used in both multiple and single versions, the latter devices providing for offset null (shown dotted).

This circuit, as can be noted, is once again the classic differential-input two-stage amplifier which is similar to the basic 741 topology. The essential difference is that the bipolar-input transistors have been replaced by an ion-implanted p-channel JFET pair,

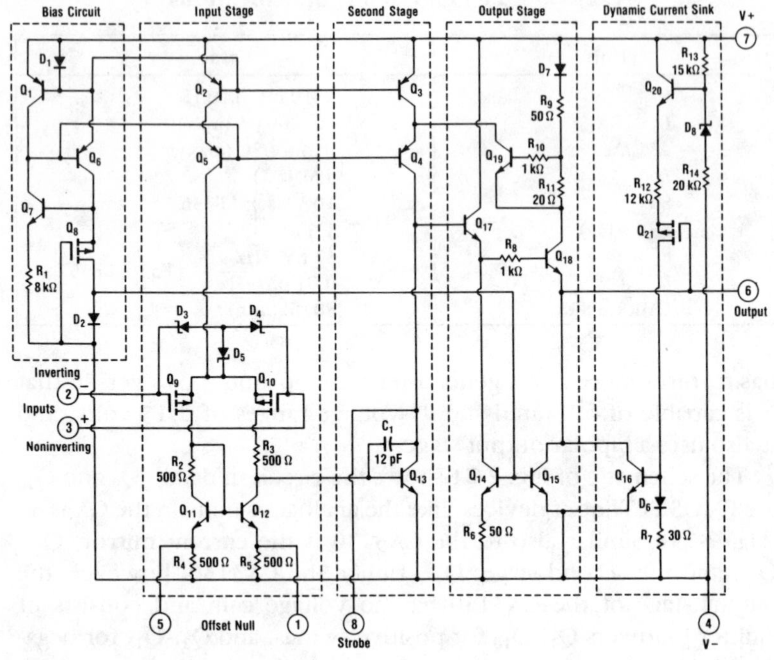


**Fig. 2-21. Schematic of the CA3140 op amp.**
*(Courtesy RCA Solid State Division)*

$J_1$-$J_2$. The general circuit topology of Fig. 2-22 is applicable to many currently available JFET general-purpose parts.

In an op-amp circuit, the largest single improvement to be gained by the ion-implanted JFET is in the input stage, where lower bias currents and higher speed can be achieved (vis-a-vis the 741). Therefore, a topology such as this can be advantageous over a 356-type circuit, as it is basically more simple (requires less die area) and, consequently, will have potentially greater yields and lower ultimate cost. Also, the lower quiescent power drain of this circuit, compared to a 356 type, makes it feasible for multiple units, whereas a dual or quad 356 would be prohibitive in terms of dissipation.

The TL081 series consists of the TL080, an externally compensated unit pinned like the 301A, and internally compensated units consisting of the TL081 (single), the TL082/083 (dual), and the TL084 (quad).

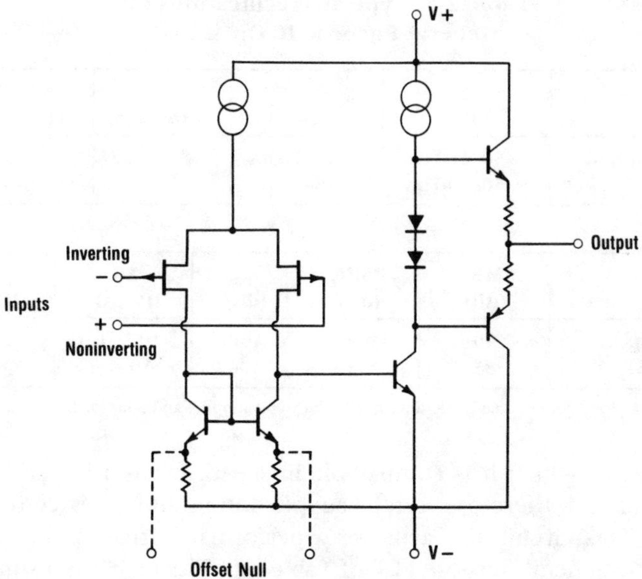


**Fig. 2-22. Simplified schematic of general-purpose JFET-input op amp(s).**

This device series, as well as a host of many others with similar specifications, could be called the general-purpose FET-input op amp. That is, it represents a FET-input device with the utility and multiple device availability of the bipolar general-purpose units. Thus, a specification comparison can be made between the categories of devices by means of the hypothetical device specifications of Table 2-6.

The TL071 series (Texas Instruments, 1977) consists of devices that are numbered similar to the TL081 series; that is, TL071 (single), TL072 (dual), and TL074/075 (quad). The specifications of the devices are similar to the TL081 series, but feature lower noise and distortion for ac uses.

### *The LF351/353/347 Series*

The LF351 series (National Semiconductor, 1977) was introduced with the LF351, a general-purpose single-unit FET-input op amp. The LF351 also uses the classic two-stage amplifier topology with only a single JFET pair used at the input (Fig. 2-22). This device is intended to be a more general-purpose unit than the

**Table 2-6. Typical Specifications for General-Purpose IC Op Amps***

| | $I_{ib}$ | $I_{io}$ | $V_{io}$ (mV) | $\Delta V_{io}/\Delta T$ ($\mu$V/°C) |
|---|---|---|---|---|
| **Bipolar** | 50–100 na | 10–20 nA | 2 | 10 |
| **FET** | 50–200 pA | 5–20 pA | 5 | 10–15 |

| | CMRR (dB) | PSRR (dB) | $A_{vo}$ (dB) | SR (V/$\mu$s) | $f_t$ (MHz) |
|---|---|---|---|---|---|
| **Bipolar** | 90 | 90 | 100 | 0.5–1 | 1–2 |
| **FET** | 85 | 85 | 100 | 10–15 | 3–5 |

*At 25°C, for a commercial-grade device with ±15-volt supplies and 2-kΩ load.

LF356, to which it is comparable in speed. It has a lower supply current and, therefore, can be supplied in multiple device form as well. This circuit also achieves a performance that is very much like the general-purpose FET of Table 2-6. The LF353 is a similarly specified dual; the LF347 is a quad.

***The LF411/412 Series***

The LF411 and LF412 devices (National Semiconductor, 1981) are single and dual units with performance similar to the LF351 and LF353, respectively. Both devices are available in "A" versions, with lower input offset.

### 2.1.6 A Broad Overview of General-Purpose Op Amps

General-purpose IC op amps are used in the greatest percentage of applications. From Table 2-6 and the discussion thus far, such a unit would be loosely defined as having a unity-gain bandwidth of 1 to 5 MHz, operation from power supplies of ±5 volts to ±20 volts without serious degradation of performance, and usually (but not always) internally compensated. This "loose" definition is fulfilled by a very great number of IC op amps, for example all of those just described, as well as many to follow. In fact, the range of devices available is so huge that it is impossible to cover all of them in any degree of detail.

As has been noted, the definition given for a general-purpose op amp is perhaps too broad, as there are many distinctions that can be made among types that fit this general definition. The following discussion and classifying tables illustrate this point, by categorization of devices into groups by either FET or bipolar, and by single, dual, and quad pinout arrangements.

Tables 2-7, 2-8, and 2-9 illustrate standard pinout arrangements for both bipolar and FET single, dual, and quad devices. As can be noted, there is an almost overwhelming array of units from which to choose. And, it should be noted, this list of devices is not 100% complete, as that is beyond reason. It does, however, list representative devices in almost every performance category and type. The part number prefix indicates the original source; but, in many cases, alternate sources are available.

The units listed in these tables will plug into the pin arrangement as noted, and function as intended. In some cases, however, there may be subtle or secondary performance changes possible such as an optimum offset null with a 100-kΩ potentiometer (as opposed to 10 kΩ, as shown) for example. The manufacturer's literature will be your final guide in such cases. Data sheets for some of the most popular devices are included in the Appendix.

The remarks columns of these tables are intended to give an "at-a-glance" assessment of the key feature of a device, beyond that of a pure general-purpose unit. The OP-07 of Table 2-7, for example, will more than meet the general-purpose specifications of Table 2-6, but also (as noted) features instrumentation-grade performance compared to the "741" general-purpose specifications for a bipolar device. The OP-07 is an extremely popular device, and is widely available. As another example, the dual CA3240 of Table 2-8 has a lower bias current than the normal general-purpose FET, and is suitable for single-supply operation as well.

In the area of quad devices, there is a double standard in pinouts, namely the "324" and "4136" pinouts. The tables reflect this; that is, the TL075 is a "4136-pinout" quad FET amplifier, with essentially general-purpose specifications (no remarks); the "324" pinout equivalent of this device is the TL074.

There are many devices listed on these standard pinout charts that actually go far beyond general-purpose performance, insofar as their exact specifications are concerned. These types, covered further in Section 2.2, may be optimized in terms of dc errors, speed, voltage, etc. The previously mentioned OP-07 is such a device, and is listed in both charts.

## Table 2-7. Single Amplifier Pinout Data

| Type No. | Pinout No. | Remarks |
|---|---|---|
| **Bipolar** | | |
| $\mu$A741 | 1 | Industry standard, general-purpose. |
| MC1456 | 1 | Low $I_{ib}$, high SR. |
| MC1436 | 1 | Low $I_{ib}$, high SR, high voltage. |
| LM343 | 1 | Low $I_{ib}$, high SR, high voltage. |
| HA2645 | 1 | Low $I_{ib}$, high SR, high voltage. |
| $\mu$A799 | 1 | Single supply. |
| $\mu$A759 | 1 | Single supply, high current. |
| MC1741S | 1 | High SR. |
| NE530 | 1 | High SR, low $I_{ib}$. |
| OP-01 | 1 | High SR. |
| AD741 | 1 | Improved dc performance. |
| OP-02 | 1 | Improved dc performance. |
| 3500 | 1 | Improved dc performance. |
| RC4131 | 1 | Medium speed. |
| RC4132 | 1 | Low power. |
| CA3193 | 1 | Instrumentation grade. |
| OP-21 | 1 | Low power, single supply. |
| MC34071(A) | 1 | Single supply, high speed. |
| **FET** | | |
| TL081 | 1 | — |
| TL071 | 1 | — |
| TL061 | 1 | Low power. |
| LF351 | 1 | — |
| $\mu$A771 | 1 | — |
| MC34001 | 1 | — |
| CA3140 | 1 | Single supply, low $I_{ib}$. |
| CA3160 | 1 | Single supply, low $I_{ib}$. |
| AD515 | 1 | Ultralow $I_{ib}$. |
| 3528 | 1 | Ultralow $I_{ib}$. |
| TL087 | 1 | Low $V_{io}$. |
| AD542 | 1 | Low $V_{io}$, low drift. |
| AD544 | 1 | Low $V_{io}$, high SR. |
| AD503 | 1 | First-generation FET devices—not all are recommended for new designs (consult the manufacturers). |
| AD506 | 1 | |
| AD540 | 1 | |
| ICL8007 | 1 | |
| 3540 | 1 | |
| OPA111 | 1 | Low $V_{io}$, low $I_{ib}$, low drift, low noise, $R_p$ = 100K. |
| OPA121 | 1 | Low $V_{io}$, low $I_{ib}$, low drift, low noise, $R_p$ = 100K. |
| OP-41 | 1 | Low $V_{io}$, low $I_{ib}$, low drift, $R_p$ = 100K. |
| AD548 | 1 | Low power, low $V_{io}$. |
| AD611 | 1 | Low $V_{io}$, high SR. |
| AD711 | 1 | Low $V_{io}$, high SR, fast settling. |

## Table 2-7 (cont). Single Amplifier Pinout Data

| Type No. | Pinout No. | Remarks |
|---|---|---|
| AD547 | 1 | Low $V_{io}$, low drift. |
| AD549 | 1 | Low $V_{io}$, ultralow $I_{ib}$. |
| CA3440 | 1 | Single supply, low $I_{ib}$, programmable/micropower. |
| CA3420 | 1 | Single supply, ultralow $I_{ib}$. |
| TLC251 (Series) | 1 | Single supply, low $I_{ib}$, low voltage, programmable. |
| TLC271 (Series) | 1 | Single supply, low $I_{ib}$, programmable. |
| LF411(A) | 1 | Low $V_{io}$. |
| LF441(A) | 1 | Low power, low $V_{io}$. |
| LF400 | 1 | High speed, fast settling. |
| MC34081(A) | 1 | High speed. |
| **Bipolar** | | |
| LM301A | 2 | $C_1 = 30$ pF, $R_1 = 5$ MΩ, $R_2 = 5$ MΩ. |
| μA748 | 2 | $C_1 = 30$ pF, $R_1 = 5$ MΩ, $R_2 = 5$ MΩ. |
| LM344 | 2 | $C_1 = 30$ pF, $R_1 = 5$ MΩ, $R_2 = 5$ MΩ, low $I_{ib}$, high SR, high voltage. |
| LM308(A) | 2 | $C_1 = 30$ pF, no null provision. |
| LT1008 | 2 | $C_1 = 30$ pF, no null provision, ultralow $I_{ib}$. |
| OP-08 | 2 | $C_1 = 30$ pF, no null provision. |
| **FET** | | |
| TL080, TL070 | 2 | $C_1 = 30$ pF, $R_1 = 5$ MΩ, $R_2 = 5$ MΩ. |
| CA3130 | 2 | $C_1 = 47$ pF, $R_1 = 100$ kΩ, $R_2 = 0$. |
| CA080 | 2 | $C_1 = 30$ pF, $R_1 = 5$ MΩ, $R_2 = 5$ MΩ. |
| **Bipolar** | | |
| LM310 | 3 | Low $I_{ib}$, high SR (follower only); $R_p$ 1–8 = 1 kΩ. |
| AD504* | 3 | Instrumentation grade; $R_p$ 1–8 = 10 kΩ. |
| AD510 | 3 | Instrumentation grade; ultralow $V_{io}$; $R_p$ 1–8 = 10 kΩ. |
| AD517 | 3 | Instrumentation grade; ultralow $V_{io}$; $R_p$ 1–8 = 10 kΩ. |
| OP-05 | 3 | Instrumentation grade; $R_p$ 1–8 = 20 kΩ. |
| OP-07 | 3 | Instrumentation grade; ultralow $V_{io}$; $R_p$ 1–8 = 20 kΩ. |
| μA714 | 3 | Instrumentation grade, ultralow $V_{io}$; $R_p$ 1–8 = 20 kΩ. |
| HA2500 (Series) | 3 | High SR, fast settling; $R_p$ 1–5 = 20 kΩ. |
| μA725/725A* | 3 | Instrumentation grade; $R_p$ 1–8 = 100 kΩ. |
| LM312 | 3 | Low $I_{ib}$; $R_p$ 1–8 = 100 kΩ. |
| LM316(A) | 3 | Ultralow $I_{ib}$; $R_p$ 1–8 = 100 kΩ. |

## Table 2-7 (cont). Single Amplifier Pinout Data

| Type No. | Pinout No. | Remarks |
|---|---|---|
| HA2600 (Series) | 3 | High SR and GBP; $R_p$ 1–5 = 100 kΩ. |
| OP-27 | 3 | Instrumentation grade, ultralow $V_{io}$, low noise; $R_p$ 1–8 = 10 kΩ. |
| OP-37 | 3 | Instrumentation grade, ultralow $V_{io}$, low noise; $R_p$ 1–8 = 10 kΩ. |
| LT1001 | 3 | Instrumentation grade; ultralow $V_{io}$, low noise; $R_p$ 1–8 = 10 kΩ. |
| LT1007 | 3 | Instrumentation grade; ultralow $V_{io}$, low noise; $R_p$ 1–8 = 10 kΩ. |
| LT1037 | 3 | Instrumentation grade; ultralow $V_{io}$, low noise; $R_p$ 1–8 = 10 kΩ. |
| LT1012 | 3 | Ultralow $I_{ib}$, $R_p$ 1–8 = 100 kΩ. |
| LM11 | 3 | Ultralow $I_{ib}$, $R_p$ 1–8 = 100 kΩ. |
| CA3493 | 3 | Instrumentation grade, ultralow $V_{io}$; $R_p$ 1–8 = 10 kΩ. |
| **FET** | | |
| $\mu$A740 | 3 | First-generation device; $R_p$ 1–5 = 10 kΩ. |
| LF355 | 3 | Low power; $R_p$ 1–5 = 25 kΩ. |
| LF356 | 3 | High performance overall; $R_p$ 1–5 = 25 kΩ. |
| LF357 | 3 | High SR and GBP; $R_p$ 1–5 = 25 kΩ. |
| OP-15 | 3 | Similar to 355 } Improved in dc and ac performance; $R_p$ 1–5 = 100 kΩ. |
| OP-16 | 3 | Similar to 356 } Improved in dc and ac performance; $R_p$ 1–5 = 100 kΩ. |
| OP-17 | 3 | Similar to 357 } Improved in dc and ac performance; $R_p$ 1–5 = 100 kΩ. |
| LT1022 | 3 | Higher performance, similar to OP-16. |
| LT1055 | 3 | Higher performance, similar to LF355. |
| LT1056 | 3 | Higher performance, similar to LF356. |
| OPA356 | 3 | Higher performance, similar to LF356. |
| OPA606 | 3 | Higher performance, similar to LF356. |

*Compensation component(s) required.

**Table 2-7 (cont). Single Amplifier Pinout Data**

**Pinout Diagram**

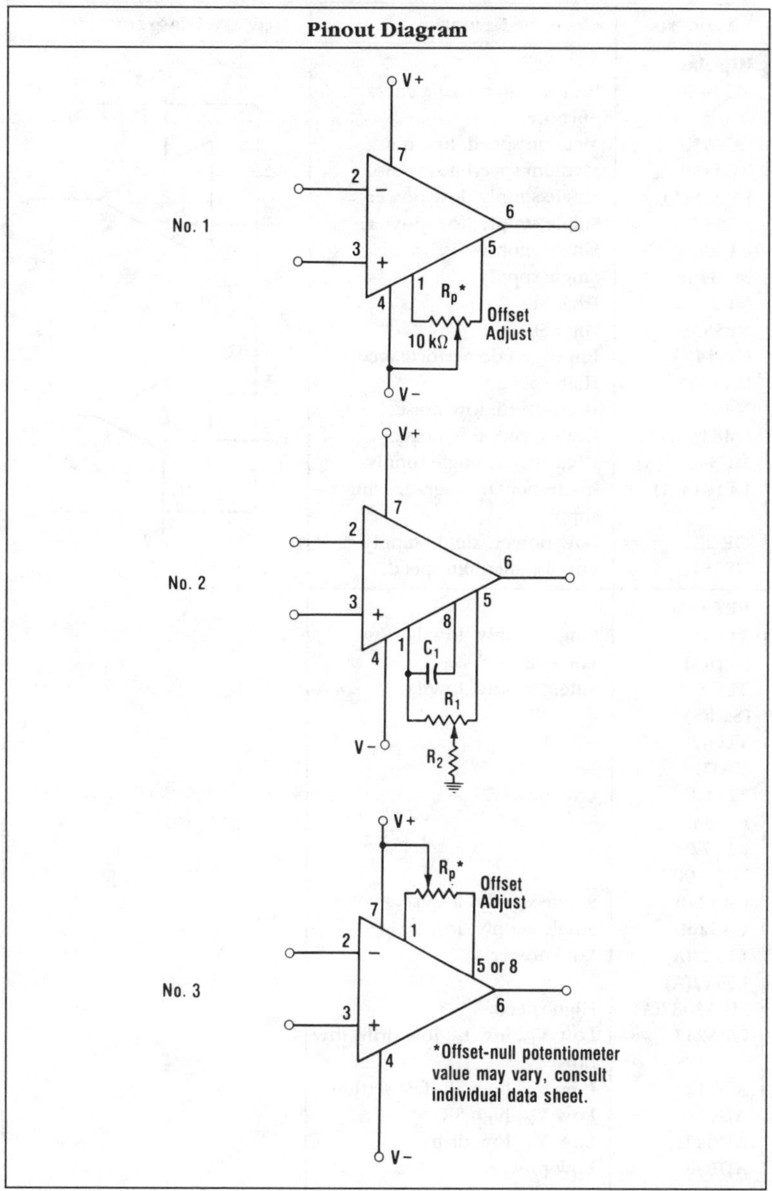

## Table 2-8. Dual Amplifier Pinout Data

| Type No. | Remarks | Diagram |
|---|---|---|
| **Bipolar** | | 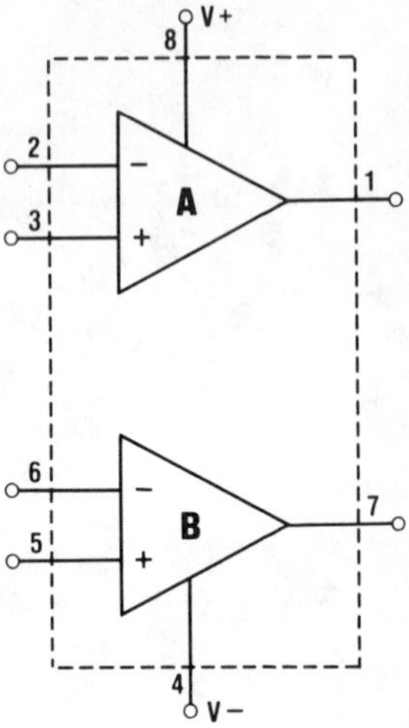 |
| MC1458 | Industry standard, general purpose. | |
| RC4558 | Medium speed, low noise. | |
| RC4559 | Medium speed, low noise. | |
| LM358(A) | Single supply, low power. | |
| NE532 | Single supply, low power. | |
| μA798 | Single supply. | |
| MC3458 | Single supply. | |
| MC1458S | High SR. | |
| NE5535 | High SR. | |
| OP-14 | Improved dc performance. | |
| HA2655 | High speed. | |
| NE5532 | High speed, low noise. | |
| LM833(A) | High speed, low noise. | |
| MC34072(A) | High speed, single supply. | |
| LT1013(A) | Instrumentation-grade, single supply. | |
| OP-221 | Low power, single supply. | |
| NE5512 | Low $I_{ib}$, medium speed. | |
| **FET** | | |
| TLC252 (Series) | Single supply, low $I_{ib}$, low voltage. | |
| TLC272 (Series) | Single supply, low $I_{ib}$. | |
| TL082 | — | |
| TL072 | — | |
| TL062 | Low power. | |
| LF353 | — | |
| μA772 | — | |
| MC34002 | — | |
| CA3240 | Single supply, low $I_{ib}$. | |
| CA3260 | Single supply, low $I_{ib}$. | |
| LF442(A) | Low power. | |
| LF412(A) | — | |
| MC34082(A) | High speed. | |
| OPA2111 | Low $V_{io}$, low $I_{ib}$, low drift, low noise. | |
| AD712 | Low $V_{io}$, high SR, fast settling. | |
| AD644 | Low $V_{io}$, high SR. | |
| AD647 | Low $V_{io}$, low drift. | |
| AD648 | Low power. | |
| OP-215 | — | |

## Table 2-9. Quad Amplifier Pinout Data

| Type No. | Remarks | Diagram |
|---|---|---|
| **Bipolar** | **324 Pinout** | 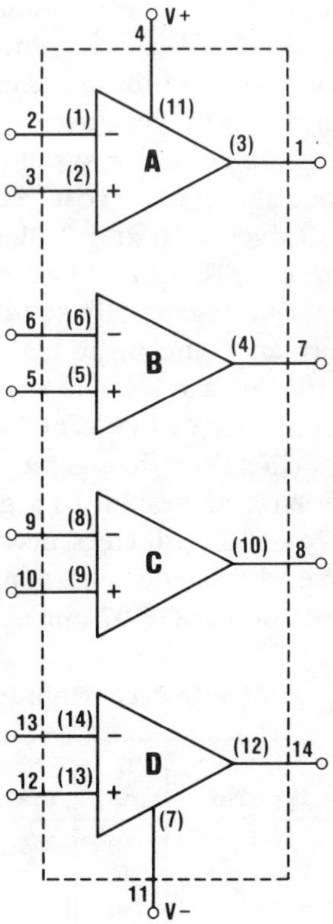 |
| LM348/349 | — | |
| MC4741 | Industry standard, general purpose. | |
| LM324(A) | Single supply, low power. | |
| SA534 | Single supply, low power. | |
| MC3403 | Single supply. | |
| RC3403A | Single supply. | |
| HA4741 | Medium speed, low noise. | |
| RC4156 | Medium speed, low noise. | |
| XR4212 | Medium speed, low noise. | |
| OP-11 | Medium speed, low noise. | |
| OP-421 | Single supply, low power. | |
| LT1014(A) | Instrumentation-grade, single supply. | |
| MC34074(A) | Single supply, high speed. | |
| NE5514 | Low $I_{ib}$, medium speed. | |
| **FET** | | |
| TL084 | — | |
| TL074 | — | |
| TL064 | Low power. | |
| LF347 | — | |
| μA774 | — | |
| MC34004 | — | |
| MC34084 | High speed. | |
| LF444(A) | Low power. | |
| TL254 (Series) | Single supply, low $I_{ib}$, low voltage. | |
| TL274 (Series) | Single supply, low $I_{ib}$. | |
| **Bipolar** | **4136 Pinout** | Note: First number denotes "324" pinout. Number in ( ) denotes "4136" pinout. |
| RC4136 | Medium speed, low noise. | |
| OP-09 | Medium speed, low noise. | |
| **FET** | | |
| TL075 | — | |

There are also, of course, a large number of pinouts that are different in some other detail, such as offset null and/or compensation. These cases are simply too numerous to document here. A number of such units are listed in the performance charts that follow.

### *High-Performance Op Amps*

Some devices offer substantial improvement over the characteristic specification of the general-purpose amplifier, although they may be plug-in replacements for a similar basic design in some instances. A case in point is the fact that the 741 design has now been refined to the point that several manufacturers offer devices with 30-nA input-bias currents and 5 μV/°C maximum drift. These units can be very useful, either for a performance upgrade of an existing circuit or system, or for new designs with good performance at moderate cost (see following caveat).

Tables 2-10 and 2-11 list a number of these devices, for both bipolar (Table 2-10) or FET units (Table 2-11). The tables themselves are generally self-explanatory; it should be noted that performance specifications are given in terms of *worst-case limits,* either as a minimum or a maximum, for a commercial temperature-range device. The guaranteed limit is, of course, the type of specification that is necessary to compile an error budget, or worst-case design, to a given system specification. Note that the "741-style" devices may or may not be performance/cost effective for *new* designs, compared to a more modern design (such as the popular OP-07 unit).

**Table 2-10. High-Performance Bipolar Op-Amp Data**

| Type No. | $I_{ib}$ (nA) | $I_{io}$ (nA) | $V_{io}$ (mV) | $\Delta V_{io}/\Delta T$ (μV/°C) | CMRR (dB) | PSRR (dB) | $A_{vo}$ (dB) |
|---|---|---|---|---|---|---|---|
| AD741L | 50 | 5 | 0.5 | 5 | 90 | 90 | 94 |
| AD301AL* | 30 | 5 | 0.5 | 5 | 90 | 90 | 98 |
| OP-02E | 30 | 2 | 0.5 | 8 | 90 | 90 | 100 |
| OP-14E | 50 | 2 | 0.75 | 8 | 90 | 90 | 100 |
| 3500E | 50 | 30 | 0.5 | 1 | 88 | 100 | 100 |

Note: All specifications are worst-case values @ 25°C (minimum or maximum, as applicable).
*External compensation component(s) required.

The charts are arranged to highlight the general areas where a particular device type is notable. For example, the FET types are also capable of excellent speed; therefore, this is listed as well as the dc specifications.

## 2.2 SPECIALIZED GROUPS OF IC OP AMPS AND THEIR CHARACTERISTICS

There are many op-amp applications that cannot be met satisfactorily or optimally by general-purpose types. Special situations may demand very high speed (either wide bandwidth or high slew rate), very low input current, high input impedance, very low offset or offset drift, high voltage or high current output, or other confining characteristics.

To complement the general-purpose op amps discussed in Section 2.1, this section discusses the more special-purpose types. The array of types presented is broad, but there is a considerable amount of overlap. In fact, several of the ICs could even be categorized in both groups. Most of these types are also standards, representative of the popular version of a particular device. For reference, the key features that govern the selection of a particular op amp within its group are listed in the respective tables. These listings are in terms of *single* devices; however, dual versions are also available in many cases.

### 2.2.1 Instrumentation-Grade Op Amps

Table 2-12 lists amplifiers that are classified as "instrumentation grade."* Generally, what this connotation implies is *precision;* that is, very high performance with regard to dc errors, very low $V_{io}$ (and drift), very high CMRR, PSRR, and $A_{vo}$. Such devices usually have low noise and low loading errors as well. In short, these devices are optimum for highly accurate dc amplification, such as may be found in instrumentation and control systems. Specifications are given in terms of worst-case limits, except as noted. The table includes both bipolar and FET devices.

The first general instrumentation-grade op amp was the 725, an externally compensated type. As can be noted with this device (and many others as well), there are two offset-voltage-drift specifications given, one without null and one after null. With many units, it should be noted that the condition of minimum drift is attained with the lowest (adjusted) offset voltage. In such cases,

---

* Do not confuse "instrumentation-grade" op amp with "instrumentation amplifier." For the latter device, see Chapter 7.

**Table 2-11. High-Performance FET Op-Amp Data**

| Type No. | $I_{ib}$ (pA) | $I_{io}$ (pA) | $V_{io}$ (mV) | $\Delta V_{io}/\Delta T$ ($\mu V/°C$) | CMRR (dB) | PSRR (dB) | $A_{vo}$ (dB) | SR ($V/\mu s$) | $f_t$ or GBP (MHz) | $t_s$† ($\mu s$ to 0.01%, 10 V Step) |
|---|---|---|---|---|---|---|---|---|---|---|
| LF356A | 50 | 10 | 2 | 5 | 85 | 85 | 94 | 10 | 4 | 1.5 |
| OP-16E | 50 | 10 | 0.5 | 5 | 86 | 86 | 100 | 18 | 6 | 3.8 |
| AD542L | 25 | 2† | 0.5 | 5 | 80 | 80 | 108 | 3 | 1† | — |
| AD544L | 25 | 2† | 0.5 | 5 | 80 | 80 | 94 | 8 | 2† | 3.0 |
| LT1022AC | 50 | 10 | 0.25 | 5 | 86 | 88 | 104 | 23 | 8.5† | 1.3 (0.02%) |
| LT1056AC | 50 | 10 | 0.18 | 4 | 86 | 90 | 104 | 12 | 6.5† | 1.5 |
| OPA606L | 10 | 5 | 0.5 | 5 | 85 | 90 | 100 | 27 | 12 | 2.1 |
| OP-41E | 5 | 1 | 0.5 | 8 | 100 | 92 | 120 | 1 | 0.5† | — |
| AD711L | 50 | 5† | 0.25 | 2 | 80 | 80 | 94 | 20† | 4† | 1.0 |

Note: All specifications are worst-case values @ 25°C (minimum or maximum, as applicable).

*$A_v$ (minimum) = 5.

†Denotes typical specification.

the manufacturer's recommended null procedure should be used, if nulling is to be used. (This is discussed further in Chapter 3.)

Note that some of the devices listed as "first" and "second" generation are quite old, and while they still may be available, a more recent device will likely be better for new designs.

With the advent of the actively trimmed op amp, input offset voltage as a major source of error has become much less of a problem. The term *active trimming* means that at some point during processing, the manufacturer adjusts the internal circuitry of the op amp for a minimum input offset-voltage condition. This is accomplished by either selectively shorting zener diode strings (popularly called "zener zapping") or burning resistor areas with a laser ("laser trimming"). The end result is the same—a device that reaches the user with an offset voltage of 100 $\mu$V or less. As can be noted from the listings, devices are available with offsets as low as 25 $\mu$V (and less).

Such a device is the Precision Monolithics OP-07 device, shown in Fig. 2-23, which is trimmed for offset by adjustment of $R_{2A}$ and $R_{2B}$ on the chip. The OP-07 is essentially a trimmed version of the earlier OP-05. A comparison of their specifications reveals what the trimming achieves in terms of error reduction—nearly an order of magnitude reduction in $V_{io}$. The OP-07 has achieved status as an industry standard device, and is widely available. The OP-27 and OP-37 are more recent (and lower noise) improvements, and are also widely second-sourced.

While instrumentation-grade op amps are generally dominated by the bipolar devices, there are exceptions. Notable devices in this table are the AD547 and OPA-111, which achieve maximum drifts of 1 $\mu$V/°C, unusual for FET-input devices.

The advent of routine production-line trimming of circuits to extremely low errors has been a second major technological advance of recent years, as it has significantly advanced the op-amp state-of-the-art, at less than prohibitive cost. While any op amp can be nulled for zero offset *individually,* nulling often adversely alters (increases) drift. Also, an op amp with a potentiometer attached is *not* one that is field replaceable without recalibration and the attendant cost/complication. A specified low-offset low-drift device simply eliminates this problem.

Chopper-stabilized op amps attack the input-voltage offset error in a different manner by using an on-chip modulation/demodulation process. This results in offset voltages on the order of 10 $\mu$V,

**Table 2-12. Instrumentation-Grade Op Amps (Bipolar and FET)**

| | Type No. | $I_{ib}$ (nA) | $I_{io}$ (nA) | $V_{io}$ ($\mu$V) | $\Delta V_{io}/\Delta T$ ($\mu$V/°C) Unnulled | Nulled | CMRR (dB) | PSRR (dB) | $A_{vo}$ (dB) |
|---|---|---|---|---|---|---|---|---|---|
| First and Second Generation | $\mu$A725E* | 75 | 5 | 500 | 2.0 | 1.0 | 120 | 106 | 120 |
| | LM308A* | 7 | 1 | 500 | 5.0 | NA | 96 | 96 | 98 ($R_L = 10$ kΩ) |
| | LM308A-1* | 7 | 1 | 500 | 1.0 | NA | 96 | 96 | 98 ($R_L = 10$ kΩ) |
| | AD504L* | 80 | 10 | 500 | 2.0 | 1.0 | 110 | 100 | 120 |
| | OP-05E | 4 | 3.8 | 500 | 2.0 | 0.6 | 110 | 94 | 106 |
| Actively Trimmed Devices | OP-07E | 4 | 3.8 | 75 | 1.3 | 1.3 | 106 | 94 | 106 |
| | $\mu$A714E | 4 | 3.8 | 75 | 1.3 | 1.3 | 106 | 94 | 106 |
| | AD OP-07E | 4 | 3.8 | 75 | 1.3 | 1.3 | 106 | 94 | 126 |
| | AD510L | 10 | 2.5 | 25 | 2.0 | 0.5 | 110 | 100 | 120 |
| | AD517L | 1 | 0.25 | 25 | 0.5 | 0.5 | 110 | 110 | 120 |
| | LH0044AC | 15 | 2.5 | 50 | 0.5 | 0.5 | 120 | 120 | 120 ($R_L = 10$ kΩ) |
| | 3510CM* | 15 | 10 | 60 | 0.7 | 0.5 | 110 | 110 | 120 |
| | OP-08E* | 2 | 0.2 | 150 | 2.5 | NA | 104 | 104 | 94 |
| | OP-12E | 2 | 0.2 | 150 | 2.5 | NA | 104 | 104 | 94 |
| | LM11C | 50 (pA) | 10 (pA) | 300 | 3.0 | NA | 110 | 100 | 100 ($R_L = 5$ kΩ) |
| | OP-27E | 40 | 35 | 25 | 0.6 | 0.6 | 114 | 100 | 120 |
| | OP-37E | 40 | 35 | 25 | 0.6 | 0.6 | 114 | 100 | 120 |
| | AD547L | 25 (pA) | 2 (pA)† | 250 | 1.0 | 1.0 | 80 | 80 | 108 |
| | OPA111B | 1 (pA) | 0.75 (pA) | 250 | 1.0 | 1.0 | 100 | 100 | 120 |
| | LT1001AC | 2 | 2 | 15 | 0.6 | 0.6 | 114 | 110 | 113 |
| | LT1007AC | 35 | 30 | 25 | 0.6 | 0.6 | 117 | 110 | 137 |
| | LT1037AC | 35 | 30 | 25 | 0.6 | 0.6 | 117 | 110 | 137 |
| | LT1012C | 150 (pA) | 150 (pA) | 50 | 1.5 | 1.5 | 110 | 110 | 102 |
| | LT1008C | 100 (pA) | 100 (pA) | 120 | 1.5 | NA | 114 | 114 | 102 |

**Table 2-12 (cont). Instrumentation-Grade Op Amps (Bipolar and FET)**

| | Type No. | $I_{ib}$ (nA) | $I_{io}$ (nA) | $V_{io}$ ($\mu$V) | $\Delta V_{io}/\Delta T$ ($\mu$V/°C) | | CMRR (dB) | PSRR (dB) | $A_{vo}$ (dB) |
|---|---|---|---|---|---|---|---|---|---|
| | | | | | Unnulled | Nulled | | | |
| **Chopper Stabilized** | ICL7650C* | 10 (pA) | 0.5 (pA) | 5 | 0.05 | NA | 120 | 120 | 120 |
| | ICL7652C* | 30 (pA) | 60 (pA) | 5 | 0.05 | NA | 110 | 110 | 120 |
| | LTC1052C* | 30 (pA) | 30 (pA) | 5 | 0.05 | NA | 120 | 120 | 120 |
| **Electrometer Grade** | AD515L | 75 (fA) | — | 1000 | 25.0 | — | 70 | 74 | 88 |
| | 3528CM | 75 (fA) | 20 (fA) | 500 | 10.0 | — | 70 | 80 | 90 |
| | AD549C | 75 (fA) | — | 1000 | 10.0 | — | 90 | — | — |
| | OPA128L | 75 (fA) | 30 (fA)† | 500 | 5.0 | — | 90 | — | 110 |
| | CA3420B | 1000 (fA) | 700 (fA) | 2000 | 4.0† | — | 80 | 75 | 86 |

Note: All specifications are worst-case values @ 25°C (minimum or maximum, as applicable), for best-grade commercial (or industrial) temperature-range units.

*External compensation component(s) required.

†Typical specification(s).

or less, for the devices shown. The ICL7652 or LTC1052, for example, achieve a typical $V_{io}$ of 5 $\mu$V, concurrent with a bias current of 30 pA, as well as other excellent specifications. The long-term drift of a chopper amplifier is, of course, an inherent minimum, as the device continually zeros itself (and never requires recalibration).

Electrometer-grade amplifiers are, in general, devices that have bias currents of 1 pA or less—a very tall order for any op amp. The AD515 is a very high performer in this regard with its 75-fA bias current, as is the 3528. The CA3420 has a 1-pA (max) bias current, but it can also work with supplies as low as ±1 volt. The recently introduced AD549 and OPA128 are high all-around performance electrometer amps.

### 2.2.2 High-Speed Op Amps

Table 2-13 lists a number of op amps that are generally classified as high-speed devices. Speed is a somewhat general parameter and can mean several things. However, these devices are characterized by exceptional performance with regard to $f_t$ (or GBP), SR, and settling time (to a rated accuracy). They may or may not have good dc parameters as well.

Since high performance in speed-related terms depends on a number of interrelated (and complex) factors, no single parameter will usually predict in a straightforward manner everything needed. The most complete and meaningful single specification in speed terms is settling time, as it indicates how quickly an amplifier will reproduce a changing step input to a rated accuracy. However, to be useful, it must be specified under controlled closed-loop conditions. Therefore, a given specification may not always match an intended use exactly. For applications such as data-conversion uses, settling time is the most meaningful parameter and may, in fact, be all that is needed.

For ease of use, Table 2-13 is broken down into the subcategories of FET and bipolar amplifier types. Most of the data are self-explanatory and the specifications given are typical. The highest-performance FET types are the AD711, LT1022, and the OPA606 series; highest-performance bipolar types are the HA2541 and HA2542.

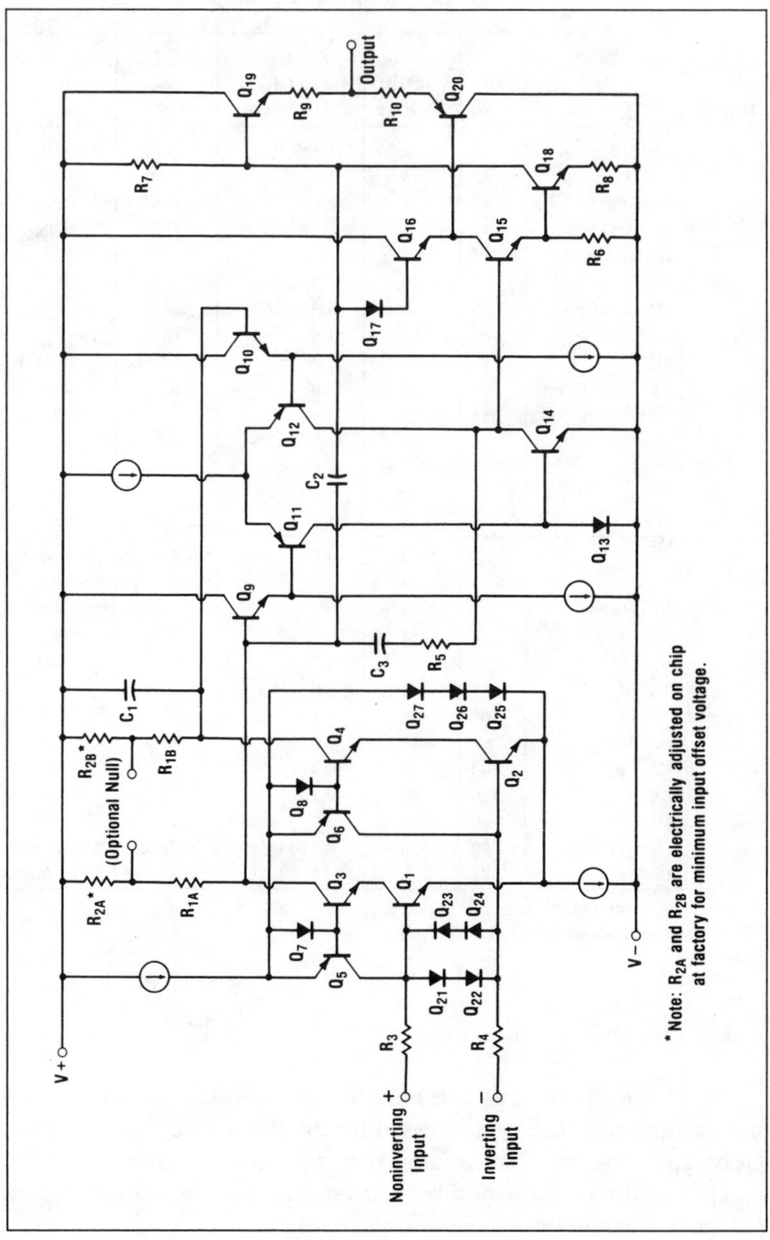


**Fig. 2-23. Schematic of OP-07 op amp.**

*(Printed with permission from Precision Monolithics, Inc.)*

**Table 2-13. High-Speed Op Amps**

| Device | Type No. | SR (V/$\mu$s) | $f_t$ or GBP (MHz) | $t_s$ ($\mu$s to 0.01%* 10 V Step) | $V_{io}$ (mV) | $I_{ib}$ (nA) |
|---|---|---|---|---|---|---|
| FET | CA3140 | 9 | 4.5 | 1.4 (0.1%) | 5 | 0.01 |
| | TL071 (Series) | 13 | 3 | NS | 3 | 0.03 |
| | TL080* | ≥12 | ≥3 | NS | 3–5 | |
| | LF351,353,347 | 13 | 4 | 2 | 1–5 | 0.05 |
| | LF356 (Series) | 12 | 5 | 1.5 | † | † |
| | LF357 (Series) | 50 | 20 | 1.5 ($A_v$=5) | † | † |
| | OP-16 (Series) | 25 | 8 | 3.8 | † | † |
| | OP-17 (Series) | 60 | 30 | 1.5 ($A_v$=5) | † | † |
| | AD711 | 20 | 4 | 1.0 | 0.25 (max) | 0.05 (max) |
| | MC34080 (Series) | 25 | 8 | 1.6 | 0.5–6 | 0.06 |
| | OPA606 | 35 | 13 | 2.1 | 0.1 | 0.005 |
| | LT1022 | 24 | 8 | 1.3 (0.02%) | 0.1 | 0.01 |
| | LT1056 | 14 | 5.5 | 1.5 | 0.07 | 0.01 |
| | LF400 | 57 | 18 | 0.4 | 4 | 0.2 |
| | LF411 | 15 | 4 | 1.8 | 0.8 | 0.05 |
| Bipolar | CA3100* | ≥25 | ≥30 | 0.6 (0.5%) | 1 | 700 |
| | LM318 | ≥70 | ≥15 | 0.8 | 4 | 150 |
| | NE5534* | ≥7 | ≥10 | NS | 0.5 | 500 |
| | MC1741S | 12 | 1 | 3 | 2 | 200 |
| | NE531* | ≥30 | ≥1 | 2.5 | 2 | 400 |
| | NE530 | 35 | 3 | 0.9 | 2 | 65 |
| | NE538 | 60 | 6 | 1.2 ($A_v$=5) | 2 | 65 |
| | HA2625* | ≥7 | ≥12 | NS | 3 | 5 |
| | HA2525* | 100 | 20 | 0.2 ($A_v$=3) | 5 | 125 |
| | AD509 | 120 | 20 | 0.2 | 5 | 125 |
| | MC34074 | 10 | 4.5 | 1.1 | 2 | 100 |
| | HA2541 | 280 | 40 | 0.1 (0.1%) | 10 (max) | 25 $\mu$A (max) |
| | HA2542 | 375 | 45 | 0.1 (0.1%) | 10 (max) | 25 $\mu$A (max) |

Note: All specifications typical, except as noted.
* External compensation component(s) required.
† See Table 2-11.

## 2.2.3 Other Op-Amp Devices

There are other considerations of application category which may not be specifically addressed by the types discussed thus far. Generally, however, an op amp that does not fall into one of the categories discussed would be quite specialized. Examples of such categories would be:

**Table 2-14. Op-Amp Parameter Optimization Chart**

| | Parameter | Category | Example | Specification |
|---|---|---|---|---|
| Input | $V_{io}$ | Chopper stabilized or Instrumentation grade. | ICL7652 or LTC1052 (Choppers) | 5 $\mu$V |
| | | | LTC1001 | 15 $\mu$V |
| | | | OP-27/37 (Instrumentation grade) | 25 $\mu$V |
| | $\Delta V_{io}/\Delta T$ | Chopper stabilized or Instrumentation grade. | ICL7652 or LTC1052 | 0.05 $\mu$V/°C |
| | $I_{ib}$ | Electrometer. | AD549 | 75 fA |
| | | MOSFET input. | CA3420 | 1 pA |
| | $I_{io}$ | Same as $I_{ib}$. | Same | |
| | CMRR | Chopper stabilized or Instrumentation grade. | LT1052 | 120 dB |
| | PSRR | Chopper stabilized or Instrumentation grade. | LH0044 | 120 dB |
| | $e_n$ | Bipolar. | OP-27/37 | 3 nV/$\sqrt{Hz}$ |
| | $I_n$ | FET. | LF356 Series | 0.01 pA/$\sqrt{Hz}$ |
| Output | $V_{om}$ | Bipolar, high voltage. | HA2645 | ±35 V @ $V_S$=±40 V |
| | $I_{om}$ | Power (TO-3). | LM12 | 12 A |
| | | Power (TO-99). | $\mu$A759 | 0.35 A |
| Dynamic | $f_t$ | Bipolar. | HA2541 | 40 MHz |
| | GBP | FET (decompensated). | OP-17 | 30 MHz |
| | | Bipolar (decompensated). | OP-37 | 65 MHz |
| | SR | Bipolar. | HA2542 | 350 V/$\mu$s |
| | $A_v$ | Chopper stabilized or Instrumentation grade. | LTC1007/1037 | 137 dB |

- **High-Voltage Op Amps ($V_s$ > 20 V)**
  MC1436
  LM343, LM344
  HA2645
  3580, 3584
- **Low-Voltage Op Amps ($V_s$ < 5 V)**
  CA3420
  TLC251
  TLC271
  LM10

- **High-Power Op Amps**
  μA759
  μA791
  LM12
  OP-50
- **Low-Power Op Amps**
  CA3078
  CA3420
  CA3440
  OP-21
  OP-22
  OP-32
  LM441
  AD548
  LP324
  LT1013/1014
  LM10
  TLC251/271
- **Preamps**
  MC1303
  μA739, μA749
- **Programmable Op Amps**
  4250
  μA776
  HA2725
  CA3078
  TL066
  CA3440
  OP-22
  OP-32
  TLC251/271

Most of the categories mentioned are self-explanatory with, perhaps, the exception of the latter—programmable op amps. A programmable op amp is one that allows user adjustment of performance characteristics, via a programming input pin. Note that since programmability implies power control, most programmable op amps can also be considered to be low-power types.

### 2.2.4 Parameter Optimization

When selecting an op amp for a particular application, very often one (or two) amplifier performance parameters will reduce to primary importance. The selection of an optimum device can then proceed, given a catalog of devices from which to choose.

As an aid to this process, the chart given in Table 2-14 will be helpful. This chart lists a number of basic op-amp performance parameters, and the category of the device that will be found to be optimum for that parameter. Then, a representative device and its specifications are listed. Although the specific device examples may change with the introduction of new types, the categories will not—thus the user generally will always know where to look.

As has been mentioned, selected data sheets for individual devices are included in the Appendix. Also included in the Appendix is a complete list of devices and their manufacturers, for complete specification information.

Further background information on specific devices may be obtained by consulting the references listed at the end of the chapter.

## 2.3 REFERENCES


1. *Analog Data Manual,* 1985. Signetics Corp., Sunnyvale, CA.
2. Bell, K., and Giolma, B. "Single-Supply CMOS Op Amps Sacrifice Nothing." *Electronic Design,* March 3, 1983.
3. Borlase, W. H., and Maidique, M. A. "Precision IC's Challenge Chopper Amplifier Designs." *Electronic Products,* December 18, 1972.
4. Cave, D. L., and Davis, W. R. "A Quad JFET Wide-Band Op Amp IC Featuring Temperature Compensated Bandwidth." *IEEE Journal of Solid State Circuits,* Vol. SC-12, No. 4, August 1977.
5. Cotreau, G., and Creed, S. "Pair of One-Chip Op Amps Streamlines the Design of Fast Analog Circuits." *Electronic Design,* February 21, 1985.
6. Davis, S. "IC Op Amp Families." *EDN,* January 20, 1978.
7. Davis, W. F. "Bipolar Op Amp Achieves JFET-Like Speeds." *Electronic Design,* September 1, 1983.
8. Davis, W. F., and Vyne, R. L. "Design Techniques for Improving the HF Response of a Monolithic JFET Operational Amplifier." *ISSC Journal,* Vol. SC-19, No. 6, December 1984.

9. "Designing With TL080 BIFET Op Amps." *Texas Instruments Bulletin CA-193,* July 1977. Texas Instruments Incorporated, Dallas, TX.

10. Dobkin, R. C. "IC Preamp Challenges Choppers on Drift." *National Semiconductor Application Note AN-79,* February 1973. National Semiconductor Corp., Santa Clara, CA.

11. _________. "LM118 Op Amp Slews 70 V/μs." *National Semiconductor Linear Brief LB-17,* September 1971. National Semiconductor Corp., Santa Clara, CA.

12. Erdi, G. "A Low Drift, Low Noise Monolithic Operational Amplifier For Low Level Signal Processing." *Fairchild Semiconductor Application Bulletin APP-136,* July 1969. Fairchild Semiconductor, Mountain View, CA.

13. _________, Schwartz, T., Bernardi, S., and Jung, W. "Op Amps Tackle Noise—and for Once, Noise Loses." *Electronic Design,* December 12, 1980.

14. _________, and Williams, J. "Precision Op Amp Serves Host of Design Needs." *Electronic Design,* September 1, 1983.

15. Fullagar, D. "A New High Performance Monolithic Operational Amplifier." *Fairchild Application Brief,* May 1968. Fairchild Semiconductor, Mountain View, CA.

16. _________. "Better Understanding of FET Operation Yields Viable Monolithic J-FET Op Amp." *Electronics,* November 6, 1972.

17. _________. "The 8007—A High-Performance FET-Input Operational Amplifier." *Intersil Application Bulletin A005,* March 1972. Intersil, Inc., Cupertino, CA.

18. Grant, D. "Low Drift Super Beta Op Amp." *Analog Dialogue,* Vol. 12, No. 1, 1978.

19. Gray, P. R. "A 15W Monolithic Power Amplifier." *ISSCC Digest,* February 1972, pp. 180–181.

20. Hearn, H. W. "Fast Slewing Monolithic Operational Amplifier." *IEEE Journal of Solid State Circuits,* Vol. SC-6, No. 1, February 1971.

21. Hoskins, K. "Autozero Chip Cuts Op Amp Offset Voltages to a Mere ±5 μV." *Electronic Design,* March 21, 1985.

22. Jung, W. G. "IC Op Amps For Audio," Parts I and II. *The Audio Amateur,* February 1973, January 1974, February 1974.

23. _________. "IC Op Amps Have Evolved." *Electronic Design,* January 9, 1978.

24. _________. "An IC Op Amp Update." *Ham Radio,* March 1978.

25. _________. "Stable FET-Input Op Amps Achieve Precision Performance," *EDN,* November 10, 1982.

26. Maidique, M. A. "A High Precision Super Beta Operational Amplifier." *ISSCC Digest,* February 1972, pp. 178–179.

27. ________. "Monolithic Operational Amplifier With 1 $\mu$V/°C Drift." *Analog Dialogue,* Vol. 5, No. 3, April-June 1971.

28. Metzger, J. "Bipolar and FET Op Amps." *Electronic Products,* June 1977.

29. Millaway, S. "Monolithic Op Amp Hits Trio of Lows." *Electronic Design,* February 9, 1984.

30. Ochi, S., and Flink, J. "LM143 Monolithic High Voltage Operational Amplifier Applications." *National Semiconductor Application Note AN-127,* April 1976. National Semiconductor Corp., Santa Clara, CA.

31. Pippenger, D., and May, D. "Put BIFETs Into Your Linear Circuits." *Electronic Design,* January 4, 1978.

32. Roberge, J. K. *Operational Amplifiers—Theory and Practice.* John Wiley & Sons, Inc., New York, 1975.

33. Russell, R. W., and Culmer, D. D. "Ion-Implanted JFET-Bipolar Monolithic Analog Circuits." *ISSCC Digest,* February 1974.

34. ________, and Frederiksen, T. M. "Automotive and Industrial Control Systems Building Blocks." *ISSCC Digest,* February 1972, pp. 116–117.

35. Schade, O. H. Jr. "CMOS/Bipolar Linear Integrated Circuits." *ISSCC Digest,* February 1974, pp. 136–137.

36. Schade, O. H. Jr. "A New Generation of MOS/Bipolar Operational Amplifiers." *RCA Review,* Vol. 37, September 1976.

37. Schade, O. Jr., and Kramer, E. "Low-Voltage BIMOS Op Amp." *ISSCC Journal,* Vol. SC-16, No. 6, December 1981.

38. Sherwin, J. "Use the LM158/LM258/LM358 Dual, Single Supply Op Amp." *National Semiconductor Application Note AN-116,* May 1974. National Semiconductor Corp., Santa Clara, CA.

39. Soderquist, D., and Erdi, G. "The OP-07 Ultra-Low Offset Voltage Op Amp." *Precision Monolithics Application Note AN-13,* December 1975. Precision Monolithics, Inc., Santa Clara, CA.

40. Solomon, J. E. "The Monolithic Op Amp: A Tutorial Study." *IEEE Journal of Solid State Circuits,* Vol. SC-9, No. 6, December 1974.

41. ________, Davis, W. R., and Lee, P. L. "A Self-Compensated Monolithic Operational Amplifier With Low Input Current and High Slew Rate." *ISSCC Digest of Technical Papers,* Vol. 12, February 1969, pp. 14–15.

42. ________, and Russell, R. W. "A New Op Amp Is Born." *Motorola Monitor,* Vol. 18, No. 1, 1970.

43. Stout, D. F., and Kaufman, M. *Handbook of Operational Amplifier Design.* McGraw-Hill Book Co., New York, 1976.

44. Sullivan, D. R., and Maidique, M. A. "High Performance IC FET-Input Op Amp." *Analog Dialogue,* Vol. 4, No. 2, December 1970.

45. Tobey, G. E., Graeme, J. G., and Huelsman, L. P. (eds.). *Operational Amplifiers—Design and Applications.* McGraw-Hill Book Co., New York, 1971.

46. Underwood, R. K. "New Design Techniques for FET Op Amps." *National Semiconductor Application Note AN-63,* March 1972. National Semiconductor Corp., Santa Clara, CA.

47. Widlar, R. J. "A Monolithic Operational Amplifier." *Fairchild Semiconductor Application Bulletin APP-105/2,* July 1965. Fairchild Semiconductor, Mountain View, CA.

48. ________. "A New Monolithic Operational Amplifier Design." *National Semiconductor TP-2,* June 1967. National Semiconductor Corp., Santa Clara, CA.

49. ________. "A Unique Circuit Design for a High Performance Operational Amplifier Especially Suited to Monolithic Construction." *Proceedings of the NEC,* Vol. XXI, October 1965, pp. 85-89. (Also available as Fairchild TP-32/2, November 1965. Fairchild Semiconductor, Mountain View, CA.)

50. ________. "Design Techniques for Monolithic Operational Amplifiers." *National Semiconductor TP-12,* August 1969. National Semiconductor Corp., Santa Clara, CA.

51. ________. "Future Trends In DC Amplifiers." *National Semiconductor TP-4,* January 1968. National Semiconductor Corp., Santa Clara, CA.

52. ________. "IC Op Amp Beats FETs on Input Current." *National Semiconductor Application Note AN-29,* December 1969. National Semiconductor Corp., Santa Clara, CA.

53. ________. "IC Op Amps Close the Performance Gap on Discretes." *National Semiconductor TP-9,* December 1968. National Semiconductor Corp., Santa Clara, CA.

54. ________. "IC Op Amp With Improved Input-Current Characteristics." *EEE,* December 1968.

55. ________. "Super Gain Transistors for ICs." *National Semiconductor TP-11,* March 1969. National Semiconductor Corp., Santa Clara, CA.

56. ________. "The Improved μA702 Wideband DC Amplifier." *Fairchild Semiconductor Application Bulletin APP-111/2,* July 1965. Fairchild Semiconductor, Mountain View, CA.

57. ________. "The LM110—An Improved IC Voltage Follower." *National Semiconductor Linear Brief LB-11,* March 1970. National Semiconductor Corp., Santa Clara, CA.

58. ________. "Working With High-Impedance Op Amps." *National Semiconductor Application Note AN-241,* February 1980.

59. ________. "Reducing DC Errors in Op Amps." *National Semiconductor Technical Paper 15,* December 1980.

60. ________, and Yamatake, M. "A 150W Op Amp Introduces New Power Techniques." *ISSCC Digest,* February 1985.

61. ________, Pease, B., and Yamatake, M. "Applying a New Precision Op Amp." *National Semiconductor Application Note 242,* April, 1980.

62. Williams, J. "Applications Considerations and Circuits for a New Chopper-Stabilized Op Amp." *Linear Technology Application Note No. 9,* March 1985.

63. ________. "Bipolar IC Op Amps Enhance Circuit Precision." *EDN,* June 9, 1983.

64. Witten, S. "Designing Junction FET Input Op Amps." *Siliconix Application Note AN74-3,* August 1974. Siliconix, Inc., Santa Clara, CA.

THREE

# *General Operating Procedures and Precautions in Using IC Op Amps*

In this final chapter of Part I, we are but one step removed from applying IC op amps in actual circuits. However, in a practical sense, this chapter may be the most important of all the introductory material because it deals with the actual problems involved in using the devices: offset adjustment methods, standard compensation methods, protection against abuses, failure mechanisms, and application pitfalls. It is recommended that this chapter be studied thoroughly before any attempts are made at building op-amp circuits, particularly if the reader has had little previous experience with linear ICs.

## 3.1 METHODS OF OFFSET ADJUSTMENT

In applications that require a high degree of dc accuracy, the residual input offset voltage of the op amp used may be a significant source of error, particularly when operating at high stage gains. These situations require that the input offset voltage be adjusted to zero (nulled) for best performance. Offset nulling can be accomplished either by using a technique recommended by the manufacturer for the particular device (termed *internal nulling* in these discussions), or by using a universally applicable *external nulling* procedure.

### 3.1.1 Internal Offset Nulling for Individual Devices

In general, the manufacturer's technique for nulling a particular device is usually optimum for that device. Unfortunately, as touched on previously, the exact method of nulling that is used varies widely from device to device, and there is no completely standard method of internal offset adjustment (although the 741 technique shown in diagram No. 1 of Table 2-7 enjoys a fair degree of usage). Other details necessary in the practical application of op amps are methods of frequency compensation and pinout arrangements. In this section, we will discuss alternate methods of internal offset nulling. Compensation techniques (when required) are by no means common and must be dealt with case by case due to the large number of devices involved.

Although most op amps are available in a variety of packages, the most popular are the 8-pin packages—the TO-99 metal can and the 8-pin, dual in-line package (MINI-DIP). For dual devices, the 8-pin MINI-DIP package is the most popular, although the limitation of eight pins sacrifices the null capability. Alternate pinout arrangements may be arrived at by consulting the manufacturers' data sheets for the particular device (see the Appendix).

Actually, the basic pinouts for single devices, shown in Chapter 2 with offset null, can be considered to be the "standard" forms of offset null—those *most often* seen. The discussions in this section supplement those basic modes, and are for the remaining applications.

The information provided in this section will be sufficient to utilize virtually any device in its basic circuit arrangement. Offset adjustments, and even compensation components, are not required in *all* circumstances and may be deleted where not necessary. Power-supply connections are obviously always required and are shown in this section for reference. Throughout the remainder of the book, however, the power connections are not shown as long as they involve standard supplies. In special circuits or nonstandard power-supply operation, the power connections will be shown in full detail.

Fig. 3-1 shows the standard op-amp pinout with dotted supply connections to illustrate this point. The V−, or ground, pin is pin 4 (for single devices) and is common for both *power* and *signal* in single-supply types. For example, a single-supply op amp might typically operate from a +15-volt (only) power supply and handle a 0- to 10-volt signal range. In such a case, pin 4 would be grounded. It is assumed, by implication, that both input and output can operate over the 0- to +10-volt range. For dual-supply op amps, the more general ±5- to ±15-volt (or other) supply ranges will be used.

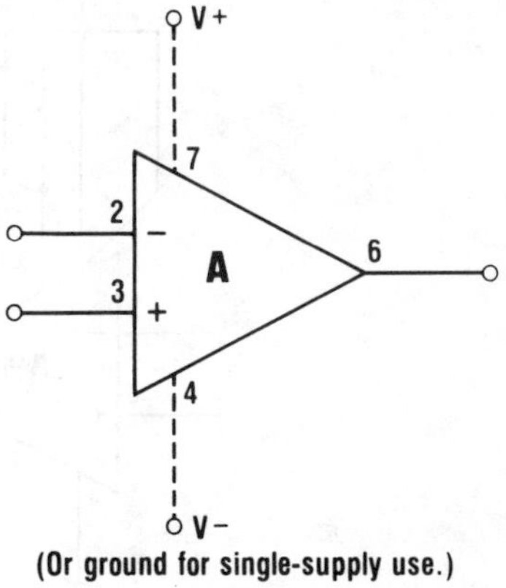

**Fig. 3-1. Standard op-amp pinout for dual- or single-supply use.**

### 3.1.2 Nulling Precision Op Amps

Although, in theory, virtually any op amp can be adjusted for a condition of zero input voltage, there are special techniques for accomplishing it when precision op amps are to be used.

A precision, instrumentation-grade op amp will most likely be a device with a low initial offset voltage and, consequently, a low offset-voltage drift. This is because these amplifiers typically exhibit a drift-change characteristic of 3.3 μV/°C/mV of change in offset voltage. Thus, in theory, when $V_{io} = 0$, zero drift is attained. It is for this fundamental reason (as well as the low offset voltage itself) that a factory-trimmed, low-offset-voltage op amp is a desirable item—two major errors are minimized simultaneously. Obviously, for operation consistent with this high-performance capability, it is desirable not to compromise the low-drift/low-offset-voltage relationship.

In general, offset null of a precision op amp should be avoided, if at all possible. This is for the simple reason that, if care is not taken, the nulling technique used may actually *increase* drift. However, where nulling of the last 50–100 μV offset of a precision op amp is necessary, one of two schemes should be used.

Fig. 3-2 illustrates a high-resolution null technique suitable for trimming 100–200 μV of offset to 1 or 2 μV. This magnitude of trim range is really about all that should be considered for null, and nulling of a precision op amp should *never* be used to compensate for other system errors.

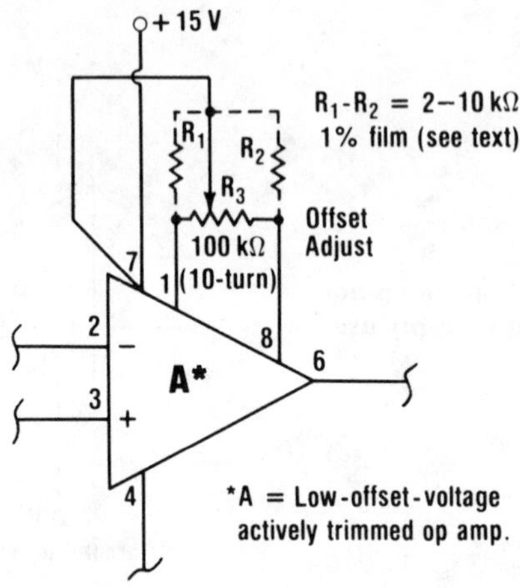


**Fig. 3-2. High-resolution null technique for precision op amps.**

In this circuit, $R_3$ is a relatively high-value, multiturn, cermet trimmer, which shunts the internal amplifier resistances of 1–10 kΩ (typically). Resistors $R_1$ and $R_2$ are 1% metal-film units, selected to yield the desired resolution and range, and will be in the range of 2–10 kΩ. (On some amplifier types, these resistors will not be necessary.) Trimmer $R_3$ should be placed as close to the amplifier as feasible for minimum noise pickup. The wiper arm of the trimmer (and $R_1$-$R_2$) should be connected directly to the V+ pin of the op amp.

Fig. 3-3 illustrates a method of offset null that will not alter the drift of a precision op amp. In this circuit, $A_1$ is a low-drift precision type and $A_2$ is a unity-gain summer. The variable current applied to $A_2$ from the offset network trims the amplified offset of $A_1$ (as well as that of $A_2$). Although the offset and drift of $A_2$ will appear at the output, as well as that of $A_1$, they will be negligible in proportion to those of $A_1$, for high $A_1$ gains. This allows a low-cost general-purpose unit to be used for $A_2$. Amplifier $A_2$ is also advantageous from the standpoint that it buffers $A_1$. This general concept is applicable to many other precision devices; not just the OP-07 and OP-27 types shown.

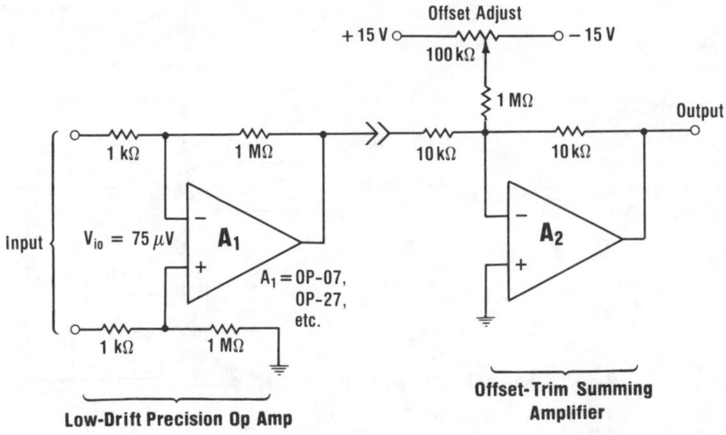


**Fig. 3-3. Offset null technique to preserve low drift of precision op amp.**

### 3.1.3 Nulling Programmable Op Amps

Op amps which have user-programmable operating currents will also require some special consideration for optimum nulling. While not all programmable op amps provide for an internal offset null, those that do may require that the value of the offset null potentiometer used be inversely proportional to the programmed current.

Fig. 3-4 shows such a case, the 3440 programmable op amp. For this specific device, the value of the null potentiometer should be as shown in the illustration. In practice, a value can be selected from the table according to the $I_{SET}$ in actual use.

### 3.1.4 More General Nulling Procedures

The generalized nulling technique shown in Fig. 3-5 allows almost any op amp to be nulled, using a voltage referred to ground. High-value resistors $R_1$ and $R_2$ connect to the balance terminals, 1 and 8 (or 1 and 5), and sink (or source) a few microamperes of current. A variable voltage applied to $R_2$ unbalances the two currents and thereby changes the offset value.

This scheme will work with a degree of precision that depends somewhat on the specific op amp employed. This is because the circuitry seen at pins 1 and 8 (or 1 and 5) varies widely from type to type. Thus, $R_1$-$R_2$ may need to be optimized for individual cases.

Fig. 3-6 shows how this basic null technique can be used with

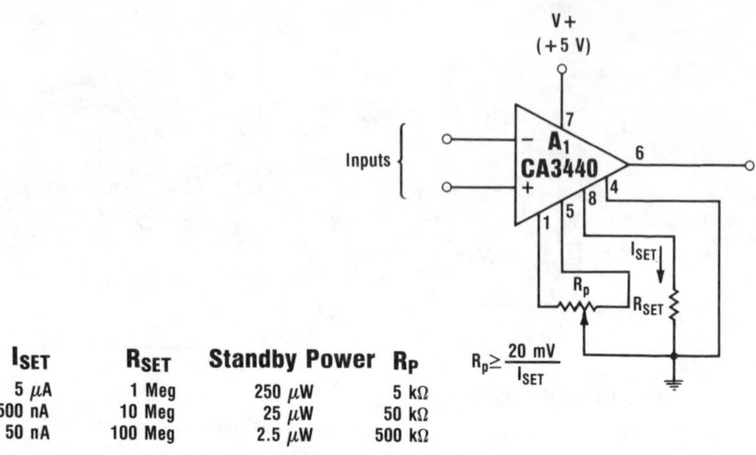


| $I_{SET}$ | $R_{SET}$ | Standby Power | $R_P$ |
|---|---|---|---|
| 5 μA | 1 Meg | 250 μW | 5 kΩ |
| 500 nA | 10 Meg | 25 μW | 50 kΩ |
| 50 nA | 100 Meg | 2.5 μW | 500 kΩ |

**Fig. 3-4. Nulling a programmable op amp.**

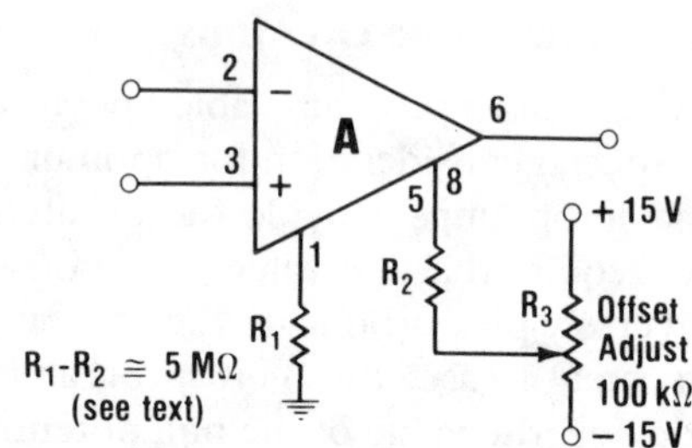


**Fig. 3-5. General-purpose null technique.**

an auto-null offset correction scheme. Here, $A_2$ is a low-drift and low-offset-voltage FET-input op-amp integrator, which stores a dc offset voltage on capacitor C. The auto-null system timing periodically closes switch S and drives the output of $A_1$ to a low voltage, essentially the offset voltage of $A_2$.

A final variation of this basic scheme is shown in Fig. 3-7, where digital control is used for offset changes. In this circuit, the DAC-08C is used as a 6-bit D/A converter, which develops a differential voltage across the 100-kΩ load resistors. The voltages are converted to currents by the 4.7-MΩ resistors, which control the offset voltage of the op amp.

This scheme is a useful one for large systems, as virtually any desired offset voltage can be stored in memory and applied to the amplifier when desired. The entire circuit can be made remote, of course, allowing correction at a transducer site, for example.

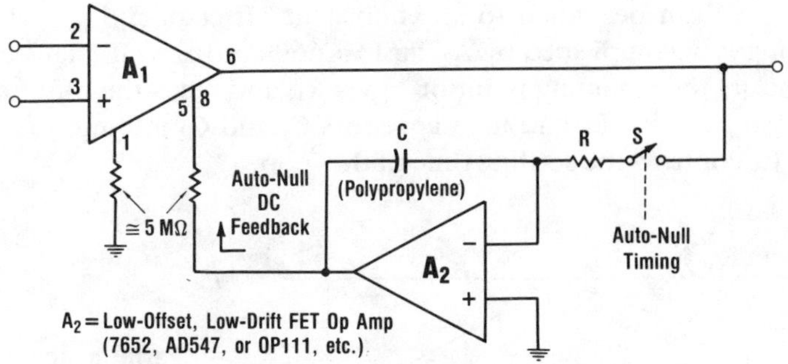


**Fig. 3-6. General-purpose auto-null technique.**

For op amps which do not have a pair of offset pins available, the "auto-null" technique can be adapted to drive the input pins (discussed in the following sections).

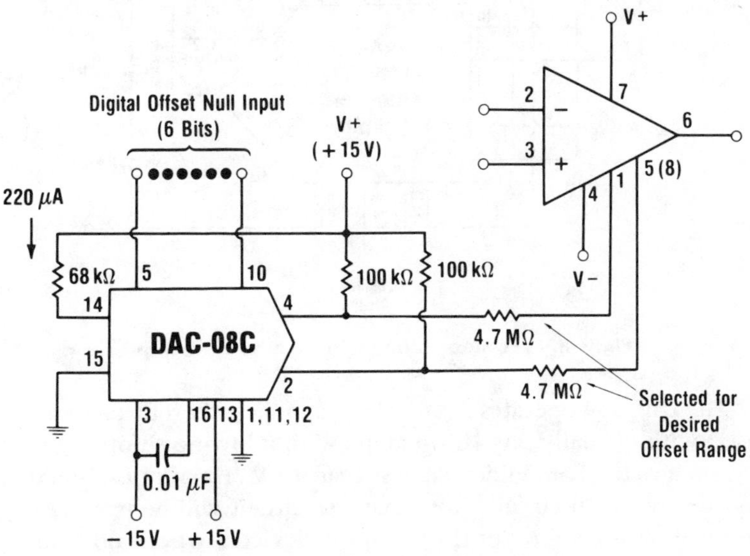


**Fig. 3-7. General-purpose, digitally controlled, offset trim technique.**

## *Precision Auto-Nulling*

There are situations where an automatic nulling scheme is operated "stand-alone," where it can continuously correct for offset and drift, as well as worst-case amplifier offset.

Fig. 3-8 illustrates how an inverting amplifier with an offset of ±10 mV can be nulled to an equivalent offset of ±5 μV. Here, nulling is accomplished by $A_2$, an LMC669 auto-zero IC. This chip monitors the summing point of $A_1$ via $R_1$, and feeds the offset correction back via $R_2$ and $R_3$. Capacitors $C_1$ and $C_2$ minimize noise, and $C_3$ controls the settling time of the loop.

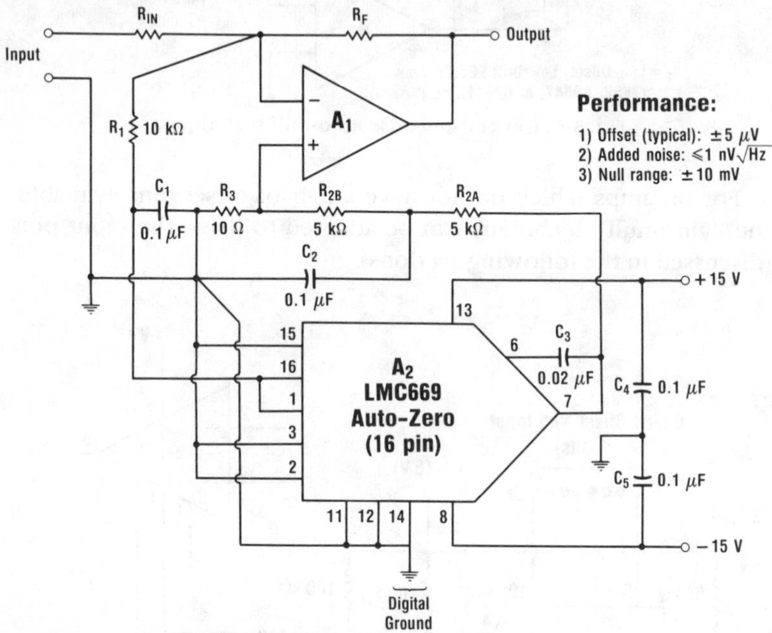


**Fig. 3-8. Precision nulling with an auto-zero chip.**

The LMC669 operates from ±15-V supplies; therefore, it can be used with virtually any IC op amp. With a low-noise op amp for $A_1$, such as the LM833 device, less than 1 nV$\sqrt{Hz}$ of noise is added by the auto-zero circuit. Note that this circuit will be particularly useful for 8-pin dual or 14-pin quad devices, where no nulling terminals are provided.

### 3.1.5 Universal External Offset Nulling Techniques

As pointed out in Section 3.1.1, the optimum method of offset adjustment for a particular op amp is the internal method provided for by the manufacturer. This method results in the condition of minimum input offset-voltage drift, which occurs when the input offset voltage is zero. Unfortunately, the mechanism for accomplishing this is not available on all op amps, and *external* methods sometimes must be used. This section describes procedures for universal external-offset adjustment at the ***input*** terminals, allowing any op amp to be nulled for an equivalent input offset voltage of zero. This allows any type which does not provide an internal offset adjustment to be used with greater precision (i.e., 308 types and many duals and quads), or it preserves the low-drift operation of certain types. The exact method used varies slightly with the type of configuration, as detailed in the following discussion.

### *Inverting Configuration*

There are actually two slightly different techniques that can be used with the inverting stage to adjust the input offset voltage. The simpler of the two is shown in Fig. 3-9A. This circuit uses the $R_1$ and $R_2$ input and feedback resistors as part of an attenuator network, in conjunction with $R_3$, to generate a variable offset voltage at the inverting input. This voltage is scaled down from the $\pm 15$ volts available from $R_4$ by the $R_3$ and $R_1 \| R_2$ divider. In this case, the ratio is 1000/1, yielding an offset range of $\pm 15$ mV. For different supply voltages, a lesser (or greater) range, or different $R_1$-$R_2$ values, $R_3$ may be rescaled by use of the equation given in Fig. 3-9A. Note that if more than one input is used, the offset range will be reduced because of the lower equivalent resistance. In this case, the more general equation for multiple inputs should be used to select $R_3$.

With just one additional resistor, the circuit of Fig. 3-9B accomplishes offset adjustment independent of the feedback elements. This circuit is generally more useful since it injects the offset voltage at the noninverting input terminal, away from the direct signal path. In this circuit, $R_3$ and $R_5$ form a simple 1000/1 divider, providing $\pm 15$ mV of offset voltage across $R_5$ with the values shown. The exact values used for $R_3$ and $R_5$ are not critical, but it is good practice to keep $R_5$ below 1000 ohms.

### *Noninverting Configuration*

A circuit for external null adjustment for low-gain, noninverting stages is shown in Fig. 3-10A. Here, a low resistance, such

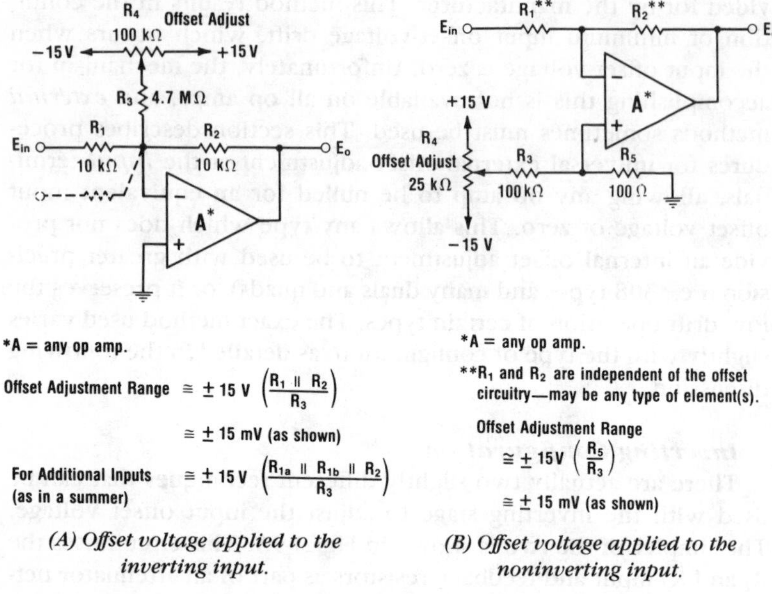


*(A) Offset voltage applied to the inverting input.*

*(B) Offset voltage applied to the noninverting input.*

**Fig. 3-9. Universal external-offset null-adjustment circuits—inverting amplifier.**

as $R_5$, is inserted in the return leg for $R_1$. The offset voltage is thus developed by the $R_3$-$R_5$ high-ratio voltage divider at this point. Since gain-determining resistors $R_1$ and $R_2$ constitute a voltage divider for the voltage at $R_5$, the voltage appearing at the inverting input will be reduced by their ratio. This yields a slightly more complicated offset-voltage-range equation. Again, the values shown yield $\pm15$ mV of adjustment voltage.

It should be understood that the gain and offset calculations for this stage are not independent. The presence of $R_5$ in series with $R_1$ alters the gain equation as shown; and, with the values shown, the error is 1% (ratio of $R_1$ to $R_5$).

At high gains, a null may be more simply achieved with what is basically a modification of Fig. 3-10A. This circuit is shown in Fig. 3-10B, which generates the offset voltage directly at the summing junction, using $R_1$ and $R_3$ as the offset divider. The reason this technique is not preferred at low gains is that the signal voltage across $R_1$ (equal to the input voltage by the theory of zero differential input) is larger at lower gains, which creates a change in the

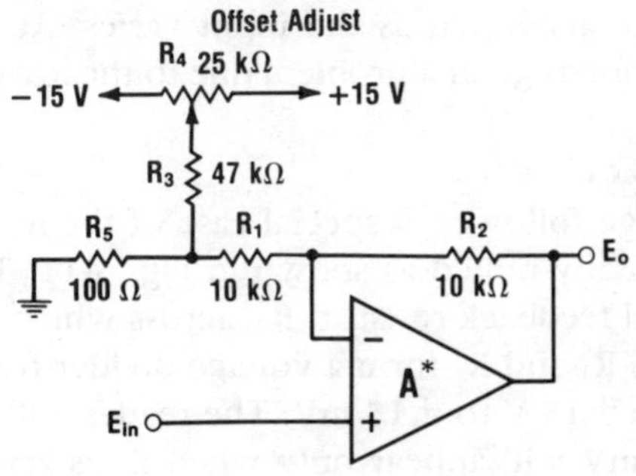


*A = any op amp.

Offset Adjustment Range $\cong \pm 15\text{ V} \left(\frac{R_5}{R_3}\right)\left(\frac{R_2}{R_1 + R_2}\right)$

$\cong \pm 15$ mV (as shown)

Precise gain of stage modified by presence of $R_5$ in series with $R_1$:

$$A_v = \frac{R_5 + R_1 + R_2}{R_5 + R_1}$$

(disregarding $R_3 \parallel R_5$, since $R_3 \gg R_5$)

*(A) Circuit for low-gain noninverting amplifier.*

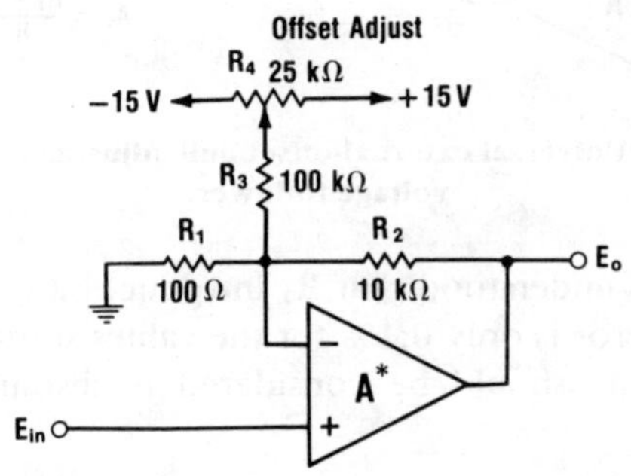


*A = any op amp.

Offset Adjustment Range $\cong \pm 15\text{ V} \left(\frac{R_1}{R_3}\right)$

$\cong \pm 15$ mV (as shown)

*(B) Circuit for high-gain noninverting amplifier.*

**Fig. 3-10. Universal external-offset null-adjustment circuits—noninverting amplifier.**

offset voltage generated as the input varies. At high gains, however, this is not as great a problem due to the lower input level.

### *Voltage Follower*

The voltage follower, a special case of the noninverting amplifier, is externally nulled as shown in Fig. 3-11. This circuit introduces a small feedback resistor, $R_2$, across which the offset voltage is generated. $R_1$ and $R_2$ form a voltage divider from the arm of $R_3$, reducing the ±15 V to ±15 mV. The reader will note that the offset of ±15 mV will appear only when $E_{in}$ is zero. As $E_{in}$ changes from zero, the offset voltage generated will also change. This may not be a serious drawback because the error it does introduce is greater at full scale, where it is still a small percentage error.

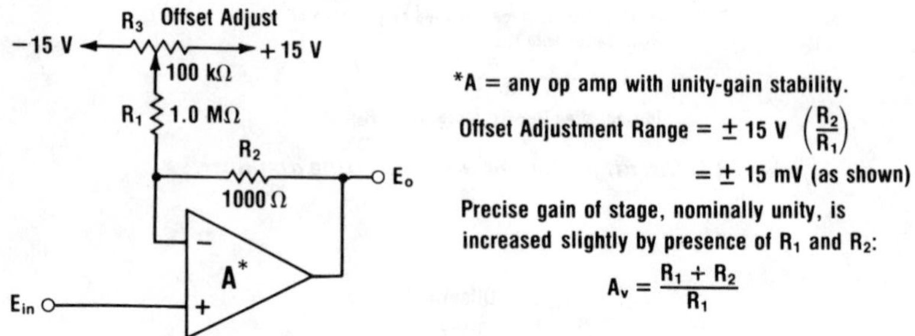


**Fig. 3-11. Universal external-offset null-adjustment circuit—voltage follower.**

It should be understood that $R_1$ introduces a small gain error, although the error is only 0.1% for the values shown. Both of the preceding points should be considered if absolute accuracy is important.

### *The Differential Amplifier*

Fig. 3-12 illustrates a useful technique for balancing a differential-amplifier configuration. The main problem with introducing an offset voltage into a differential circuit is that it must be done *without* disturbing the differential signal balance.

In this circuit, high-ratio divider $R_3$-$R_5$ provides a variable potential at the common end of $R_2'$, which is normally grounded. After being further divided by $R_1'$ and $R_2'$, this voltage serves to null the amplifier. A very small $R_5$ is chosen so that it will be a

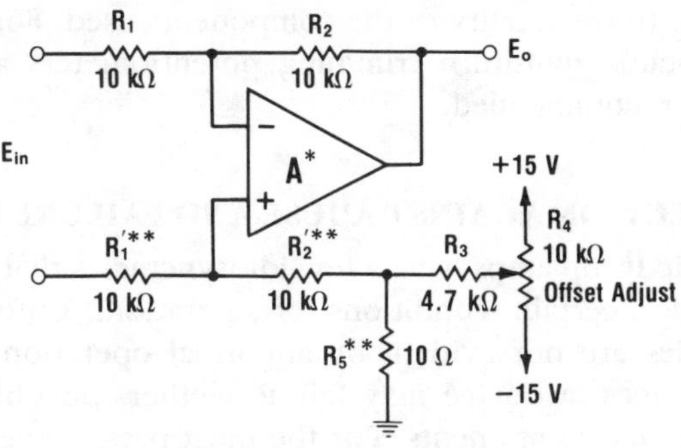


*A = any op amp.

Offset Adjustment Range $\cong \pm 15\text{ V}\left(\frac{R_5}{R_3 + R_5}\right)\left(\frac{R_1'}{R_1' + R_2'}\right)$

$\cong \pm 15$ mV (as shown, if $R_{source} \ll R_1'$)

**Presence of $R_5$ introduces a common-mode error, since

$$\frac{R_2}{R_1} \neq \frac{R_2' + R_5}{R_1'}.$$

If necessary, trim $R_2'$ such that

$$\frac{R_2}{R_1} = \frac{R_2' + R_5}{R_1'}.$$

**Fig. 3-12. Universal external-offset null-adjustment circuit—differential amplifier.**

small percentage of $R_2'$, thereby minimizing the deviation from the ideal bridge ratio. In demanding situations, $R_2'$ may need trimming to ensure that

$$\frac{R_2}{R_1} = \frac{R_2' + R_5}{R_1'}$$

(A 10-meg resistor connected across $R_2'$ will provide this, for the values chosen.) With the values shown, the offset voltage range is ±15 mV.

In all of the preceding discussions, it is assumed that the ±15-volt power supplies used to generate the offset voltages are stable potentials. If they are not, it should be obvious that they will introduce noise into the system in proportion to their deviation from perfection. A noisy or relatively unstable power supply may be dealt with by conventional decoupling techniques (see Chapter 4).

In the same light, the stability of the offset adjustment is also proportional to the quality of the components used. For best performance, stable multiturn trimming potentiometers and stable resistors are recommended.

## 3.2 PROTECTION AGAINST ABUSE AND FAILURE MODES

Monolithic IC op amps have a few idiosyncracies that can cause trouble under certain conditions of operation. Unfortunately, these troubles are not evident by abnormal operation alone; in certain instances, a device may fail altogether, possibly causing damage to other components. For the most part, however, these failure modes are predictable and, if taken into account beforehand, there should be no problems in applying the devices. In this section, we will discuss such failure conditions and give positive treatment for their prevention.

### 3.2.1 Input Limitations

Failure in IC op amps can be induced in the input stage in two general ways: (1) by exceeding the differential input rating, or (2) by exceeding the common-mode rating. Of these two, the differential input rating is the parameter that is most susceptible to abuse; therefore, it will be discussed in some detail.

#### *Differential Input Breakdown*

One of the easiest methods of inducing failure in an unprotected op-amp input stage is to exceed its differential input voltage rating. When this happens, one or the other of the differential input transistors (depending on the relative polarity of the input voltage) will go into emitter-to-base zener breakdown. This is shown in Figs. 3-13A and B. An npn-input differential pair is actually equivalent to a pair of back-to-back 7-volt zener diodes, as shown in Fig. 3-13B. The anodes of these two diodes are the (+) and (−) input terminals of the op amp.

Whenever the difference between the input terminals exceeds ±7 volts, these emitter-base diodes will break down (as will any zener) and conduct a current which will be limited only by the external source resistance. If the source impedance feeding both inputs is low, the current can rise to destructive levels very quickly. Currents greater than 50 mA will cause permanent failure—usually a junction short. Currents lower than this, while not causing junction damage, are still very undesirable because they

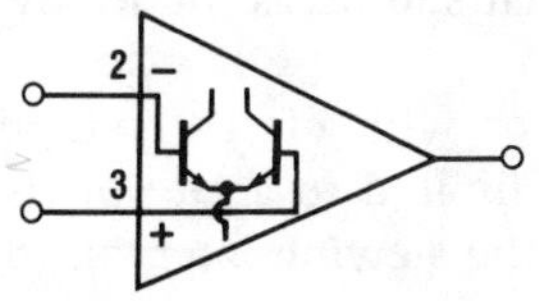


*(A) Unprotected differential input stage.*

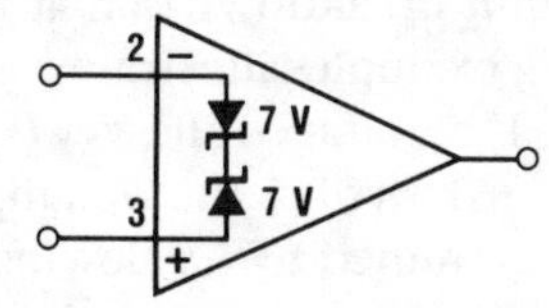


*(B) Zener equivalent to differential input stage.*

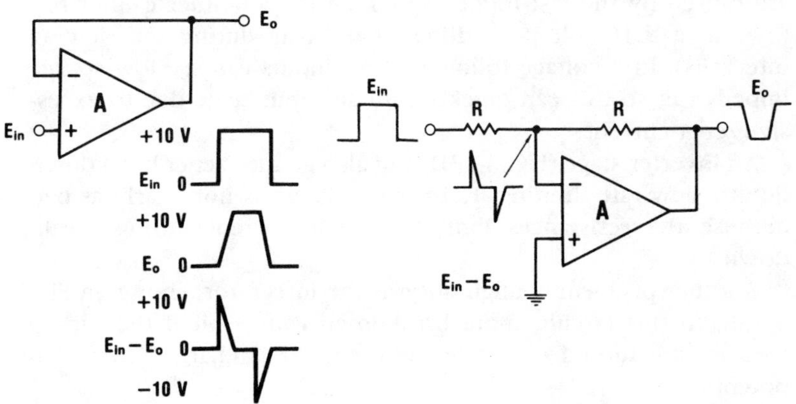


*(C) Differential input voltage in a slew-rate-limited voltage follower.*

*(D) Differential input voltage in a slew-rate-limited inverter.*

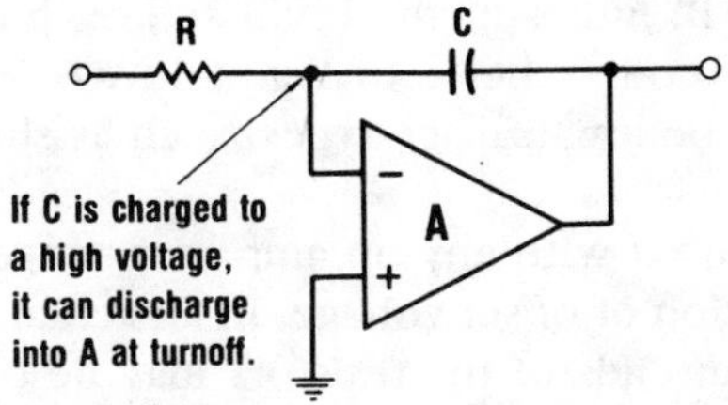


*(E) Differential input voltage in a charged integrator at turnoff.*

**Fig. 3-13. Illustration of possible causes of differential input breakdown in unprotected op amps.**

can cause permanent changes in such dc input parameters as bias, offset current, gain, and noise; i.e., failure of a "soft" variety.

Obviously, in order to prevent this type of breakdown, we must anticipate it beforehand. However, the possible sources of input breakdown can at times be somewhat subtle, as the following examples illustrate.

If a voltage follower (Fig. 3-13C) is driven with a fast step input of +10 volts, the op-amp output will be limited to a rate of rise determined by its slewing ability. During the slewing interval, the amplifier is not operating in a normal closed-loop mode where the differential input voltage is zero; rather, a large differential error voltage exists, as shown by the waveform sketches. These spikes are caused by the fast input rise which the amplifier cannot follow, so a ±10-volt peak differential exists during the slewing interval(s). In a voltage follower, both inputs can see low source impedance, so this can quickly ruin an input stage due to excessive zener current.

An inverter stage (Fig. 3-13D) can also go into zener breakdown due to slew-rate limiting. Here the problem is not nearly as bad because the resistances limit the zener current during breakdown.

Another problem configuration is the integrator, shown in Fig. 3-13E. In this circuit, input breakdown can result if the supply voltages are turned off while the capacitor is charged to a high potential.

The solution to all of the preceding is quite simple. It involves an input clamping network to limit the differential input voltage and some series resistance to limit the current during clamping. Two methods of accomplishing overvoltage protection are shown in Figs. 3-14A and B. Either method will suffice, but the reverse-connected parallel-diode scheme of Fig. 3-14A is less expensive. The diodes should be low-leakage types, such as shown in either Fig. 3-15A, B, or C.

The resistances used with any op amp may range up to 10 kΩ with little degradation of offset voltage. In an actual working configuration, the input ends of the resistors may be considered the inputs of the op amp and used accordingly. Generally, for best dc accuracy, equal resistance will be desirable in each leg. However, in many instances it will be possible to eliminate one or possibly both resistors, as long as input or feedback resistances limit the clamping-diode current. Such an example would be the circuit of Fig. 3-13D, which could be protected just by the addition of the diodes, using the input and feedback resistances as current limiters.

The reader may wonder if perhaps these points on differential

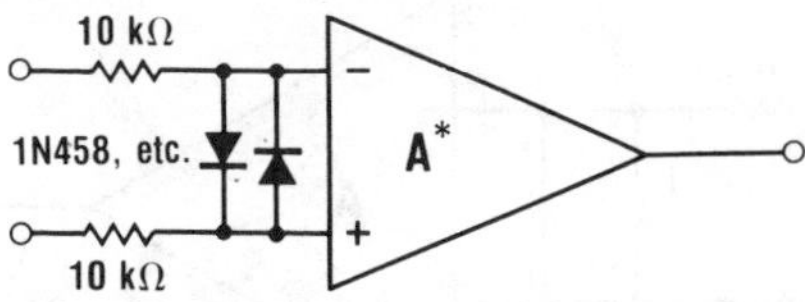


*(A) Preferred overvoltage protection for unprotected op amps.*

*A = any op amp with an unprotected input stage; e.g., 709, 725, etc.

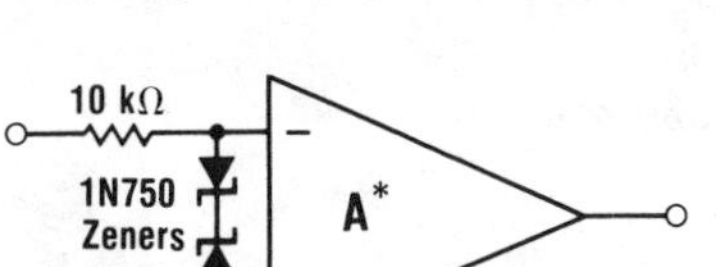


*(B) Alternate overvoltage protection for unprotected op amps.*

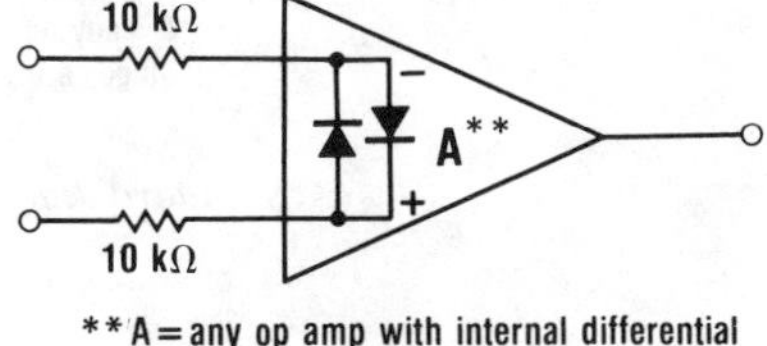


**A = any op amp with internal differential clamping diodes; e.g., 308, 318, 504, OP-27, etc.

*(C) Overcurrent protection for voltage-clamped op amps.*

**Fig. 3-14. Differential input overvoltage/ overcurrent protection methods.**

input protection are not belabored, since many IC op amps of newer design do not have restrictive input-voltage limitations. This is certainly true for op amps, such as the 101, 741, and the JFET-type families, but it is by no means a universal practice with other designs. The 725, for example, has an unprotected input stage, and, if driven into breakdown, the devices' low-noise capability will be compromised. It is of prime importance, particularly in high-performance dc and low-noise amplifiers, to prevent differential input breakdown because it can seriously degrade performance.

Many instrumentation amplifiers now have the differential clamping diodes built into the device. In the absence of internal resistance, amplifiers such as these must be used with external resistors to prevent possible excessive current in the diodes. This is illustrated in Fig. 3-14C. Beware of the exact input configuration—types such as the OP-07 are fully protected with series resistance and shunt diodes. Low-noise types will likely *not* have the series resistance—for example, the OP-27/37 (as shown in Fig. 3-14C). With the latter types, use some resistance of a minimal value for protection, with lowest added noise (i.e., $\simeq$200 ohms).

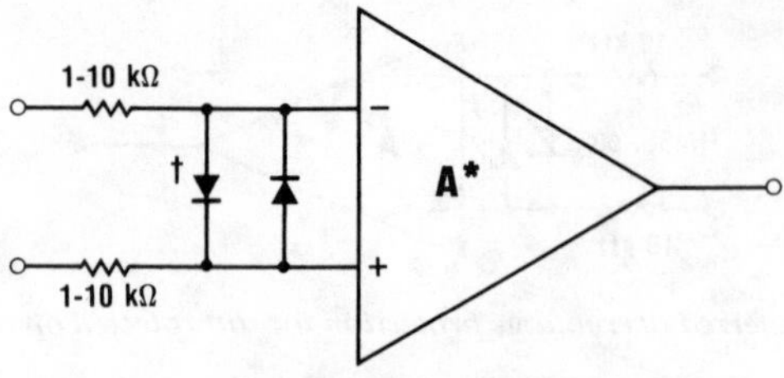


*(A) Standard diode clamp (desired).*

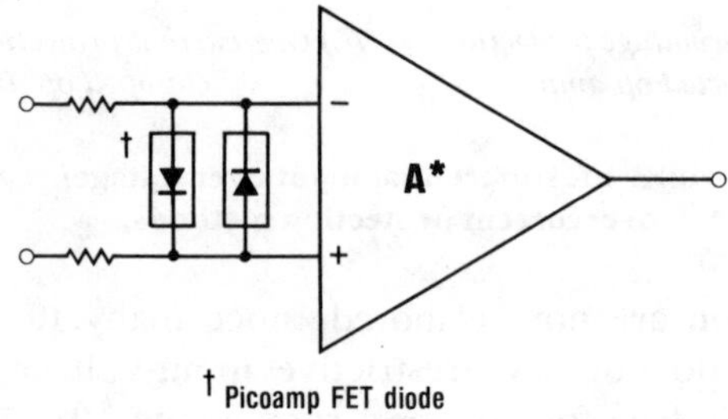


*(B) FET clamp pair; ID100, PAD10, or PAD100.*

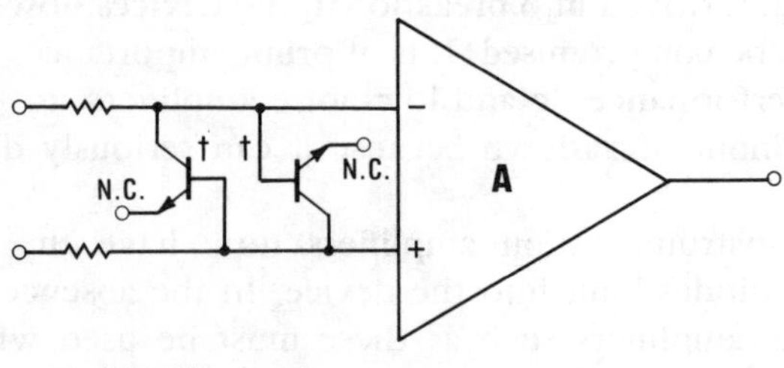


*(C) Diode-connected transistors.*

**Fig. 3-15. Preferred overvoltage protection for unprotected op amps.**

By and large, user-added clamping will likely be of the form shown in Fig. 3-14A, and using low-leakage diodes. Three variations on this theme are shown in Fig. 3-15, using alternate types of diodes for the clamping.

The reader can use the type most appropriate in terms of cost versus performance.

### *Common-Mode Input Breakdown*

Another cause of input stage failure is a possible destructive input current flow, due to exceeding the common-mode input-voltage range of the device. Although the 101 and 741 families are notably free of differential input breakdown, they can under certain circumstances fail from common-mode voltage abuse.

This failure mechanism is illustrated by Fig. 3-16A, which shows the input stage of a 301-type op amp. In a 301A op amp, both inputs are the base terminals of npn transistors, the collectors of which go to V+. In normal operation, the input terminals will always be negative with respect to V+, so the collector-base junctions of these transistors are never forward biased. If, however, V+ is removed from the op amp with a positive potential remaining on pins 2 and 3, the transistors will then conduct through the collector-base junction into the V+ line. If the source impedance seen at either pin 2 or pin 3 is sufficiently low, the resulting current may be high enough to destroy the input transistors. Such a condition may occur when the input to a system is left connected and the power is removed. Another example is with a capacitive source charged to a high potential at the time of power turnoff. If the capacitance is greater than 0.1 μF, the discharge current from the capacitor can also destroy the input transistors.

The solution to the preceding is rather simple. It involves the use of series input resistors to limit the worst-case fault current to 10 mA or less (Fig. 16B). Since 301A input characteristics are relatively unaffected by input resistances up to 10 kΩ, the resistors may be this high with no side effects. For general use, however, 1000 ohms per side will be adequate. The resistors need only be connected to those inputs which see potentially low impedances or excessive voltage. If it can be guaranteed that neither of these conditions will occur under any combination of input and power, then the resistors can be deleted.

Before concluding, it should be noted that this type of failure mechanism is not peculiar to only 301-type op amps. The exact

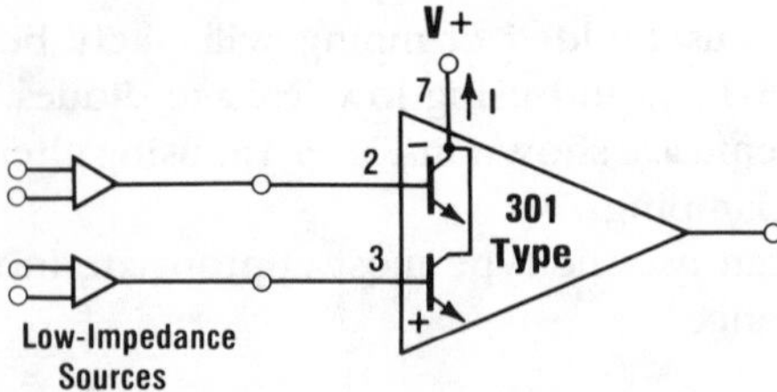


*(A) Reason for common-mode input failure in 301-type op amps.*

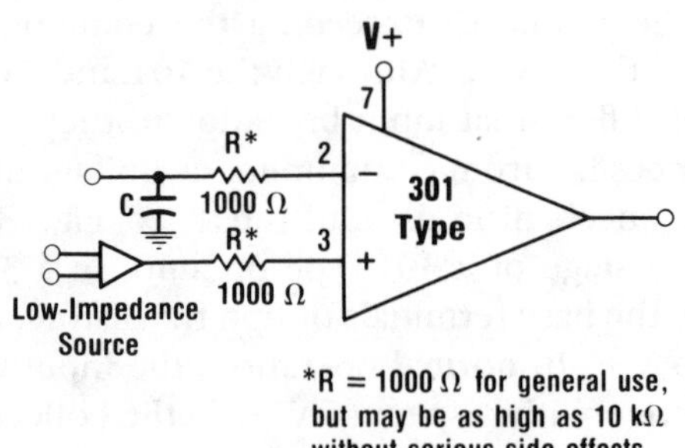


*(B) Protection against common-mode input failure in 301-type (or any) op amp.*

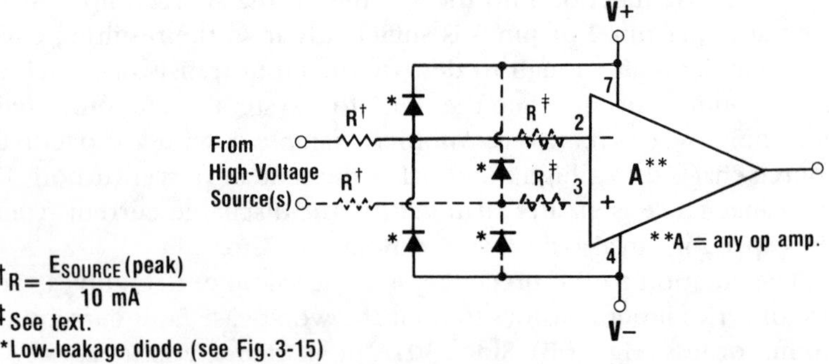


*(C) Protection against excessive source-voltage peaks for any op amp.*

**Fig. 3-16. Common-mode input protection methods.**

internal circuitry of 301A (and 741) types will vary from one manufacturer to another. It is best to check the particular devices you intend to use, either by a simple ohmmeter check (pins 2 and 3 to pin 7) or by measuring fault current in a series resistor under the stress conditions.

In general, be wary of any sort of potential common-mode fault, for any amplifier type, particularly with supply sequencing. For example, some currently available JFET input devices can be damaged by loss of the V− supply, with an input applied. A simple series resistor of several thousand ohms can prevent grief from this problem (see Fig. 3-16B).

Another device type that must be protected against common-mode overvoltage is the LM324. This op amp (and its family members) must not be stressed with an overvoltage which can go below V− (or ground), or an internal diode will conduct. This can cause burnout, latch-up, or both. The cure is restriction of the input common-mode voltage for negative swings. A clamp diode with the anode connected to V− will restrict the voltage, and a series resistor ($\simeq$10 kΩ) between the diode and the (+) or (−) input will then protect the amplifier (see Fig. 3-16C).

Again, it is wise in general to be wary of possible power sequencing phenomena in any op-amp circuit, regardless of the device. When operating from sources of voltage which may have peaks in excess of the op-amp power supplies, the protection schemes of Fig. 3-16C can be used to protect any op amp. The diodes clamp the inputs to a level such that the source-voltage peaks cannot exceed V+ and V− by more than the forward-biased voltage drops of the diodes. The series resistor limits the input current, either in the diode (R†) or the op amp (R‡). Schottky diode clamps are preferred for single-supply amplifiers, such as the LM324 discussed previously. If the amplifier being protected features low bias current, low-leakage diodes are obviously preferable.

### *Latch-up*

A phenomenon that sometimes occurs in op amps is called *latch-up*. Latch-up occurs most often in voltage-follower stages where the output swing is equal to the input and the op amp is driven to a high peak level. If the bias levels of the input stage are not well in excess of the maximum peak-to-peak voltage swing that the input terminals must undergo, the input stage can saturate on the peaks. When saturation occurs, a normally inverting stage no longer inverts; thus, what was negative feedback becomes positive feedback. With positive feedback, the stage will then remain in saturation, and thus it is said to be *latched up*.

This is illustrated in Fig. 3-17A, an example of an op-amp input stage. To use a specific example, assume a minimum common-mode range of +8 volts, and an output swing as high as +14 volts. If a signal greater than +8 volts were applied to such an op amp connected as a follower, it is quite possible that the op amp may latch up due to saturation of $Q_2$. Once latch-up occurs, the positive feedback can usually be broken only by removal of power.

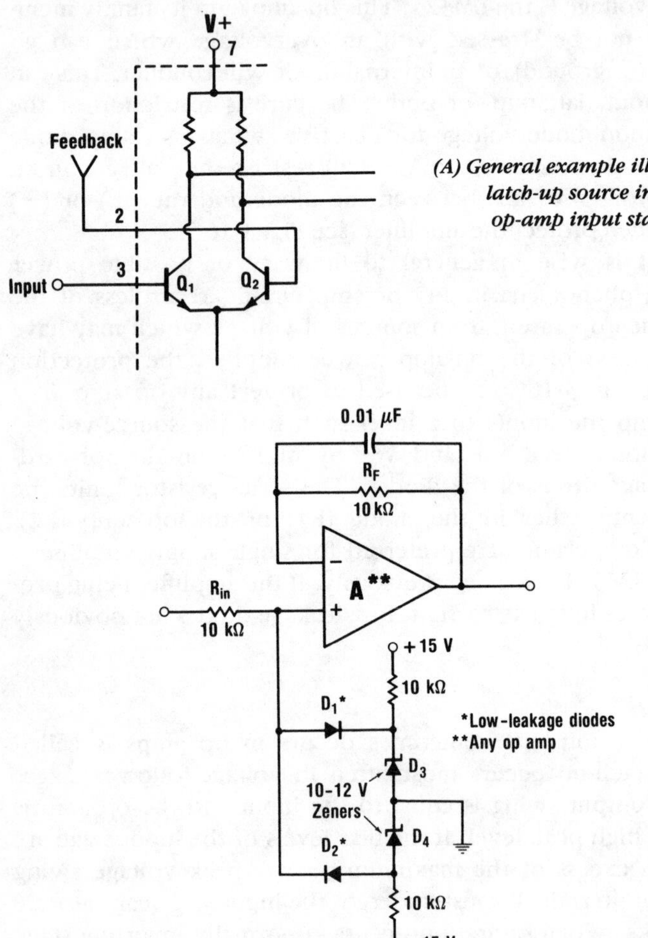


*(A) General example illustrating latch-up source in an op-amp input stage.*

*(B) Elimination of latch-up (any op amp) by input limiting.*

**Fig. 3-17. Illustration of latch-up in IC op amps, and method of elimination.**

Aside from the "nuisance" disadvantages of latch-up (the necessity of removing power to restore normal operation), there is a more serious potential hazard with the phenomenon. If the circuit of Fig. 3-17A uses an npn bipolar pair as shown, and is driven beyond its common-mode range, it will latch up on the positive peak. If an input signal from a low-impedance source is assumed, the input signal at pin 3 need go only 7 volts below the saturation level before $Q_1$ will go into emitter-base zener breakdown. This will almost surely ruin the device because both inputs are connected to low-impedance sources in this state. Latch-up may not even be triggered by the input signal. Any positive transient on the output line which is fed back to $Q_2$ can trigger it, and, once in the latch-up condition, the circuit will stay that way until power is removed.

There are remedies for the latch-up problem, which can be either specific to a given device or general in nature. Perhaps general advice is best, as it can always be applied to any device.

Since latch-up will occur only at the extreme limits of input common-mode range, any method that reduces the common-mode input swing will eliminate (or lessen) the latch-up problem. Therefore, latch-up in an amplifier prone to it will have little likelihood of occurrence when the amplifier is operated at gains ***higher*** than unity or in cases when the input levels are safely ***less*** than that voltage which triggers latch-up.

A circuit that takes advantage of this latter point can be used to prevent latch-up with any op amp; such a circuit is shown in Fig. 3-17B. Since latch-up will occur at only one common-mode limit, only one diode is (usually) needed—diode $D_1$ or $D_2$. The (lightly) biased zener diodes, $D_3$ and $D_4$, simply provide a defined input limit. The sum of the zener voltage and the forward voltage of $D_1$ ($D_2$) should be less than the maximum safe input to the op amp.

Some general-purpose JFET-input op amps are subject to latch-up and/or state reversal, as are some single-supply bipolar types, for negative common-mode swings.

### 3.2.2 Output Short-Circuit Protection

Some early op amps (such as the 709) did not incorporate full-time current limiting in the output stage. Although a 709 will survive a short of a few seconds duration, a sustained short to ground, V+, or V− will result in destruction of the device.

Any amplifier type that does not incorporate current limiting may be protected against short circuits by a low value resistor connected in series with the output as shown in Fig. 3-18. If this resistor is connected within the feedback loop as shown, its presence will have little effect on performance, except for the obvious drop in output voltage that it entails (10% with a 2000-ohm load).

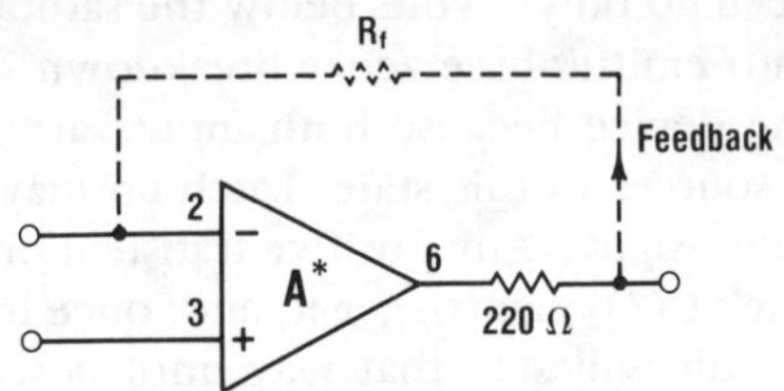


*A = any op amp with an unprotected output stage.

In addition to preventing destruction of the amplifier due to shorted loads, the series resistance will enhance stability by isolating capacitive loads.

**Fig. 3-18. Output short-circuit protection.**

A side benefit of this approach is the extra stability it provides with capacitive loads. For this reason, the series resistor may be helpful even if the amplifier used does have current limiting.

### 3.2.3 Supply-Voltage Protection

***Reversal of Polarity***

Because of their internal construction, ICs must always be operated with the specified supply-voltage polarity. If the voltages ever become reversed, even momentarily, destructive currents will flow through the normally reverse-biased isolation diodes of the IC chip. This is a point to be wary of in any IC op amp, with no exceptions. Positive protection against this may be provided by connecting a rectifier diode in the negative supply lead as shown in Fig. 3-19A.

Protection for a number of amplifiers in a system or on a printed-circuit card may be provided by connecting a single pair of power diodes in reverse across the supply leads as shown in Fig. 3-19B. The diodes used should have a current capacity greater than the fuse or short-circuit limit of the supplies used. When a

*(A) Reverse-polarity protection for a single amplifier.*

$C_1$, $C_2$ = 100 µF low ESR electrolytics

*(B) Reverse-polarity protection for a number of amplifiers in a system.*

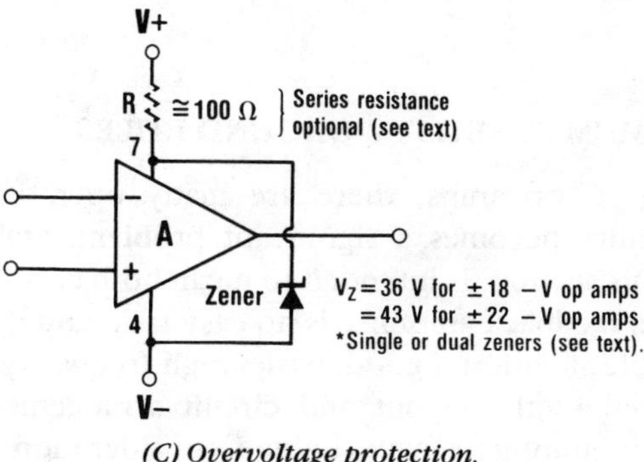


*(C) Overvoltage protection.*

**Fig. 3-19. Supply-voltage protection methods.**

supply reversal or transient condition attempts to force opposite polarity on the amplifiers, $D_1$ and $D_2$ clamp the supplies into the limit mode (or blow the fuse) and nothing is damaged. This is a good feature to incorporate into any but the very smallest system—breadboards, in particular, since they are often disconnected and reconnected. In addition, the entry point of the card should preferably be ac bypassed, using 100-µF (or similar) *low ESR* electrolytics. This should be done for each supply, as shown.

***Overvoltage***

Commercial-grade IC op amps are generally specified for a total operating voltage of 36 volts (±18 volts), while full-temperature-range devices can operate up to 44 volts (±22 volts). These limits on supply voltage should not be exceeded even for brief durations. If there is a likelihood that the supply may exceed 36 volts (or 44 volts), a voltage clamp in the form of a zener diode should be used across the terminals as shown in Fig. 3-19C. This zener will be nonconducting with normal ±15-volt (30-volt) potentials. For higher maximum-rated supplies, choose the zener voltage closest to but not more than the total maximum—for example, for ±22 volts, use a 43-volt zener.

As another alternative, two individual zeners can be employed, one for each supply. Note that if a pair of 18-volt zeners are connected as shown in Fig. 3-19B ($D_1$, $D_2$), they will provide *both* reverse and overvoltage protection. They should be high-wattage types (5 watts or more).

The series resistance in Fig. 3-19C is not absolutely necessary, as long as the supply is fused or current limited; thus, it is shown as optional.

## 3.3 OPTIMUM STABILITY GROUND RULES

In using IC op amps, there are many operating conditions where stability becomes a significant problem. Stability, as discussed in this section, is intended to mean both ac and dc stability. Stabilizing a feedback amplifier is no easy task, and it can turn into an impossible situation if good, basic, high-frequency practices are not followed in the layout and circuit arrangement. Similarly, low-level dc amplifiers have unique considerations for stability also.

### 3.3.1 AC Stabilization Techniques

***Layout and Bypassing***

Of utmost importance to stability is a compact, minimum-lead-length layout. The leads to the input and compensation terminals (if used) of any op amp should be as direct as possible with a minimum of conductor length and proximity to other signal leads. Ground paths should have low resistance and low inductance. In wideband circuits, it is good practice (if possible) to use a ground plane on printed-circuit cards to obtain a high-quality ground.

Supply leads to amplifiers should be bypassed at least once for each PC card with good low-inductance capacitors, such as 0.01-$\mu$F to 0.1-$\mu$F disc or chip ceramics, stacked films, or other similarly rated types, as shown in Fig. 3-20A. If the amplifiers used have a GBP above 10 MHz, more thorough bypassing will be necessary. In such cases, a set of local rf-quality bypasses, such as shown in Fig. 3-20A is recommended for *each* amplifier. For good measure, a pair of low-ESR electrolytic bypasses for each PC card (or every five devices) should also be used, as shown in Fig. 3-19B. General-purpose devices are less critical as to bypassing, but they cannot be ignored. It is good practice, even with general-purpose devices, to include at least one set of bypasses for every five devices, or at least one set per card, regardless of the number of ICs used. With larger numbers of amplifiers, or longer supply runs, more local bypassing may be needed.

### *Input Capacitance Compensation*

Stray input capacitance in feedback amplifiers can also lead to stability problems. Fig. 3-20B illustrates the nature of input-related capacitance problems. At the input of an op amp there will always be a few picofarads of stray capacitance, $C_s$, which consists of the amplifier input capacitance plus wiring capacitance (*including* the source, if $R_{in} \simeq 0$). From the feedback path through $R_f$, this capacitance represents a potential phase shift at a corner frequency of

$$f = \frac{1}{2\pi R_f C_s}$$

The problem is particularly noticeable if $R_f$ is large, since this moves the $R_f$-$C_s$ frequency downward into a region where it can add to the amplifier phase shift. The result of these cumulative phase shifts may be enough to cause oscillation.

A simple solution is to keep $R_f$ low, which forces the $R_f$-$C_s$ frequency upward beyond the amplifier limit. But this is not always practical, since $R_f$ may need to be high for gain reasons.

A more general solution is to use a compensation capacitor, $C_f$, across $R_f$, which in effect makes $C_f$-$R_f$ and $R_{in}$-$C_s$ a frequency-compensated divider. It may not be possible to readily determine the exact value of $C_f$ to satisfy this relation, since $C_s$ is not precisely known. $C_f$ is best determined by experiment; or, if the amplifier

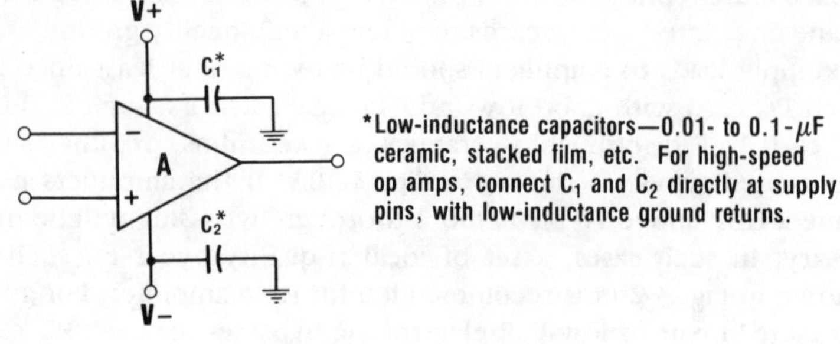


*(A) Power-supply bypassing.*

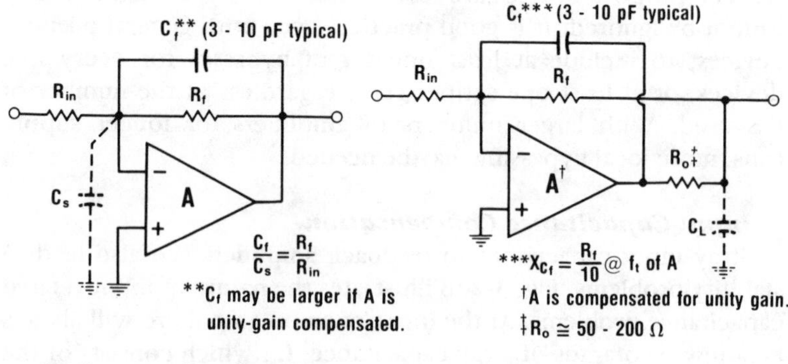


*(B) Compensation of stray input capacitance.*

*(C) Compensation of stray output capacitance.*

**Fig. 3-20. Stability enhancement techniques.**

used is compensated for unity gain, $C_f$ may be greater than the formula relation to whatever degree desirable. In practice, values of 3–10 pF are typical with feedback resistors of 10 kΩ.

### *Output Capacitance Compensation*

On many IC op amps, the phase shift caused by stray output load capacitance and the amplifier output resistance can be troublesome if the output capacitance is much more than 100 pF. By adding a series output resistance, $R_o$, the load capacitance is isolated from the amplifier (Fig. 3-20C). Feedback is taken after this resistor, thus compensating for its dc loss. The feedback capacitor, $C_f$, is then added to reduce the gain of the loop at high frequencies. $C_f$ is chosen so that its reactance is one tenth (or less) that of $R_f$ at the unity-gain frequency of the amplifier used. The amplifier must be compensated for unity gain, since $C_f$ reduces high-frequency gain to unity. Again, typical values for $C_f$ fall in the 3–10-pF region. $C_f$ can be (optionally) increased, so long as bandwidth does not suffer.

#### *Other Instability Sources*

The preceding by no means represents all of the sources of possible stability problems in IC op amps—just those that are most likely to occur. Other problem areas are more specific, either in terms of the devices themselves or in terms of circuits in which the devices are used. For example, source resistance may be a problem with some devices. In stubborn cases of instability, when other alternatives have been exhausted, amplifier overcompensation may be the solution. If extra bandwidth or optimum slew rate are not necessary, it is usually wise to overcompensate an op amp. For example, this may be accomplished on 301A-type amplifiers by increasing the compensation capacitors by a factor of 10 (or by the ratio necessary to obtain stability). Of course, this general point applies in principle to any amplifier with user-adjustable compensation.

High-current stages can cause instability by feedback through the power supplies. Such stages should be decoupled to localize their circulating currents. Similarly, ground-return paths for high-level and high-gain stages should be kept separate.

In general, many of the stability ground rules simply amount to good practices in layout and construction. These rules should obviously be followed for consistently stable results. Always consult the manufacturer's literature for specific recommendations and for more thorough background on specific op amps.

### 3.3.2 DC Stability Considerations

The stabilization of dc amplifiers can take on a number of meanings. Already discussed were considerations for minimizing offset voltage and its drift; in this section, some further considerations are undertaken.

#### *Guarding*

When very-low ($\leq 1$ pA) input currents are involved, such as with electrometer amplifiers, consideration of leakages on the order of a few picoamperes becomes part of the design process.

Consequently, every low-level signal-path point must be carefully controlled for potential leakage and noise, which can easily contaminate the desired signals.

A technique that aids the control of excess leakage current is *guarding*. Guarding is simply the reduction of the potential difference between two conductors by bootstrap techniques. Its value can easily be assessed by the input current versus voltage graph of Fig. 3-21.

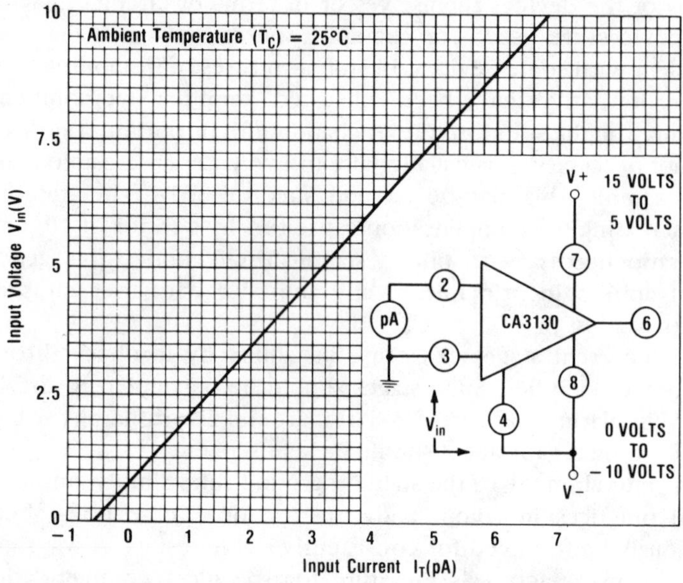


**Fig. 3-21. CA3130 input current change with common-mode voltage.** *(Courtesy RCA Solid State Division)*

This curve is for the CA3130 op amp, which can operate with input pins 2 and 3 at a potential equal to pin 4. Clearly, when the difference between these pins is low (or zero), the leakage (bias) current falls to very low levels—below 1 pA. This illustrates, in principle, the concept of guarding, as the current flow is reduced when the potential difference is zero. This assumes, of course, that the amplifier has appropriately low bias currents to begin with.

With a standard 8-pin TO-99 op-amp package, leakage would normally be excessive because of large potentials at pins adjacent to pins 2 and 3 (pin 4 is at −15 volts, for example). Low-bias-current amplifiers usually employ the metal can as a shield, or guard, and bring it out to pin 8. With some amplifiers designed to work with both inputs at the V− (ground) potential, pin 4 can effectively be used as a guard (such as the CA3130 example).

A printed-circuit layout for the IC must realize the full capability of guarding, and should be laid out as shown in Fig. 3-22. This bottom view of a guarded TO-99 IC shows how the guarding completely surrounds pins 2 and 3, preventing leakage. This is done on *both* sides of the PC board.

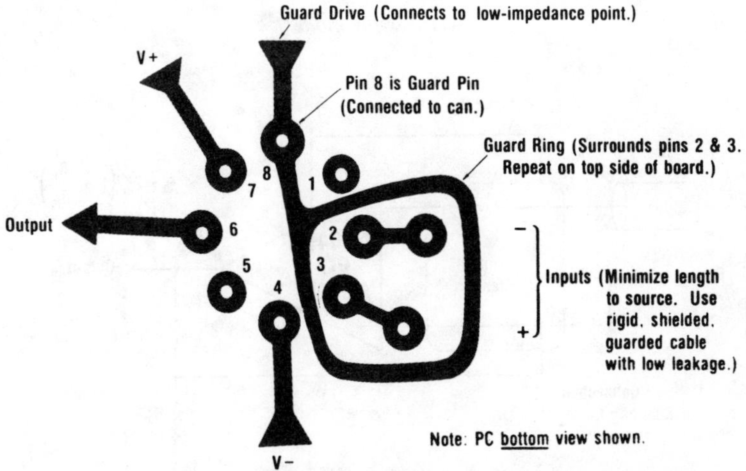


**Fig. 3-22. Printed-circuit-board guard techniques for low-input-current amplifiers.**

The signal is connected to pin 2 (or 3), as appropriate, by rigid shielded guarding cable having low leakage. The signal path to the source should be minimized in length, and, preferably, the cable should have Teflon® insulation.

The amplifier connection to be used will govern the connection of the guard ring, and will be similar to the circuits shown in Fig. 3-23 for an electrometer-grade op amp.

In Fig. 3-23A for an inverting amplifier, the guard ring is connected to ground, since pin 3 is grounded for this mode of use. The series protection resistor ($R_P$) is appropriate for sources that can supply currents of 0.5 mA or more (200 kΩ for a 100-volt

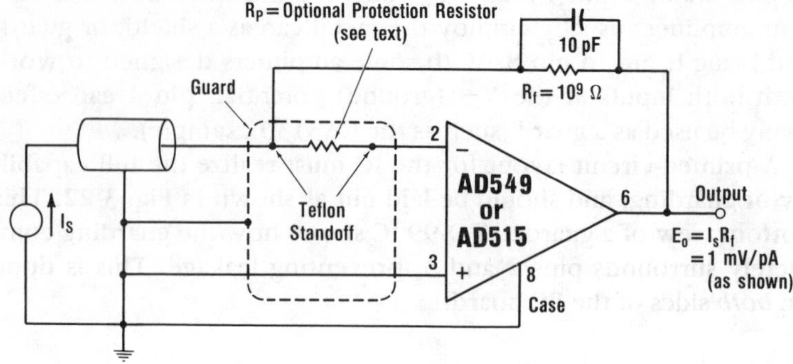


*(A) Inverting mode.*

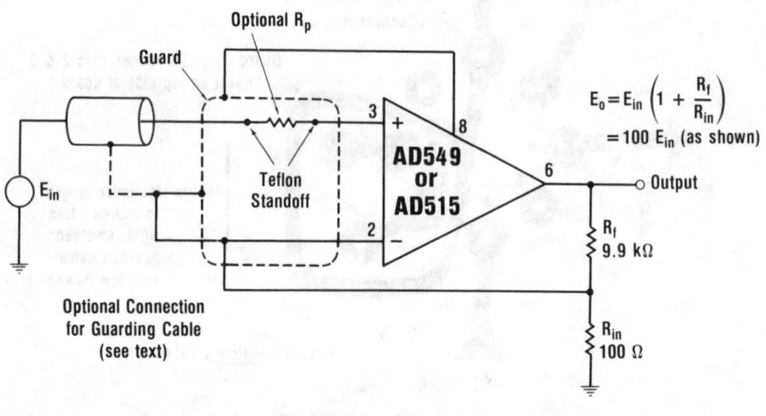


$$E_o = E_{in}\left(1 + \frac{R_f}{R_{in}}\right) = 100\ E_{in}\ \text{(as shown)}$$

*(B) Noninverting mode.*

**Fig. 3-23. Guarded input circuitry using an electrometer op amp.**

overvoltage, for example). With $R_f = 10^9$ ohms, this circuit converts current to voltage with a 1-V/nA sensitivity (or 1 mV/pA).

A noninverting connection is shown in Fig. 3-23B and, in this case, the guard is bootstrapped by the feedback voltage applied to pin 2 from $R_{in}$-$R_f$. To maintain low input capacitance, the signal cable can be bootstrapped as well, as indicated. For the very lowest possible deterioration of the signal at the ultrahigh impedance node, a low-loss guarded standoff, such as Teflon®, can be used. This standoff should terminate the sensitive IC lead, plus the input and feedback component wires (and a protection resistor, if used).

Since this requires special handling of the PC card assembly, the simple trace guard is more often used. In Fig. 3-23, the use of such an optional standoff is shown, for both configurations.

Just about all of the components associated with the input of an electrometer amplifier must be carefully chosen. PC boards must be cleaned carefully and sealed against humidity and dirt. Teflon® or other comparable-quality insulation is preferable, if not necessary.

Feedback resistors must be chosen for stability, low-voltage coefficient, as well as tolerance—a difficult order for high-value types. Capacitors, if used, should be of Teflon®, polystyrene, or polypropylene dielectrics, as few others will have the low leakage required.

The amplifier can be optimized also by minimizing self-heating, which will lower the bias currents. This can be realized by using only the lowest required supply voltages, heat-sinking the device package, and minimizing the load.

Other IC op-amp types which can be useful in electrometer circuits are the AD545, the CA3420, the OPA104, and the OPA111. While the specific pinouts of all these devices are not completely identical, they will all benefit from the application of the above general principles.

Further information on electrometers and low-current-level circuitry can be found in the references at the end of the chapter, and the circuit discussions in the applications section.

## 3.4 REFERENCES


1. Brokaw, A. P. "Analog Signal Handling for High Speed and Accuracy." *Analog Dialogue,* Vol. 11, No. 2, 1977.
2. __________. "An IC Amplifier User's Guide to Decoupling & Grounding." *Electronic Products,* December 1977.
3. __________. "Designing Sensitive Circuits? Don't Take Grounds for Granted." *EDN,* October 5, 1975.
4. Demrow, R. "Op Amps as Electrometers." *Analog Dialogue,* Vol. 5, No. 2, 1971.
5. Dobkin, R. C. "Universal Balancing Techniques." *National Semiconductor Linear Brief LB-9,* August 1969. National Semiconductor Corp., Santa Clara, CA.

6. Grant, D., and Wurcer, S. "Avoiding Passive-Component Pitfalls." *Analog Dialogue,* Vol. 17-2, 1983.
7. Jung, W. G., and Marsh, R. "Picking Capacitors, Parts I and II," *Audio,* February and March 1980.
8. Morrison, R. *Grounding & Shielding Techniques in Instrumentation,* 2nd Ed., John Wiley & Sons, Inc., 1977.
9. Motchenbacker, C. D. "Protect Your Transistors Against Turn-on or Testing Transient Damage." *Electronics,* December 6, 1971.
10. Patstone, W. "Designing Femtoampere Circuits Requires Special Considerations." *EDN,* July 1, 1972.
11. Pease, R. "Bounding, Clamping Techniques Improve Circuit Performance." *EDN,* November 10, 1983.
12. Rich, A. "Shielding and Guarding." *Analog Dialogue,* Vol. 17-1, 1983.
13. Manufacturers' Data Sheets:

| Manufacturer | Device | Description |
|---|---|---|
| Analog Devices | AD549 | Precision, Low-Power, FET-Input Electrometer OP Amp |
| | AD545 | Precision, Low-Drift, FET-Input Op Amp |
| Burr-Brown | OPA111 | Low-Noise, Precision Op Amp |
| | OPA104 | Ultralow Bias Current, |
| | OPA128 | Low-Drift, FET-Input Op Amp |
| RCA Solid State Division | CA3130 | BIMOS Op Amp (see also CA3160, 3260, 3140, 3240) |
| | CA3420 | Low-Supply-Voltage, Low-Input-Current, BIMOS Op Amp |

# II

# *Op-Amp Applications*

FOUR

# *Voltage Regulators, References, and Power Supplies Using Op Amps*

## 4.1 IC VOLTAGE REGULATION

Voltage regulators using IC operational amplifiers as control elements can make very attractive designs in some respects as compared to ICs designed specifically for the purpose of voltage regulation. The advantage of IC op amps is in their inherently high open-loop gain (100,000 typical) which, under closed-loop conditions, can provide regulation percentages as good as 0.01% if sufficient care is taken in layout and construction, and if consideration is given to the significant contributions to error.

### 4.1.1 Basic Elements of an Op-Amp Voltage Regulator

The four basic elements of a series-type op-amp voltage regulator are shown in Fig. 4-1. In this functional diagram, a reference voltage source produces a constant output voltage, $V_{ref}$, under varying conditions of temperature and input voltage, $+V_{in}$. The reference voltage is applied to the noninverting input of op amp $A_1$. The output of $A_1$ drives the series-pass transistor, $Q_1$. The output of $Q_1$ is the voltage, $E_o$, which is developed across the load resistance, $R_L$. Voltage $E_o$ is also fed back through the feedback network, $\beta$, causing a fraction of $E_o$, or $\beta E_o$, to appear at the inverting input of $A_1$. Recalling one of the laws of ideal op-amp behavior, zero differential-input voltage will be present at the inputs of $A_1$;

thus, voltage $\beta E_o$ will be exactly equal to $V_{ref}$, since the infinite open-loop gain of $A_1$ will adjust $E_o$ until this equality is satisfied. The feedback mechanism of the active components acts to generate an output voltage that is a multiple of the input voltage, $V_{ref}$. The relationship is simply

$$E_o = \frac{V_{ref}}{\beta}$$

where

$\beta$ is the attenuation of the feedback network.

The three active components may be thought of as reference, control, and power, represented by $V_{ref}$, $A_1$, and $Q_1$, respectively. The functions of $V_{ref}$ and $A_1$ are obvious enough, and $Q_1$ is necessary to handle load currents beyond the basic output capability of the op amp.

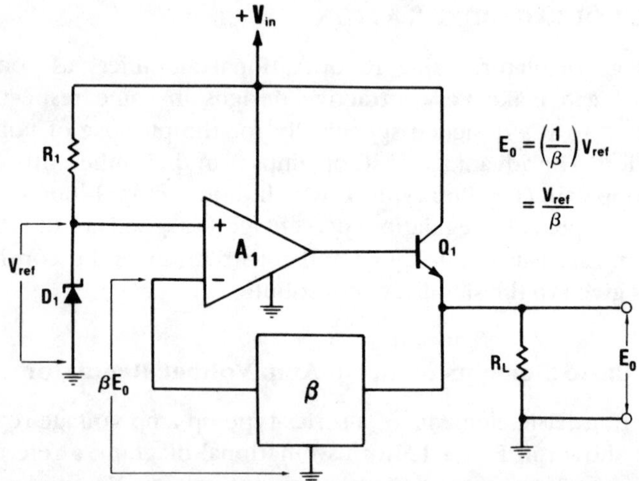


**Fig. 4-1. Functional diagram of a series-type voltage regulator.**

Within the concept of this basic configuration, voltage regulators of practically any voltage and/or current capability can be constructed by using appropriate elements for $V_{ref}$, $A_1$, $Q_1$, and $\beta$. The concept can be extended to include negative voltages by using complementary devices to perform the functions of $-V_{in}$, $-V_{ref}$, and $Q_1$. Therefore, to design a regulator using an op amp for $A_1$, we must have as a minimum the element for $V_{ref}$, and a suitable feedback network, $\beta$. These elements can be either a custom circuit assembly, or integral to an IC reference or regulator chip. The discussions of this chapter will illustrate the various options available towards creating a stable reference and/or power source. To begin the discussion of op-amp voltage regulators, we will examine the first element, the reference voltage source.

### 4.1.2 Requirements of the Reference Voltage Source

The prime requirement of any reference voltage source is that it must maintain an accurate and stable output voltage with varying conditions of input voltage, time, temperature, and loading.

It is first appropriate to look in a broad sense at various devices which are appropriate for use as references. Generally, these consist of voltage-regulator diodes (zeners), or complete integral regulator circuits which produce a reference voltage. Given a stable and accurate voltage, a current can always be derived, when necessary.

To give some perspective as to the general performance level being considered here for "reference" quality, it should be understood that this implies total regulation and/or stability to small fractions of a percent. For example, one of the most-common reference applications is with d/a and/or a/d converters. And, a 12-bit converter, for instance, has a basic resolution of 0.025%. Some, if not a major portion, of such a converter's inherent capability would be lost if the reference used in it were not capable of a performance better than the 0.025% figure; i.e., a lower percentage of error—this includes drift, regulation errors, and so on. The importance of this becomes even more apparent when temperature variations are considered, as TC (temperature coefficient) is the largest single source of reference error.

In light of the previous information, it is not at all unusual to speak of reference-source regulation (source and/or load induced) errors in the range of 0.001% to 0.01% (10 to 100 ppm), TCs of 1 to 20 ppm/°C, and long-term drifts of 10–20 ppm per month, or even year. These specifications generally should be maintained even with load-current variations, input-voltage fluctuations, temperature variations, and time itself. A reference is, after all, what its name implies, the quantity to which all input or output levels of an a/d or d/a converter are referred. If the reference is variable or uncertain, a converter can measure to an accuracy no better than that of the reference, regardless of the resolution, or *suggested* accuracy, of the converter. So, it is important that a converter's reference circuit be consistent with its intended use. If 12-bit accuracy is needed, you need a good reference. On the other hand, if only 8-bit performance is required, the reference burden is eased quite considerably.

Table 4-1 lists the key performance features of a number of popular devices which are suitable for use as converter references. Listed are such parameters as output voltages, current rating, output impedance, TC, input voltage and stability, and general comments on applicability.

In the table there are two broad categories—reference diodes and integral reference circuits. Classes A and B are diodes, as is class C in a sense. Class-C units, however, are reference diodes on a thermally stabilized substrate and, as such, exhibit some of the lowest temperature coefficients (TCs) of all those shown. Device classes D and E are integral "three-terminal" reference circuits which can be simply plugged into a circuit to deliver a rated output voltage.

Further details of performance for the various devices are covered in their individual discussions which follow. The general conditions of this table summarize the specifications for the best 0–70°C range device, for all those listed. Consult specific data sheets for full details (see chapter references).

## 4.2 BASIC DEVICE HOOKUP AND DETAILS OF OPERATION

This section takes a brief look at each reference device, and shows what it takes to operate it and how it is packaged. In some cases, as for instance that of integral references, this information alone is all that is needed to apply the device and achieve the specified rated performance.

### 4.2.1 Classification A: Temperature-Compensated Reference Diodes

The 1N821–1N829 and 1N4611–1N4611C series of devices are actually combinations of zener and conventional diodes, hooked

## Table 4-1. Performance Features of Reference Devices

| Device | Classification (1) | Output Voltage (V) | $I_z$ or $I_o$ (mA) | $R_z$ or $R_o$ (Ω) | TC (as noted) | Input Voltage (V) | Long-Term Stability (as noted) | Comments |
|---|---|---|---|---|---|---|---|---|
| 1N821–829 series | A | 6.2 ± 5% | 7.5 ± 0.01 | 15 | 5 ppm/°C (max) (1N829A) | ↑ | — | |
| PRD series (CODI) | A | 6.2 ± 5% | 7.5 ± 0.01 | 15 | 1 ppm/°C (max) (25–45 °C) | | 5 to 60 ppm/year | |
| 1N4611–4611C series | A | 6.6 ± 5% | 2 ± 0.2 | 75 | 5 ppm/°C (max) (1N4611C) | determined by user application | — | |
| AD589L | B | 1.235 + 1.2% − 2.8% | 50 μA–5 mA | 0.6 | 25 ppm/°C (max) | | — | micro-power |
| LM385BX (2) | B | 1.24 to 5.3 | 10 μA–20 mA | 0.4 | 30 ppm/°C (max) | | 20 ppm/ 1000 hrs (typ) | micropower, programmable |
| LT1004-1.2C (2) | B | 1.235 ± 0.3% | 8 μA–20 mA | 0.2 | 20 ppm/°C (typ) | | 20 ppm/ 1000 hrs (typ) | micro-power |
| ICL8069A | B | 1.23 + 1.63% − 2.4% | 50 μA–5 mA | 1 | 10 ppm/°C (max) | | — | micro-power |
| LM336B (3) | B | 2.49 ± 2% | 0.4 to 10 | 0.2 | 50 ppm/°C (max) | | 20 ppm (typ) | adjustable |
| LT1009C | B | 2.5 ± 0.2% | 0.4 to 10 | 0.2 | 25 ppm/°C (max) | ↓ | 20 ppm/ 1000 hrs (typ) | adjustable |

**Table 4-1 (cont). Performance Features of Reference Devices**

| Device | Classification (1) | Output Voltage (V) | $I_z$ or $I_o$ (mA) | $R_z$ or $R_o$ (Ω) | TC (as noted) | Input Voltage (V) | Long-Term Stability (as noted) | Comments |
|---|---|---|---|---|---|---|---|---|
| LM329A | B | 6.9 ± 5% | 0.6 to 15 (max) | 0.8 | 10 ppm/°C (max) | determined by user application | 20 ppm (typ) | |
| LM3999 | C | 6.95 ± 5% | 1 (typ.) | 0.6 | 5 ppm/°C (max) | ↓ | 20 ppm/ 1000 hrs (typ) | |
| LM399 | C | 6.95 ± 5% | 1 (typ.) | 0.5 | 2 ppm/°C (max) | ↓ | 20 ppm/ 1000 hrs (typ) | |
| LM399A | C | 6.95 ± 5% | 1 (typ.) | 0.5 | 1 ppm/°C (max) | ↓ | 20 ppm/ 1000 hrs (typ) | |
| AD580M | D | 2.5 ± 0.4% | 10 | 1 (max) | 10 ppm/°C (max) | 4.5 to 30 | 100 ppm (typ) | |
| MC1403A | D | 2.5 ± 1% | 10 | 0.91 (max) | 25 ppm/°C (max) | 4.5 to 40 | — | |
| REF-01E | D | 10 ± 0.3% | 10 | 0.8 (max) | 8.5 ppm/°C (max) | 13 to 33 | — | adjustable |
| REF-02E | D | 5 ± 0.3% | 10 | 1 (max) | 8.5 ppm/°C (max) | 8 to 30 | — | adjustable |
| REF-10A | D | 10 ± 0.3% | 10 | 8 (max) | 8.5 ppm/°C (max) | 13 to 33 | 50 ppm/ 1000 hrs (max) | adjustable |

## Table 4-1 (cont). Performance Features of Reference Devices

| Device | Classification (1) | Output Voltage (V) | $I_z$ or $I_o$ (mA) | $R_z$ or $R_o$ (Ω) | TC (as noted) | Input Voltage (V) | Long-Term Stability (as noted) | Comments |
|---|---|---|---|---|---|---|---|---|
| REF-05A | D | 5 ± 0.3% | 10 | 1 (max) | 8.5 ppm/°C (max) | 8 to 33 | 100 ppm/ 1000 hrs (max) | adjustable |
| AD581L | D | 10 ± 0.05% | 10 | 0.5 (max) | 5 ppm/°C (max) | 13 to 30 | 25 ppm/ 1000 hrs (typ) | series or shunt |
| AD584L | D | 10 ± 0.05%<br>7.5 ± 0.05%<br>5.0 ± 0.05%<br>2.5 ± 0.05% | 10 | | 5 ppm/°C (max) | $V_{out}$ + 2.5V to 30V | 25 ppm/ 1000 hrs (typ) | series or shunt, programmable |
| LT1019AC-5 (4) | D | 10 ± 0.05%<br>5 ± 0.05%<br>2.5 ± 0.05% | 10 | 0.05 (max) | 5 ppm/°C (max) | $V_{out}$ + 1.5V to 40V | — | series or shunt, heated operation option, adjustable |
| LT1021CC-5 (4) | D | 5 ± 0.05% | 10 | 0.1 (max) | 20 ppm/°C (max) | $V_{out}$ + 2V to 40V | 15 ppm/ 1000 hrs (typ) | series or shunt, adjustable |
| LH0070-2 | E | 10 ± 0.05% | 5 | 0.2 | 0.04% (max) | 13 to 33 | 500 ppm/yr (max) | |
| LH0071-2 | E | 10.24 ± 0.05% | 5 | 0.2 | over temp | 13 to 33 | 500 ppm/yr (max) | |

**Table 4-1 (cont). Performance Features of Reference Devices**

| Device | Classification (1) | Output Voltage (V) | $I_z$ or $I_o$ (mA) | $R_z$ or $R_o (\Omega)$ | TC (as noted) | Input Voltage (V) | Long-Term Stability (as noted) | Comments |
|---|---|---|---|---|---|---|---|---|
| AD2700L | E | +10 ± 0.025% | 10 | 0.05 (max) | 3 ppm/°C (max) | +13 to +18 | 100 ppm/ 1000 hrs (max) | |
| AD2701L | E | −10 ± 0.025% | 10 | 0.05 (max) | 3 ppm/°C (max) | −13 to −18 | 100 ppm/ 1000 hrs (max) | |
| AD2710L | E | +10 ± 0.01% | 10 | 0.05 | 1 ppm/°C (max) | +13.5 to 16.5 | 25 ppm/ 1000 hrs (typ) | |
| REF10KM | E | +10 ± 0.05% | 10 | 0.2 (max) | 1 ppm/°C (max) | +13.5 to 35 | 50 ppm/ 1000 hrs (typ) | adjustable |
| REF101KM | E | +10 ± 0.05% | 10 | 0.2 (max) | 1 ppm/°C (max) | +13.5 to 35 | 25 ppm/ 1000 hrs (typ) | adjustable |

Notes:

(1) A = Temperature-compensated reference diode.
B = Active temperature-compensated reference diode (IC).
C = Type B, with thermally stabilized substrate.
D = Three-terminal monolithic IC.
E = Hybrid IC, actively trimmed to high accuracy.

(2) See also the −1.2 and −2.5 versions of the LM385BX and LT1004.

(3) See also the −5.0 version of the LM336.

(4) See also the −2.5 and −10 versions of the LT1019AC and LT1021CC.

up in series. This cancels the respective diode reverse-and-forward TCs, and thus achieves a nominal net temperature coefficient of zero. Lowest maximum TC of the JEDEC registered "1N" series of devices is 5 ppm/°C, as typified by the 1N821 and 1N4611C. The TC of a reference diode is critically dependent upon its current, however, and this current must be closely regulated. This is particularly true in the case of 1N829 types, but somewhat less so in the 1N4611 series units. Very low TCs, on the order of 1 ppm/°C, and very high long-term stability are achieved by the "PRD" series (proprietary units by CODI).

No schematic of these devices is shown, as they are represented by the familiar zener symbol. They are packaged in standard packages.

### 4.2.2 Classification B: Active Temperature-Compensated Reference Diodes

The National Semiconductor LM329 is like other temperature-compensated diodes in that it has a low nominal TC. Various performance grades are available, down to 10 ppm/°C typical drift, and are represented by suffixes "A." However, a major difference in this type of device is the fact that the dynamic impedance is very low, typically less than one ohm. Also, the temperature coefficient is relatively independent of current. These features are in actuality achieved by an IC which does, in fact, comprise the "diode." The zener diode is a "subsurface" type—one which is inherently more stable than surface zeners.

The LM329 is packaged in a TO-92 format. The schematic symbol used (not shown) is that of a zener, although the IC circuit actually consists of many components.

The National Semiconductor LM336 (Fig. 4-2) is also an active zener type of reference, but it is synthesized with a low-voltage architecture, with a reference voltage of 2.5 volts, based on the band-gap voltage of silicon. The TC is somewhat higher than the subsurface zener type, but is still of general reference quality. This type of reference finds wide use with lower-voltage supplies; i.e., 5 volts or less. Note that the device has a third terminal, which is used for adjustment purposes. When trimmed, as in Fig. 4-2B, for a voltage of 2.490 volts, the TC will be at minimum, typically 15 ppm/°C. Trimming is optional and can be deleted. Another form of trim uses the potentiometer with the diodes shorted out, and also adjusts the terminal voltage. This allows the initial tolerance of the unit to be adjusted out, without changing the basic TC.

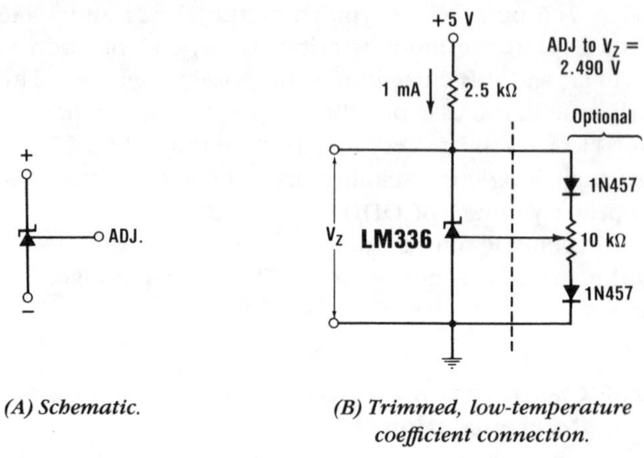


*(A) Schematic.*

*(B) Trimmed, low-temperature coefficient connection.*

**Fig. 4-2. LM336 zener diode (TO-92 package).**

The LM336 is a popular series; a 5-volt version is also available, the LM336-5.0. Also, related to the LM336 is the Linear Technology LT1009 series, a 2.5-V reference with a 0.2% tolerance and a lower guaranteed TC (see Table 4-1).

The Analog Devices AD589 is another band-gap reference-diode type that is very popular. It has a lower voltage reference (1.235 V) and can operate at microampere-current levels. This diode type allows very low-power circuits to be built, yet they can still operate at higher currents as well, with low TC. Variations of this diode type feature a wider current range—the Linear Technology LT1004-1.2 (8 μA–20 mA); lower TC—the Intersil ICL8069A (10 ppm/°C); and programmability—the National Semiconductor LM385BX series (1.24 V to 5.3 V). The diodes of this classification feature maximum utility with good performance. In general, with the exception of the LM385 series, they are not adjustable.

### 4.2.3 Classification C: Active Temperature-Stabilized Reference Diodes

This series includes the LM3999, LM399, and LM399A. The National Semiconductor LM3999 is similar to the LM329 insofar as it is an active, temperature-compensated, sub-surface, zener reference diode. However, it goes a step further in that it also incorporates a temperature stabilizer (or heater) which keeps the internal substrate temperature at a constant 85°C. This minimizes sensitivity to external temperature changes, allowing a 5 ppm/°C drift to be achieved.

As indicated by the schematic in Fig. 4-3, the diode and heater have a common terminal (1); the device is contained in a TO-92 package. The heater operates from 9- to 36-volt supplies, typically consuming 360 mW of power in 25°C (still) air. Initial heater surge current is typically 140 mA, and warm-up time is 5 seconds.

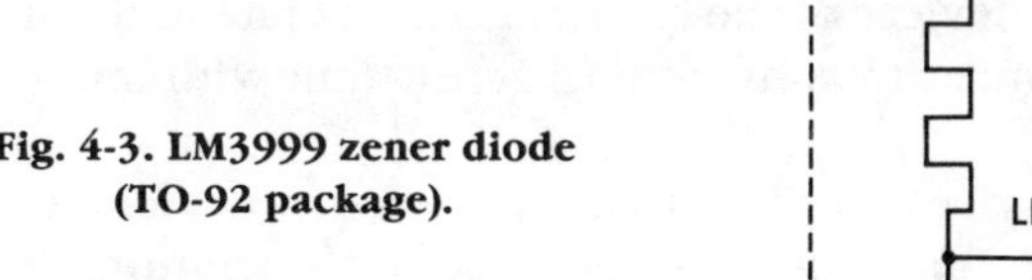


**Fig. 4-3. LM3999 zener diode (TO-92 package).**

The National Semiconductor LM399 and LM399A (Fig. 4-4) are electrically similar to the LM3999, but feature improved performance and flexibility. These devices are packaged in an insulated TO-46 casing, using a four-lead configuration. This provides electrical isolation of the heater and diode common terminals.

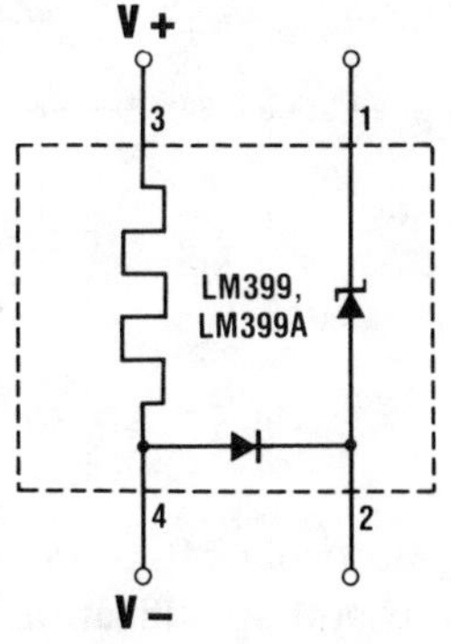


**Fig. 4-4. LM399 and LM399A zener diodes (TO-46 package).**

The LM399 provides a TC of 2 ppm/°C maximum, while the LM399A has a maximum TC of only 1 ppm/°C. Typically, both devices have a TC of 0.3 ppm/°C. Power consumption of both units is typically 255 mW at 25°C, and warm-up time is 3 seconds. A selected version of the LM399A is available for lowest long-term drift, the LM399AH-50.

### 4.2.4 Classification D: Three-Terminal Reference ICs

"Three-terminal" references are complete unregulated-input/regulated-output circuits within a single IC, with a common (ground) terminal. For application, they are used in their "series mode," simply connected between the unregulated input and the load. Selection from this class of devices can be made for a specified voltage, voltage tolerance, TC, and/or long-term stability. In addition, selected devices of the class can also operate in the shunt mode, which is similar to conventional zeners (but with improved stability).

The Analog Devices AD580 is an IC reference device configured for three-terminal voltage-in, voltage-out operation. It accepts an unregulated input voltage in the range of +4.5 volts to +30 volts, and produces a +2.5-volt (±0.4%) reference voltage. Its load-current range is up to 10 mA, and the TC is 10 ppm/°C (max). Long-term stability is 100 ppm/year, typically. This device, like the others of this class, uses a voltage-reference principle based on the band-gap voltage of silicon. A minimum component hookup for applying the AD580 is shown in Fig. 4-5. It is supplied in a 3-pin TO-52 package.

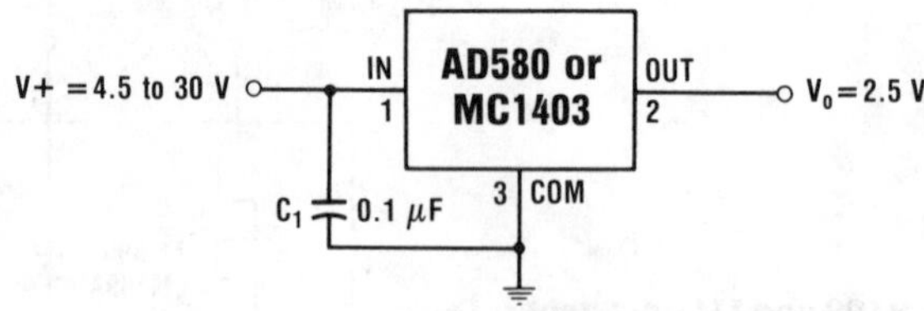


**Fig. 4-5. AD580 and MC1403 monolithic ICs.**

The Motorola MC1403 is a three-terminal IC reference, with operation and specifications that are similar to the AD580. It is packaged in an 8-lead "MINIDIP" package. It may also be applied as noted in Fig. 4-5.

The Precision Monolithics REF-01 (Fig. 4-6A) is a three-terminal IC reference which operates somewhat similar to the AD580 in basic concept, but with some important differences. The device is packaged in an 8-pin TO-99 package. (Note: This particular pinout is a de facto standard, common to many devices.) The REF-01 is a 10-volt output device, with an input range of 12 to 40 volts. Initial output tolerance is ±0.3% for the REF-01E, which also has an 8.5 ppm/°C (max) TC.

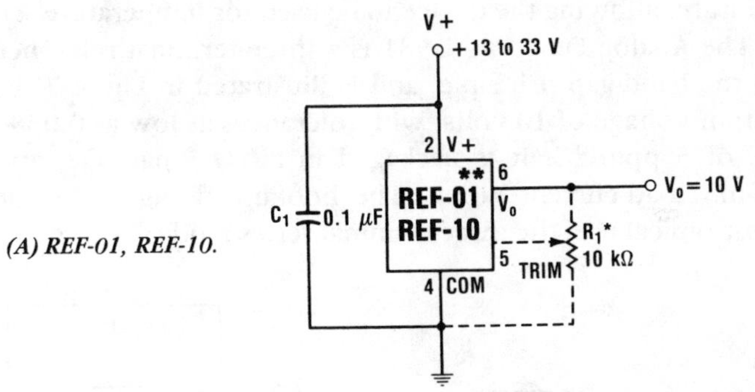


*(A) REF-01, REF-10.*

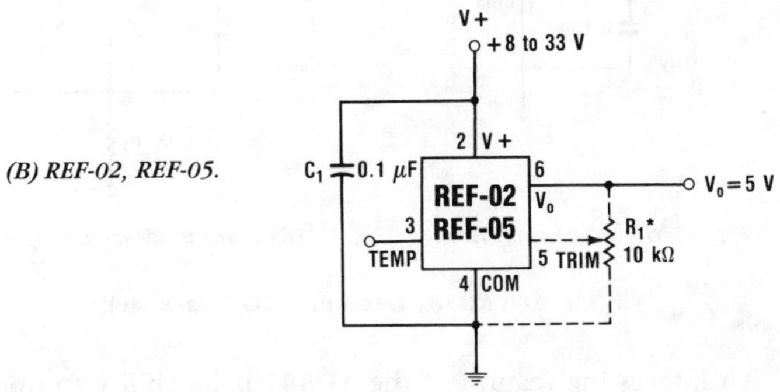


*(B) REF-02, REF-05.*

**Fig. 4-6. REF-01–REF-10 series (TO-99 package).**

As may be noted from the hookup in Fig. 4-6A, this device has an additional terminal, labeled TRIM (pin 5). If left open, the output voltage tolerance is as mentioned above. However, at the user's option, trim pot R1 may be used to set the output anywhere within the range of ±3% about nominal. Its load-current rating is 10 mA. The REF-10 is similar to the REF-01, but has better long-term stability.

The Precision Monolithics REF-02 (Fig. 4-6B) is similar to the REF-01, except that its output voltage is a nominal 5 volts, and its input range is 7 to 40 volts. It is trimmed up to ±6% (when desired) in a like manner, and has a similar load rating. The REF-05 is similar, with a tighter long-term stability. One additional pin is provided with the REF-02 (and REF-05), however, termed the TEMP (pin 3). This pin provides an output voltage proportional to temperature, allowing the device to be used for temperature sensing.

The Analog Devices AD581 is a three-terminal reference based on the band-gap principle, and is illustrated in Fig. 4-7. It has an output voltage of 10 volts, with tolerances as low as 0.05%, and a TC of 5 ppm/°C. It is packaged in a TO-5 package, and has a 10-mA load-current rating. The hookup of Fig. 4-7A shows its most typical use, the three-terminal series mode.

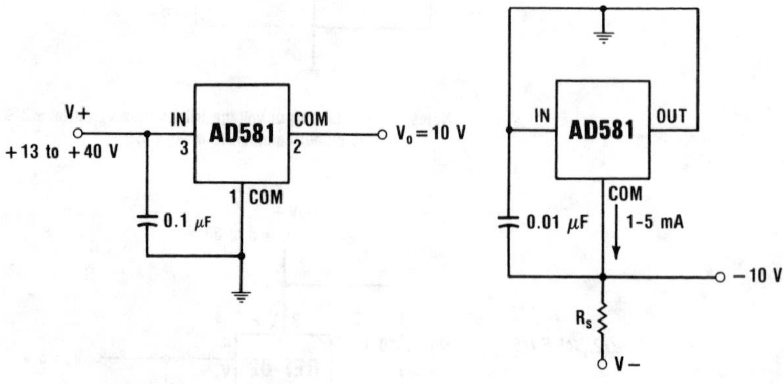


*(A) Series-mode operation.*

*(B) Shunt-mode (zener) operation.*

**Fig. 4-7. AD581 reference (TO-5 package).**

An interesting feature of the AD581 is its ability to operate in two-terminal zener mode, as shown in Fig. 4-7B. With this feature, the device can be used for negative outputs, as well as positive. Device zener current should be maintained in the range of 1 to 5 mA, by selection of $R_S$.

A device related to the AD581 is the Analog Devices AD584,

shown in Fig. 4-8. This unit has similar performance in terms of accuracy, but adds pin-programmed voltage outputs of +10 V, +7.5 V, +5 V and +2.5 V, as illustrated in Table 4-1. In addition to the four series-mode operation settings, the AD584 can also be operated in the shunt mode, at voltages of 5, 7.5, and 10 volts (not explicitly shown, but generally similar to the AD581).

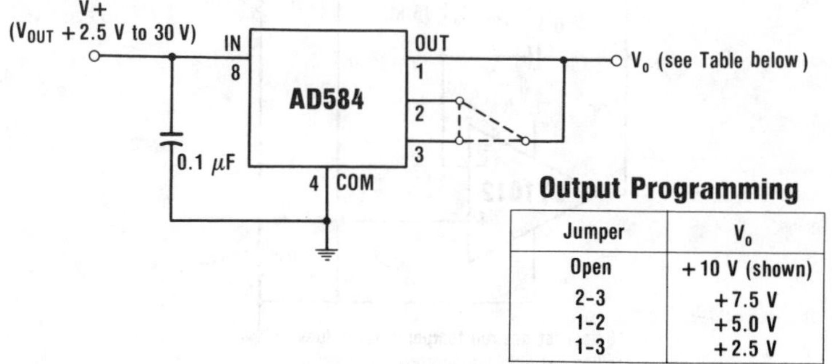


**Output Programming**

| Jumper | $V_0$ |
|---|---|
| Open | +10 V (shown) |
| 2-3 | +7.5 V |
| 1-2 | +5.0 V |
| 1-3 | +2.5 V |

**Fig. 4-8. AD584 reference (TO-99 package).**

The Linear Technology LT1019 is a three-terminal reference generally similar to the REF-01 and REF-02, and available in output voltages of 2.5, 5, and 10 volts. It features an output tolerance of 0.05%, and a TC as low as 5 ppm/°C. In addition, an optional heated mode is available for drifts as low as 2 ppm/°C, and it can be operated in both the series and shunt modes.

For standard series-mode operation, the LT1019 is connected the same as in the REF-01 (Fig. 4-6A), since the basic pinout is the same. For the option of heated-mode operation, the LT1019 can be connected as shown in Fig. 4-9. This configuration is a 10-V (±0.05 V) circuit using the LT1019AC-10, and the device will stabilize at 70°C. With the use of heater stabilization, the circuit will have a temperature drift of less than 2 ppm/°C.

The Linear Technology LT1021 is another three-terminal reference generally similar to the LT1019, but without the on-chip heater. It is available in output voltages of 5, 7, and 10 volts, and a TC as low as 5 ppm/°C. In addition, long-term stability is optimized with the 7-V unit, featuring a 7 ppm/1000 hrs (typ) drift. The LT1021 can also be operated in a shunt mode, for the 7- and 10-V output versions.

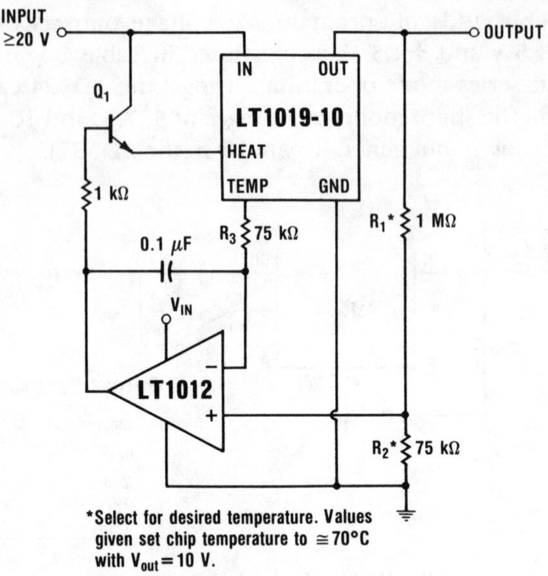


**Fig. 4-9. LT1019 reference (8-pin package) in heated mode.**
*(Courtesy Linear Technology Corp.)*

## 4.2.5 Classification E: Actively Trimmed, Hybrid IC References

This class of reference devices is comprised of three-terminal references which operate in a *functional* sense similar to the three-terminal monolithic ICs. There are, however, significant advantages of this class of device in terms of performance. This is due to active circuit trimming (by laser, or similar means) that allows both a precise adjustment of final output voltage and a minimum TC to be achieved. This group of devices allows the greatest measure of "as delivered" accuracy, and a minimum of user design attention. Generally speaking, these devices will deliver the tightest output-voltage tolerances, the lowest TCs, and the best long-term stability.

The National Semiconductor LH0070 and LH0071 are devices which deliver 10- and 10.24-volt (±0.05%) output voltages, respectively. Load current is 5 mA and TC is specified as 0.04% (max) over the device's temperature range (for the −2 suffix).

Input-voltage range is 13 to 33 volts and long-term stability is specified as 500 ppm/year. They are packaged in TO-5 packages; a minimum-component hookup circuit is diagrammed in Fig. 4-10.

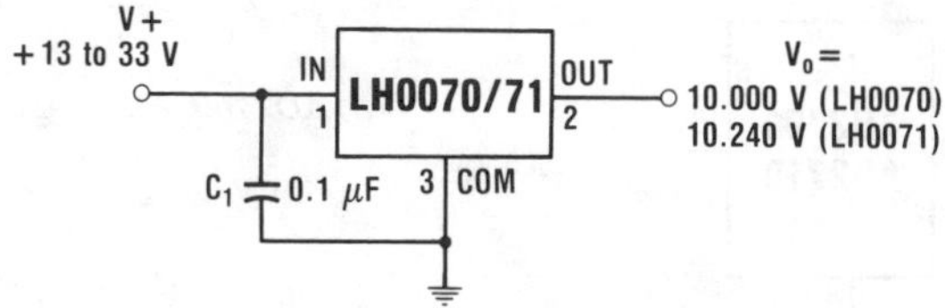


**Fig. 4-10. LH0070,71 reference (TO-5 package).**

The Analog Devices AD2700 series of references delivers output voltages of +10 volts, −10 volts, and +10 volts, in devices AD2700, AD2701, and AD2710, respectively. These units are packaged in a 14-pin DIP package.

Initial accuracy for the AD2700 and AD2701, respectively, is +10 volts and −10 volts (±0.025%) and the load current is 10 mA. The input-voltage range is 13.5 to 16.5 volts, the long-term stability is 100 ppm/1000 hours, and the devices have a 3 ppm/°C (max) TC. The AD2710 improves the +10-V output accuracy to ±0.01%, the TC to 1 ppm/°C, and the stability to 25 ppm/1000 hours.

Minimum-component hookup circuits are as noted in Fig. 4-11. Although the minimum hookup arrangement is three-terminal, an optional output trim may be used with any device to adjust the output voltage to ±30 mV, when desired.

The Burr-Brown REF10 and REF101 series of references offer high-accuracy +10-V outputs, at ±0.05% for the KM (TO-99 package) grades. They both feature 1 ppm/°C (max) TC, with 50 and 25 ppm/1000 hour stabilities for the REF10 and REF101, respectively. They both basically use the "REF-01" pinout (Fig. 4-6A) and can be optionally trimmed when necessary.

## 4.3 APPLICATIONS OF REFERENCES

In this section, the various devices just discussed are illustrated in a variety of reference applications, which serve to bring out the various points of consideration in optimum reference designs. This will begin with the most simple of circuits and progress through more advanced designs.

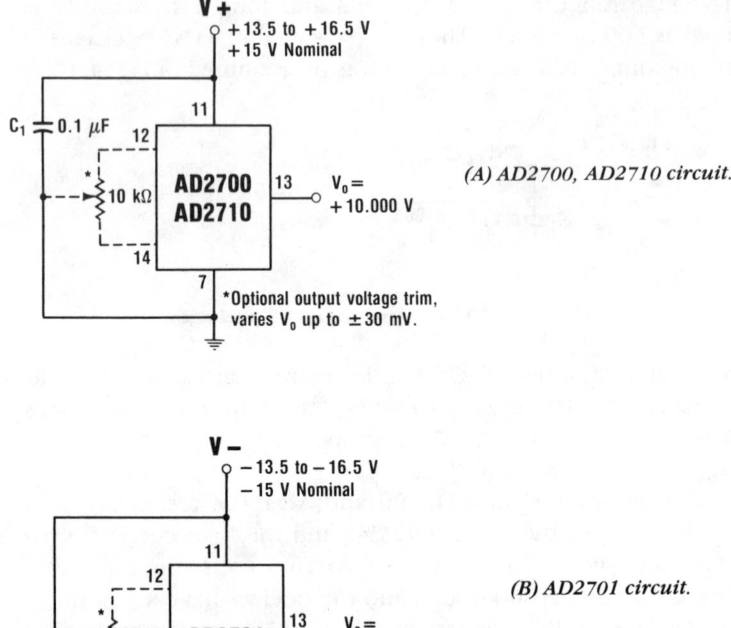


*(A) AD2700, AD2710 circuit.*

*(B) AD2701 circuit.*

**Fig. 4-11. AD2700-series reference (14-pin DIP package).**

### 4.3.1 Simple Reference Sources

The most basic reference source has traditionally been the temperature-compensated diode. The circuit can, in its most simple form, consist of only an input resistor to the supply line and the reference diode itself. The output voltage is taken from across the diode with suitable attention given to minimizing load-current variations.

A modern and very practical form of such a circuit is shown in Fig. 4-12A using the LM329 reference diode. Because of its very low dynamic impedance of 0.8 ohm, the LM329 is attractive in such a circuit, as the low impedance minimizes errors due to input voltage variations, load variations, and feed-resistor drift. The

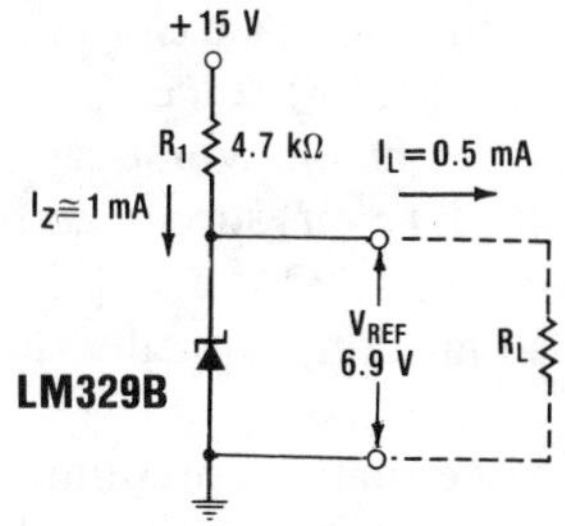


**TYPICAL PERFORMANCE**

| | | |
|---|---|---|
| Temp. Coefficient: | 20 ppm/°C(MAX) × 45°C | = 900 ppm |
| Input Regulation: | 425 μV/V | = 62 ppm |
| Load Regulation: | 0.0035%($\Delta I_L \leq 2.5\%$) | = 35 ppm |
| Long Term Stability: | 20 ppm | = 20 ppm |
| | | 1017 ppm |
| Total Error, 0-70°C: | | ≤1017 ppm<br>or ≅ 0.1%<br>suitable ≤ 8 bits |

*(A) Using the LM329.*

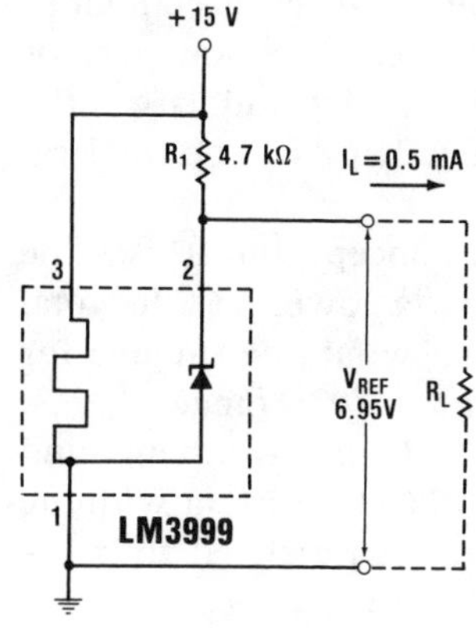


**TYPICAL PERFORMANCE**

| | | |
|---|---|---|
| Temp. Coefficient: | 5 ppm/°C(MAX) × 45°C | = 225 ppm |
| Input Regulation: | 467 μV/V | = 70 ppm |
| Load Regulation: | 0.0037%($\Delta I_L \leq 2.5\%$) | = 37 ppm |
| Long Term Stability: | 20 ppm | = 20 ppm |
| | | 352 ppm |
| Total Error, 0-70°C: | | ≤352 ppm<br>or ≤0.035%<br>suitable ≤ 9 bits |

*(B) Using the LM3999.*

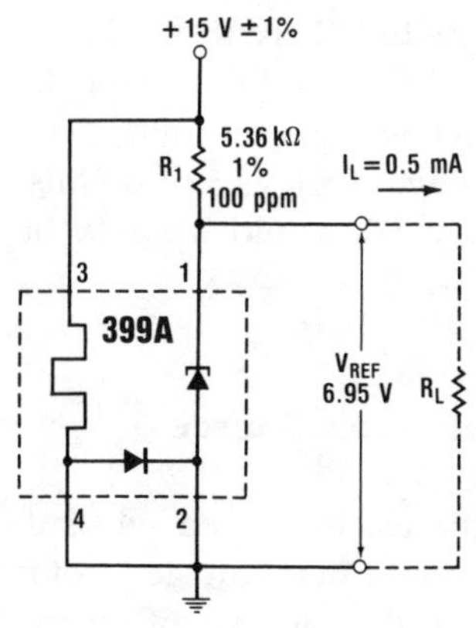


**PERFORMANCE (worst case)**

| | | |
|---|---|---|
| Temp. Coefficient: | 1 ppm/°C(MAX) × 45°C | = 45 ppm |
| Input Regulation: | 280 μV/V × 0.15 V | = 6 ppm |
| Load Regulation: | 0.00087%($\Delta I_L \leq 25\%$) | = 26 ppm |
| Long Term Stability: | 20 ppm | = 20 ppm |
| | | 97 ppm |
| Total Error, 0-70°C: | | ≤97 ppm<br>or ≤0.0097%<br>suitable ≤ 11 bits |

*(C) Using the LM399A.*

**Fig. 4-12. Simple reference sources.**

resulting major error of such a circuit will be the diode's basic TC, as will be seen from the following discussion of performance.

Included in Fig. 4-12A is a tabulation of the major worst-case errors. The conditions assumed are a 15-volt (±1 volt) supply and a 0.5-mA load current which varies no more than 25%. The performance figures are tabulated in worst-case fashion; typically, it will be appreciably better.

From the error budget, it is fairly easy to see that the predominant error is due to TC, which comprises about 90% of the total error. The total error in a 0–70°C temperature range (a change of 45°C from 25°C) is about 0.1%. This would restrict this circuit to converters of 8-bit, or less, accuracy, if the reference source error is to be held to ¼ LSB or less. Note that by and large, the temperature-induced change in reference voltage will be the largest error for most of these circuits.

The circuit of Fig. 4-12B is similar in concept, but it uses the temperature-stabilized LM3999 to achieve a lower TC. Its total error is 352 ppm or less (0.035%), which would be suitable for converter accuracies of 9 bits or less, over a 0–70°C range.

The circuit of Fig. 4-12C is optimized for high performance and uses the LM399A with its 1 ppm/°C (max) TC. Consistent with the high-performance goal, the supply voltage is restricted to ±1% and a stable resistor type is used for $R_1$. With these measures, this circuit is able to achieve a worst-case error of less than 190 ppm over this range, which is suitable for 11-bit accuracy.

Errors which these analyses have not included are those due to initial voltage tolerance of the diodes, and the total effects of the load, $R_L$. In typical use, the initial tolerance would be trimmed out by using only a fraction of the output voltage, such as 5 or 6 volts, for example. The divider resistors thus used would have to be included in the overall error budget.

### 4.3.2 The "Bootstrapped-Diode" Reference Source

Although the simple reference circuits just described will suffice for many applications, quite often a prescribed voltage level is required (such as 10 volts), and at a current output level of several milliamperes. This is usually beyond the capabilities of simple unbuffered diode circuits so additional support circuitry is required. An example of this form of circuit is shown in Fig. 4-13.

This circuit has appeared in the literature in many forms (see references) and is quite popular as a voltage-reference source. This is due to its high performance without undue complexity. The circuit "bootstraps" the reference diode, because the diode is used not only to regulate its output voltage, but also to set and maintain its own current, a prerequisite for a low TC. This current depends primarily on the reference voltage itself and, thus, will be stable, which lends further overall stability to the circuit. The reference voltage output can be taken directly from $D_1$ for high-impedance loads, or from the output of the amplifier, where it is scaled upward by the $(R_2 + R_3)/R_2$ ratio. This output appears at a low impedance and can supply appreciable current without the basic or scaled reference voltages being affected. It is, however, potentially less stable than the basic output across the diode, unless the resistor ratio, $R_3/R_2$, is designed to also be stable with temperature (see component suggestions in Fig. 4-13).

So that the output can assume only one polarity to ensure start-up, the amplifier is operated from a single supply voltage. Once the diode has broken down, the negative feedback loop sets the output voltage as:

$$V_o = V_{REF} \left( \frac{R_2 + R_3}{R_2} \right)$$

$R_1$ sets the current in the diode, since the voltage drop across it is simply the difference between the output and reference voltages. Since these two voltages are stable, this stabilizes the current.

The key to effective TC performance is the correct choice of components suitable to the required performance. The diode used should, of course, be consistent with the desired TC, and representative types are listed. $R_2$ and $R_3$ are most critical from the standpoint of their *tracking* TC, and their absolute value (or TC) is of lesser importance. $R_4$ is selected to be equal to $R_2 \parallel R_3$, for minimum bias-current error. The amplifier used should be capable of low drift, on the order of 5 $\mu$V/°C, or less. (This is equivalent to 1 ppm/°C or less for a 6-volt reference diode.)

The selection of $R_1$ is governed by the diode used. Thus, three values are shown, for use with the LM329, 1N4611, or 1N829 (or 7.5 mA) types, respectively. This will yield nominal currents of 1, 2, and 7.5 mA, respectively. For 7.5-mA use, additional current can be supplied by $R_5$ to ease the burden on the op amp.

For the 1N4611 and, in particular, the 1N829-type units, the value of $R_1$ should be trimmed to optimize (minimize) the overall TC of the circuit. This has the effect of adjusting the TC of the reference diode about its nominal zero point, and will serve to minimize the *net* drift errors of the circuit, including the op amp and resistors $R_1$, $R_2$, and $R_3$. $R_2$ and $R_3$ are trimmed for output voltage calibration. The trim process is interactive to an extent.

An example of this type of circuit, designed for reduced TC from the resistor divider, is shown in Fig. 4-13B. This circuit also includes a TRIM potentiometer for calibration, to effectively cancel the diode's ±5% tolerance. This hookup uses a thin-film resistor array which features low TCR (tracking), in like values of 2K (kilohms). The array has 5 ppm/°C tracking specification, and the

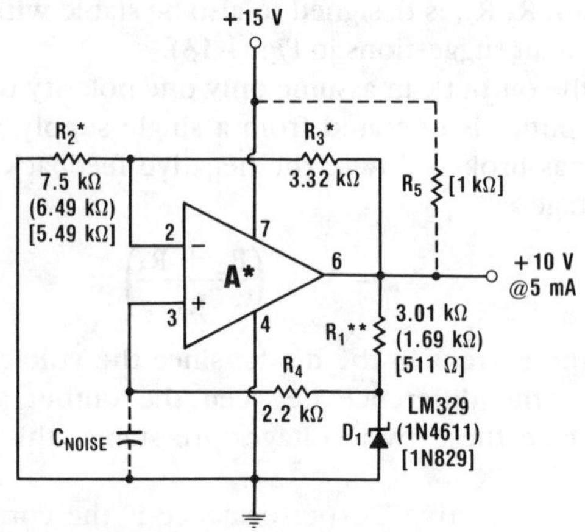


*Trim ratio to calibrate output voltage.
**Trim to minimize temp drift (1N4611, 1N829).

**Suggested Component Combinations**

| Overall TC | $D_1$ | A | $R_2/R_3$ | $R_1$ |
|---|---|---|---|---|
| <50 ppm<br><25 ppm<br>10-15 ppm<br>≅5 ppm<br>(or less) | 329A or 329B<br>1N4611B or 1N827<br>1N4611B or 1N827<br>1N4611C or 1N829<br>"PRD" or LM399A | 741,301A<br>OP-07C<br>OP-07C<br>OP-07C | 10 ppm/°C tracking<br>5 ppm/°C tracking<br>2 ppm/°C tracking<br>1 ppm/°C tracking<br>or less | 100 ppm<br>↑<br>optimize<br>see text<br>↓ |

*(A) Basic circuit.*

**Fig. 4-13. Basic bootstrapped**

7 values are connected to allow a gain trimming pot to be connected across one portion. This portion of the divider is scaled to provide a gain of 10/6.9 (for a 10-volt output) with a ±5% trim. The pot is used in the potentiometric mode for minimum error, and its TC is effectively reduced by the ratio of resistances (25/2). Overall, this circuit should have a TC of appreciably less than 25 ppm/°C, yet it uses no exotic parts.

Because of the foregoing interrelated factors which influence the overall TC, it is difficult to give specific component recommendations for these circuits, particularly for the very low TCs. Below drifts of 20 ppm/°C, the circuit must be considered as a complete thermal system and designed from both a mechanical and electrical standpoint accordingly. It is this type of approach,

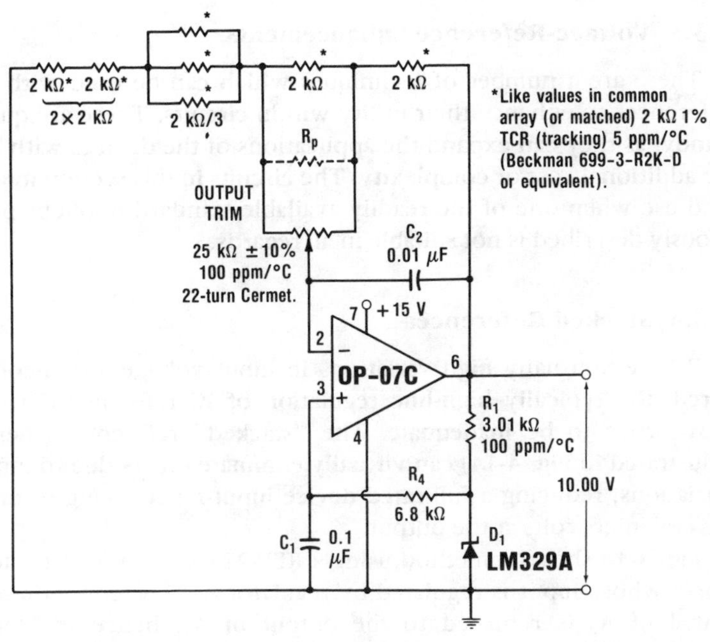

*(B) With trimming.*

**reference source.**

with this form of circuit, which does, in fact, result in the high performance of the actively trimmed hybrid references of Classification "E" (Table 4-1), and in references of lab-standard quality, such as the CODI "Certavolt" (see references).

For those who desire to build and characterize a circuit of this type, some general guidelines are given in the figure, but these should only be considered as starting points. It should be understood that only through extensive testing and trimming, after component qualification, will the highest performance be reached.

For negative forms of this reference circuit, the diode is reversed, pin 7 of the op amp is grounded, and the unregulated negative input is applied to pin 4. Resistor $R_5$ (when used) would then be returned to pin 4, instead of pin 7.

Various other reference-circuit schemes have been devised to attack specific problems or sources of error which arise in varying applications. The following series of "applications" discussions illustrate some of these techniques.

### 4.3.3 Voltage-Reference Enhancements

There are a number of techniques which can be used with IC references to enhance their utility within circuits. These are quite handy, as they can expand the applications of the devices with little additional cost or complexity. The circuits in this section might find use when one of the readily available standard hookups previously described is not suitable in all regards.

### 4.3.4 Stacked References

Where unusually high variations in input voltage are encountered, the typically high-line regulation of IC reference devices may prove to be inadequate. The "stacked" reference scheme (illustrated in Fig. 4-14) can virtually eliminate errors due to input variations, reducing a full-rated device input-range swing to only several microvolts at the output.

Fig. 4-14 shows a method using a REF-02 ($A_1$) as a 5-volt reference, whose input is regulated by regulator $A_2$. The common terminal of $A_2$ is returned to the output of $A_1$, hence the term "stacked." $R_1$ provides a bleed path for the common-leg current of $A_2$. Note that only $A_1$ need be a precision device in this circuit, the $A_2$ function is relatively noncritical. In fact, even ordinary

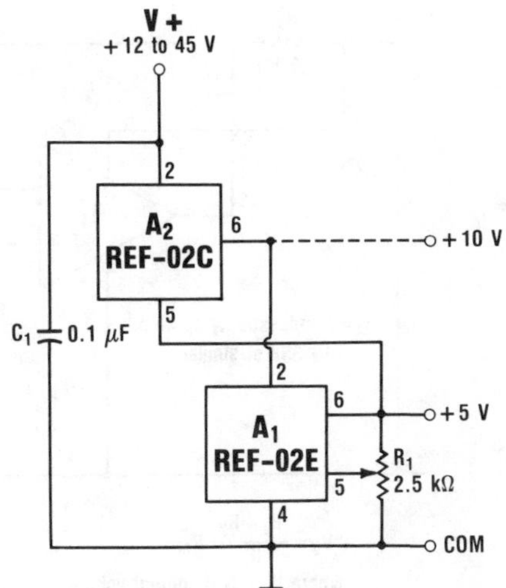


**Fig. 4-14. Stacked references.**

three-terminal regulators can be adapted for use as in this circuit $A_2$ (LM309, etc.).

When using the REF-01, REF-02 units (or similarly pinned devices), making $R_1$ a variable potentiometer, as shown, can also trim the output voltage. If trim is not needed, a fixed resistor should be used for $R_1$. Again, the device used for $A_2$ is relatively noncritical, and it allows a bonus output to be available (+10 V, in this case).

### 4.3.5 Shunt-Regulated References

With some additional support circuitry, IC references can operate as two-terminal shunt regulators, as shown in Fig. 4-15. Here a low-power, low-voltage reference, the AD589, is used in conjunction with the single-supply op amp, $A_1$. The negative feedback loop formed around $A_1$ programs the output-terminal voltage, $V_o$, to a multiple of the reference voltage. $R_s$ is a series input resistor chosen to supply the necessary load current, plus the current in $A_1$ and $A_2$ (less than 500 $\mu$A), and, as shown, is suitable for 3-mA loads.

Note that this circuit can "float," as it has no necessarily common point. For low-power, low-voltage applications, a single-supply op amp, such as the 358, should be used.

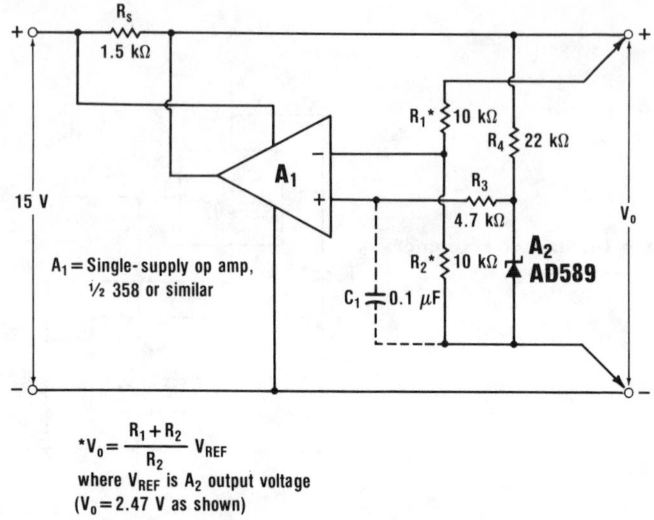


**Fig. 4-15. Shunt-regulated reference.**

### 4.3.6 Precision, Buffered Reference Sources

With a combination of circuit techniques, it is possible to design buffered references where the total error is reduced to only the TC and long-term drift of the reference diode itself. This is accomplished through combinations of bootstrapping, reference-device buffering, and input preregulation.

In Fig. 4-16, an LM399A IC reference is connected in a bootstrapped configuration which multiplies its output voltage by the product of the two ratios:

$$\left(\frac{R_A + R_B}{R_B}\right)\left(\frac{R_D}{R_C + R_D}\right)$$

For the ratio values chosen, the output is 10 V, the op amp gain is 2, and the values are much simplified. This allows $R_A$ and $R_D$ to be contained in a common array, for best stability. Note that for best results, output voltage should be sensed directly at the load terminals, as shown. Input voltage variations are absorbed by $A_2$, a low-cost 5-volt three-terminal regulator which can also supply up to 25 mA of output. $R_3$ and $D_1$ are used for start-up purposes.

Here, because the reference diode has such a low inherent TC,

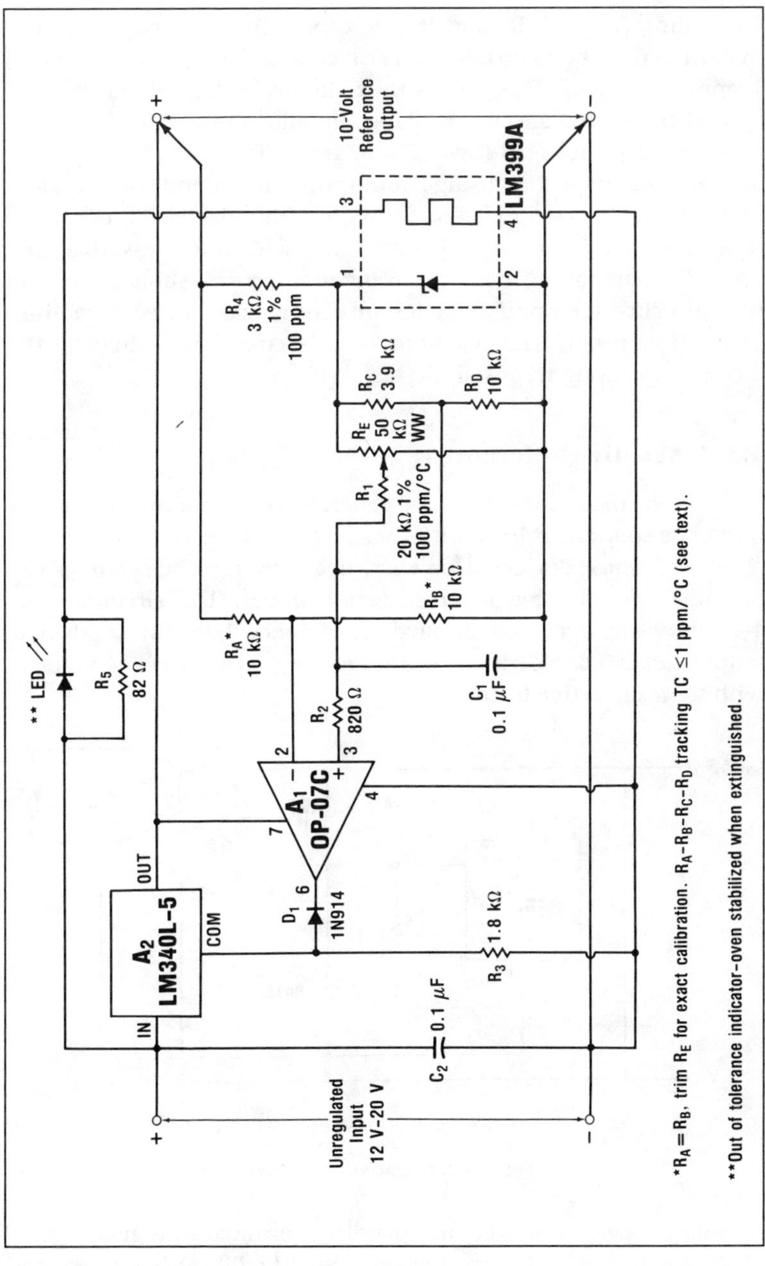


**Fig. 4-16. Low TC precision reference.**

the scaling resistors $R_A$ and $R_D$ are especially important to a low overall TC. They should have tracking TCs of (preferably) 1 ppm/°C or less. This means they should be high-stability wirewound units of Evanohm or Manganin alloys in a common array, or a similarly packaged low-TC film array. Trimming for a 10-volt output is accomplished using a multiturn wirewound trimmer, $R_E$.

A low-drift op amp is used for $A_1$, to minimize its TC contribution. The amplifier type specified has a drift of less than 0.2 ppm/°C. This, along with the remaining errors, should yield an overall drift of 2 ppm/°C or less if care is taken in construction. The ultimate performance limit of this circuit will then be the diode's basic long-term stability.

### 4.3.7 Negative References

Thus far, the discussions have generally been in terms of positive reference sources, at least with those circuits involving three-terminal IC reference devices. However, negative reference sources can be implemented using reversed zener diodes. The circuit in Fig. 4-17 shows how a three-terminal IC reference (basically, a positive-output device) can be used to produce negative reference voltages with some attractive features.

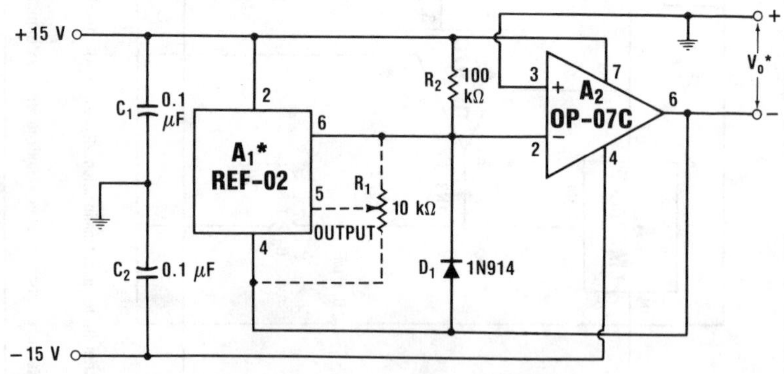


**Fig. 4-17. Negative reference.**

A basic negative source, using a three-terminal IC regulator ($A_1$), is shown in Fig. 4-17; in this case, an REF-02. Other types and voltage outputs can be used as well, with no modification of the basic principles of operation. In this circuit, the op amp supplies the output current. $A_1$ biases the op amp input-to-output at a voltage equal to its output voltage; in this case, 5 volts. With the + input of the op amp referred to ground, the output of the op amp is negative, or −5 V.

Potentiometer $R_1$ serves its normal trim purpose (when desired) and the output current of the circuit is determined by the op amp. In this case, the output current is 5 milliamperes. Overall TC of this circuit is less than 10 ppm/°C, or basically that of the TC of the reference IC used. $R_2$ and $D_1$ are used for start-up and protection purposes.

An extension of the negative reference idea is illustrated in Fig. 4-18. This circuit can be used to divide a given reference voltage into equal (or unequal) smaller voltages. The outputs are then bipolar tracking reference voltages, from low-impedance sources.

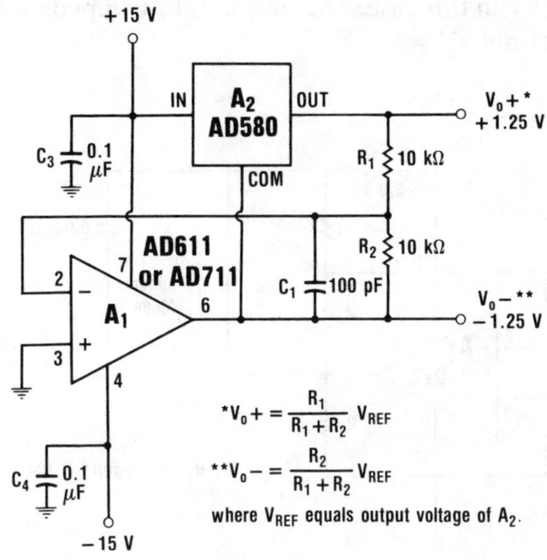


**Fig. 4-18. Bipolar tracking reference source.**

With a given reference IC output voltage, such as 2.5 volts, from the AD580 shown in Fig. 4-18, resistors $R_1$ and $R_2$ can be adjusted for any desired positive or negative output voltage, with an absolute-value sum equal to the IC's output voltage. Positive (+) output current is determined by $A_2$, while negative (−) output is set by $A_1$. For balanced tracking of the two reference voltages, $R_1$-$R_2$ should be a ratio-tracking pair, preferably in a single array.

### 4.3.8 Trimmable, Buffered Zener Reference

One of the basic problems that often arises in the application of voltage references is a means of adjustment for the calibration of system gain or scale factors. While generally any reference can be adjusted by a voltage divider, this usually gives rise to additional error factors, due to divider resistances. Also, a voltage divider is unbuffered so it sacrifices output impedance and it cannot, of course, furnish a higher voltage, if this is desired.

The circuit of Fig. 4-19 illustrates a means of trimming a zener diode type of reference source an equal (+/−) amount about its nominal level, with the output appearing in low-impedance form. Here, an AD581 (or other zener-type reference) is operated within the negative-feedback path of $A_1$, with the zener current set by $R_5$. If the (+) input of $A_1$ were grounded, the output of the circuit would simply be 1% less than the voltage of the reference diode; that is, −9.9 V in this case. The output is low impedance, with the current determined by $A_1$.

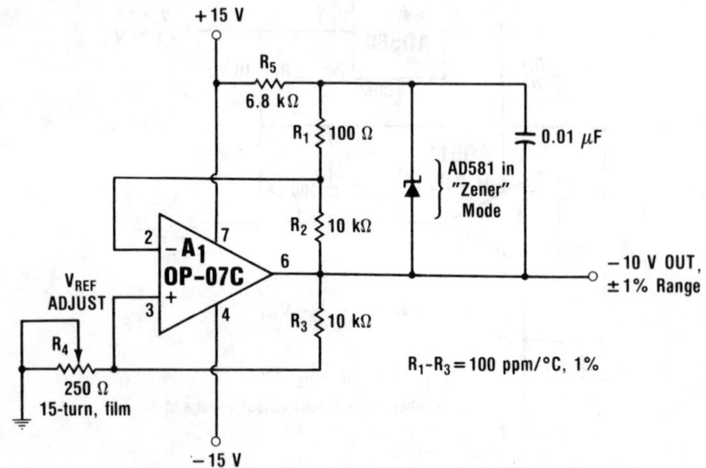


**Fig. 4-19. Adjustable reference source.**

In the circuit of Fig. 4-19, $R_1$ and $R_2$ reduce the voltage from the AD581 by 1%, which produces a −1% trim limit. Positive feedback through $R_4$ to $R_3$ allows the op amp to bootstrap the output by the amount of voltage across $R_3$. Resistor $R_3$ can then be trimmed for exact calibration, with range limits of ±1% about the nominal voltage.

The circuit can generally be used with virtually any reference type of zener, and is applicable to positive outputs by reversing the diode (and feeding it from −15 volts).

## 4.4 CURRENT REGULATORS

Current regulators are similar to voltage regulators in that they use the same basic elements of control to accomplish their function. The difference lies in their configuration and method of final application. For most of the following current-regulating circuits discussed, a stable voltage reference as an input is assumed.

The basic elements of a current regulator are illustrated in Fig. 4-20. There is a reference source ($V_{ref}$), an amplifier ($A_1$), and a feedback network ($\beta$), which consists in this circuit of $R_{sense}$ and $R_{load}$. At this point, it appears that there is no difference between this circuit and the basic voltage regulator of Fig. 4-1. Actually, with a given value for $R_{load}$, there would be no difference. $R_{load}$ and $R_{sense}$ would form a feedback network to define some value of $\beta$, and the loop would generate a stable output. However, the main difference in a current regulator is that it is the *current* rather than the voltage that we wish to maintain constant. In the circuit of Fig. 4-20, it will be noted that if $V_{ref}$ is constant, if $R_{sense}$ is a fixed value, and if $A_1$ is an ideal amplifier, the current, $I_L$, flowing in $R_{sense}$ (and thus $R_{load}$) will remain constant at a value

$$I_L = \frac{V_{ref}}{R_{sense}}$$

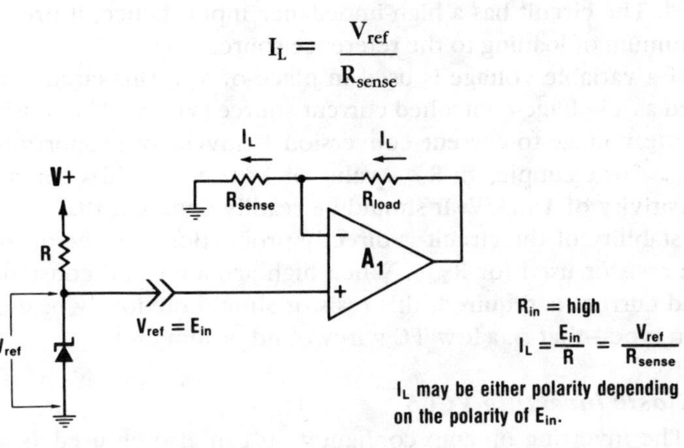


Fig. 4-20. Basic noninverting VCCS.

This is true regardless of the value of $R_{load}$. The basic concept of a current regulator, then, is the maintenance of a definite value of current flowing in a variable load impedance, with independence from the load variations, temperature changes, and supply-voltage changes.

There are many different types of regulators, which attain various levels of performance. The performance attainable using IC op amps in current regulators can be nearly perfect if sufficient attention is paid to the significant sources of error.

Op-amp parameters which tend to be important to current regulators are offset-voltage drift, bias current, CMRR, and PSRR. With modern amplifiers, virtually all of these factors can be made negligible in a wide range of applications. The very-low-input-current FET-input amplifiers allow currents to be regulated down to microampere levels and less, while buffers and/or power op amps extend the output range to amperes.

### 4.4.1 Low-Power, General-Purpose Current Regulators

#### *Basic Noninverting VCCS*

With the assignment of component values, the circuit of Fig. 4-20 can serve as a highly useful, general-purpose, low-power current regulator. It can supply current of either polarity through $R_{load}$, up to the rated output current and voltage of the amplifier used. The circuit has a high-impedance input; hence, it presents a minimum of loading to the reference source.

If a variable voltage is used in place of $V_{ref}$, this circuit can be used as a voltage-controlled current source (VCCS). The sensitivity of the voltage-to-current conversion is inversely proportional to $R_{sense}$. For example, an $R_{sense}$ value of 1 kΩ will yield a conversion sensitivity of 1 mA/V. It should be readily apparent that the overall stability of the circuit is directly proportional to the quality of the resistor used for $R_{sense}$. When high accuracy and constancy of load current is required, this resistor should obviously be a precision type—that is, a low-TC wirewound or film unit.

#### *Basic Inverting VCCS*

The inverting op-amp configuration can also be used as a current regulator and a VCCS. This is illustrated in Fig. 4-21. In this circuit, input current $I_{in}$ is equal to load current $I_L$; therefore,

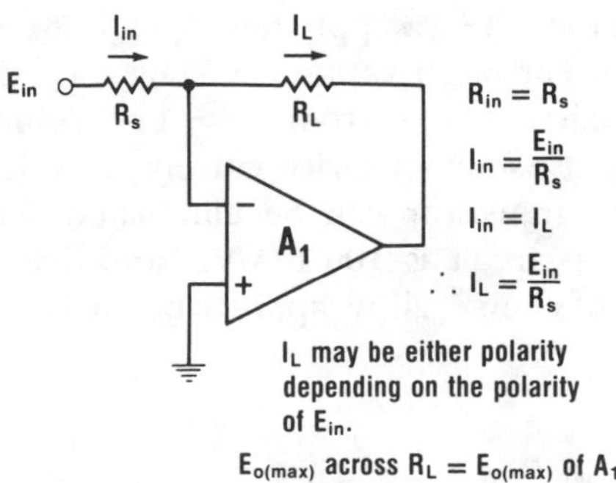


**Fig. 4-21. Basic inverting VCCS.**

$$I_L = \frac{E_{in}}{R_s}$$

Resistor $R_s$ determines the sensitivity of the voltage-to-current conversion as in the previous circuit, but in this circuit it also determines the input resistance, which is equal to $R_s$. This circuit can also supply current of either polarity up to the rated output current and voltage of the amplifier. In addition, it has the advantage that more voltage is available to drive $R_L$ because one end of $R_L$ is at ground potential, whereas in the circuit of Fig. 4-20, one end of $R_L$ is at the same potential as the input voltage, which subtracts from the voltage available to drive $R_L$.

## 4.4.2 Boosted-Output Current Regulators

General-purpose op amps, such as the 741 and the more precise OP-07, are limited to a maximum output current of ±5 mA. Power types, such as the 759 and the OP-50, can handle currents from 50 mA, while a power device like the LM12 can handle from 1 A. Low-current types can be boosted to higher currents, however, by adding current-gain transistors within the feedback loop, a step which can often be cost-effective.

### *Noninverting VCCS With Boosted Output*

A basic circuit illustrating a boosted-output current regulator is shown in Fig. 4-22. This circuit is a 100-mA-output noninverting VCCS with bidirectional current capability in load resistance $R_L$. For positive input voltages, $Q_1$ conducts, $Q_2$ is turned off, and current flows from the +15-V supply through $Q_1$, $R_L$, and $R_1$ (the 5 Ω sensing resistor). For negative input voltages, $Q_2$ conducts, $Q_1$ is turned off, and current flows from the −15-V supply through $Q_2$, $R_L$, and $R_1$. If only single-polarity current flow is required, the opposite-polarity transistor can be eliminated. The conversion sensitivity of this circuit is 100 mA/V. Note that higher-current transistors will, of course, allow appreciably higher currents.

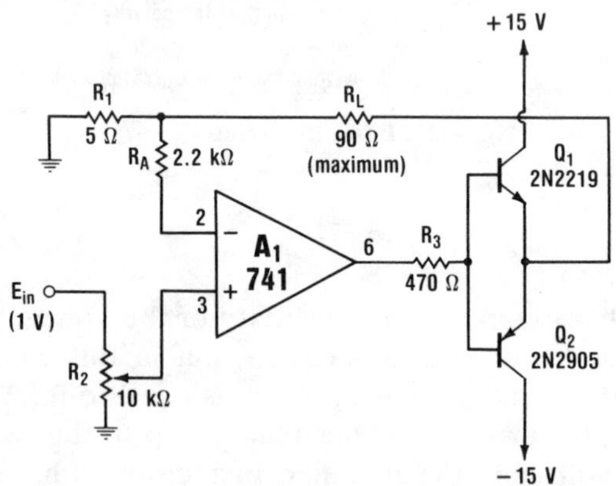


**Fig. 4-22. Noninverting VCCS with boosted output.**

In this circuit (as well as many others), it should be appreciated that sensing resistor $R_1$ should be a stable type. Therefore, some means of calibration, which will *not* alter $R_1$, is desirable and should be used. This can best be accomplished by reducing the value of $R_1$ slightly, and introducing a scaled potentiometer for $R_2$, as shown.

There are other advantages to reducing the voltage across $R_1$, such as minimizing the self heating. A 741-type is shown for $A_1$, but an OP-07 will enhance stability at low input voltages.

### *Inverting VCCS With Boosted Output*

An inverting VCCS with output boosted to 100 mA is illustrated in Fig. 4-23. One undesirable characteristic of the basic inverting configuration is the fact that the input current is equal to the load current, as diagrammed in Fig. 4-21. This is obviously a drawback for high-current use because the signal or reference might have to supply currents as high as 100 mA. However, there is a method of circumventing this problem known as *current scaling,* which provides an effective gain between the input current, $I_{in}$, and the load current, $I_L$.

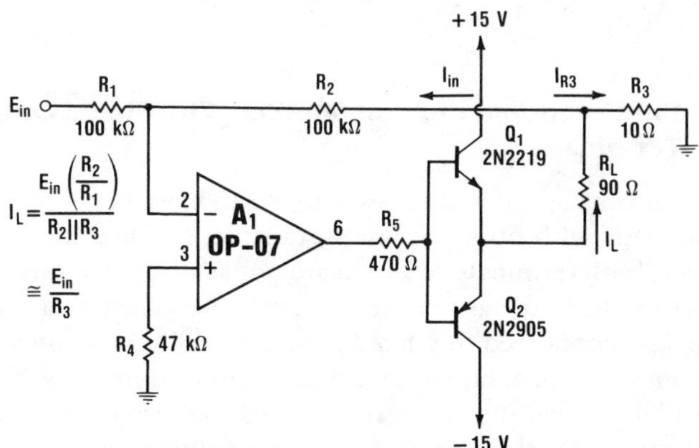


**Fig. 4-23. Inverting VCCS with boosted output.**

The circuit of Fig. 4-23 utilizes the current-scaling technique, which works as follows: A voltage level of $-1$ V at the input will create a current of 10 $\mu$A in $R_1$ and, because of feedback action, this same current must also flow through $R_2$. Therefore, the junction of $R_2$ and $R_3$ will be at a potential of $+1$ V, which means that $R_3$ will also have 1 V across it. Hence, the current in $R_3$ will be

$$I_{R_3} = \frac{1\ \text{V}}{10\ \Omega}$$

$$= 100\ \text{mA}$$

Thus, $I_{R_3}$ has been effectively scaled upward by the ratio of $R_2/R_3$, which in this case is 10,000/1.

The total load current, $I_L$, flowing in $R_L$ consists of the two currents, $I_{in}$ and $I_{R_3}$, so the actual ratio of load current to input current will be

$$\frac{I_L}{I_{in}} = \frac{I_{R_3} + I_{in}}{I_{in}}$$

$$\cong \frac{I_{R_3}}{I_{in}}$$

The circuit can be calibrated most conveniently (if necessary) via $R_1$. $R_1$, $R_2$, and $R_3$ should be stable types. This circuit can also be made bidirectional by the use of complementary buffer transistors for $Q_1$ and $Q_2$. Note the practicality of controlling amperes of current in $R_3$ (given the appropriate booster transistor) for high ratios of $R_2/R_3$.

### 4.4.3 Unidirectional Current Sources With Single Output Terminals

The current sources discussed thus far have been capable of driving loads with both terminals available, or "floating." Often, however, both terminals of a load are not available for this type of connection. For various reasons, it may be necessary that one load terminal be connected to a fixed potential—either supply voltage, for example, or ground. In fact, these applications may be more commonplace than the two-terminal (floating) drives. Also, a large number of applications do not require a bidirectional output current—the current may always remain the same polarity, and may even be fixed in magnitude. There is a wide range of circuits that fall into these general categories, some with even more subtle options of their own, which will be discussed in this section.

#### *Current Sources With Loads Referred to More-Positive Supply Potentials*

*Positive-Input VCCS*—Fig. 4-24 illustrates a circuit that partially resembles the noninverting buffered VCCS of Fig. 4-20, but with one very important difference. In Fig. 4-24, load resistance $R_L$ has been moved to the collector circuit of $Q_1$, which places $R_L$ *outside* the feedback loop for $A_1$. In operation, this circuit maintains the $Q_1$ emitter current constant as it passes through $R_1$ by comparing it to input voltage $E_{in}$. In this respect, the circuit of Fig. 4-24 is the same as an npn buffered version of the Fig. 4-22 circuit. In Fig. 4-22, however, the total emitter current of $Q_1$ (collector current plus base current) passes through $R_1$ and $R_L$. The situation is slightly different in Fig. 4-24 since $R_L$ is in the collector circuit, but the current that is regulated is the emitter current. Therefore, there is an error due to the base current of $Q_1$, which does not flow through $R_L$. This error is caused by the finite $h_{FE}$ of $Q_1$, and the difference between the current sampled and the current actually flowing in $R_L$ will be $1/h_{FE}$ of the current in $R_1$. This error is minimized by using a high-gain transistor (or a Darlington) for $Q_1$.

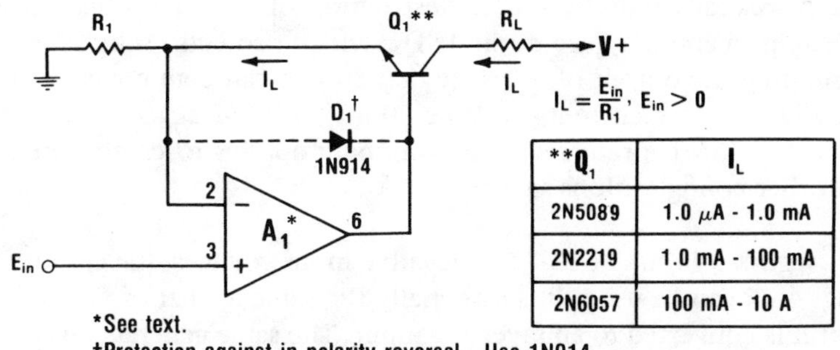


| **Q1 | IL |
|---|---|
| 2N5089 | 1.0 μA - 1.0 mA |
| 2N2219 | 1.0 mA - 100 mA |
| 2N6057 | 100 mA - 10 A |

*See text.
†Protection against in polarity reversal. Use 1N914 for general-purpose; use 1N458A for precision, low-current.

**Fig. 4-24. Basic positive-input VCCS; load referred to V+.**

The V+ supply to $R_L$ and $Q_1$ can be either the +15-V line, or it can be as high as the $V_{CB}$ rating of the transistor used, as long as the maximum power dissipation of the transistor is not exceeded. The V+ parameter is not limited by the op amp in this circuit, and the circuit responds to positive voltages at $E_{in}$.

Since this circuit does not load the reference source, it is a very useful type of current regulator. $R_1$ and $Q_1$ are selected for the desired operating current. $Q_1$ is selected primarily from the standpoint of power and voltage rating. Typical recommendations for $Q_1$, for different ranges of $I_L$, are listed in the chart in Fig. 4-24.

A 741 is generally adequate for $A_1$, since $A_1$ must be a unity-gain-compensated type due to the 100% feedback through $Q_1$. For very low values of $I_L$, the input bias current of $A_1$ (which subtracts from $I_L$) can become a significant percentage of $I_L$, thus causing a significant error. For $I_L$ values below about 10 μA, low-input-current bipolar types, such as the OP-07, OP-27, etc., or FET types, such as the OPA111, AD547, or AD542 may be necessary for better accuracy. Single-supply types would include the 3130 and 3140, the TLC271, the OP-20 and LM10, and the 358 or 324 selections. Note: These general recommendations for amplifier types may be applicable to the following circuits as well.

One possible trouble with this circuit is the failure of $Q_1$ if the input voltage is reversed from its normal operating polarity. Since $Q_1$ can conduct only a unidirectional current, a reversal of the input voltage to $A_1$ will result in an opposite-polarity, saturated, output voltage from $A_1$. This may be sufficient to cause emitter-base breakdown and possible destruction of $Q_1$. This situation is easily prevented by use of diode $D_1$, which protects $Q_1$ by reverse clamping the output of $A_1$. Note that $D_1$ can cause an error due to leakage. At low current levels ($\leq 10\ \mu A$), a low-leakage type such as a 1N458A is preferred. This comment applies to clamp diodes of other configurations as well.

*Negative-Input VCCS*—For negative input voltages, the circuit of Fig. 4-25 can be used. It is essentially the same as that of Fig. 4-24, but it is converted to an inverting input. The same general considerations that were true for $A_1$ and $Q_1$ in Fig. 4-24 also apply in this circuit. Since $Q_1$ is biased at near ground potential, this circuit has a larger available output swing to drive $R_L$. The circuit has the same basic disadvantage as the circuit of Fig. 4-23, however, in that $I_{in} \approx I_L$, restricting (practical) operation to small input currents.

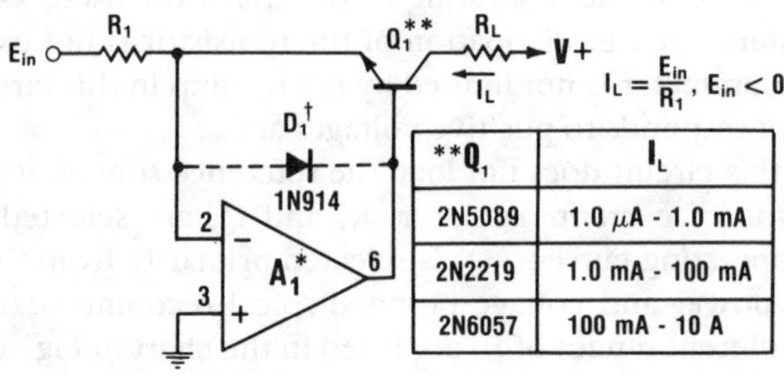


| **$Q_1$ | $I_L$ |
|---|---|
| 2N5089 | 1.0 μA - 1.0 mA |
| 2N2219 | 1.0 mA - 100 mA |
| 2N6057 | 100 mA - 10 A |

* See text.

† Protection against input polarity reversal.

**Fig. 4-25. Basic negative-input VCCS; load referred to V+.**

This disadvantage can be overcome by using the current-scaling technique, as shown in Fig. 4-26. In this circuit, resistor $R_3$ forces $Q_1$ to conduct a much heavier current than the feedback current. The operation is similar to the circuit of Fig. 4-23, with the gain between $I_{in}$ and $I_L$ determined by the $R_2/R_3$ ratio; and $R_1$ can be trimmed for calibration.

*Improving the Accuracy of the Output Current*—The circuits discussed thus far are all somewhat imprecise, as was pointed out initially. There are several remedies that can be used to minimize the inaccuracies, and basically all are schemes to minimize that

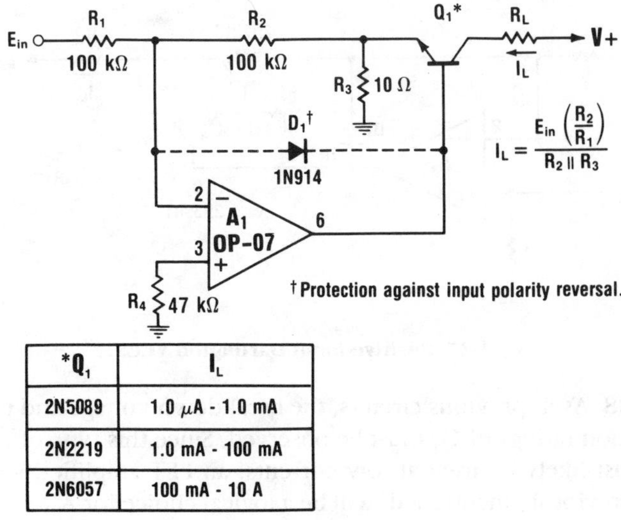


| $^*Q_1$ | $I_L$ |
|---|---|
| 2N5089 | 1.0 μA - 1.0 mA |
| 2N2219 | 1.0 mA - 100 mA |
| 2N6057 | 100 mA - 10 A |

**Fig. 4-26. Negative-input VCCS with current gain.**

portion of the regulated current which does not reach the load. This problem can become rather severe, particularly at high currents where the gain of the transistor falls off, and inaccuracies can reach several percent.

An example of one remedy that can be used is to simply substitute a Darlington device for the single transistor used in any of the previous circuits. Such a circuit is shown in Fig. 4-27. In this circuit, $Q_1$ can be a single-package Darlington amplifier, such as the MPS-U45, or it can be two discrete transistors wired as a Darlington pair (shown as $Q_1$ and $Q_2$). The use of a Darlington amplifier will improve the accuracy by the additional gain, and it will suffice for all but the most stringent applications.

For high current-accuracy applications, an FET can be used, as shown in Fig. 4-28. With the use of an FET buffer, either a JFET (Fig. 4-28A) or a MOSFET (Fig. 4-28B) type, the error current is greatly reduced. This lowers the drive current to the FET (which is, in fact, the error component) to the reverse-bias gate leakage of the particular device used.

In Fig. 4-28A, the JFET used must have a minimum zero-bias drain current ($I_{DSS}$) that is equal to or higher than the desired load current, $I_L$. For a 2N4867, the minimum $I_{DSS}$ is 0.4 mA, which is also, therefore, the maximum $I_L$ for the device in the circuit of

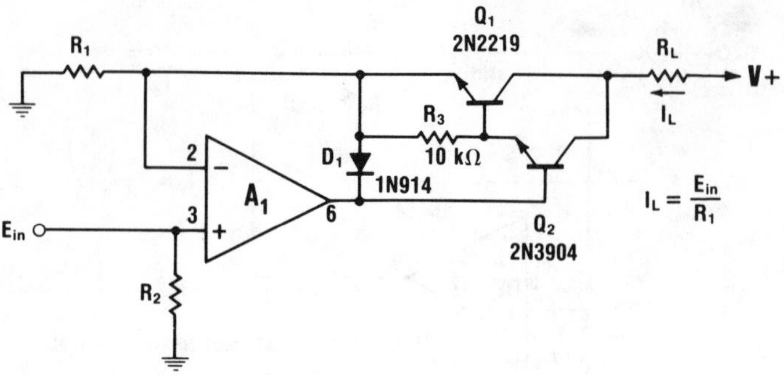


**Fig. 4-27. Positive-input Darlington VCCS.**

Fig. 4-28. As in previous circuits, the breakdown voltage and power dissipation ratings of $Q_1$ must be observed. Since this type of circuit will most likely be used at low currents, an FET amplifier, such as those previously mentioned, will be a logical choice for $A_1$.

The MOSFET version of Fig. 4-28B is quite similar, but it uses an n-channel device for $Q_1$, as shown. This device will be suitable for currents up to about 100 mA, limited actually by power dissipation. Its attractiveness lies in its simplicity and very wide dynamic range.

Many applications can require both a small bias-current error and a high $I_L$. In addition to the MOSFET buffer, requirements may also be satisfied by an FET/power-transistor combination, as shown in the circuit of Fig. 4-29. In this circuit, the FET drives the npn transistor as a source follower, and the combination is equivalent to a power FET. The breakdown voltage is limited to that of the device that has the lower rating, which in this circuit is the FET, $Q_1$. The power rating is determined primarily by $Q_2$, and the circuit can be used at any current level up to 50 mA with the devices shown. While this version is more complex than the MOSFET, it may be attractive for reasons of availability.

In all of the preceding VCCS circuits with improved accuracy, the general considerations regarding the op amp used should follow the points that were discussed for the circuit of Fig. 4-24, since they are modifications of this basic configuration. In all cases, $R_1$ is selected to provide the desired output current with a full-scale input voltage in accordance with the equation,

$$I_L = \frac{E_{in}}{R_1}$$

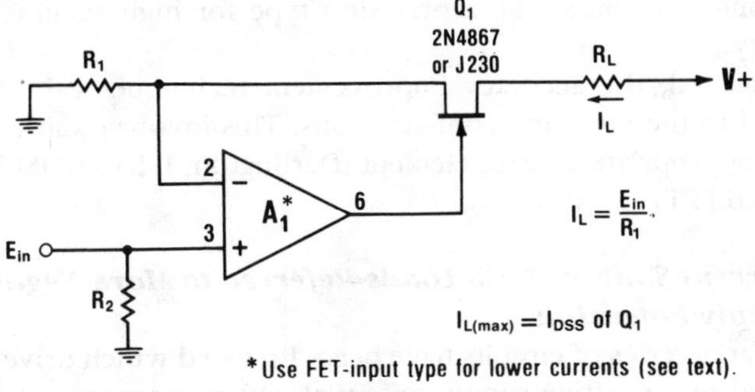

* Use FET-input type for lower currents (see text).

*(A) JFET.*

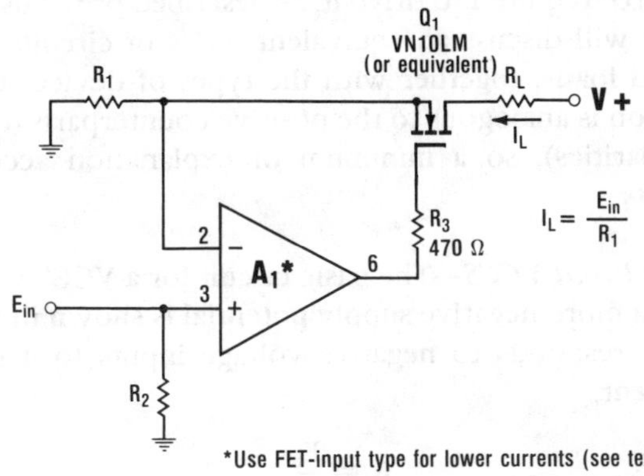

*Use FET-input type for lower currents (see text).

*(B) MOSFET.*

**Fig. 4-28. Positive-input FET VCCS.**

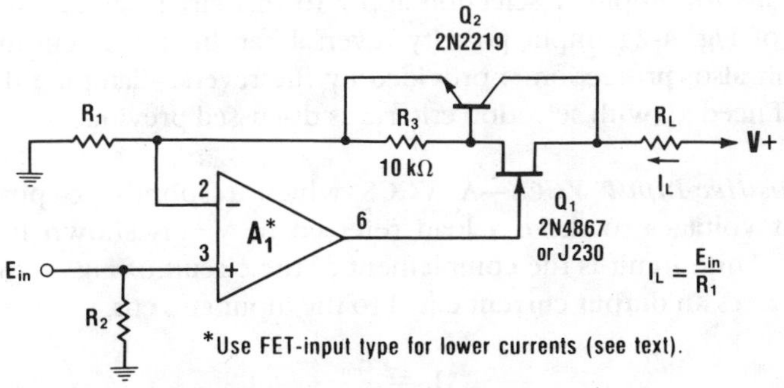

*Use FET-input type for lower currents (see text).

**Fig. 4-29. Positive-input buffered-FET VCCS.**

$R_1$ should, of course, be a precision type for high accuracy and stability.

If desired, the accuracy improvement techniques can also be applied to the inverting configurations. This involves substitution of the appropriate control element (Darlington, JFET, MOSFET, or buffered FET).

### *Current Sources With Loads Referred to More-Negative Supply Potentials*

So far, a series of circuits have been discussed which drive loads referred to a positive supply potential. Often, however, the load may be referred to a negative supply potential, with a constant-current source required to drive it, as described previously. In this section, we will discuss an equivalent series of circuits for negative-referred loads, together with the types of devices used. Circuit operation is analogous to the positive counterparts (except for voltage polarities), so a minimum of explanation accompanies these circuits.

*Negative-Input VCCS*—The basic circuit for a VCCS with a load referred to a more-negative supply potential is shown in Fig. 4-30. This circuit responds to negative voltage inputs to generate an output current,

$$I_L = \frac{E_{in}}{R_1}$$

which, in general principle, is exactly the same as the circuit of Fig. 4-24. In the circuit of Fig. 4-30, however, $Q_1$ is a pnp transistor, which allows $R_L$ to be returned to a V− potential. The same considerations for amplifier selection apply to this circuit as for the circuit of Fig. 4-24. Input polarity reversal can be a problem in this circuit also; protection is provided by the reverse-clamping diode, $D_1$, if needed (with selection criteria as discussed previously).

*Positive-Input VCCS*—A VCCS which responds to positive input voltages to drive a load referred to V− is shown in Fig. 4-31. This circuit is the complement of the circuit of Fig. 4-25 and generates an output current equal to the input current,

$$I_L = \frac{E_{in}}{R_1}$$

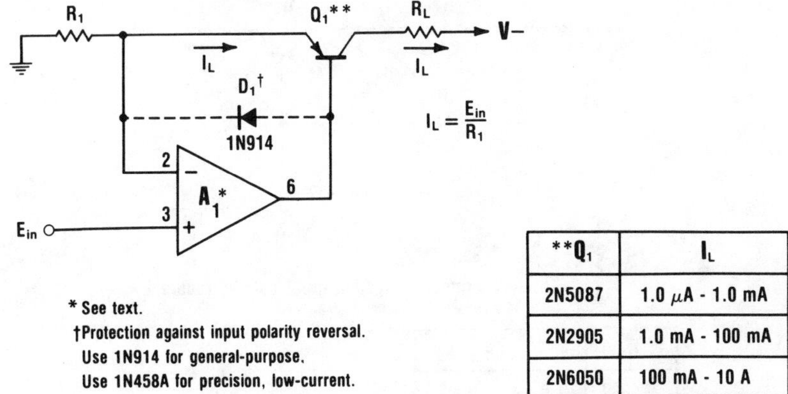


| **$Q_1$ | $I_L$ |
|---|---|
| 2N5087 | 1.0 μA - 1.0 mA |
| 2N2905 | 1.0 mA - 100 mA |
| 2N6050 | 100 mA - 10 A |

**Fig. 4-30. Basic negative-input VCCS; load referred to V−.**

This circuit can be modified to provide a current "gain," as shown in Fig. 4-32.

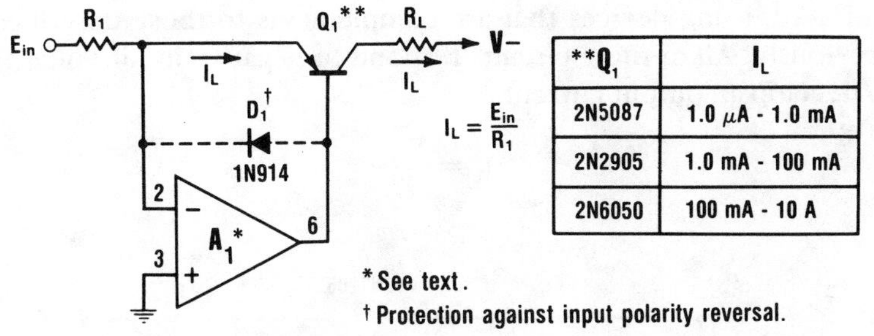


| **$Q_1$ | $I_L$ |
|---|---|
| 2N5087 | 1.0 μA - 1.0 mA |
| 2N2905 | 1.0 mA - 100 mA |
| 2N6050 | 100 mA - 10 A |

**Fig. 4-31. Basic positive-input VCCS; load referred to V−.**

*Improved Accuracy VCCSs for Negative-Referred Loads*—The three basic configurations of Figs. 4-30, 4-31, and 4-32 are subject to the same inaccuracies as those described for the circuits of Figs. 4-24, 4-25, and 4-26. This problem is solved for the circuits of Figs. 4-30, 4-31, and 4-32 in a similar manner as for the circuits of Figs. 4-24, 4-25, and 4-26—by the use of higher-current-gain control elements, such as Darlington pairs, JFETs, MOSFETs, and buffered FETs, arranged to handle opposite polarity currents.

Three circuits that eliminate this problem for negative-referred loads are shown in Figs. 4-33, 4-34, and 4-35. These circuits oper-

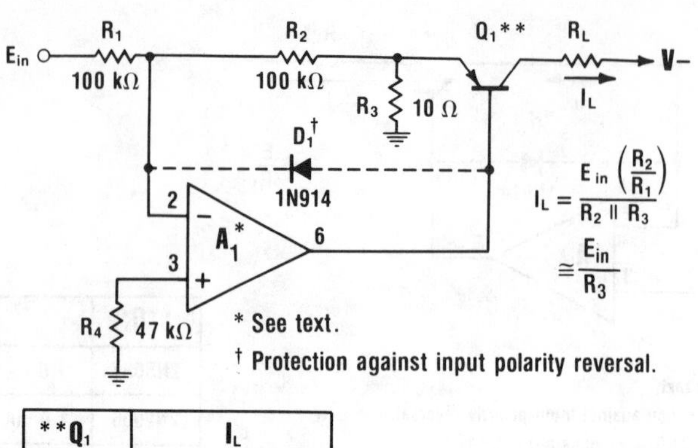


$$I_L = \frac{E_{in}\left(\frac{R_2}{R_1}\right)}{R_2 \parallel R_3} \cong \frac{E_{in}}{R_3}$$

| **Q1 | $I_L$ |
|---|---|
| 2N5087 | 1.0 µA - 1.0 mA |
| 2N2905 | 1.0 mA - 100 mA |
| 2N6050 | 100 mA - 10 A |

**Fig. 4-32. Positive-input VCCS with current gain.**

ate essentially the same as their counterparts in Figs. 4-27, 4-28, and 4-29, using devices that are complements to those described previously. All of these circuits respond to negative input voltages to develop an output current,

$$I_L = \frac{E_{in}}{R_1}$$

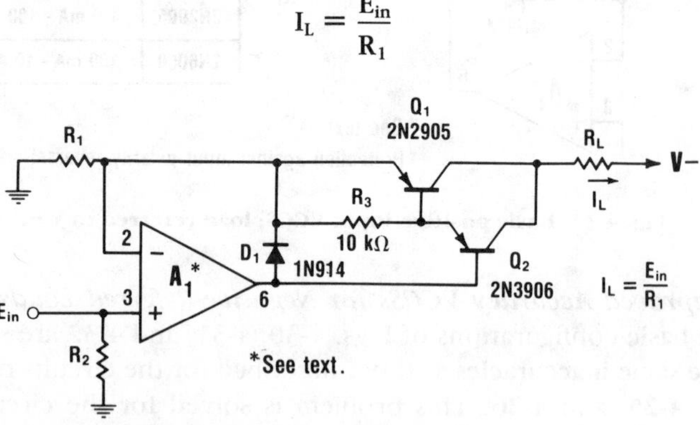


**Fig. 4-33. Negative-input Darlington VCCS.**

### 4.4.4 Bidirectional-Output VCCS

A very interesting category of VCCS is the most general-purpose type of all, and is known as a bidirectional-output VCCS, or

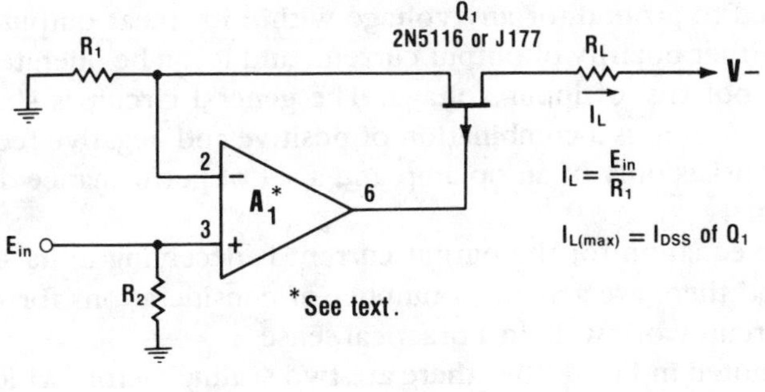


*(A) JFET.*

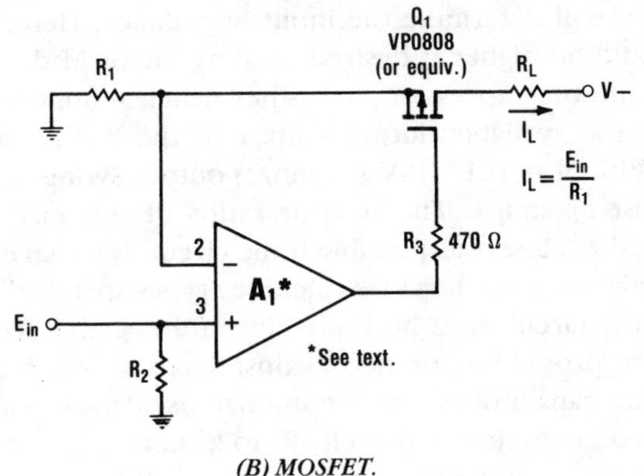


*(B) MOSFET.*

**Fig. 4-34. Negative-input FET VCCS.**

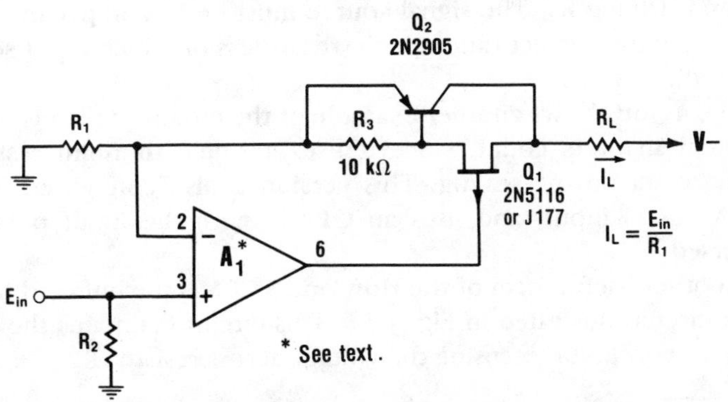


**Fig. 4-35. Negative-input buffered-FET VCCS.**

Howland circuit.* This type of current source will drive a load referred to ground (or any voltage within its linear output range) with either polarity of output current, and it can be operated from either polarity of input voltage. The general circuit is shown in Fig. 4-36. It uses a combination of positive and negative feedback, and consists only of an op amp and a set of performance-defining resistors.

The equation for the output current is deceiving in its simplicity, and there are actually a number of considerations for making this circuit work well, in a practical sense.

As noted in Fig. 4-36A, there are two scaling factors which govern operation. Resistor $R_2$ can be selected for the desired output current; then the scaling factor, K, can be used to set $R_1$, which essentially will determine the input impedance. Here, K is unity, but it could be higher if desired. Scaling factor M does not enter directly into the expression, but is nevertheless important. As M is lowered, the available output voltage at the $R_2$-$R_4$ node is maximized. With M = 0.1, 10-V (or more) output swings are available from most op amps. The resistor ratios of this circuit must be maintained as closely as possible if the circuit is to have a high output impedance, thus the close-tolerance types specified.

Since the circuit uses both positive and negative feedback, the user must provide assurance against latch-up or oscillation. A small-value capacitor, $C_f$, will minimize oscillation tendencies, as will the load capacitance from $R_2$-$R_4$ to ground.

The example shown in Fig. 4-36 uses an FET-input OPA111 op amp and delivers a 1-mA load current ($I_L$) with a +10-V input. The sense of input polarity can be reversed if desired, by grounding $R_1$ and driving $R_2$. The signal source must be low impedance, to guarantee the correct ratio match (regardless of which input sense is used).

Fig. 4-36B shows another example of the circuit, which is more efficient in parts count, since it uses a single thin-film resistor array to set up the scaling. This version is also configured for a ±5-V (max) input, and uses an OP-07, with the input polarity reversed.

Another useful form of the Howland VCCS is the buffered feedback circuit illustrated in Fig. 4-37. This circuit maintains the current, $I_L$, constant by sensing the voltage across resistor $R_5$.

---

* See references at the end of the chapter.

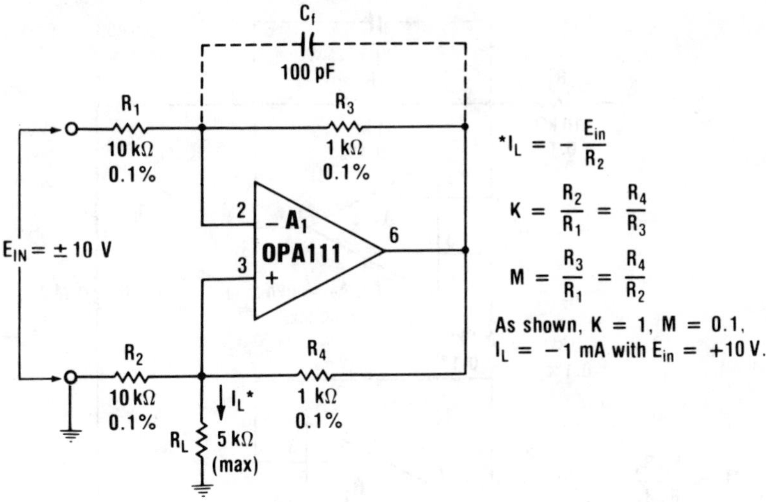


*(A) Using discrete resistors.*

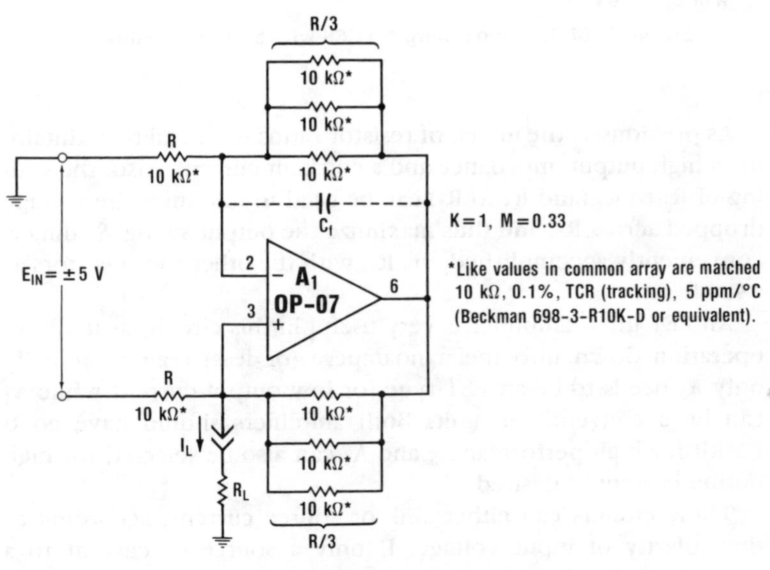


*(B) Using resistor network.*

**Fig. 4-36. General-purpose bidirectional-output VCCS.**

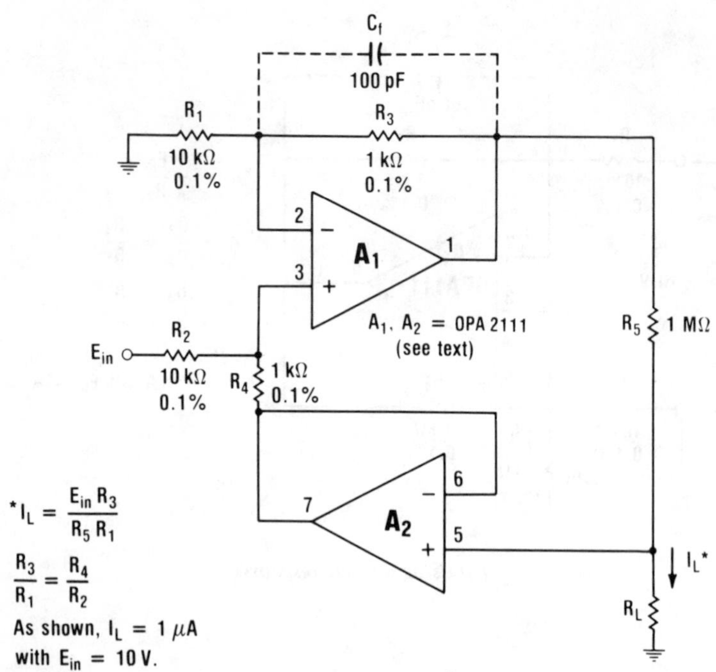


**Fig. 4-37. Bidirectional-output VCCS with buffered feedback.**

As previously, the match of resistor ratios is critical to maintaining a high output impedance and a constant current. Also, the scaling of $R_4$ to $R_2$ (and $R_3$ to $R_1$) can be used to minimize the voltage dropped across $R_5$, and thus maximize the output swing. Scaling is conveniently accomplished, via $R_5$, with the other four resistors as a stable set.

An FET-input amplifier is very useful in this circuit, as it allows operation down into the nanoampere (or less) region. Actually only $A_2$ needs to be an FET type for low output current while $A_1$ can be a conventional unit. Both amplifiers should have good CMRR for high performance, and $A_1$ can also be selected for high output current, if desired.

These circuits can either sink or source current, according to the polarity of input voltage. If only a source of current to a ground-referred load is needed, they can also be operated from a single positive supply, using a 3240, 358, TLC272, or other single-supply type.

## 4.4.5 Current Source With Integral Reference

An interesting and useful variation of current source is the type using an integral reference-voltage source, as shown in Fig. 4-38. In this circuit, the reference-voltage source "floats" and the output current drives a load referred to ground.

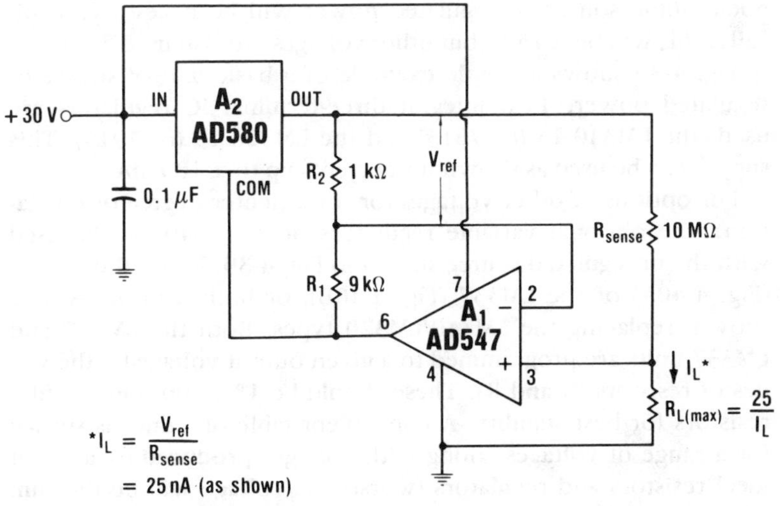


**Fig. 4-38. Current source with integral reference.**

Here, $A_1$ is a low-bias-current FET-input op amp with low offset voltage and drift. $A_2$ is a 2.5-V reference-source IC which provides a stable 2.5 V between its COM and OUT terminals. $R_1$ and $R_2$ divide the output of $A_2$ to 0.25 V, the voltage labeled $V_{ref}$.

Because of the zero input differential of the op amp, the circuit will adjust so that 0.25 V also appears across $R_{sense}$. This stable voltage determines the output current, according to the relations shown.

The control of very low currents is a problem in op-amp circuits because it implies a very-low-bias-current op amp, and a high-value resistor. The 25-pA bias current of the AD547 simplifies the current aspect of the problem, and dividing the reference down to 0.25 V limits the size of $R_{sense}$. A useful advantage of this circuit is that it can also very easily be trimmed, by making $R_1$-$R_2$ a potentiometer.

The main errors of this circuit will be due to the PSRR and CMRR of $A_1$-$A_2$. For the units specified, this will be on the order of 80 dB, or 0.01%.

### 4.4.6 A Basic Regulated Power Source

For most of the circuits within this chapter and the rest of the book, some source of regulated power will be necessary. Typically, this will be ±15 V, but other voltages are also used.

Fig. 4-39 shows a simple example of a basic ±15-V source of regulated power. Two integral three-terminal IC regulators are used, the LM340-15 (or 7815) and the LM320-15 (or 7915). This supply can be used as shown for currents up to ±100 mA.

For options of other voltages (or for a tighter degree of regulation), the optional variable regulators of Fig. 4-40 can be used with the unregulated source shown in Fig. 4-39. Either the LM317 (Fig. 4-40A) or the LM337 (Fig. 4-40B), or both, can be used as shown, replacing the LM340/LM320 types. Both the LM317 and LM337 units are programmed to a given output voltage by the values of resistors $R_1$ and $R_2$. These should be 1%, 100 ppm/°C film resistors for best stability. A convenient table of values is shown for a range of voltages, along with voltages produced by a set of ideal resistors and regulators (worst-case voltage will be the sum of resistor and regulator errors).

A list of references for further background information on references and regulators used for voltage and current circuitry is provided at the end of the chapter.

## 4.5 REFERENCES


1. Accardi, L. "Self-Stabilized Zener Insures Constant Current in Op-Amp Voltage Reference." *Electronic Design,* December 21, 1972.

2. __________. "Superstable Reference Voltage Source." *EDN,* October 1, 1971.

3. *Applications Manual for Computing Amplifiers,* 2nd. ed. Philbrick/Nexus Research, Dedham, MA, 1966, 1984.

4. *Applications of OEI Products.* Optical Electronics, Inc., Tucson, AZ, 1971.

5. Birman, P. *Power Supply Handbook.* Kepco, Inc., Flushing, NY.

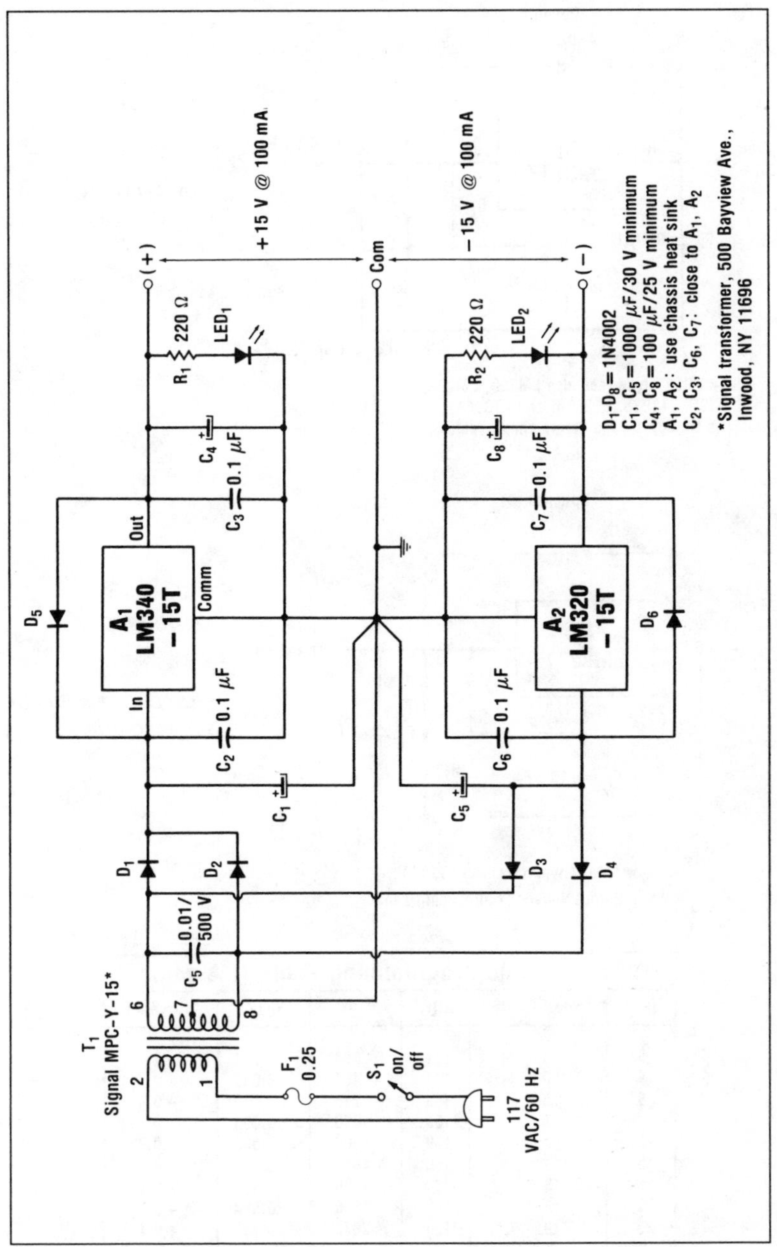


**Fig. 4-39. Basic ±15-volt source.**

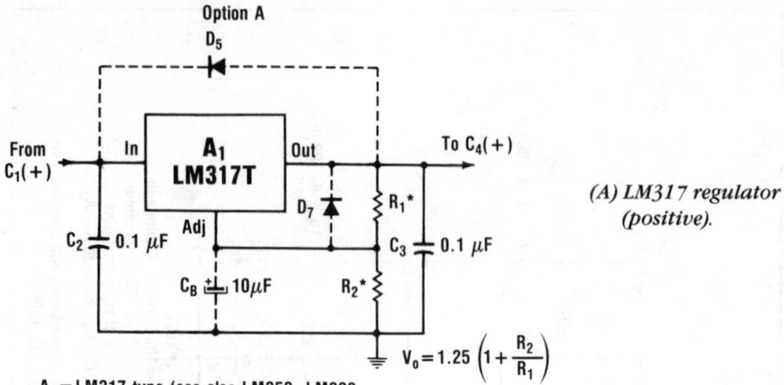


*(A) LM317 regulator (positive).*

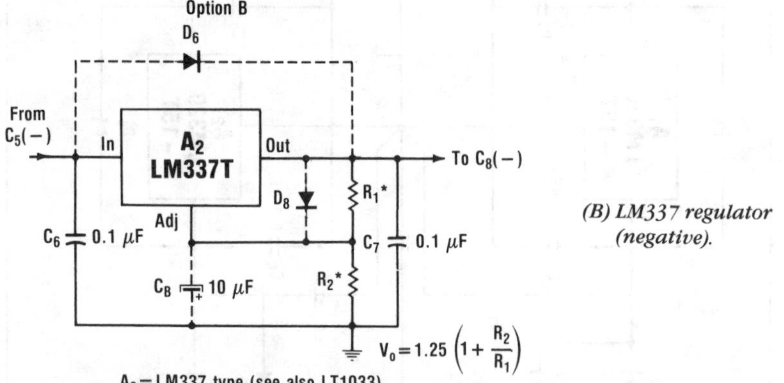


*(B) LM337 regulator (negative).*

***Output Voltage Programming Table (1% values)**

| $V_0$ (V) | ideal $R_2/R_1$ | $R_1(\Omega)$ | $R_2(\Omega)$ | $V_0$ (V) | Error (%) |
|---|---|---|---|---|---|
| 5 | 3 | 121 | 365 | 5.021 | +0.4 |
| 6 | 3.8 | 121 | 464 | 6.043 | +0.7 |
| 8 | 5.4 | 133 | 715 | 7.970 | −0.4 |
| 10 | 7 | 150 | 1050 | 10.000 | 0 |
| 12 | 8.6 | 150 | 1300 | 12.083 | +0.7 |
| 15 | 11 | 150 | 1650 | 15.000 | 0 |
| 18 | 13.4 | 133 | 1780 | 17.980 | −0.1 |
| 20 | 15 | 150 | 2260 | 20.083 | +0.4 |
| 25 | 19 | 150 | 2870 | 25.167 | +0.7 |

**Fig. 4-40. Variable voltage regulators.**

6. Brokaw, A. P. "A Simple Three-Terminal IC Bandgap Reference." *IEEE Journal of Solid State Circuits,* Vol. SC-9, No. 6, December 1974.
7. Clayton, G. B. *Operational Amplifiers.* Butterworth, Inc., Toronto, Ontario, Canada, 1971.
8. *DC Power Supply Handbook.* Hewlett-Packard Co., Palo Alto, CA, 1970.
9. Dobkin, R. C. "Don't Forget A/D Reference Stability." *EDN,* June 20, 1977.
10. ________. "High Stability Regulators." *National Semiconductor Linear Brief LB-15,* January 1971. National Semiconductor Corp., Santa Clara, CA.
11. ________. "IC Voltage Reference Has 1 PPM Per Degree Drift." *National Semiconductor Application Note AN-161,* June 1976. National Semiconductor Corp., Santa Clara, CA.
12. ________. "On Chip Heater Helps To Stabilize Monolithic Reference Zener." *Electronics,* September 16, 1976.
13. ________. "References For A/D Converters." *National Semiconductor Application Note AN-184,* July 1977. National Semiconductor Corp., Santa Clara, CA.
14. Giles, J. N. *Fairchild Semiconductor LIC Handbook.* Fairchild Semiconductor, Mountain View, CA, 1967.
15. Goldfarb, W. "Single-Supply Reference Source Uses Self-Regulated Zener." *Electronics,* June 7, 1973.
16. Gunn, J. "10 Volt (±1 mV) References." *Analog Dialogue,* Vol. 10, No. 1, 1976.
17. Harold, P. "Solid-State DC Voltage Standards Rival Standard Cell Performance." *EDN,* April 4, 1985.
18. Henderson, B., and Glasser, S. "Constant Voltage Diodes Are Getting Sharper." *Electronic Products,* March 18, 1974.
19. Henry, T. "Analysis and Design of the Op Amp Current Source." *Motorola Application Note AN-587,* October 1972. Motorola Semiconductor Products, Inc., Phoenix, AZ.
20. Jung, W. G. "Applications of the AD580." *Analog Dialogue,* Vol. 9. No. 2, June 1975.
21. ________. "IC Regulated Power." *The Audio Amateur,* April 1974.

22. __________. "Precision Reference Source Eliminates Power Supply Regulation Errors." *Electronic Design,* September 1, 1975.

23. __________. "Precision Reference Source Features Minimum Errors." *EDN,* August 5, 1976.

24. __________. "Programmable Voltage Reference Is Stable, Yet Simple." *EDN,* November 5, 1975.

25. __________. "Simple Regulator Circuits Provide Tracking Bipolar Reference Voltages." *Electronic Design,* July 5, 1976.

26. Kennedy, E. J. "Temperature Stable Supply Uses IC as Error Amplifier." *Electronic Design,* November 8, 1969.

27. Kesner, D. "Choose Regulator Op Amps With Care." *Electronic Design,* June 21, 1969.

28. __________. "Regulators Using Operational Amplifiers." *Motorola Application Note AN-480,* October 1970. Motorola Semiconductor Products, Inc., Phoenix, AZ.

29. Koep, K., "Solid State Voltage Standards." *Instruments and Control Systems,* October 1972.

30. Kuijk, K. E. "A Precision Reference Voltage Source." *IEEE Journal of Solid State Circuits,* Vol. SC-8, No. 3, June 1973.

31. Miller, W. "Single Amplifier Current Sources." *Analog Dialogue,* Vol. 1, No. 1, April 1967.

32. Nelson, C. "Monolithic Regulator Specifications." *Teletype Semiconductor Application Note,* 1971. Teledyne Semiconductor, Mountain View, CA.

33. Pouliot, D. "Build Precision Reference Sources." *Electronic Design,* July 5, 1976.

34. Ritmanich, W. "Bandgap Voltage Reference Applications." *Proceedings of EE Times IC Applications Seminar,* 1976.

35. Routh, W. S. "An Applications Guide for Operational Amplifiers." *National Semiconductor Application Note AN-20,* February 1969. National Semiconductor Corp., Santa Clara, CA.

36. Scott, M. "Circuit Building Blocks for Voltage Regulators." *Fairchild Application Note,* 1968. Fairchild Semiconductor, Mountain View, CA.

37. Sevastopoulos, N., and Moyer, J. "Micropower References Stays Stable." *EDN,* September 5, 1978.

38. Shah, M. J. "Stable Voltage Reference Uses Single Power Supply." *Electronics,* March 13, 1972.

39. Sheingold, D. H. "Impedance and Admittance Transformations Using Operational Amplifiers." *The Lightning Empiricist,* Vol. 12, No. 1, January 1964. Philbrick Researches, Inc., Dedham, MA.

40. Shirley, F. R., and Vanderlosk, L. "A Stable Voltage Reference Uses Only Six Components." *Electronic Design,* January 18, 1968.

41. Stitt, R. M. "Precision Voltage References Outgrow Stabilizing Ovens." *Electronics,* November 17, 1983.

42. Tobey, G. E., Graeme, J. G., and Huelsman, L. P. (eds.). *Operational Amplifiers—Design and Applications.* McGraw-Hill Book Co., New York, 1971.

43. Widlar, R. J. "New Development in IC Voltage Regulators." *IEEE Journal of Solid State Circuits,* Vol. SC-6, No. 1, February 1971.

44. Williams, J. "Don't Bypass the Voltage Reference That Best Suits Your Needs." *EDN,* October 5, 1977.

45. *Product Catalog, PRD Series Precision Reference Diodes,* CODI Corp., Fairlawn, NJ.

FIVE

# *Signal-Processing Circuits*

The broadest application area imaginable is one that is ideally suited to IC op amps. This is the general area of signal processing, where a circuit "operates" on an input signal to create a new form of output signal, altered in some manner from the original. Included in this area are such common functions as rectification, clipping, logarithmic or antilogarithmic conversions, phase shifting, filtering, and many others. The basic functions of dc and ac amplification are signal processing in a general sense, but since they are covered in a later chapter, they will not be detailed here except in special examples. Signal processing, as treated in this chapter, encompasses those "specialty" functions which, although common circuit techniques, do not fall into any other category.

## 5.1 THE PRECISION DIODE

A popular and useful signal-processing circuit is the precision-diode circuit, shown in a number of variations in Fig. 5-1. The first, in Fig. 5-1A, is basically a voltage follower with a diode inside the feedback loop to confine its operation to half of the input signal swing, thus accomplishing half-wave rectification. $D_1$ is the rectifier diode, which conducts on input potentials higher than zero, forming a unity-gain replica of the input during the

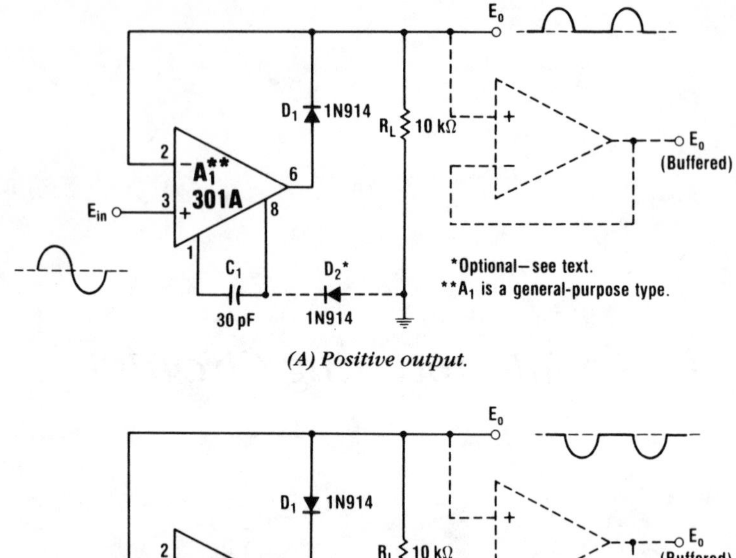


*(A) Positive output.*

*(B) Negative output.*

**Fig. 5-1. Precision-**

positive half cycle. During the negative half cycle, $D_1$ switches off and the output is zero.

The nonlinearity and temperature sensitivity of $D_1$ are not seen in the output waveform, because they are removed by feedback around the op amp. Furthermore, the circuit will *not* have a minimum forward-threshold voltage, as does a conventional diode, because the high amplifier gain will automatically adjust the voltage drive to the diode so that the rectified output is accurately maintained—even at millivolt input levels.

The limitations of the circuit are defined by the op amp used and are speed related. Since the feedback loop is broken when $D_1$ switches off, the amplifier alternates between half cycles of open- and closed-loop operation. The time it takes for $A_1$ to settle into a stable output condition after $D_1$ turns on is limited by the slew-

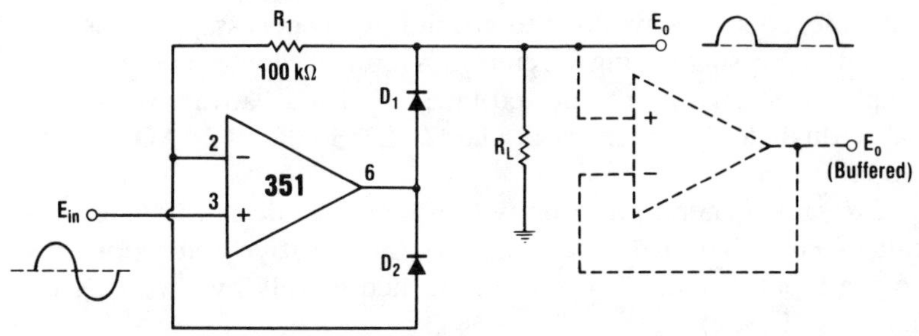


*(C) "Semiprecision" circuit—positive output.*

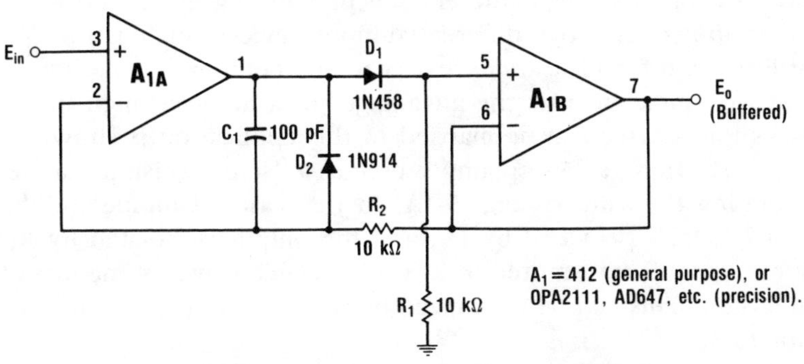


*(D) Precision buffered circuit—positive output.*

**diode circuits.**

ing rate and bandwidth of the amplifier, particularly since $A_1$ must be compensated for unity gain because it operates as a voltage follower when $D_1$ is on.

The 301A op amp is shown for this application because it can be clamped (optionally) at pin 8. This aids high-speed operation by eliminating internal saturation. Other unity-gain-stable op amps can also be used without clamping, but with some speed sacrifice. FET types are generally preferred because of their high slew rates, which will allow operation to well beyond 10 kHz. General-purpose bipolar types such as the 301A or 741 will operate in this circuit up to about 1 kHz with full output. Beyond this, a faster-slewing type should be used, such as one of the general-purpose FET types.

A lesser problem of the circuit is its high output impedance during the inactive period. The output and $D_1$ are floating during this time, so a resistive load to ground ($R_L$) is necessary if it is not provided by succeeding circuitry. A buffer can be added quite simply (as shown by the dotted lines), and is an attractive option for a dual device such as a TL082, LF353, LF412, AD712, or MC34082, etc.

The same general philosophy is also applicable to the sister circuit of Fig. 5-1B, arranged to rectify the negative-going portions of the input signal. This is accomplished merely by reversing $D_1$ (and $D_2$, if used).

A word of caution: Do not attempt to use this circuit with low-differential-input-voltage op amps such as the OP-27, 318, or 308, since the input-voltage differential is high during the period when the loop is open. High-differential-input devices, such as the 741 or FET-input families, are necessary to prevent input breakdown.

An example of how the advantages of a modern, high-speed, FET-input device can be married to this application is shown in Fig. 5-1C. Here, a 351 op amp is used as a "semiprecision" diode, operating the same as Fig. 5-1A. In this case, clamping for the "off" state is provided by $D_2$, and the output is isolated by $R_1$. Some signal is transmitted to $R_L$ via $R_1$, which causes some loss of precision. This can be minimized by maintaining $R_1$ large in relation to $R_L$.

When a dual FET-input device is available, the version shown in Fig. 5-1D is very attractive, as it allows high-precision operation. Here, the rectification is performed by $D_1$-$R_1$, with $R_1$ analogous to $R_L$ of the previous circuit. A second-stage voltage-follower buffer isolates the feedback through $D_2$-$R_2$, and allows precision unity-gain rectification. None of the components used are highly critical, and a general-purpose FET-input device, such as the 412, allows good performance up to 10 kHz (or more). The high-dc precision units listed allow accurate rectification down to millivolt levels.

For negative outputs, simply reverse the diodes. The diodes used should be low-leakage types—$D_1$, in particular.

## 5.2 PRECISION CLIPPERS

Such a basically useful circuit as the precision rectifier, described in Section 5.1, naturally has many variations. In this case, they are quite numerous and all are very useful.

### 5.2.1 Simplified Series Clipper

One very slight variation of the circuits in Fig. 5-1 turns them into precision clippers. This is shown in Fig. 5-2. The circuit works by varying the output base line during the period when diode $D_1$ is off. Since $D_1$ will conduct whenever the noninverting input is higher than the inverting input, it follows that the input threshold need not necessarily be ground (as in the circuits of Fig. 5-1). By varying the input reference, a clipper (biased rectifier) may be made. This is accomplished merely by returning $R_L$ to a voltage other than ground. In the circuit of Fig. 5-2A, which is a positive-peak clipper, the output will be a replica of those portions *lower* than the reference voltage. The circuit requires a low-impedance source for reference, particularly if clamp $D_2$ is used. Note that one-half of a dual device can suffice as a buffer for $V_{ref}$, if desired. For example, high-speed clipping can be accomplished using a 353, TL082, etc.

A negative-peak clipper can be implemented merely by reversing $D_1$ (and $D_2$, if used), as shown in Fig. 5-2B. Note that in both cases there is no restraint on the reference voltage as long as it is within the input range of $A_1$. The output will follow the input whenever the input exceeds $V_{ref}$. External loading on $R_L$ has the effect of reducing the reference voltage due to voltage division, so the circuit may require an output buffer amplifier. General-purpose duals, such as the 1458, 4558, or 358, are useful here.

### 5.2.2 Self-Buffered Series Clipper (Linear OR Gate)

In situations where the output impedance of the simple series clipper is a disadvantage, an alternate form of clipper is available (Fig. 5-3). In this circuit, the output is always in a low-impedance state because one feedback loop is always closed (around either $A_1$ or $A_2$). The two diodes perform a gating function for the input signals, allowing whichever input is greater to pass to the output; thus the terminology, "linear OR." If $E_{in}$ is greater than $V_{ref}$, $A_1$-$D_1$ will be on. When $E_{in}$ falls below $V_{ref}$, $A_1$ turns $D_1$ off, and $A_2$-$D_2$ provide the output signal, $V_{ref}$. The inputs are interchangeable, due to the complete symmetry of the circuit.

Because both amplifiers are switching as the input signal goes through threshold, the response time of this circuit is slowed considerably with general-purpose amplifiers, although they may be

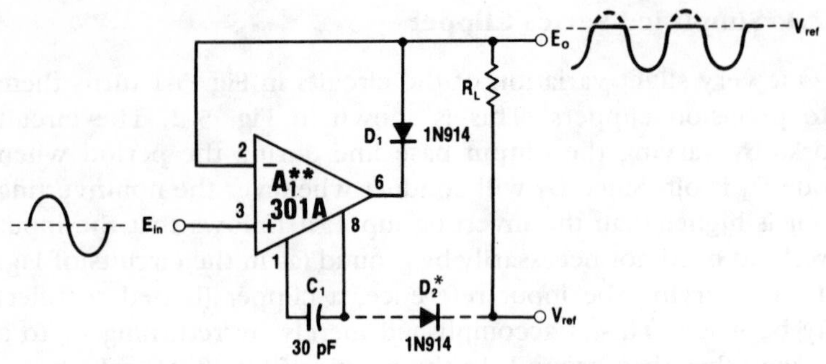


*(A) Positive-peak clipper.*

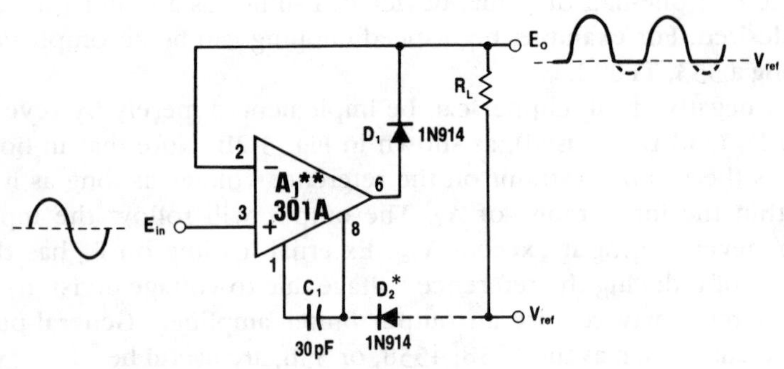


*(B) Negative-peak clipper.*

**Fig. 5-2. Precision clippers.**

useful below 100 Hz. The faster-slewing FET units called for in Fig. 5-3, however, give good performance to above 10 kHz. A positive clipping action can be obtained by reversing $D_1$ and $D_2$, and returning $R_1$ to +15 V. This circuit is useful when the clipping potential must be precise, when a low-impedance reference source is not available, or when a low-output impedance is necessary. It is capable of very low offset errors if a precision dual device and low-leakage diodes are used. Note also that it can be used with single-supply types of devices, such as the 3140, 358, LT1013, etc.

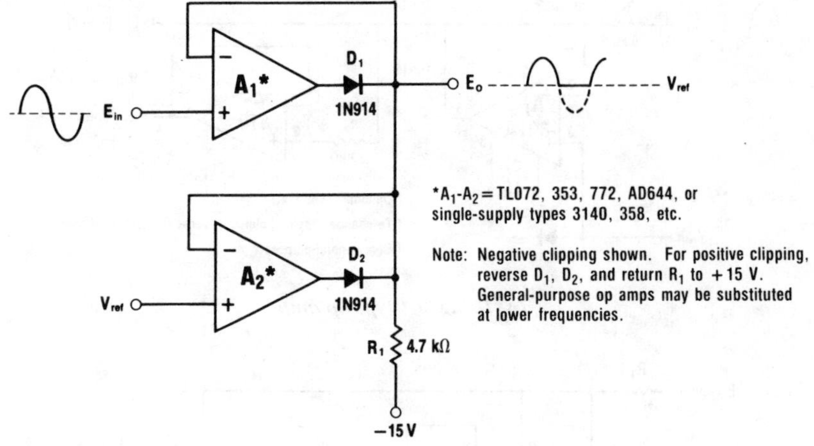


**Fig. 5-3. Self-buffered series clipper.**

### 5.2.3 Shunt Clipper

A shunt clipper can be formed by yet two more variations on the basic precision diode. This is diagrammed in Fig. 5-4. When, in Fig. 5-4A, $E_{in}$ is above the reference voltage applied to the noninverting input of $A_1$, $D_1$ will be reverse biased, and $E_{in}$ will pass unaffected through $R_1$ to the output. When a negative peak of $E_{in}$ exceeds $V_{ref}$, $A_1$ drives $D_1$ on and absorbs the input current from $R_1$, clamping the output at the $V_{ref}$ level.

The same considerations for speed apply to this circuit as to the basic one of Fig. 5-1A. An antisaturation clamp ($D_2$) is useful to minimize the clipper response time. If an antisaturation clamp is used (with 301 types), it requires a low source impedance from $V_{ref}$. A clipper of opposite polarity can be constructed by reversing $D_1$ (and $D_2$, if used).

Due to the high output impedance (essentially the value of $R_1$), this clipper will usually require a buffer amplifier. A 310 voltage follower, or follower-connected FET op amp, will work well with no additional sacrifice in operating speed. One of the previously mentioned FET duals will also work well here as $A_1$ and will allow much higher $R_1$ values, up to 1 MΩ.

A more efficient hookup is shown in Fig. 5-4B, using a dual op-amp device. This version offers a buffered output, with a mini-

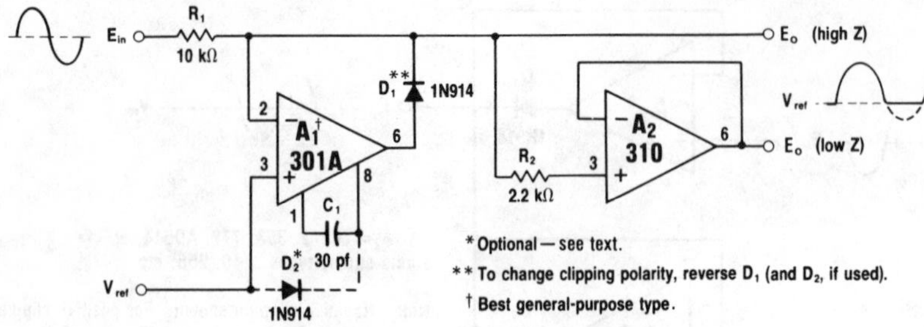


*(A) Using 301-type op amp.*

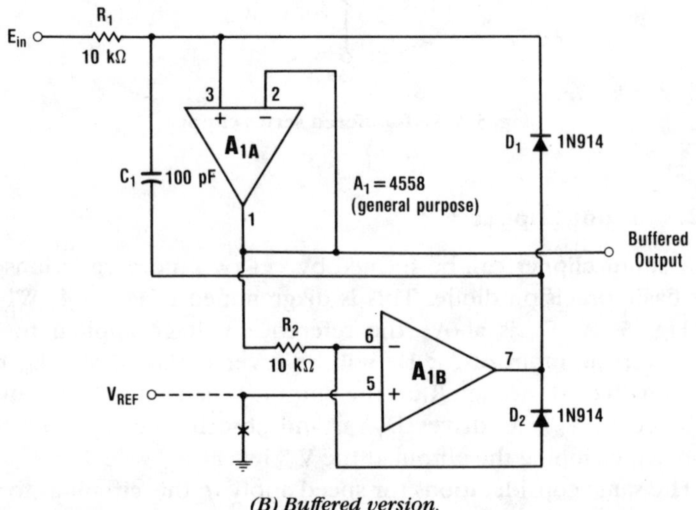


*(B) Buffered version.*

**Fig. 5-4. Shunt clipper.**

mum of complexity. Here $V_{ref}$ is shown grounded, but it can be a variable voltage if desired, even from a high-impedance source. No components are critical to operation, and general-purpose types are shown. As previously, reverse the diodes for opposite polarity, and use precision op amps and low-leakage diodes for best results.

## 5.3 DC RESTORER

Yet another useful variation of the ideal rectifier is the dc restorer or peak clamper of Fig. 5-5. This circuit charges $C_1$ to a voltage that clamps the input waveform peak at $V_{ref}$. When $E_{in}$ swings negative (below $V_{ref}$), $A_1$ drives $D_1$ on, forming a low-impedance path to charge $C_1$. After the peak has passed and $D_1$ is turned off again, the charge on $C_1$ is retained, since the only bleed paths are the input currents of $A_1$ and $A_2$. This gives the circuit a fast-attack, slow-decay response to input level changes. Faster recovery between level changes is accomplished by a shunt resistance to ground ($R_B$), with some sacrifice in clamping precision. As with the precision diode, this circuit is responsive into the millivolt region, due to the high gain of $A_1$ driving $D_1$. The circuit, as shown, clamps negative peaks to $V_{ref}$, but positive-peak clamping is possible by reversing $D_1$. The value of $C_1$ may be changed to accommodate different input rates and pulse widths.

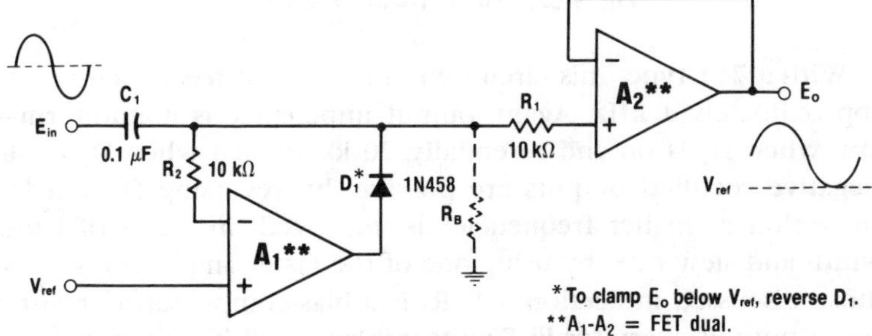


**Fig. 5-5. Dc restorer (peak clamper).**

Note that since $A_2$ sees a high source impedance, it should be a low-bias-current type (1 nA or less). A FET dual type will work well in this circuit. Note also that $E_{in}$ must be taken from a low source impedance to charge $C_1$, and $D_1$ should be a low-leakage unit for best results.

## 5.4 HALF-WAVE RECTIFIER

Fig. 5-6 is a half-wave rectifier that produces an inverted half-wave replica of the input signal. For negative input signals at $R_1$, the output is positive, forward biasing $D_1$ and closing a negative feedback loop through $R_2$. This produces an inverted gain of almost exactly one, since $R_1$ and $R_2$ are closely matched.

For positive input signals, the amplifier output is negative and $D_1$ is off. $D_2$ is on in this case, applying negative feedback to the summing point and clamping the op-amp output to −0.6 V. This clamped-output swing aids in minimizing response time, because it prevents amplifier saturation.

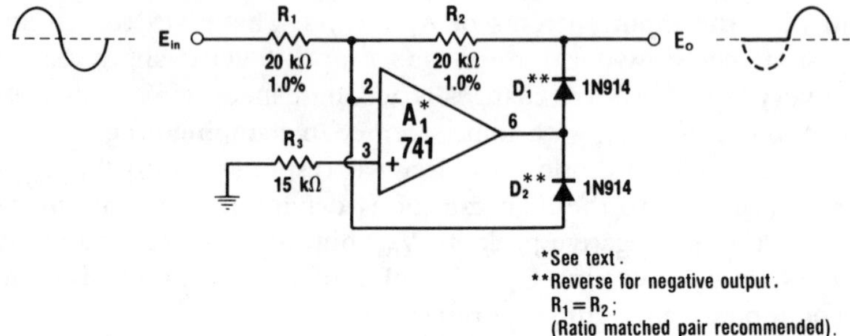


**Fig. 5-6. Basic half-wave rectifier.**

With a 741 type, this circuit will be useful at frequencies up to approximately 1 kHz. Again, output impedance is nonuniform—low when $D_1$ is on and essentially 20 kΩ (or $R_2$) when $D_1$ is off. Negative rectified outputs are possible by reversing $D_1$ and $D_2$. Operation at higher frequencies is improved, in terms of bandwidth and slew rate, by using one of the faster amplifiers, such as those discussed in Section 5.1. $R_3$ is a bias-compensation resistor and is not necessary for FET units (replace it with a direct connection to ground). Best dc precision is obtained with a precision bipolar op amp, such as an OP-07 or OP-27 type.

With this circuit, and many of its derivatives that follow, the absolute values of $R_1$ and $R_2$ are not important, as only their *ratio* determines the accuracy of rectification. Therefore, they can be a pair from a resistor array with close *matching* tolerances, which is typically more economical than low absolute tolerances, as well as being more package-efficient.

### 5.4.1 High-Speed Rectifier

Many applications demand a rectifier having a good fidelity at frequencies far above a few kilohertz. This is best achieved with fast-slewing, wide-bandwidth, amplifier configurations. The general-purpose 301A-type amplifier has an alternate form of compensation, termed *feedforward,* which is very useful in inverting applications; it allows rectification with low errors at frequencies up to 100 kHz. Since this circuit is an inverting configuration, feedforward compensation can be used as shown in Fig. 5-7. In this case, $C_1$ is the compensation capacitor. The circuit can be modified for a negative-going output by reversing $D_1$ and $D_2$. Due to the higher operating frequencies involved, oscillations are a potential problem. Therefore, a good layout and careful attention to power-supply bypassing are important considerations (see Chapter 3).

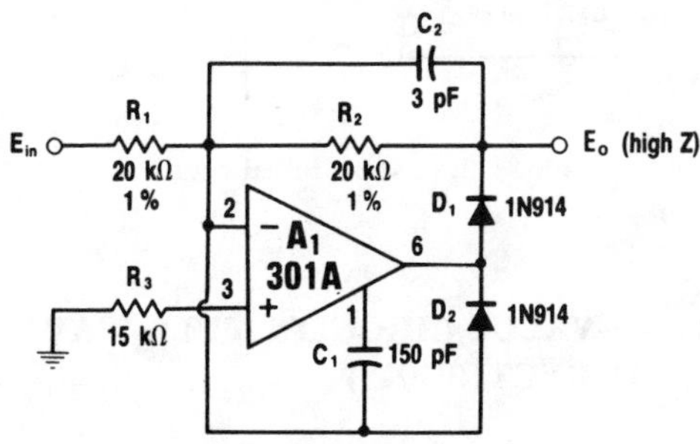


**Fig. 5-7. Fast half-wave rectifier.**

## 5.4.2 Rectifier With Buffered Output

Both of the previous rectifier circuits work in a similar fashion (except for response time), and they both have the slight problem of a nonuniform output impedance between alternations. For applications where this could be troublesome, the circuit of Fig. 5-8 can be used. This circuit contains a 310-type, high-speed, voltage follower ($A_2$) within the feedback loop of $A_1$; the function of $A_2$ is to maintain a low output impedance. During positive output alternations, $D_1$ and $R_4$ rectify the signal; $A_2$ "follows" this signal, returning feedback to the input through $R_2$. On opposite alternations, $D_1$ is off and the feedback loop of $A_2$ is closed through $D_2$. However, the output terminal is maintained at a low impedance by $A_2$, which sees a ground-level signal from $R_4$ during this period. As with previous circuits, opposite output polarity is obtainable by reversing $D_1$ and $D_2$. The buffering property of $A_2$ can be applied to either of the previous half-wave rectifier circuits—Fig. 5-6 or Fig. 5-7—with equally good results. Also, one of the previously mentioned wide-bandwidth, fast-slewing, FET dual units will work well in this circuit. Note that $R_1$-$R_2$ should be ratio matched; the remaining resistors are not critical.

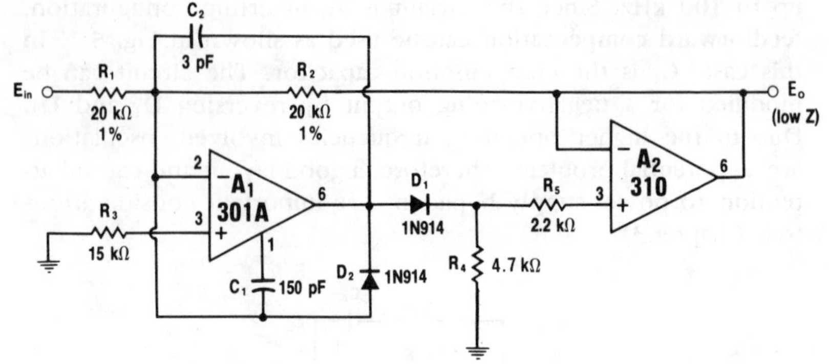


**Fig. 5-8. Fast buffered rectifier.**

## 5.5 ABSOLUTE-VALUE CIRCUITS (FULL-WAVE PRECISION RECTIFIERS)

An absolute-value circuit, or full-wave precision rectifier, can be implemented by summing the output of a half-wave rectifier and its input with the proper phase and amplitude relations. Such a circuit in its basic form is shown in Fig. 5-9. This circuit will be the starting point for a number of other absolute-value circuits which have evolved from this basic form.

In this circuit, $A_1$ is an inverting rectifier similar to that of Fig. 5-6. The output from $A_1$ is added to the original input signal in $A_2$ (a summing mixer), with the signal amplitude and phase relations shown. Negative alterations of $E_{in}$ result in no output at $E_1$ due to the rectification. $E_{in}$ feeds $A_2$ through a 20-kΩ resistor, and $E_1$ feeds $A_2$ through a 10-kΩ resistor. The net effect of this scaling is that, for equal amplitudes of $E_{in}$ and $E_1$, $E_1$ will provide twice as much current into the summing point. This fact is used to advantage here, as the negative alternation of $E_1$ produces twice the input current of that caused by the positive alternation of $E_{in}$. This causes a current of precisely half the amplitude which $E_1$ alone would generate due to the subtraction of $E_{in}$. It is the equivalent of having $E_1$ feed through a 20-kΩ input resistor and having $E_{in}$ nonexistent during this half cycle, and it results in a positive-going output at $A_2$. During negative alternations of $E_{in}$, $E_1$ is absent and $E_{in}$ produces the alternate positive output swing which, in summation, produces the desired full-wave rectified response. As before,

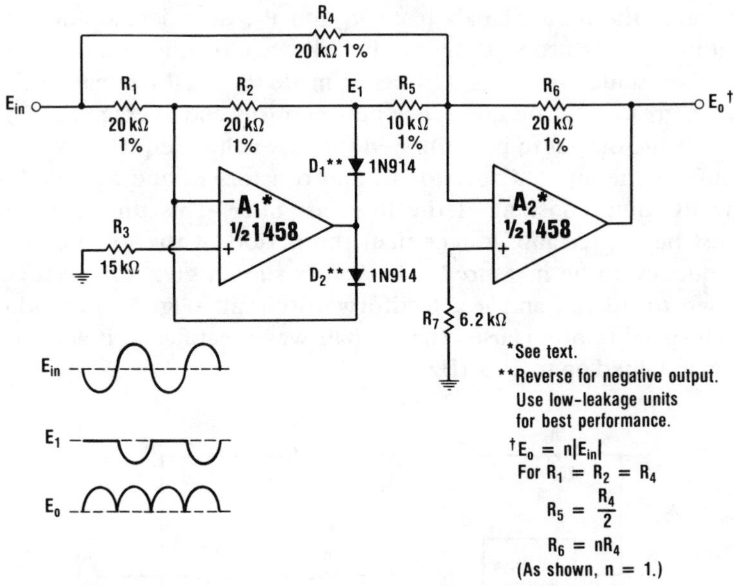


**Fig. 5-9. Basic absolute-value circuit (full-wave precision rectifier).**

operation with the opposite output polarity is possible by reversing $D_1$ and $D_2$.

The general-purpose dual-741 type indicated can be replaced with higher-speed units or FET types, as appropriate. The relationships between resistors for proper circuit operation are noted in the illustration, and may be satisfied best by a single network. Note that resistor $R_6$ can be used as an overall gain trim, or for scaling to net gains of ($n$) other than unity, as is shown by the equations. If impedances are raised, FET input units, such as the 3240, TL082, etc., are effective. With high impedances, low-leakage diodes are suggested.

### 5.5.1 Filtered Rectifiers

The previous full-wave rectifier can, with a minimum of complexity, be converted to a version having a dc-averaged output, as shown in Fig. 5-10. This circuit is nearly identical to the full-wave circuit described before; any one of the half-wave circuits already described (Figs. 5-6, 5-7, and 5-8) can be used as the first stage. The addition of $C_1$ across $R_3$-$R_5$ converts $A_1$ to a low-pass filter that averages the input signals from $R_1$ and $R_2$, thus developing a dc output voltage proportional to the full-wave rectified input. Resistor $R_5$ is added in series with $R_3$ to make the total feedback resistance greater by the difference between rms and average ac. This allows the output to be calibrated for a dc voltage equal to the rms value of the input. Capacitor $C_1$ and resistors $R_3$ and $R_5$ establish the RC time constant of the low-pass filter. This time constant must be appreciably longer than the period of the lowest input frequency to be measured. The values shown give 1% accuracy down to 20 Hz, and if a feedforward circuit (Fig. 5-7) or other high-speed option is used for the half-wave rectifier, this accuracy is maintained up to 100 kHz.

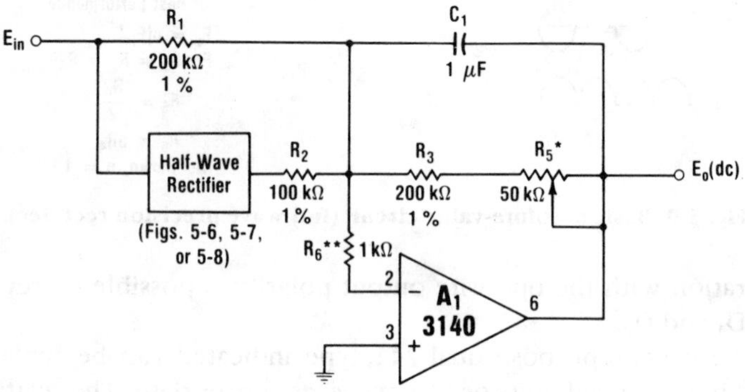


**Fig. 5-10. Filtered full-wave rectifier.**

The circuit can be used with either polarity of input rectifier, observing the appropriate polarity of $C_1$ (if it is an electrolytic type). Note that with a FET-input type for $A_1$, relatively high resistance values can be used without degradation, which allows $C_1$ to be proportionally smaller. This also allows $C_1$ to be a nonpolarized unit, with better relative results. A general-purpose FET unit is shown; for highest dc precision, an OPA111, AD547, or OP-41 would be useful.

The remaining circuits of this section illustrate alternate means of achieving the absolute-value function, with emphasis on such areas as simplicity, number of precision components, accuracy, etc. In general, almost all of them provide an output of the form

$n|E_{in}|$, where $n$ is either positive or negative unity, but can also be a scaler in some cases. The circuits as shown will be in a form where $n$ is positive, providing a positive-going rectified output. Unless stated otherwise, they can be arranged for negative outputs by reversing the diode(s) used.

Inasmuch as there has been considerable discussion in the literature on precision rectifiers, the reader will find the bibliographic references at the end of the chapter useful toward more insight on these techniques.

### 5.5.2 Equal-Value Resistance, Absolute-Value Circuit

The circuit of Fig. 5-11 is a form originated by Lloyd* to counter the unequal resistor values necessitated in the circuit of Fig. 5-9. In the Fig. 5-11 circuit, all resistor values are exactly equal, for an $n$ of unity. Although there is no reduction in the number of values, they can be economically realized by a common array. Scaling is accomplished by adjustment of $R_1$; as shown here, $n$ is 2.

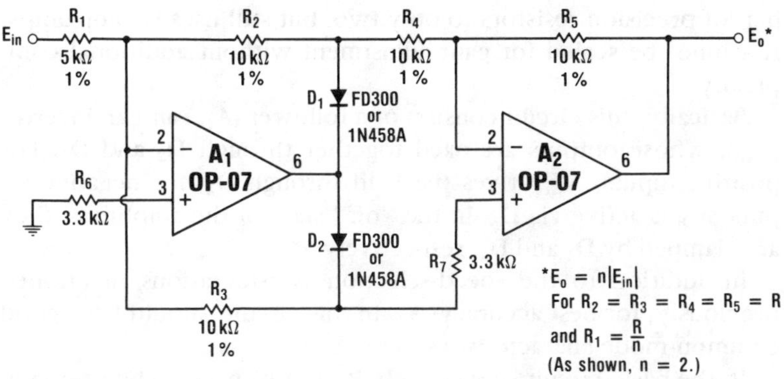


**Fig. 5-11. Equal-value resistance, absolute-value circuit.**

Op-amp selection in this example is for maximum low-frequency and dc accuracy. The OP-07 devices have input offsets of 150 $\mu$V or less, depending on the grade used, which optimize the dynamic range. The low bias currents of 2 nA (typical) minimize the

---

* See reference 19 at the end of the chapter.

requirement for matched dc source resistances; however, the compensation values are still shown. Low-leakage diodes must also be used (such as those shown), otherwise these advantages are lost. An OP-27 can also be substituted directly, with excellent dc accuracy and greater speed. Of course, the dual versions of these units, the OP-207 and OP-227, are also useful.

Speed is not a virtue of the circuit as shown, and it is only accurate to about 2 kHz. Better high-frequency (as well as excellent dc) performance can be obtained with high SR and GBP devices, such as the 356 types, the OPA-606, the AD711 or AD547, or the LTC1022 FET devices, or similar duals. These comments are basically true for all of the absolute-value circuits illustrated here, so it need not be repeated and the reader is encouraged to optimize them accordingly.

### 5.5.3 Improved Absolute-Value Circuit

Fig. 5-12 shows a circuit of a form described originally by Smither.* It is improved in the sense that it minimizes the number of precision resistors to only two, but still uses two op amps. It cannot be scaled for gain adjustment without additional complexity.

Basically, this circuit consists of a follower ($A_{1A}$) and an inverter ($A_{1B}$), whose outputs are ORed together through $D_2$ and $D_3$. For positive inputs, $A_{1A}$ drives the load through $D_2$; for negative inputs, $A_{1B}$ is active via $D_3$. In the "off" state of the amplifiers, they are clamped by $D_1$ and $D_4$, respectively.

In addition to the speed-selection considerations mentioned previously, for best accuracy, $A_1$, in this circuit, should have good common-mode characteristics.

In the basic circuit shown, only $R_1$ and $R_2$ need to be precision types, since $R_3$ does not enter into the gain expression. An alternate form of the circuit trades one more precision resistor for increased input impedance.

In the standard circuit shown, the input impedance is set at a constant value by $R_1$, which may not be a problem if the source impedance is sufficiently low. However, for high-impedance sources, the circuit can be modified simply by connecting $R_1$

* See reference 24 at the end of the chapter.

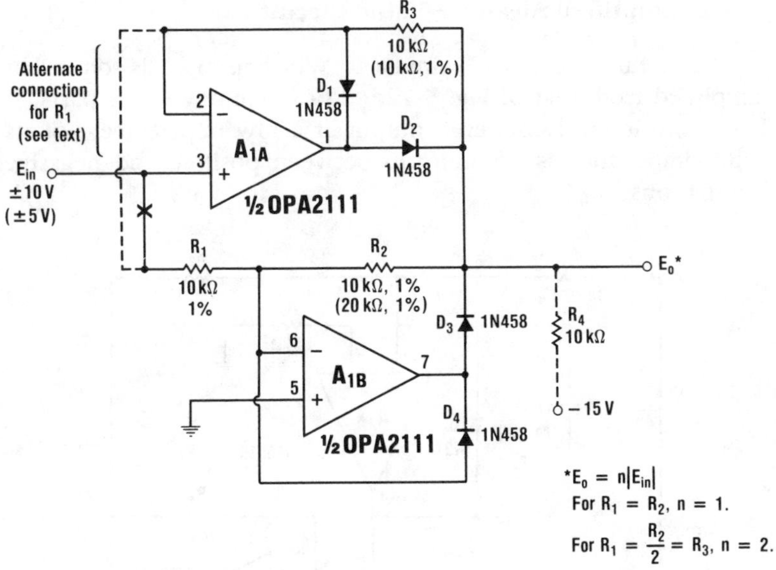


**Fig. 5-12. Improved absolute-value circuit.**

with $D_1$-$R_3$, as shown by the dashed line. This causes $A_{1A}$, rather than the source, to drive $R_1$, and the source sees only the low bias current of $A_{1A}$.

For this alternate form, the gain characteristics are no longer so simply defined. For positive inputs ($A_{1A}$ active), the gain is $1 + R_3/R_1$. For negative inputs ($A_{1B}$ active), the gain is $R_2/R_1$. To make these ratios the same, $R_2$ must be twice $R_1$, thus the 20-kΩ alternate value shown in Fig. 5-12. All three resistors should be precision types, or, preferably, ratio-matched. Note that for this alternate form, $n = 2$, so the maximum input is ±5 V.

For either form of this circuit, a pull-down resistor ($R_4$) may be found useful at high frequencies. This is an option, however, and the real necessity is governed to a large degree by the characteristics of the load.

A precision FET-input type is used as shown for the amplifiers, along with low-leakage diodes. If less precision is required, a more general-purpose unit can be used, along with conventional diodes. When this circuit is used in the alternate form, with the OPA2111, input current will be, typically, 2 pA.

### 5.5.4 Simplified Absolute-Value Circuit

The circuit of Fig. 5-13, similar to Wincentsen's,* is somewhat simplified from that of Fig. 5-12, and can employ fewer parts. It has a continuously buffered output at a low impedance, but its input impedance is nonuniform between positive and negative alternations.

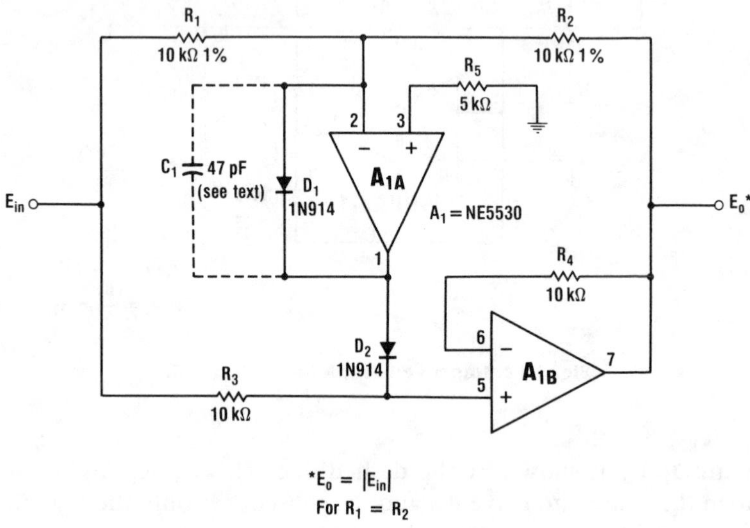


**Fig. 5-13. Simplified absolute-value circuit.**

For positive inputs, $A_{1A}$ and $D_2$ are inactive, and $A_{1B}$ operates as a simple follower to the output. For negative inputs, $A_{1A}$ becomes active, driving $A_{1B}$ through $D_2$, and with a gain defined by $R_2/R_1$. For positive inputs, the input impedance is $R_1$, while for negative inputs, it becomes $R_1$ in parallel with $R_3/2$. For best accuracy, then, the driving source impedance should be low, to prevent nonuniform loading errors. For best gain accuracy, $R_1$ and $R_2$ should be ratio-matched.

It should be noted that this circuit employs $A_{1A}$ and $A_{1B}$ in cascade, with overall feedback for negative inputs. Because of this factor, dynamic stability is a potential problem, and the circuit may oscillate with some op amps. $C_1$ is employed to control the

* See reference 32 at the end of the chapter.

effective circuit bandwidth to minimize this problem, and should be adjusted to suit the particular op amps used, if different than those indicated.

The 5530 is a high-slew-rate device, which makes this circuit useful up to 100 kHz. For these (or other bipolar input-stage) devices, bias-compensation resistors $R_4$ and $R_5$ are used; they are not necessary for FET devices and can be replaced by "shorts." FET-input units which can perform well in this circuit are the AD712, the MC34082, and a pair of OPA606s or LTC1022s.

### 5.5.5 Single-Amplifier Absolute-Value Circuit

A number of related circuits, which are used to achieve the absolute-value function, can be realized with only a single op amp. Since, for economy, many will want to use only a single amplifier, three variations on this theme are shown in Fig. 5-14. It is interesting that these circuits, while very simple, are still capable of very high precision if certain operating conditions are met. Virtually all of these circuits depend upon resistor ratios for their operation. While discrete sets of low-tolerance resistances can be used, ratio-matched array types are again preferred (and will likely be more economical).

The first of the circuits is shown in Fig. 5-14A, and is called a "T" configuration because of the T arrangement of scaling resistors $R_1$, $R_2$, and $R_3$. In this circuit, $R_2$ and $R_3$ set the gain for positive inputs, and $R_1$-$R_2$ for negative inputs. Note that the load impedance *must* be high in regard to the resistances specified to preserve accuracy in the positive input range when $R_2$ and $R_3$ serve as a simple input attenuator and $D_1$ is off. During negative inputs, $D_1$ is on and the output is buffered.

As can be noted from the scaling relations, *n* must always be less than unity for this circuit (as well as those to follow). The problem of the unbuffered output and signal-level loss can be optionally corrected by a buffer, as shown by the dashed lines. The other major constraint on the circuit is a nonuniform input impedance, which is $R_2$ for negative input swings and $R_2 + R_3$ for positive swings.

The version shown here is optimized for dc accuracy and uses the LM11, a low offset-voltage and bias-current op amp. It also uses low-leakage rectifier diodes, and an (optional) offset-null potentiometer. The offset null should only be used where necessary; the preferable method of offset control is to use a specified low-offset device.

Other precision op amps useful here are the OP-07 and OP-27, the LT1012, and such FET types as the OPA111 or AD547. Note that if a buffered output is desired, a precision low-offset dual FET device will be useful.

The "Pi" configuration circuit shown in Fig. 5-14B is very similar in operating principle, but arranges the scaling resistances in the form of a pi attenuator. In this form, $R_3$ and $R_2$ set the gain for positive inputs, while $R_2$ and $R_1$ do the same for negative inputs. A similar set of restraints on output impedance and input loading is true for this circuit, and *n* must be scaled less than unity. Of course, it can also be buffered at the output, at the cost of an additional amplifier. The circuit example for this version uses a general-purpose op amp and two diodes, and it has a more modest dc performance.

The "ladder" circuit configuration of Fig. 5-14C uses the resistances in a series-connected ladder and develops the output across

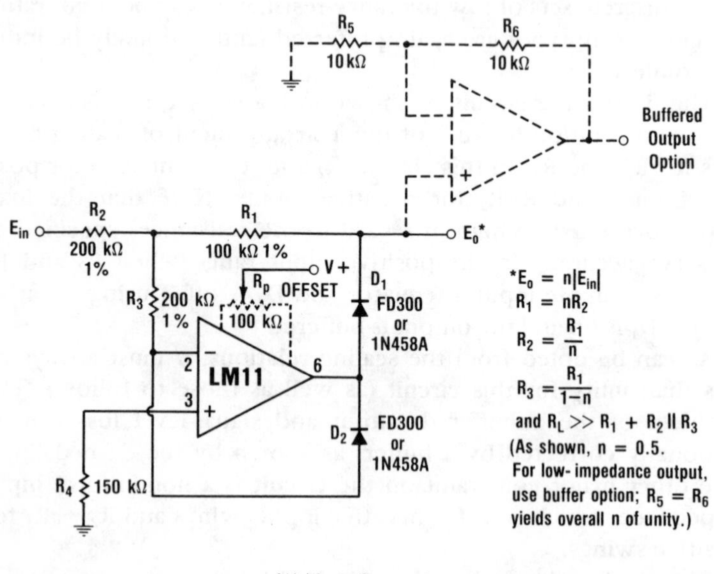


*(A) T configuration.*

**Fig. 5-14. Single-amplifier**

$R_3$. Note that $R_3$ may, in fact, be a *load* resistance; however, if this is to be the case, it should be a precise and stable type. This is simply because the scaling relations depend on this resistor, as well as on $R_1$-$R_2$.

Here, $R_1$-$R_2$-$R_3$ set the gain for positive inputs, while $R_1$ and $R_2$ do the same for negative inputs. Note that unlike the previous two circuits, there is no feedback returned to the summing point for

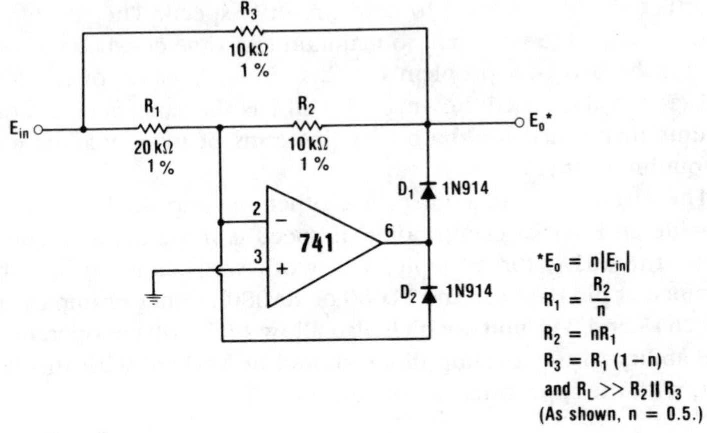


*(B) Pi configuration.*

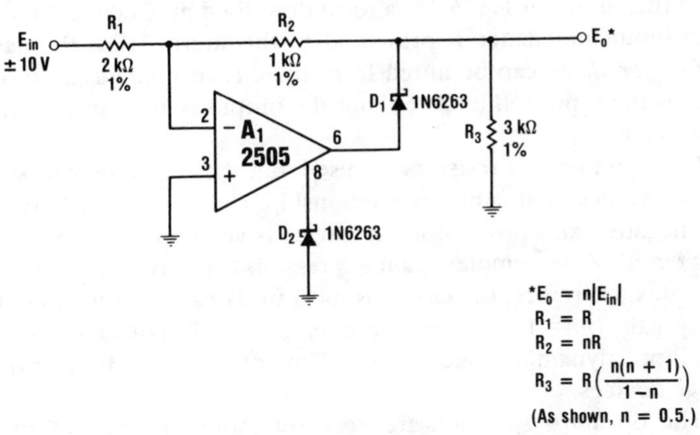


*(C) Ladder configuration.*

**absolute-value circuits.**

the input state when $D_1$ is off; therefore, $A_1$ operates open-loop. This would cause saturation of $A_1$ and subsequent recovery-time problems, were it not for $D_2$, a diode used to clamp the swing of $A_1$ in the off state. Also, because $A_1$ is allowed to see a large differential input swing during this period, it must have a high differential input rating.

The circuit example shown is optimized for speed using the 2505 (or 2605), both which are high GBP and SR op amps. Also, Schottky diodes are used to help preserve speed. The resistance values are held low as well, to maintain low time constants. Loading can be less of a problem for this circuit because of the low resistance values used; however, it still has the same general constraints mentioned for the others (in terms of input loading and output impedance).

The circuit can, of course, use other op-amp devices if they provide an internal clamp point. If speed is not a major requirement, the 301A, for example, will work well, as it can also be clamped at pin 8 (as can the TL080 or CA080). Other examples are the 2645 and 344 units, which also allow high-voltage operation. The ability to use a clamp diode should be verified with this circuit, with the appropriate amplifier.

### 5.5.6 Unity-Gain Self-Buffered Absolute-Value Circuit

In the circuit of Fig. 5-15, a form described by Graeme,* a very high input impedance is presented to the source, thus the name *self-buffered.* As can be noted from the circuit configuration, not only is the input self-buffered, but the output is also continuously buffered by $A_{1B}$.

Four gain-scaling resistors are used, but they are only necessary in two values, with a binary relationship. Gain switching between the negative and positive input ranges is accomplished by $A_{1A}$ in this circuit. The complete gain expression for both ranges is quite complex; therefore, the circuit is most practical as shown here for unity gain ($n = 1$). Due to the common-mode voltages applied, the input dynamic range is limited to $\pm 5$ V, for $\pm 10$-V-output-range devices.

The op amps specified are very-low-input-current FET units,

---

* See reference 10 at the end of the chapter.

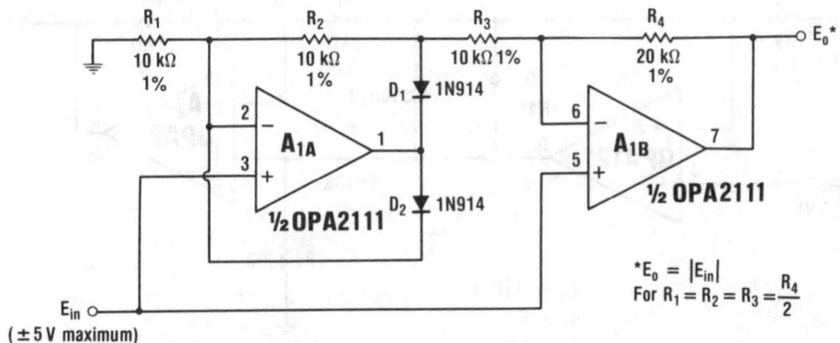


**Fig. 5-15. Unity-gain self-buffered absolute-value circuit.**

and therefore present very low loading to the source—4 pA or less (at 25°C). Bipolar-input amplifiers that could also be used include a pair of OPA27s ($I_{ib}$ = 15 nA typical). All the amplifiers mentioned are useful up to several kilohertz, but a pair of OPA606s can operate to over 100 kHz.

### 5.5.7 Programmable-Gain Self-Buffered Absolute-Value Circuit

The circuit of Fig. 5-16 is also a self-buffered type, of a form also attributed to Graeme.* This circuit requires only three precision resistors, and has the virtue that its gain can be easily programmed by a suitable choice of values.

Here, $A_1$ performs the function of input buffering and, in conjunction with $D_1$ and $D_2$, gain switching. For positive signal inputs, $D_2$ is on and a follower-with-gain loop is closed around $A_2$ and $A_1$. For negative inputs, $D_1$ is on and $A_1$ drives $A_2$ as an inverting amplifier.

The gain, *n,* may be set up for various values greater than unity, with a relative degree of freedom. In the example shown, *n* is set equal to 2, a scaler which results in convenient ratios of resistances, realizable from a single common array.

In the example shown, a FET-input amplifier is specified for $A_1$, which reduces the input bias current to 1 pA or less. $A_2$ need not have such low bias currents, as it is driven from $A_1$; thus, a bipolar unit can be employed—the OPA27. Of course, if ultralow bias

* See reference 8 at the end of the chapter.

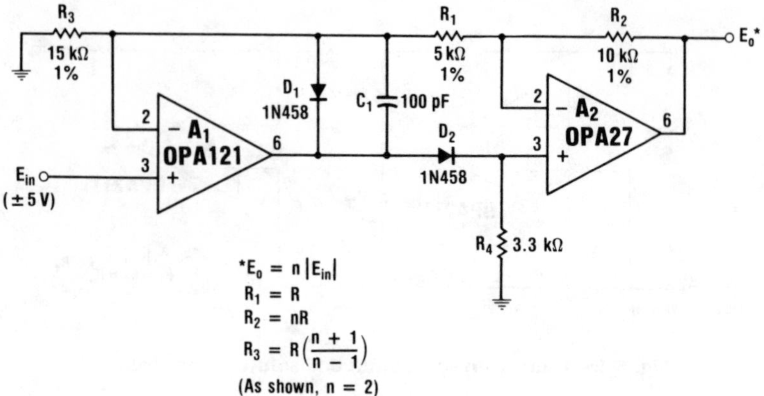


**Fig. 5-16. Programmable-gain self-buffered absolute-value circuit.**

current is not a major criterion, a bipolar type can be used for $A_1$ as well. As shown, the dc errors of this configuration are very low as both $A_1$ and $A_2$ are very-low-offset devices.

Since $A_1$ and $A_2$ operate in cascade for positive inputs, capacitance $C_1$ is suggested to control potential oscillations. It should be optimized in accordance with the specific gain and amplifier types used.

### 5.5.8 Simplified Unity-Gain Self-Buffered Absolute-Value Circuit

The circuit of Fig. 5-17* has the same self-buffered properties as the previous two examples, but it trades circuit flexibility for simplicity. It is limited to gains of only unity, but, as such, requires only one precision-resistor pair, $R_1$ and $R_2$.

The 4558, a general-purpose dual-amplifier type, is illustrated. This unit will suffice up to several kilohertz for 10-V outputs, and has reasonable dc accuracy also. Other suitable dual devices are the 4559, OP-14, LT1013, and 358, or you can use one of the dual FET types, such as the 3240, MC34082, TL082, 353, 412, or 772. The FET units will extend the frequency range to nearly 100 kHz; precision FET duals such as the OPA2111, the AD647, or the AD712 allow the best performance in both dc and ac senses.

---

* See reference 11 at the end of the chapter.

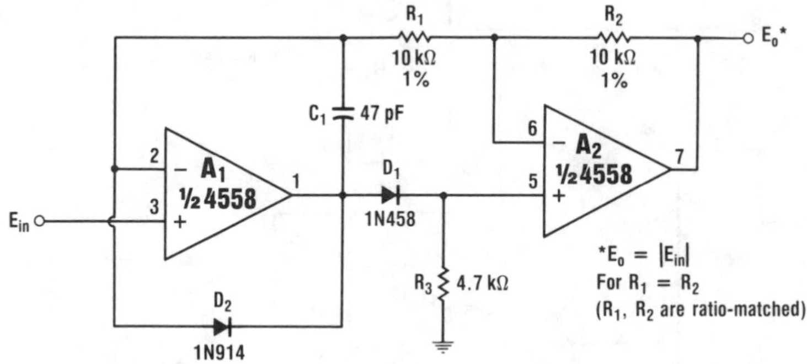


**Fig. 5-17. Simplified unity-gain self-buffered absolute-value circuit.**

## 5.5.9 "Diodeless" Absolute-Value Circuit

The circuit of Fig. 5-18 is a type illustrated by Dobkin,* and is unique in that it requires no rectifier diodes for its operation, and uses a pair of general-purpose op amps.

This circuit takes advantage of the fact that internally the 301A op amp can sink larger currents at the pin-8 node than it can source. Thus, when two devices such as $A_1$-$A_2$ have these pins wired together, the device in which pin 8 is the lower of the two will command control.

In this circuit, $A_1$ is connected as a follower and $A_2$ as a unity-gain inverter, and feedback is taken from the output of *one* device. For negative inputs, $A_1$ is active; for positive inputs, $A_2$ is active, and the output is equal to $-|E_{in}|$.

Note that due to the unique nature of this type of rectification, it is not possible to obtain positive range outputs (unless an inverting buffer is added, of course). It should also be noted that the input impedance is nonuniform; it is $R_1$ for positive inputs and a very high value for negative inputs.

## 5.5.10 Single-Supply Absolute-Value Circuits

It is a somewhat surprising, yet very useful, fact that a variety of absolute-value/precision-rectifier circuits can be implemented

* See reference 3 at the end of the chapter.

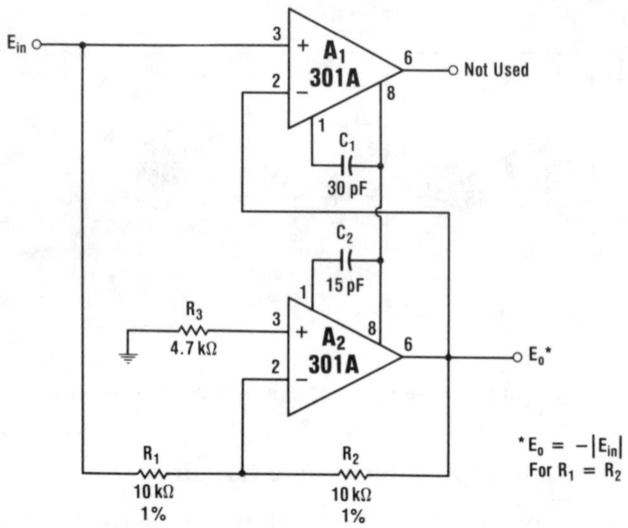


**Fig. 5-18. "Diodeless" absolute-value circuit.**

with certain op amps, using only a single power supply. Obviously, such circuits have great value in battery-powered or other simple-powered instruments, so a few of them are presented here.

A very simple single-supply absolute-value circuit is shown in Fig. 5-19. This circuit was derived from one of the earliest absolute-value circuits, which was designed back in the era of vacuum-tube op amps.*

In this circuit, $A_1$ is an inverting unity-gain amplifier for negative inputs when $D_1$ is off. For positive inputs, $D_1$ turns on and couples the input to the noninverting (+) input of $A_1$, and the circuit acts as a follower. A flaw in the operation of the circuit is the voltage drop across $D_1$. When on, $D_1$ introduces an error for positive input signals. This error can be minimized somewhat by maintaining $R_3$ at a large value and trimming it out for near full-scale inputs with $R_4$. However, because $D_1$ is *not* an ideal rectifier, the circuit does have appreciable error at low levels and should best be used at high input levels (±5 V to ±10 V).

The key to single-supply operation is the fact that $A_1$ can operate with the (+) and (−) inputs at ground potential (V−, or pin-4 potential). The 3140, 3160, and 3130 are FET-input devices which

* See reference 17 at the end of the chapter.

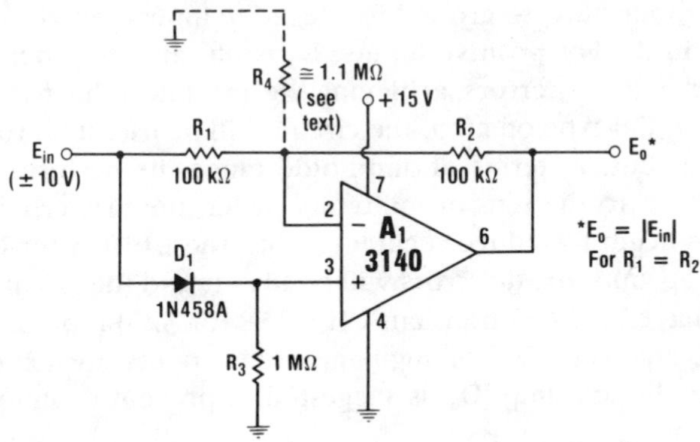


**Fig. 5-19. Simple single-supply absolute-value circuit.**

can accomplish this feat. Bipolar-input types are the 358 and 324 families, the 759, the LM10, the OP-20, and the OP-21.

A circuit that is very well suited to battery operation is shown in Fig. 5-20. This circuit takes advantage of the very low (1 mA) supply current of the LT1013 or the 358 op amp.

In this circuit, $A_1$ performs as an inverter for negative inputs, as before. However, in this circuit, $D_1$ and $A_{1B}$ clamp the noninvert-

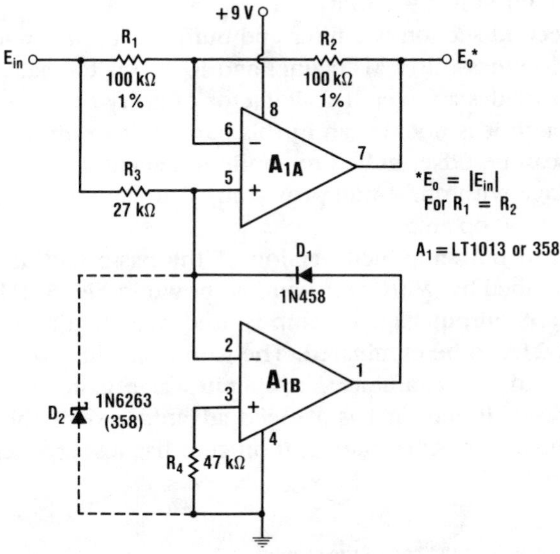


**Fig. 5-20. Battery-powered absolute-value circuit.**

ing (+) input of $A_1$ to ground for negative inputs, by virtue of the current in $R_3$. For positive inputs, $D_1$ is off, and $A_{1A}$ operates as a follower with low errors, achieving the absolute-value function.*

With a 358-type op amp, the circuit will be useful up to several kilohertz. But, in terms of amplitude range, its performance extends down to the tens of millivolts, when the bias errors of the op amps begin to reduce accuracy. With the LT1013 (or LT1014) type of op amp, the dc errors will be lower, and the circuit will be well behaved for (−) transients. If a 358 (or 324) type of op amp is used, and very fast or high-amplitude inputs are expected, a Schottky-diode clamp ($D_2$) is suggested to prevent common-mode problems.

The circuit of Fig. 5-21A is capable of even greater accuracy than the two previous ones, if certain conditions are fulfilled. It employs a dual-FET amplifier, the 3240, and has several features which can enhance its usefulness.

The $A_1$ stage of this circuit can be recognized as a ladder-type single-amplifier absolute-value circuit (as is Fig. 5-14C) arranged to employ a single-supply op amp, similar to the operation described by H. A. Wittlinger in the RCA CA3140 data sheet ("Applications"). In this section of the circuit, $n$ is 0.5; thus, an absolute value of +4 V peak is developed across $R_3$ for a ±8-V input to $R_1$. Maximum input is ±9 V to $R_1$.

The second section is a filter and buffer amplifier, where $R_6$-$C_1$ provide low-pass filter averaging, and $R_4$-$R_5$ set the gain of $A_{1B}$ to 2. This provides an overall scale factor of unity ($n$ times $m$ equals 1). Although it is not shown in this particular example, the 3240 op amp can be offset-nulled for optimum accuracy (using the 14-pin package with offset-null provision). The circuit will also work with the 3260 op amp.

A related but simplified version of the basic rectifier has also been described by Wittlinger,† and is shown in Fig. 5-21B. Here, a 3130 CMOS output-stage op amp is used, which allows the diode of Fig. 5-21A to be eliminated. The gain is calculated in a similar fashion, and this example also operates at a gain of 0.5; it can handle ±9-V inputs. It has a speed advantage over the previous circuit, however, since the 3130 op amp has less frequency compensation.

---

* See reference 26 at the end of the chapter.

† See reference 33 at the end of the chapter.

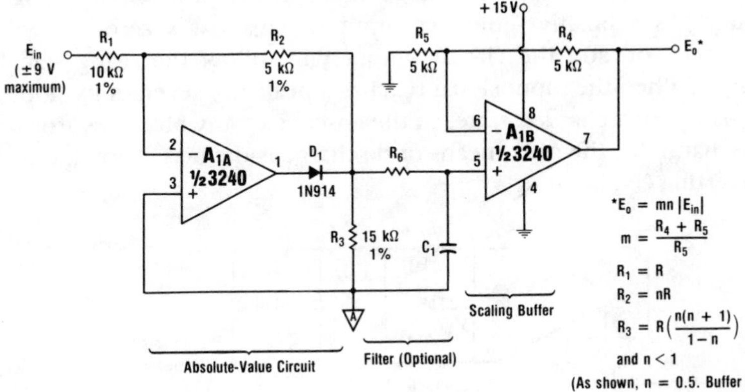


*(A) With filter and buffer.*

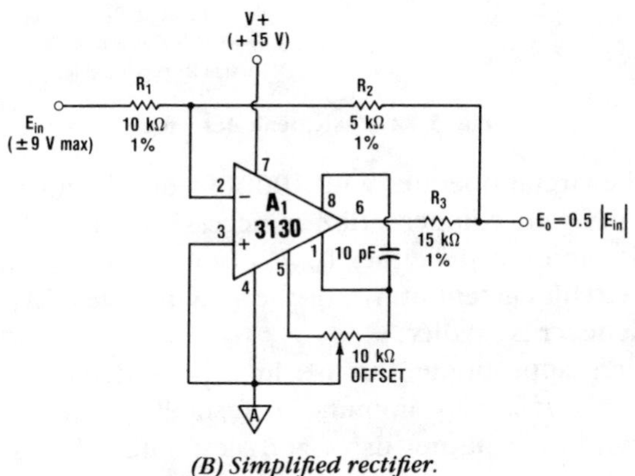


*(B) Simplified rectifier.*

**Fig. 5-21. Single-supply absolute-value circuits.**

## 5.6 PEAK DETECTORS

Peak detectors measure the maximum value of a fluctuating voltage. In this section, we will examine some op-amp circuits for accomplishing this function.

### 5.6.1 Basic Peak Detector

The basic peak detector (using an op amp) is shown in Fig. 5-22. This circuit operates like an ideal diode, but with the addition of $C_1$ to store a dc voltage equal to the peak input-voltage value. As a positive-going ac input signal crosses zero, $A_1$ will drive $D_1$ on and the circuit output will follow the rising signal slope. When the input signal reaches a peak and reverses its slope, $C_1$ is left in a charged state. In the absence of any bleed resistor to discharge $C_1$, the only means of discharge is the bias current at the inverting (−) input of $A_1$.

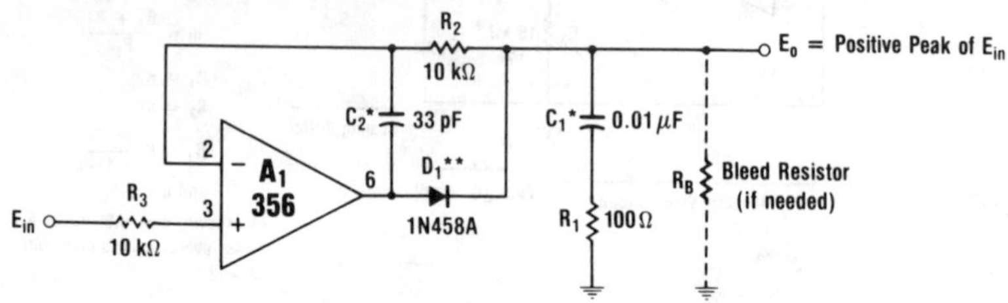


**Fig. 5-22. Basic peak detector.**

Since the circuit operates with 100% feedback when the loop is closed, unity-gain compensation is necessary for $A_1$. The rate of voltage rise across $C_1$ is either $I_{max}/C_1$ (volts/second, where $I_{max}$ is the short-circuit current of $A_1$), or the slewing rate of $A_1$ (volts/second), whichever is smaller.

Amplifiers appropriate for use in a peak detector should be types that are relatively immune to instability due to capacitive loading, with high output drive and slew rate. They should also have a high differential-input voltage rating and low input currents. One type that fits these requirements quite well is the 356, which is shown in the circuit of Fig. 5-22.

The capacitor used for $C_1$ should be a type that has low leakage and low dielectric absorption, such as polystyrene or polypropylene. The diode used for $D_1$ should also be a low-leakage type.

Capacitor $C_2$ can be adjusted to minimize settling time, and the optimum value will vary somewhat according to the value of $C_1$. In the circuit as shown, settling to within 0.1% of a 10-V peak is accomplished in 10 $\mu$s or less.

Resistor $R_2$ is used to protect the input of $A_1$ against dangerously large discharge currents from $C_1$ when the power supply is switched off; $R_2$ is particularly necessary if $C_1$ is 0.1 $\mu$F or greater. $R_3$ is (optionally) included to balance the effect of $R_2$ for bipolar amplifier types.

$C_1$ can be reset with a switch, either manually or electronically via a normally off transistor. Alternately, if a fast-attack, slow-decay response is tolerable, connecting a bleed resistance ($R_B$) across $C_1$ will also deplete the charge. Peak reading of negative-polarity signals is possible by reversing $D_1$. A post amplifier in the form of a second FET-input voltage follower will usually be necessary for this circuit, since the output impedance is quite high (as shown).

### 5.6.2 Buffered Peak Detector

When extremely long holding times are required of a peak detector, the bleedoff leakage of $C_1$ becomes important. Therefore, to minimize this discharge current, $C_1$ is usually buffered (Fig. 5-23) by a low-input-current voltage follower, $A_2$. The 356 type used here will require an input current of only about 200 pA (leakage drain current from $C_1$), so the voltage discharge rate for $C_1$ will be

$$\frac{\Delta V}{\Delta t} = \frac{I}{C}$$
$$= \frac{2 \times 10^{-10}\ A}{1 \times 10^{-8}\ F}$$
$$= 0.2\ V/s$$
$$= 20\ \mu V/ms$$

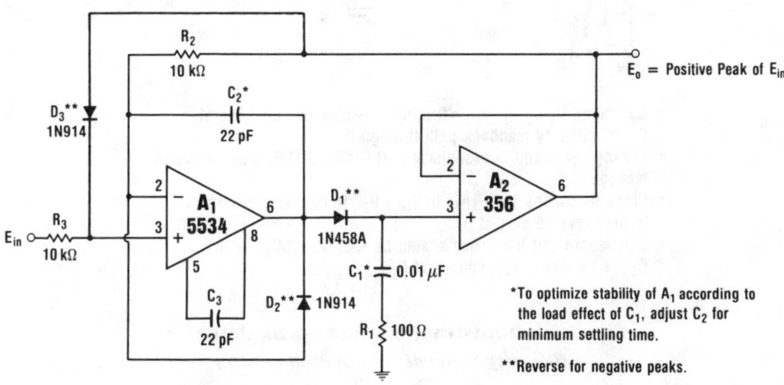


Fig. 5-23. Buffered peak detector.

In practice, diode leakage will most likely be greater than 200 pA, and so will be dominant. If very-low-leakage operation is critical, the clamp diode, $D_3$, is used, which limits the bias across $D_1$ to just over 1 volt. This will lower the leakage current substantially (and also protect $A_1$'s input stage). Note that only $D_1$ is required to be a low-leakage diode.

Amplifier $A_2$ in this circuit provides the input bias current for $A_1$ as well as the output current. $R_2$ is included here to allow $A_1$ to be clamped in the off state by $D_2$, resulting in faster recovery. $R_2$, $R_3$, and $C_2$ perform functions here similar to those described for Fig. 5-22. This circuit will acquire a 10-V peak signal to an accuracy of 0.1% or less in 10 μs.

### 5.6.3 Bootstrapped-Diode Peak Detector

The circuit of Fig. 5-24 obviates the need for ultralow-leakage diodes for the peak detection.* This is accomplished by reducing the reverse bias on the diode in the off state, and thus reducing its leakage.

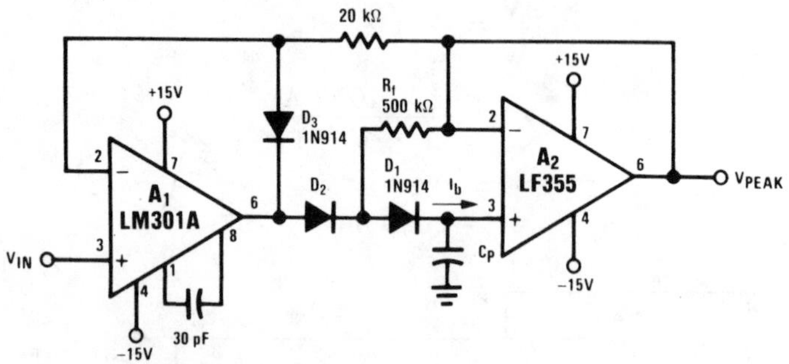


- By adding $D_1$ an $R_f$, $V_{D_1} = 0$ during hold mode. Leakage of $D_2$ provided by feedback path through $R_f$.
- Leakage of circuit is essentially $I_b$ (LF155, LF156) plus capacitor leakage of $C_p$.
- Diode $D_3$ clamps $V_{OUT}$ ($A_1$) to $V_{IN} - V_{D_3}$ to improve speed and to limit reverse bias of $D_2$.
- Maximum input frequency should be $<< 1/2\pi R_f C_{D_2}$ where $C_{D_2}$ is the shunt capacitance of $D_2$.

**Fig. 5-24. Bootstrapped-diode peak detector.**
*(Courtesy National Semiconductor Corp.)*

* See reference 4 at the end of the chapter.

In this circuit, $A_1$ and $A_2$ operate somewhat similarly to the circuit of Fig. 5-23, with the main difference being in the detector circuit. $D_1$ charges the storage capacitor, in series with $D_2$—both being ordinary diodes. In the off state of $D_1$ and $D_2$ (input below the peak level), $D_2$ sees a large reverse bias, but $D_1$ sees only the voltage drop across $R_f$, which is developed from the leakage of $D_2$. This reduces the reverse bias on $D_1$ by as much as two orders of magnitude, and allows it to function effectively as a low-leakage unit.

To take advantage of this low leakage, $A_2$ should be an FET-input device such as the 355 or 356 op amp, and the storage capacitor should be a plastic-film type, such as polystyrene, polypropylene, or Teflon®. Note that dual FET-type op amps, such as the 412, can be used effectively in this circuit, with some speed advantages.

Fig. 5-25 illustrates an alternate form of bootstrapped-diode peak detector, using a low-leakage (and low-cost) n-channel JFET, $Q_1$, as the peak detector diode. $C_H$, a polypropylene unit, is the hold capacitor. A second n-channel JFET ($Q_2$) allows the charge on $C_H$ to be dumped with a TTL LOW signal at the $\overline{\text{RESET}}$ line. With this line HIGH, positive peaks are detected and held. With the AD647, leakage from $C_H$ during hold will be no more than 25 pA; with the AD648, this will be reduced to even lower levels.

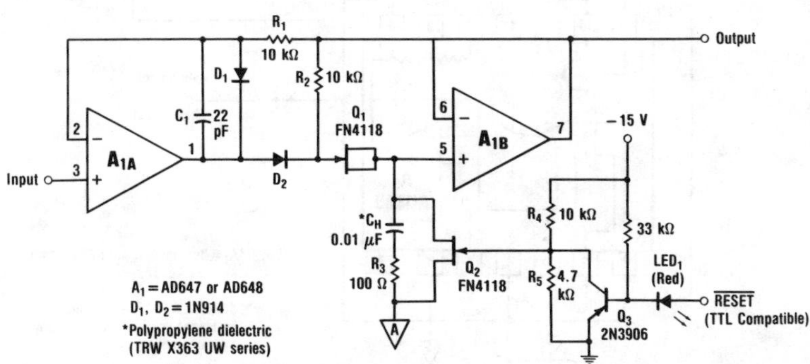


**Fig. 5-25. Bootstrapped-diode peak detector with reset.**

As shown, the circuit detects positive peaks, and the leakages of $Q_1$-$Q_2$ are consistent with op amps that have bias currents as low as a few picoamperes. Lower leakages are available, using the "A" version of $Q_1$-$Q_2$. For negative peak detection, reverse $D_1$-$D_2$, and substitute p-channel JFETs, such as 2N5116s or J270s, for $Q_1$-$Q_2$. A p-channel JFET can be driven directly from TTL for resetting.

### 5.6.4 Programmable Positive/Negative Peak Detector

Most peak detectors using op amps and precision rectifier circuits must necessarily be rewired to accommodate different peak polarities. The precision positive/negative peak detector of Fig. 5-26 does not have this disadvantage, and can be logic programmed for response to different input peak polarities.

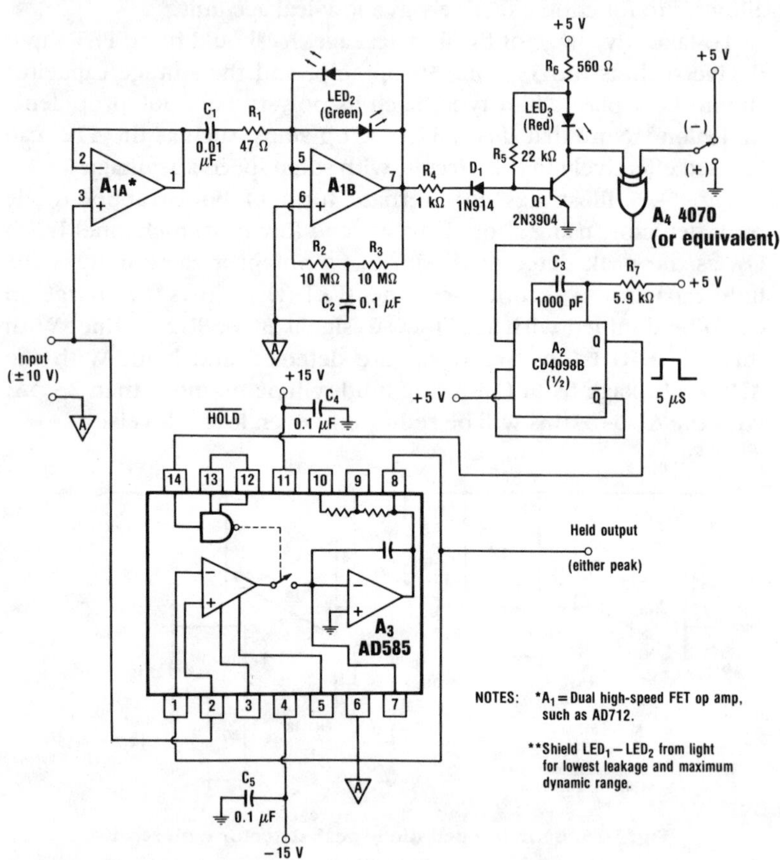


**Fig. 5-26. Programmable positive/negative peak detector.**

This circuit employs an AD585 sample/hold circuit (see section following) to act as a peak detector, and will sample the input when the $\overline{\text{HOLD}}$ line is HIGH. The upper portions of the circuit process the input, converting input peaks to logic transitions at the collector of $Q_1$. The XOR gate functions as a programmable inverter, triggering the 5-$\mu$s one-shot formed by $A_2$. Selection of the appropriate input peak is accomplished quite simply, via the XOR-gate polarity selector.

The 5-$\mu$s pulse from $A_2$ switches the AD585 to its sample mode when this line is HIGH, and, thereafter, the sample/hold reverts to the hold mode. With the timing of this sample pulse adjusted to coincide with the input peaks, a precision S/H function is accomplished.

### 5.6.5 Peak-to-Peak Detector

The circuit of Fig. 5-27 detects both positive and negative peaks and, in so doing, accomplishes peak-to-peak detection.*

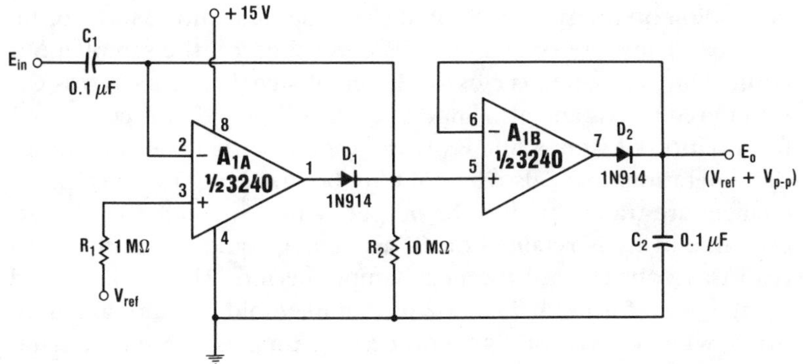


**Fig. 5-27. Peak-to-peak detector.**

In this circuit, $A_{1A}$ is a dc restorer which clamps the most-negative peak of the input to the potential, $V_{ref}$. $A_{1B}$ then detects the most-positive peak of the voltage across $R_2$ and stores it on $C_2$. The voltage on $C_2$ is then equal to $V_{ref} + V_{p\text{-}p}$, where $V_{p\text{-}p}$ is the peak-to-peak input voltage.

Note that with the 3240 single-supply op amp, $V_{ref}$ can range all the way to ground, enhancing the versatility of the circuit. Other single-supply op amps can be used, but the low input currents of the 3240 (or two 3140s) allow long time constants to be most easily realized.

---

* See reference 27 at the end of the chapter.

Although a single-supply version is shown, the circuit can easily be adapted for bipolar-supply uses. Note that additional buffering may be required at either input or output (or both).

## 5.7 SAMPLE-AND-HOLD CIRCUITS

A sample-and-hold circuit is one having two basic operational states; one in which an input signal is sampled and then transmitted to the output, and a second in which the last value sampled is held until the input is sampled again.

### 5.7.1 Basic Operation

The basic circuit operation is diagrammed in Fig. 5-28. In Fig. 5-28A, an input signal ($E_{in}$) is applied to electronic switch $S_1$ and, depending on the state of $S_1$, will either be transmitted to $C_1$ or be blocked. The state of switch $S_1$ is controlled by the sample/hold control line. When $S_1$ is closed, the input signal appears across $C_1$, is buffered by $A_1$, and also appears at the output. If $S_1$ is connected for a period of time while $E_{in}$ is varying, as with an ac waveform, the operation may also be said to be *tracking* $E_{in}$; any input changes are transmitted to the output. When $S_1$ is opened, the latest value of $E_{in}$ is retained on $C_1$ as a charge, and $A_1$ continues to read this voltage until the next sample period. This is illustrated by the waveforms in Fig. 5-28B. Sample/hold circuits are used for a wide variety of signal-processing functions, such as data-handling systems, analog/digital interfaces, etc. Perhaps the most important function is as the input interface before a successive approximation a/d circuit (an application to be illustrated).

### 5.7.2 Fast Sample/Hold Circuit

The circuit of Fig. 5-29 almost directly follows that of Fig. 5-28A. The JFET switches provide the sample/hold control function, $A_1$ serves as an input buffer to drive the hold capacitor, and $A_2$ serves as an output buffer to isolate the capacitor from external loads.

In this circuit, the dc errors will be a sum of the errors of $A_1$, the switches and capacitors, and $A_2$. Individual offset adjustments are necessary for both $A_1$ and $A_2$ to minimize the accumulation of these errors.

The circuit is a relatively fast sample/hold type, because the set-

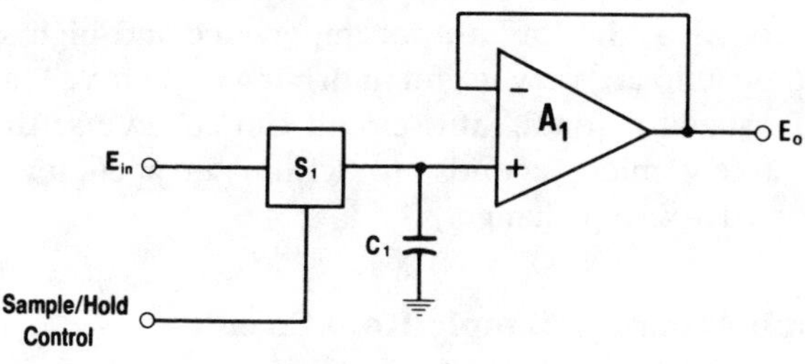


*(A) Basic circuit.*

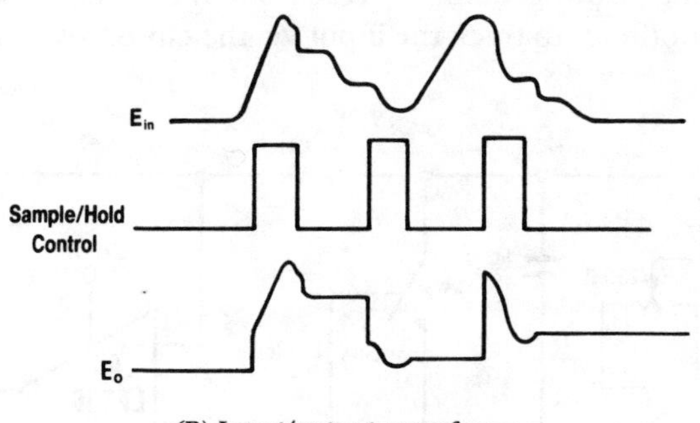


*(B) Input/output waveforms.*

**Fig. 5-28. Sample/hold operation.**

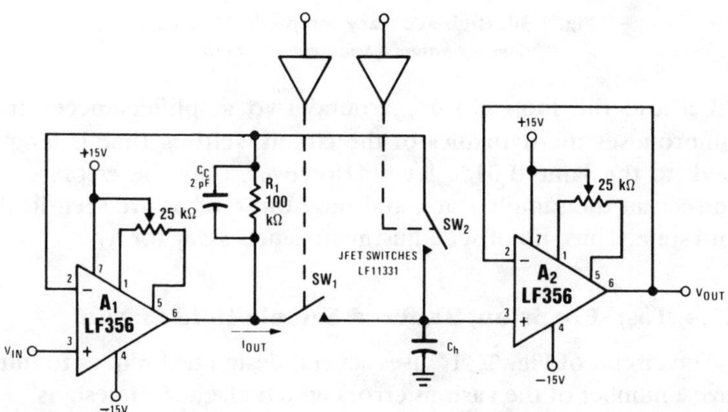


**Fig. 5-29. Fast sample/hold circuit.** *(Courtesy National Semiconductor Corp.)*

tling dynamics of $A_1$ and $A_2$ are kept separate, and are thus not complicated. Also, the low output impedance and high slew rate of the 356 op amp are very useful in driving capacitive loads. With a 1000-pF capacitor for $C_h$, this circuit can achieve settling times of under a few microseconds, to within an accuracy band of 0.01%, for a 10-V step change.

### 5.7.3 High-Accuracy Sample/Hold Circuit

The circuit of Fig. 5-30 is arranged slightly different from that of Fig. 5-29, and closes the control loop around both $A_1$ and $A_2$ during the sampling. During hold, $A_2$ reads out the voltage held on $C_h$, while $A_1$ continues to track the input via the closed switch, $SW_2$.

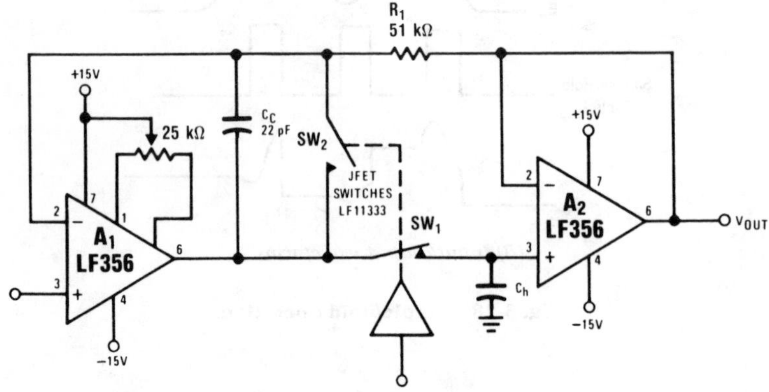


**Fig. 5-30. High-accuracy sample/hold circuit.**
*(Courtesy National Semiconductor Corp.)*

Because the loop closure around two amplifiers necessarily compromises the dynamics of the circuit, settling time is lengthened (to the same 0.01% level). However, only the errors of $A_1$ are seen in the sample state, and only those of $A_2$ are seen in the hold state. Thus, no offset adjustment is necessary for $A_2$.

### 5.7.4 Fast-Precision, Buffered Sample/Hold Circuit

The circuit of Fig. 5-31* uses several design techniques to minimize a number of the various errors which plague S/H designs.

* See reference 31 at the end of the chapter.

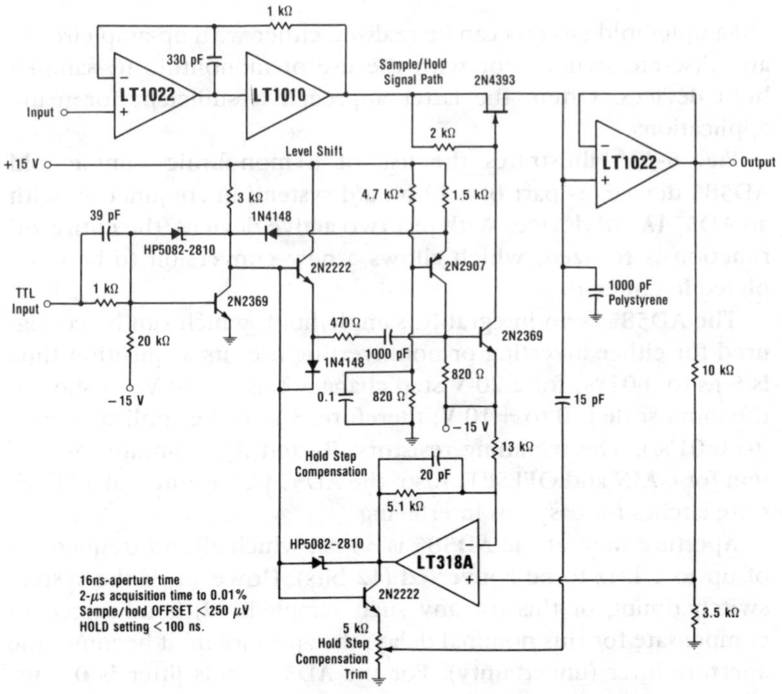


**Fig. 5-31. Fast-precision, buffered sample/hold circuit.**
*(Courtesy Linear Technology Corp.)*

For high-speed acquisition, the LT1010 monolithic unity-gain power buffer is used to drive the storage capacitor. With the 100-mA (or more) current available from the LT1010, an acquisition time limit is effectively placed upon the two buffer amplifiers; namely, the LT1022s. The acquisition time of the circuit is 2 μs to 0.01%, for a 10-V step change.

The discrete component circuitry is used as a high-speed level shift, and also to provide a hold step compensation. This allows the error of the 2N4393 JFET switch, upon entering hold, to be minimized by trim.

The 16-ns aperture time of this sample hold will allow frequencies of up to 2 kHz or more to be sampled, for a ½ LSB of error in a 12-bit system.

### 5.7.5 Monolithic Sample/Hold as Part of a 12-Bit A/D System

Sample/hold circuits can be realized either with op-amp circuits and discrete switches or with the use of monolithic IC sample/hold devices. Often, the latter approach is sufficient for many applications.

Fig. 5-32* illustrates the use of a monolithic sample/hold AD585 device, as part of a 12-bit a/d system, in conjunction with an AD574A a/d device. With just two active devices, the entire a/d function is realized, which allows a new conversion to be completed in under 40 $\mu$s.

The AD585 is an integral IC sample/hold, which can be configured for either inverting or noninverting use. Its acquisition time is 5 $\mu$s to 0.01%, for a 20-V step change; 3 $\mu$s for 10 V. As shown, the input scale is 0 to +10 V; therefore, 3 $\mu$s is the applicable time (to 0.01%). The trimming resistors, $R_1$ and $R_2$, calibrate the system for GAIN and OFFSET. Also, the AD574A has internal TTL tristate latches for easy bus interfacing.

Aperture time of the AD585 is 35 ns, which allows frequencies of up to 1 kHz to be converted (12 bits). However, if the system switch timing of this (or any such sample/hold) is advanced to compensate for this nominal delay, the aperture limit becomes the aperture jitter (uncertainty). For the AD585, this jitter is 0.5 ns, allowing the frequency range to be extended to about 70 kHz. (*Note:* It is the responsibility of the system designer to control timing to the required range.)

## 5.8 LIMITERS

In a signal chain, it is often necessary to symmetrically limit or "bound" the peak amplifier output. It is usually desirable to do this before the amplifier reaches saturation voltage, because saturation almost always involves a long recovery time before the amplifier can operate linearly once again. Many applications require peak limiting for other reasons—such as, for example, to interface the op-amp output to digital logic circuits. The positive or negative limit required for digital logic is usually much lower than the typical op-amp voltage swing.

---

* See reference 14 at the end of the chapter.

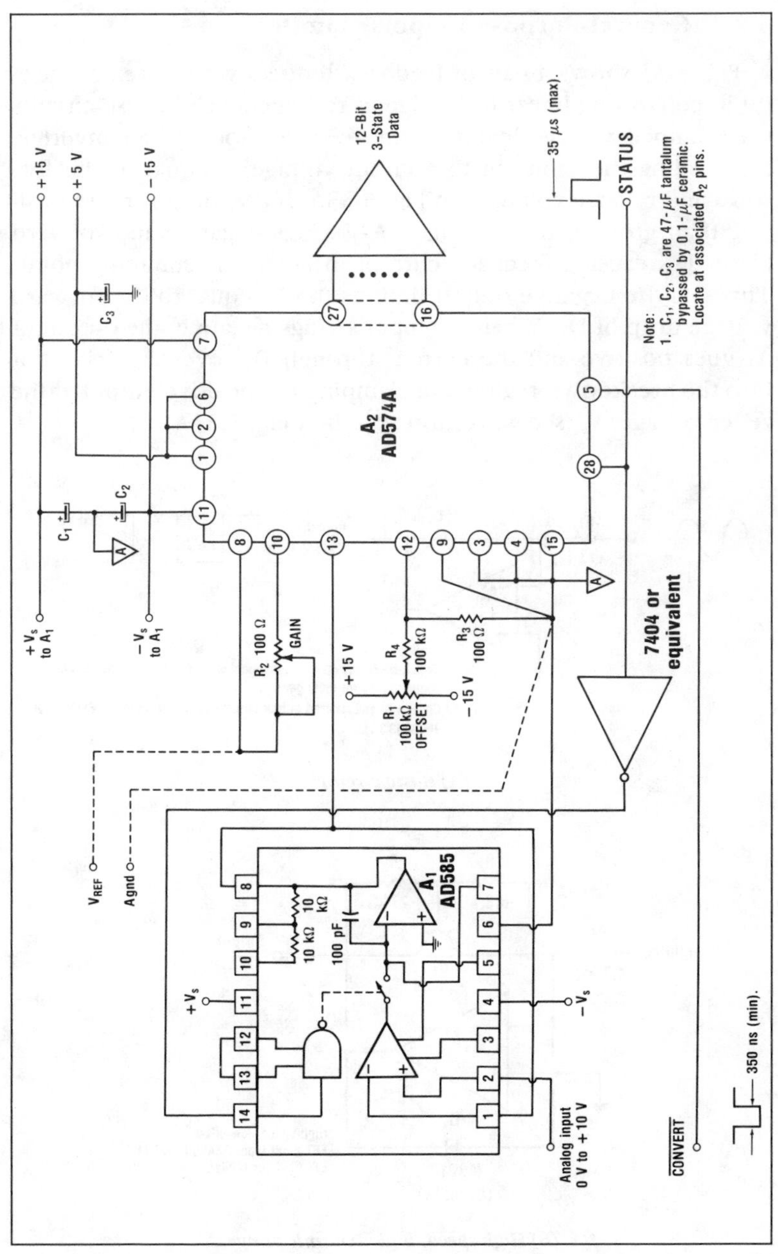


**Fig. 5-32. S/H as part of 12-bit a/d system.**

## 5.8.1 General-Purpose Unipolar Limiter

Fig. 5-33 shows a pair of feedback limiters which are very useful in converting linear input signals to logic levels. These circuits use clamped voltage levels in the feedback loop of an inverter, establishing the limits of the output voltage as equal to the forward and reverse voltages. In Fig. 5-33A, if the input is more positive than ground, the output of $A_1$ swings negative and forward biases $D_1$, causing feedback current in $D_1$ to the summing point. This sets the negative output limit as 0.6 V, equal to the forward voltage drop of $D_1$. When the input swings negative, the output of $A_1$ goes positive and the current through $D_1$ reverses, driving it into the breakdown region and clamping the positive output at the zener voltage, $V_z$ (see waveform sketch in Fig. 5-33A).

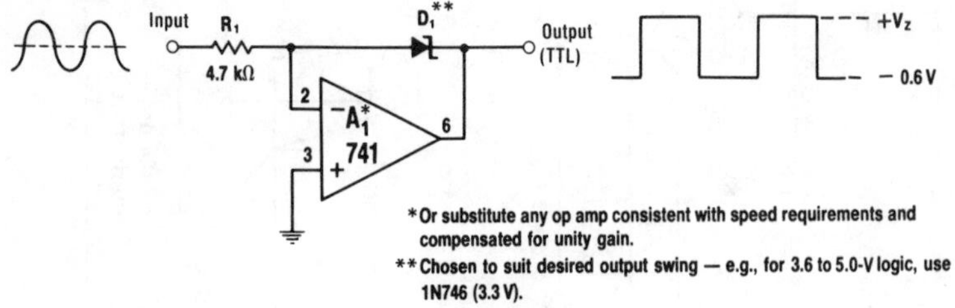


*(A) Basic circuit.*

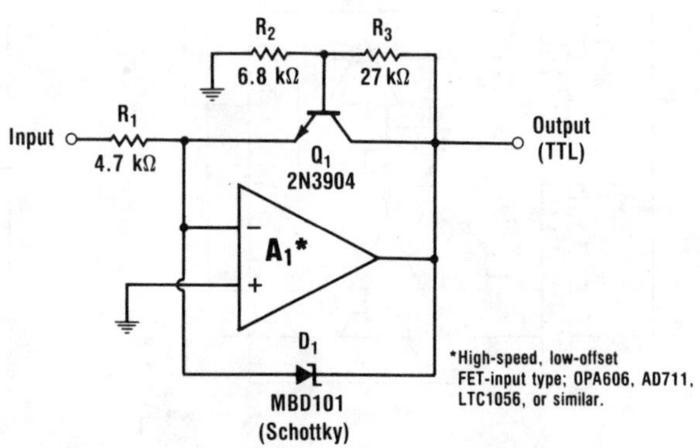


*(B) High speed, high dynamic range.*

**Fig. 5-33. Unipolar feedback limiters.**

Note that the circuit will have a high gain at low input levels, because $D_1$ will have a high reverse impedance below the breakdown voltage. This tends to amplify low-level inputs, but only up to the level of $V_z$, where the output is clamped by the breakdown of $D_1$. As a result, the circuit maintains the output swing at the limit of the diode over a rather wide range of input amplitudes.

The output voltage is equal to that of the diode used and may be reversed by reversing $D_1$. Due to the essentially 100% feedback through the low impedance of $D_1$, the op amp used must be compensated for unity gain.

The basic circuit of Fig. 5-33A is speed limited, due to the relatively high-function capacitance of zener diodes. This limits the effectiveness to frequencies below about 100 Hz.

Fig. 5-33B is speed optimized, and uses a low-capacitance $V_{BE}$-multiplier feedback scheme, with a Schottky reverse clamp. It also delivers TTL output levels, but is useful to 100 kHz or more. Either circuit should use a low-offset FET-input amplifier, for maximum low-level sensitivity.

### 5.8.2 Bipolar Zener Limiter

The unipolar limiters of Fig. 5-33 are actually special cases; the more general form of a zener limiter is shown in Fig. 5-34.

This circuit, shown in basic form in Fig. 5-34A, limits the op-amp output swing in either direction to the value of $V_z + V_f$, where $V_z$ and $V_f$ are the zener- and forward-breakdown voltages of $D_1$ and $D_2$. With matched zener voltages, the positive and negative limiting levels will be symmetrical. For example, using 10-V zeners for $D_1$ and $D_2$, the output swing of $A_1$ will be limited at slightly over 10 volts (10.0 + 0.6 volts); this allows a linear ±10-V swing, but still avoids saturation of $A_1$. Below the breakdown potentials of $D_1$ and $D_2$, the amplifier operates as if they were not there, with gain determined by $R_1$ and $R_2$ in the conventional fashion. The degree of limiting, or "hardness," is determined by the ratio of the zener network impedance to the resistance of $R_2$. If the value of $R_2$ is high in relation to the combined impedances of $D_1$ and $D_2$, the limiting will be sharp with little rounding; typical $R_2$ values of 10 kΩ or above give very sharp limiting.

A weak point of the basic form of this circuit lies in its sensitivity to zener leakages. With higher values for $R_2$, this sensitivity

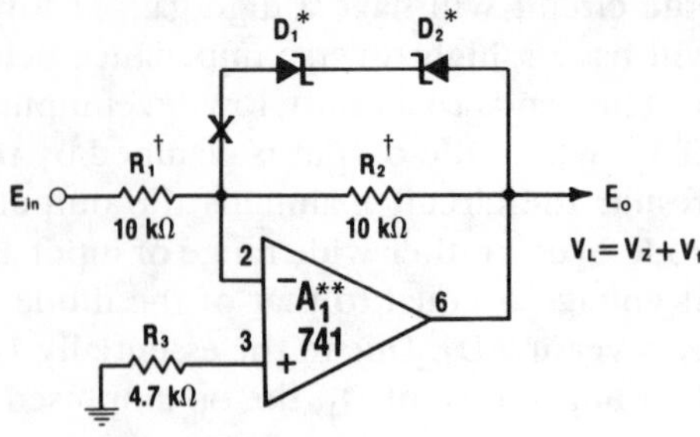


*Matched pair to suit required limit voltage—e.g., 1N758 for $E_0 = \pm 10V$.
**Any op amp compensated for unit gain.
†Selected according to gain requirement.

*(A) Basic circuit.*

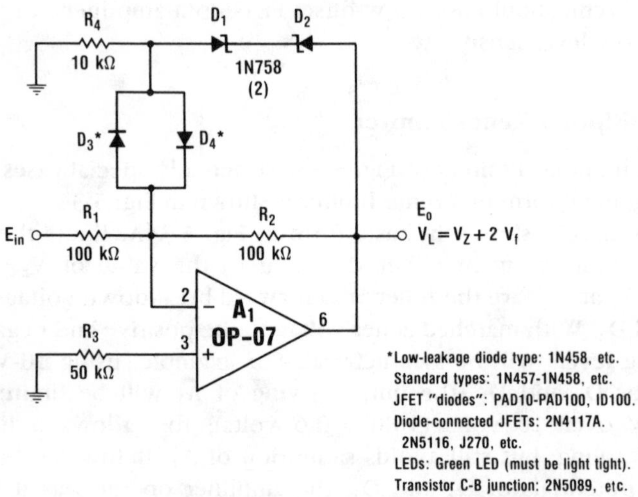


*Low-leakage diode type: 1N458, etc.
Standard types: 1N457-1N458, etc.
JFET "diodes": PAD10-PAD100, ID100.
Diode-connected JFETs: 2N4117A. 2N5116, J270, etc.
LEDs: Green LED (must be light tight).
Transistor C-B junction: 2N5089, etc.

*(B) Low-leakage "T" decoupling.*

**Fig. 5-34. Bipolar zener limiters.**

becomes a limitation. This problem can be eliminated by decoupling the few microamperes of leakage possible in conventional zeners, as shown with the "T" decoupling circuit in Fig. 5-34B. Here $R_4$ is added to bleed any diode leakage to ground, developing a relatively low voltage across $D_3$ or $D_4$ of only a few microvolts. With such a low voltage across these low-leakage diodes, the leakage error coupled into the summing point of $A_1$ will be only a few picoamperes (or less).

A variety of low-leakage diode options are listed in the notes of Fig. 5-34B. Among the better performers are the JFET units, which can actually have leakages of 1 pA or less. LEDs are also effective, but should be well shielded from light inputs to prevent photo currents. This technique of leakage decoupling is applicable to a gamut of limiting and clipping circuits related to this one, and is compatible with amplifiers with bias currents down to 1 pA (not just the 1-nA OP-07 shown). Further examples of these points follow in later applications.

### 5.8.3 Bipolar Limiter With Adjustable Level

One drawback in designing symmetrical zener limiters is that zener diodes generally have imprecise breakdown voltages and, hence, are difficult to match. If accurate symmetry is desired, one solution is the circuit of Fig. 5-35 which uses a matched pair of monolithic diodes and has an adjustable limit voltage.

A typical pair of monolithic IC transistors will have emitter-base breakdown voltages that match within a few percent or less. Also, the temperature coefficient (TC) of the matched pair will be small, typically less than 1 mV/°C. In Fig. 5-35A, a small fraction (B) of the op-amp output voltage ($E_o$) is fed to $D_1$-$D_2$ through $R_4$. Resistor $R_4$ is adjusted to set the limit voltage, which can range from zero to about 10 volts. $R_4$ should have a relatively low impedance (typically 5 kΩ), so that this resistance will not appreciably affect the output voltage as the output current changes. Fig. 5-35B shows how the matched zener pair can be generated from a CA3093 chip. With the values shown, $V_L \simeq \pm 8.4$ V, and the zeners track very well with temperature. Other possible units are the CA3046, CA3086, and the CA3096. Leakage currents can be decoupled if necessary, as shown in Fig. 5-34B.

### 5.8.4 Transistor Junction Limiter

For very high speed in a general-purpose limiter, transistor emitter-base junctions can be used effectively, because they have a low capacitance. Such a limiter configuration is illustrated in Fig.

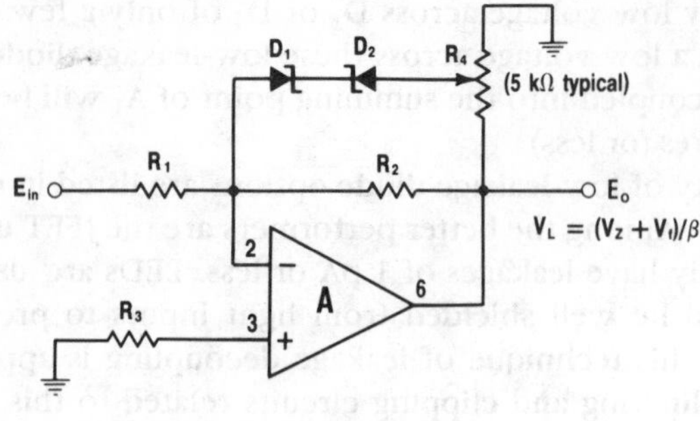


*(A) Basic circuit.*

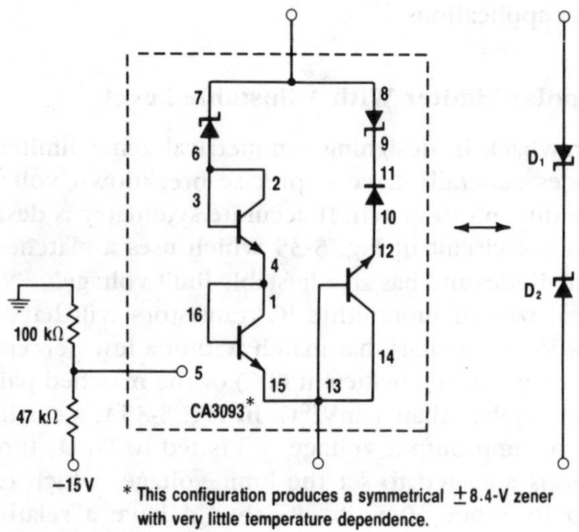


*(B) Synthesizing matched zeners.*

**Fig. 5-35. Adjustable bipolar limiter.**

5-36. The 2N3906 transistors provide a limit voltage, $V_L$, of about ±7 V; again, this value can be varied by adjusting $R_4$, and leakage decoupling added, if needed.

## 5.8.5 Zener Bridge Limiter

A single zener diode is sometimes used for symmetrical limiting by placing it in a diode bridge, so that it conducts on both polarities of the output swing. This is illustrated in Fig. 5-37. The diodes

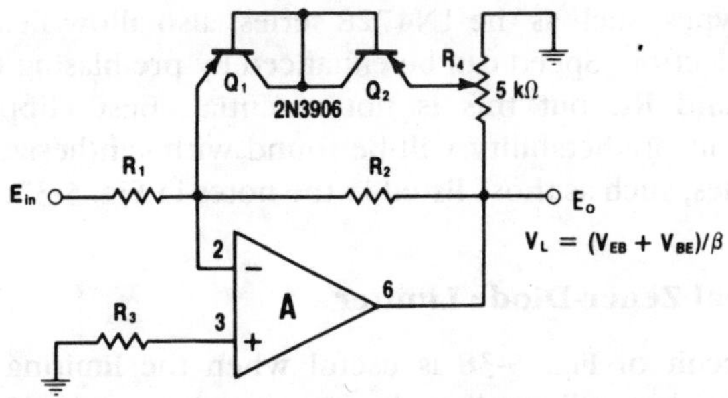


**Fig. 5-36. High-speed limiter using transistor junctions.**

can, in some cases, be ordinary general-purpose types, such as the 1N914, unless very precise symmetry is required; monolithic diode bridges will provide more accurate symmetry, if needed. For the purposes of lowest leakage, $D_1$ and $D_3$ should be low leakage, while $D_2$ and $D_4$ can be conventional types.

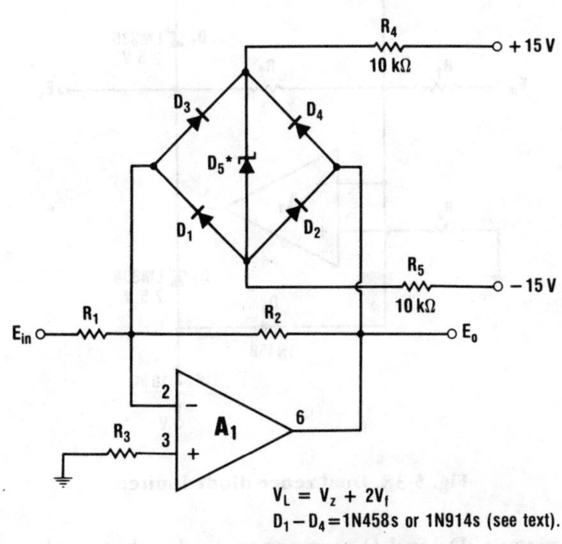


**Fig. 5-37. Diode bridge limiter using one zener.**

A very hard limit characteristic can be obtained by using a synthesized zener such as the LM329 (6.9 V), but the more general-purpose types, such as the 1N4728 series, also allow flexibility in voltage selection. Speed can be enhanced by pre-biasing the zener with $R_4$ and $R_5$, but this is not essential. Best clipping- and voltage-limit predictability will be found with synthesized monolithic diodes, such as those listed in the notes in Fig. 5-37.

### 5.8.6 Dual Zener-Diode Limiter

The circuit of Fig. 5-38 is useful when the limiting is to be asymmetric, but still predictable; or when it must be sharp with reasonably high speed.

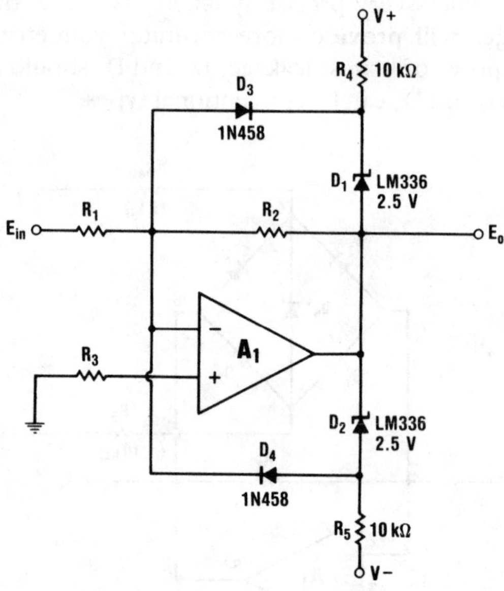


**Fig. 5-38. Dual zener-diode limiter.**

In this circuit, $D_1$ and $D_2$ are zener diodes that are biased "on" by resistors $R_4$ and $R_5$, respectively, These diodes conduct continuously, and thus need not be particularly fast. They can be monolithic zeners, for best stability.

The output is limited in its negative swing when $D_3$ conducts at a level of $V_z$ ($D_1$) + $V_f$ ($D_3$). It is limited in the positive direction when $D_4$ conducts at a level of $V_z$ ($D_2$) + $V_f$ ($D_4$). Note that if $R_1$

and $R_2$ are relatively large resistances, it will be advisable to make $D_3$ and $D_4$ low-leakage units, such as 1N458As, or another low-leakage type.

### 5.8.7 "T" Feedback Limiter

In very-high-impedance applications, such as when using electrometer-type op amps, it will be necessary to use special means to control leakages when limiting is employed. The circuit of Fig. 5-39 is useful for this task, and is another example of the general "T" leakage decoupler of Fig. 5-34B.

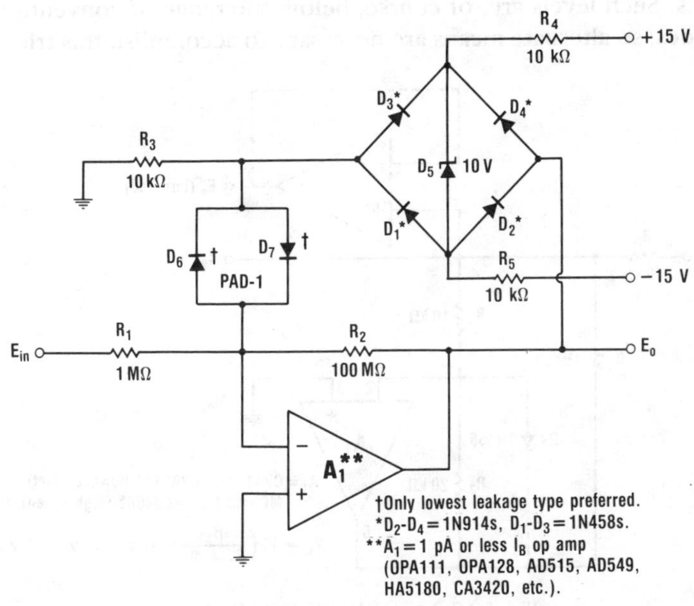


**Fig. 5-39. "T" feedback limiter to decouple leakage.**

This limiter is in "T" form, in order to decouple the leakage of bridge diodes $D_1$-$D_4$ to ground through $R_3$. $R_3$, being a relatively low resistance, allows only a few millivolts of bias to be applied to low-leakage types $D_6$ and $D_7$, greatly enhancing their usefulness. Although two ultralow-leakage types are listed ($I_L = 1$ pA), their actual necessity will depend on the temperature range of the circuit, relative impedances seen at $R_1$-$R_2$, and the bias current of $A_1$.

Note, also, that the value of $R_3$ can be minimized to bleed off leakage from $D_1$-$D_3$, and $D_6$-$D_7$ can be selected from the variety of low-leakage units mentioned previously.

The "T" decoupling technique for leakage is not limited to bridge-type feedback circuits, and any of the previously described circuits can be modified to use it. In the example shown, a 10-V zener provides limiting at just under ±12-V output.

### 5.8.8 Low-Voltage Shunt Limiter

The circuit of Fig. 5-40 is a useful one when the signal level must be limited at a very low level, such as several hundred millivolts. Such levels are, of course, below the range of conventional diodes, so alternate means are necessary to accomplish this trick.

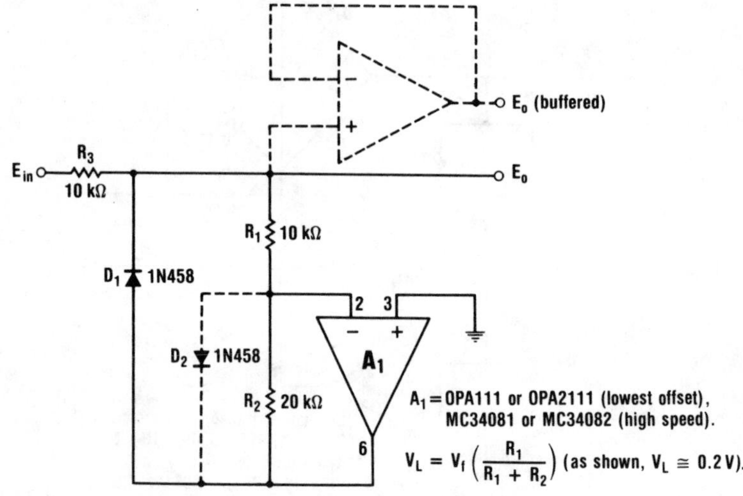


**Fig. 5-40. Low-voltage shunt limiter.**

In this circuit, a variation of a form by Schlesinger,* the op amp is used to amplify and level-shift the threshold of a conventional diode. The limit voltage is of the form shown, and can be adjusted to be any fraction of the diode voltage. Although a conventional diode is shown for $D_1$, low-drift zeners (such as the LM113 or AD589 types) can be used if temperature stability is required.

* See reference 21 at the end of the chapter.

If desired, $R_1$-$R_2$ can be a linear potentiometer, for continuous adjustment.

$D_1$ can be back-to-back-connected diodes for symmetrical clipping, in which case, the shunt antisaturation clamp, $D_2$, will not be necessary. Note that this circuit divides down the signal level, even when below the threshold, because of the feedback through $R_2$. It is unbuffered, as shown, but dual-FET types can be used (as shown optionally).

## 5.9 DIODE BRIDGE CIRCUITS

### 5.9.1 Input Current Limiter

One of the more precise types of signal limiters is formed by passing the signal through a diode bridge, as shown in Fig. 5-41. Here, diodes $D_1$ through $D_4$ are biased at a continuous current by $R_4$ and $R_5$, which are equal-value resistors connected to equal voltages. If $D_1$ through $D_4$ are all closely matched in characteristics, the dc potential at the bridge output ($D_1$-$D_3$ junction) can be connected directly to the summing point of $A_1$.

Note that there is no direct path for the signal to take through

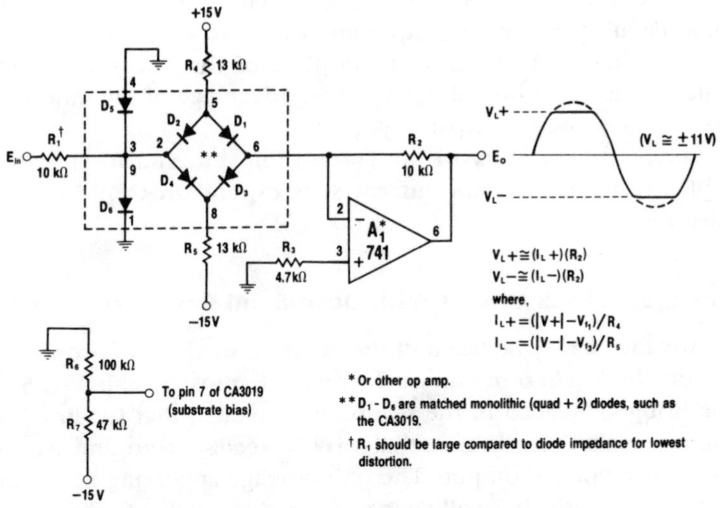


Fig. 5-41. Input current limiter.

the bridge to the summing point; instead, signal variations are transmitted to $A_1$ as variations in the currents supplied by $R_4$ and $R_5$. Limiting occurs when the input current in $R_1$ exceeds that capable of being transmitted by the bridge assembly. The bridge "opens," with $D_2$ and $D_3$ reverse biased on positive swings and $D_1$ and $D_4$ reverse biased on negative swings.

In this state, the entire current available from $R_4$ or $R_5$ is diverted through $D_1$ and $D_3$ on signal peaks, and the current from the opposite member ($R_5$ or $R_4$) is then passed through $D_1$ or $D_3$ to the summing point. In this manner, the bridge "peak limits" the input current to $A_1$ and so controls the output voltage of $A_1$.

The operation of a current-mode limiter such as this is very smooth and fast, with no "glitches" or overshoots as in voltage-switching limiters. This type is a good choice when the most abrupt limiting possible is desired. Diode nonlinearities can, however, introduce some distortion in the input signal, and this effect can be minimized by making $R_1$ large in relation to the diode impedance. As an example, the values shown give less than 0.5% distortion. The op amp used should be compensated for unity gain because, when the bridge opens, the feedback is nearly 100%.

The limiting-current threshold of the circuit is set by the current in the bridge. With a given value of current flowing in $R_4$-$R_5$, the input and output voltages are set by the choices of $R_1$ and $R_2$, with the usual considerations for gain. The values shown are for clipping at $\pm 11$ volts of output and can, of course, be rescaled if desired. Higher-speed op amps in place of the 741 (e.g., an FET unit such as the 356) will take greater advantage of the inherently fast bridge-switching capability.

For best dc stability of the current limit, $R_4$ and $R_5$ should be replaced by stable 1-mA current sources, and an OP-07 op amp used for $A_1$.

### 5.9.2 Feedback Circuit With Dead-Band Response

Advantage may be taken of the "current depletion" property of a diode bridge in other unusual circuits. For instance, in Fig. 5-42, the bridge is placed in the feedback loop such that for low-level input signals there is essentially 100% feedback around $A_1$, and very little voltage output. The only voltage appearing at the output is the relatively small swing required to modulate the current in the bridge as it passes the feedback current.

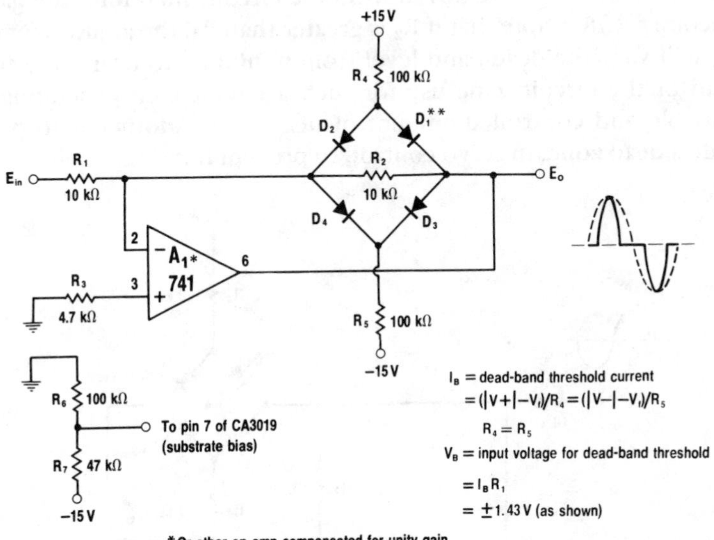


**Fig. 5-42. Dead-band circuit.**

When the input current in $R_1$ becomes greater than the allowable current, however, the bridge opens and the output voltage "jumps" to a new level determined by $R_2$, and the input is amplified by the ratio of $R_2/R_1$ in a linear manner.

This feature gives the circuit a *dead-band* property for low levels, but when the bridge current is exceeded, the circuit amplifies normally. The current and voltage threshold at the input are determined as before, by selecting $R_1$ for an input current equal to the bridge current at the threshold level.

### 5.9.3 Variable Dead-Band Circuit

The previous circuit is easily modified to provide a controlled amount of dead-band response, as shown in Fig. 5-43. With $R_2$ shorted, this circuit is the same as Fig. 5-42. For low-level input signals, increasing the value of $R_2$ raises the voltage developed across it, which raises the dead-band level. The circuit gain for small signals is then $(R_2 \parallel R_3)/R_1$. For large signals, the bridge opens and $R_2$ is effectively removed from the circuit; therefore, the gain becomes $R_3/R_1$. Note that if $R_2$ is greater than $R_3$, the adjustment of $R_2$ will vary the dead-band level from minimum to nearly the full gain of the circuit. One use for such a circuit is in generating a variable and controlled amount of distortion. Another is to provide a dead zone, in servo control, to prevent hunting.

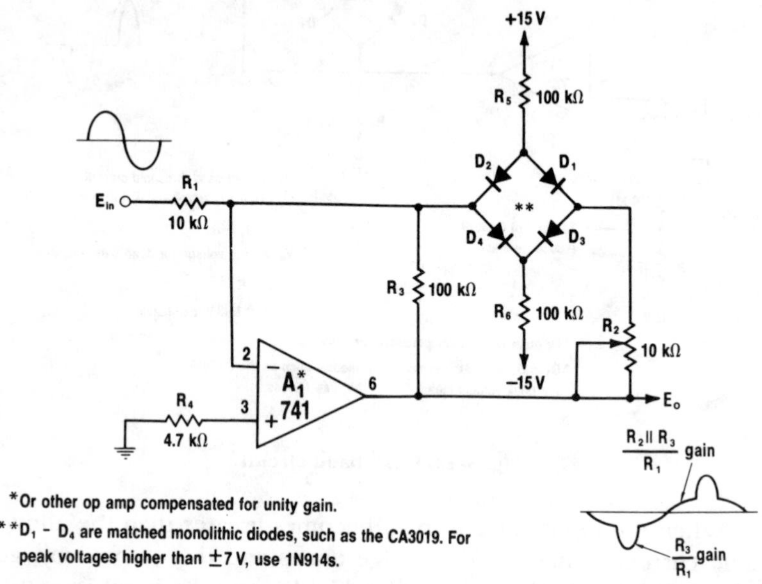


**Fig. 5-43. Variable dead-band circuit.**

## 5.10 VOLTAGE FOLLOWER WITH VARIABLE OFFSET (PRECISION LEVEL SHIFTER)

In some circuits, it is necessary to shift the dc operating point between stages without losing the signal level or loading the source. A circuit which accomplishes this is the variable-offset voltage follower of Fig. 5-44.

Diodes $D_1$ and $D_2$ provide a bias of ±1.23 V about the output potential of $A_1$. Since the tap of $R_1$ is at the same potential as $E_{in}$, adjustment of $R_1$ allows the output to be varied by the diode drop of $D_1$ or $D_2$ (about ±1.23 V). $D_1$ and $D_2$ have negligibly small impedances by comparison with $R_2$ and $R_3$, so there is little dynamic signal loss across this network and the gain is very close to

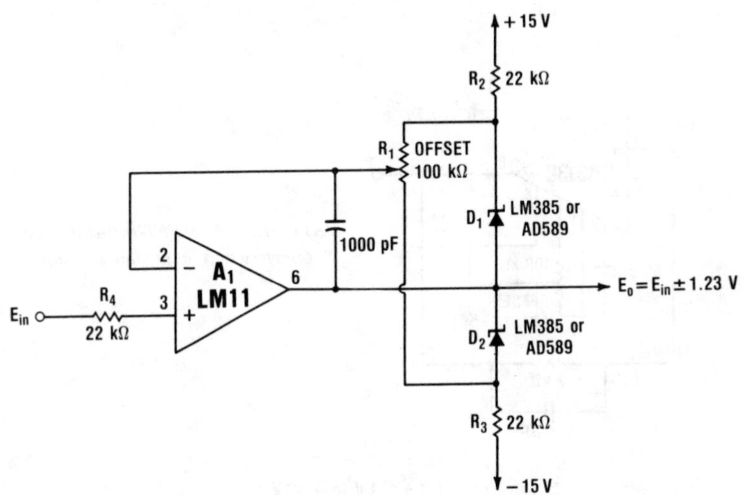


**Fig. 5-44. Variable-offset voltage follower.**

unity. Furthermore, the diodes listed are reference types with low-temperature coefficients. Thus, the temperature drift will be low (less than 100 $\mu$V/°C). Alternate lower-input-current forms of this circuit can use an FET-input op amp, such as the AD547 or AD548.

## 5.11 SINGLE-SUPPLY VOLTAGE FOLLOWERS

Some currently available op amps have the unique ability to operate from single power supplies while processing wide-range analog signals. Two good examples of this capability are the single-supply followers of Fig. 5-45.

In Fig. 5-45A, the op amp is a 3130, which can accept inputs in the range of zero to approximately +12 V, while operating as a precision voltage follower. Offset null is accomplished by the potentiometer, as shown, for applications requiring the lowest dc errors. Unity-gain compensation is provided by the 56-pF capacitor. Pin-for-pin amplifiers which also work in this circuit are the 3160 and 3140 (no compensation required), and the 3240, which functions similarly. All of these types have low bias currents due to their MOSFET input stages.

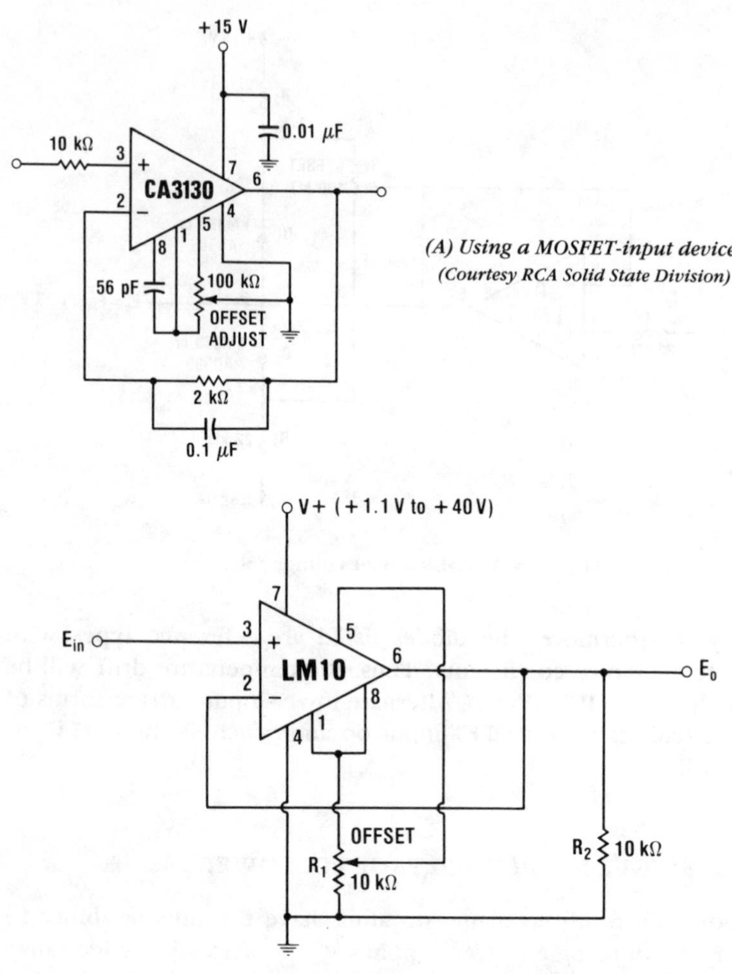


*(A) Using a MOSFET-input device.*
*(Courtesy RCA Solid State Division)*

*(B) Using a bipolar-input device.*

**Fig. 5-45. Single-supply voltage followers.**

The 3420 functions very much the same in this circuit (with no compensation), and at a highly reduced-output swing. It has, however, the unique advantage among the types mentioned of having an input bias current of less than 1 pA, and operation from supplies as low as 2 volts.

In Fig. 5-45B, a bipolar-input-stage device, the LM10, is shown used as a single-supply voltage follower. Although the input current is higher than that of an FET device (10 nA), it will generally have a lower offset voltage and offset-voltage drift. Also, the LM10 device is unique because of its extremely broad power-supply range, even down to 1.1 V, suitable for a single cell.

The input range of this circuit is from ground to 1 V less than the supply voltage, and the output range is comparable. Offset null is optional ($R_1$), but the output pulldown resistance, $R_2$, should be used to enhance the voltage swing. The load for this circuit must be referred to ground for it to work properly.

Other single-supply-compatible voltage-follower op amps are the 358 and 324 series, the LT1013/LT1014, the 3403 and 3458 types, the OP-20, the OP-21 and OP-32, and the 759. While these do not plug directly into either circuit shown, they have various other application advantages.

## 5.12 SIGN CHANGER

An interesting op-amp configuration is one that allows the choice of 0° or 180° phase reversal. This is the sign-changer circuit of Fig. 5-46.

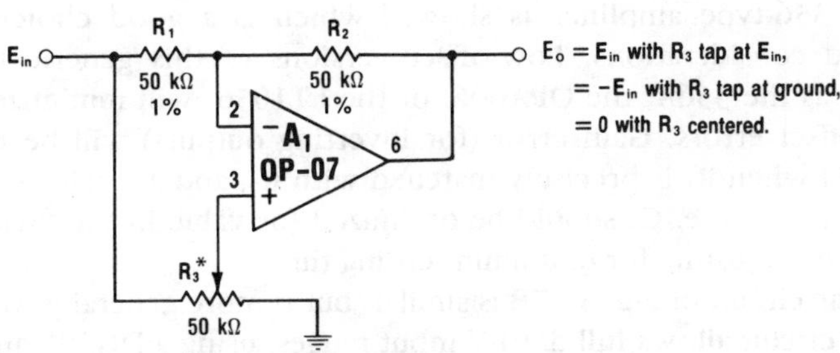


**Fig. 5-46. Sign-changer circuit with variable output.**

The circuit can be best understood by visualizing the operational state at various positions of $R_3$. If the arm of $R_3$ is at ground, the circuit is an inverter, providing a 180° phase shift. If the arm is at $E_{in}$, $E_{in}$ will appear at both inputs of $A_1$; therefore, both ends of $R_1$ are at the same potential, which implies no current in $R_1$. If this is true for $R_1$, it must also be true of $R_2$; therefore, $R_2$ will have a potential of $E_{in}$ on both ends, which makes $E_o = E_{in}$. Positions of $R_3$ at intermediate settings provide gains that are reduced from these two extremes. For instance, at the midpoint of $R_3$, the circuit will be recognized as a differential amplifier with a voltage of $E_{in}/2$ at each input. In this case, the output will be zero, because the circuit rejects the common-mode input signal (in proportion to the CMRR of $A_1$). $R_3$ may be calibrated for a linear variation from +1 through zero to −1, if desired. The best common-mode error will generally be found in bipolar-input types such as the OP-07 or OP-27. Other suitable bipolar devices are the LM11 and LT1007. FETs include the AD542 and the AD711, the 356 types, and the OPA111.

The circuit can also be used with a switch to select the choice of phase, as shown in Fig. 5-47, which allows remote programming. The basic operation of this circuit can be understood by referring to Fig. 5-47A. Here, the logic-programmable analog switch, $A_2$, connects the (+) input of the 356-type amplifier, $A_1$, to either ground or the input voltage. $A_2$ is a CMOS-array analog switch, the CD4007. The analog input range is 0 to +10 V, and the output is either 0 to +10 V or 0 to −10 V, with logic as indicated.

A 356-type amplifier is shown, which is a good choice for speed considerations. Low-offset versions of this generic type, such as the 356A, the OPA606, or the LT1056, will minimize the dc offset errors. Gain error (for inverting outputs) will be minimized when $R_1$ is precisely matched with $R_2$, and a single array is preferable here. $C_3$ should be optimized for value in the final circuit, by adjusting for minimum settling time.

The circuit of Fig. 5-47B is similar, but is more general purpose. This circuit allows full ±10-V input ranges, using a DG305 analog switch. Again, the best accuracy is achieved with the better types of the 356 amplifier, with $R_1$ and $R_2$ well matched (to 0.01% or better).

Both of these circuits are capable of very high precision (to 0.01%), and are also excellent in a utility sense. They can be used for sign control in a/d and d/a conversion, precision ac references, and balanced modulation and demodulation. A possible application caveat is the change in circuit input impedance, between the noninverting and inverting states. It is therefore best fed from a low-impedance source, for best results.

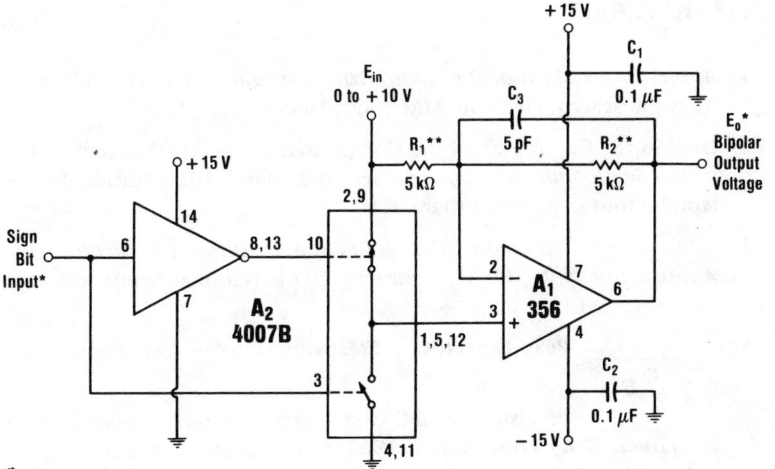

***Control Table**

| Sign Bit | Output Voltage |
|---|---|
| 0 | 0 to +10 V |
| 1 | 0 to −10 V |

$E_o+ = E_{in}$

$E_o- = -E_{in}\left(\frac{R_2}{R_1}\right)$

$^{**}R_1 = R_2$ (match to required accuracy).

*(A) Unipolar input range (0 to +10 V).*

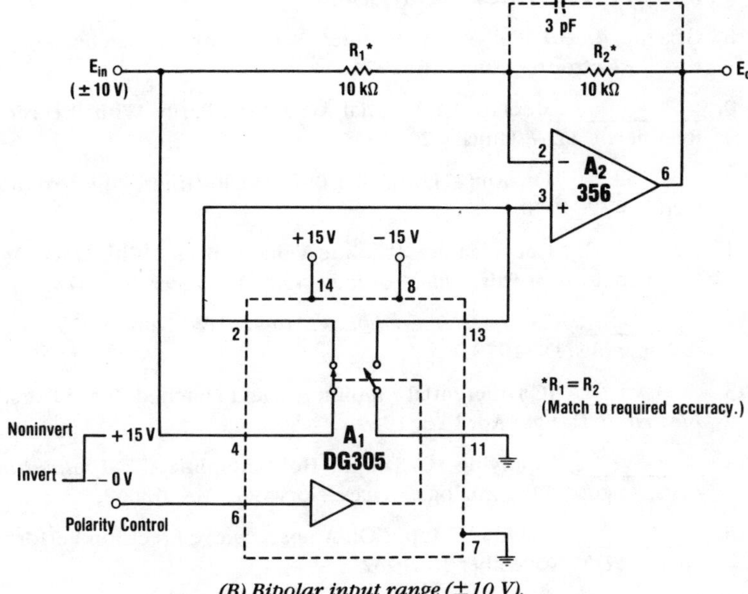


*(B) Bipolar input range (±10 V).*

**Fig. 5-47. Switch-selectable sign changers.**

## 5.13 REFERENCES


1. *Applications Manual for Computing Amplifiers,* 2nd ed. Philbrick/ Nexus Research, Dedham, MA, 1966, 1985.
2. Dobkin, R. C. "Feedforward Compensation Speeds Op Amp." *National Semiconductor Linear Brief LB-2,* March 1969. National Semiconductor Corp., Santa Clara, CA.
3. _________. "Op Amp Circuit Collection." *National Semiconductor Application Note AN-31,* February 1970. National Semiconductor Corp., Santa Clara, CA.
4. _________. "Peak Detectors." *Electronic Engineering Times,* June 17, 1974.
5. _________. "Precision AC/DC Converters." *National Semiconductor Linear Brief LB-8,* August 1969. National Semiconductor Corp., Santa Clara, CA.
6. Dudley, R. L., and Laing, V. L. "A Self-Contained, Hand-Held Digital Multimeter—A New Concept in Instrument Utility." *Hewlett-Packard Journal,* November 1973.
7. Frederiksen, T. M. "Multiple-Function LICs." *Proceedings of Electronic Products Magazine LIC Seminar,* March 1973.
8. Graeme, J. G. "Full-Wave Rectifier Needs Only 3 Matched Resistors." *Electronics,* August 8, 1974.
9. _________. "Measure Differential AC Signals Easily With Precision Rectifiers." *EDN,* January 20, 1975.
10. _________. "Op Amps Form Self-Buffered Rectifier." *Electronics,* October 12, 1970.
11. Jung, W. G. "Get Accurate Absolute Value Outputs With Only One Matched Resistor Pair." *Electronic Design,* December 20, 1978.
12. _________. *IC Converter Cookbook.* Howard W. Sams & Co., Inc., Indianapolis, IN, 1978.
13. _________. "IC Differential Amplifiers Yield Matched Zener Pairs." *Electronic Design,* April 26, 1969.
14. _________. "Applying IC Sample Hold Amplifiers." *Application Note,* August 1985, Analog Devices, Norwood, MA, 02062.
15. _________. "Stable FET-Input Op Amps Achieve Precision Performance." *EDN,* November 10, 1982.
16. Korn, G. A., and Korn, T. M. *Electronic Analog and Hybrid Computers,* 2nd ed., McGraw-Hill Book Co., New York, 1972.

17. *K2 Amplifier Series Applications Manual.* G. A. Philbrick Researches, Dedham, MA, 1956.

18. Lloyd, A. G. "Ideal Rectifiers." (Letters) *Electronic Design,* December 20, 1967.

19. __________. "Ideal Rectifier Uses Equal Value Resistors." *Electronic Design,* June 21, 1967.

20. Pease, R. A. "Bounding, Clamping Techniques Improve Circuit Performance." *EDN,* November 10, 1983.

21. Schlesinger, E. R. "Active Zener Circuit Provides Clipping Adjustable Down to Almost Zero." *Electronic Design,* April 26, 1977.

22. Sheingold, D. H. (ed.). *Non-Linear Circuits Handbook.* Analog Devices, Inc., Norwood, MA, 1974.

23. Smith, J. I. *Modern Operational Circuit Design.* John Wiley & Sons, Inc., New York, 1971.

24. Smither, M. A. "Improved Absolute Value Circuit." *EEE,* March 1969.

25. Tobey, G. E., Graeme, J. G., and Huelsman, L. P. (eds.). *Operational Amplifiers—Design and Applications.* McGraw-Hill Book Co., New York, 1971.

26. Vogler, D. L. "BI-FET Devices Improve Absolute Value Amplifier." *Electronics,* May 12, 1977.

27. __________. "Precision Peak-to-Peak AC-DC Converter Built With Single-Supply Op Amp Circuit." *Electronic Design,* November 22, 1977.

28. Widlar, R. J. "Designing With Super-Beta Transistor Op-Amp ICs." *EEE,* December 1969, February and March 1970.

29. __________. "IC Op Amp Beats FETs on Input Current." *National Semiconductor Application Note AN-29,* December 1969, National Semiconductor Corp., Santa Clara, CA.

30. __________. "Monolithic Operational Amplifiers—The Universal Linear Component." *National Semiconductor Application Note AN-4,* April 1968. National Semiconductor Corp., Santa Clara, CA.

31. Williams, J. M. "Application for a New Power Buffer." *Application Note 4,* September 1984, Linear Technology Corp., Milpitas, CA, 95035.

32. Wincentsen, R. J. "Absolute Value Circuit Uses Only 5 Parts." *EDN,* November 1, 1972.

33. Wittlinger, H. A. "Precision Op Amp Rectifier Eliminates Diodes for Economy, Simplicity." *Electronic Design,* July 26, 1984.

34. Manufacturers' Data Sheets:

| | | |
|---|---|---|
| Analog Devices | AD547 | Operational Amplifier |
| | AD711 | Operational Amplifier |
| | AD712 | Operational Amplifier |
| | AD548 | Operational Amplifier |
| | AD549 | Operational Amplifier |
| | AD648 | Operational Amplifier |
| | AD585 | Sample Hold |
| | AD574A | A/D |
| | AD589 | Reference |
| Burr-Brown | OPA111 | Operational Amplifier |
| | OPA128 | Operational Amplifier |
| | OPA2111 | Operational Amplifier |
| | OPA356A | Operational Amplifier |
| | OPA606 | Operational Amplifier |
| | OPA121 | Operational Amplifier |
| | OPA27 | Operational Amplifier |
| Fairchild Semiconductor | FD300 | Diode |
| | FJT1100 | Diode |
| Harris Semiconductor | HA2505 | Operational Amplifier |
| | HA5180 | Operational Amplifier |
| Linear Technology Corporation | LT1022 | Operational Amplifier |
| | LT1012 | Operational Amplifier |
| | LT1056 | Operational Amplifier |
| | LT1013 | Operational Amplifier |
| | LT1014 | Operational Amplifier |
| | LT1010 | Buffer |
| | LT1004 | Reference |
| Motorola Semiconductor Products | MC34080 Series | Operational Amplifier |
| National Semiconductor | LF400 | Operational Amplifier |
| | LF411 | Operational Amplifier |
| | LF412 | Operational Amplifier |
| | LM10 | Operational Amplifier |
| | LM11 | Operational Amplifier |
| | LM329 | Reference Diode |
| | LM336 | Reference Diode |

| Manufacturer | Part | Description |
|---|---|---|
| Precision Monolithics, Inc. | OP-07 | Operational Amplifier |
| | OP-27 | Operational Amplifier |
| | OP-41 | Operational Amplifier |
| RCA | CA3019 | Diode Array |
| | CA3046 | IC Array |
| | CA3086 | IC Array |
| | CA3093 | IC Array |
| | CA3130 | Operational Amplifier |
| | CA3140 | Operational Amplifier |
| | CA3420 | Operational Amplifier |
| | CD4007 | CMOS Inverter |
| Siliconix | DG305 | Analog Switch |
| | PAD-1 | Diode |
| | 2N4117 | JFET |

SIX

# *Logarithmic and Multiplier Circuit Techniques*

## 6.1 LOGARITHMIC CONVERTERS

Some of the most useful circuit techniques when using op amps are related to the use of logarithms: dynamic range compression, raising to powers or taking roots, rms conversion, and multiplier/divider circuits are only a few examples. Virtually all of these operations depend on the conversion of a linear voltage or current into a corresponding logarithmic value.

The heart of the analog logarithmic converter is the simple bipolar transistor. It is a highly useful fact that the base-emitter voltage, $V_{BE}$, of a silicon transistor is logarithmically related to the collector current over an extremely wide range. Using well-matched, monolithic transistors and IC op amps, it is possible to build log converters with a typical dynamic range of 5 decades ($10^5$ or 100 dB), or more.

### 6.1.1 Transistor Logarithmic Characteristics

Since the pn junction is the basis of all the log converters to be discussed, some review of the basic mechanism is in order. The $V_{BE}$ of a silicon planar transistor for a $V_{CB}$ of 0 V can be expressed as

$$V_{BE} = \frac{\kappa T}{q} \log_e\left(\frac{I_C}{I_O}\right)$$

where

$\kappa$ = Boltzmann's constant ($1.38 \times 10^{-23}$ J/K),
T = absolute temperature (Kelvin),
q = charge on an electron ($1.6 \times 10^{-19}$ coulomb),
$I_C$ = collector current,
$I_O$ = theoretical reverse saturation current ($10^{-13}$ A @ 27°C; will vary with processing).

Stating the equation in terms of $\log_{10}$, it becomes

$$V_{BE} = \frac{2.3\,\kappa T}{q} \log_{10}\left(\frac{I_C}{I_O}\right)$$

At T = 27°C, the $\kappa T/q$ term is 26 mV, which yields

$$V_{BE} = 60 \text{ mV} \log_{10}\left(\frac{I_C}{I_O}\right)$$

This reduced expression may be stated quite simply. At 27°C, the $V_{BE}$ of a silicon transistor will increase about 60 mV for each decade increase in $I_C$.

As stated previously, this relationship holds true over a wide range of voltages for low-leakage, high-gain transistors. The lower end of the dynamic range is limited (practically) by the op amp, while the upper end departs from logarithmic operation due to logging element ohmic-resistance drops above $10^{-2}$ to $10^{-3}$ A; exact figures depend on the specific device.

### 6.1.2 Basic Transistor Logger

The circuit of Fig. 6-1, called a *transdiode* configuration, is one of the basic logarithmic converters. The collector current of transistor $Q_1$ is equal to the input current and, since the base of $Q_1$ is grounded, the output of $A_1$ is the $V_{BE}$ of $Q_1$.

This circuit maintains the collector and base voltages of $Q_1$ equal through the virtual ground at the input of $A_1$; thus, the effects of collector-voltage changes on the logging operation are eliminated. Performance limits are governed by the transistor used and, more often, by the op amp; at very low log currents, the input current of $A_1$ becomes an appreciable error factor.

The circuit of Fig. 6-1 provides the closest approximation to actual logarithmic operation. This simple form, however, has two

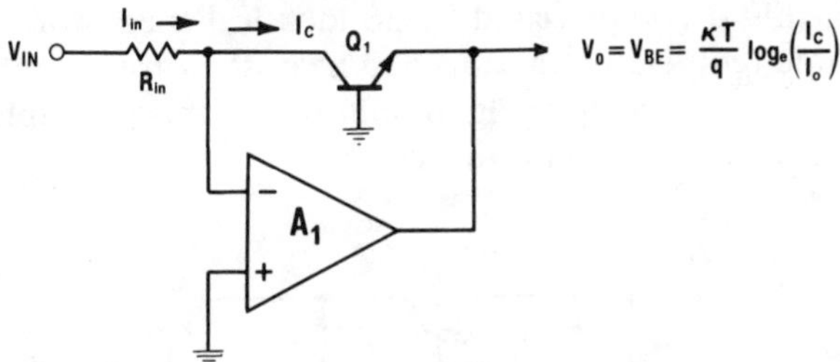


**Fig. 6-1. Transdiode logarithmic converter.**

important drawbacks: first, there are temperature-dependent components that can introduce error. Second, the relatively small (60 mV) output change per decade is inconvenient—some value such as 1 V/decade is usually preferred. From the equations of Section 6.1.1, the temperature-affected parameters are seen to be $\kappa T/q$, which is a scaling factor, and $I_o$, which is a reference factor; both of these may be compensated by additional circuit elements.

The temperature coefficient of $\kappa T/q$ amounts to $+0.3\%/°C$. This can be compensated as shown in Fig. 6-2 by using a thermistor ($R_2$) to track the changes in output voltage. Resistor $R_1$ determines the logarithmic scale; with the values shown, the scale factor is about 1 V/decade. Thermistor $R_2$ should be selected to have a temperature coefficient of $+0.3\%/°C$, to compensate the $\kappa T/q$ factor. As will be noted from the circuits which follow, there are standard wirewound units available with the correct TC, which serves for compensation.

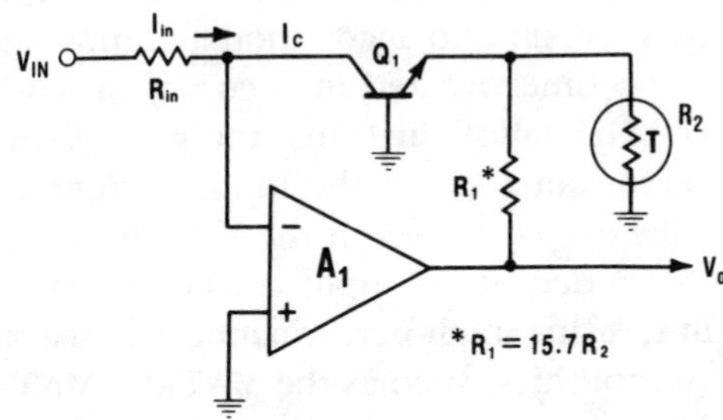


**Fig. 6-2. Log converter with temperature compensation of $\kappa T/q$.**

The $I_o$ factor in the equations also changes with temperature. In practice, this is compensated by an identical transistor operating at the same temperature and having a constant (reference) collector current. The circuit of Fig. 6-3 illustrates this principle, which will be detailed further in actual circuits.

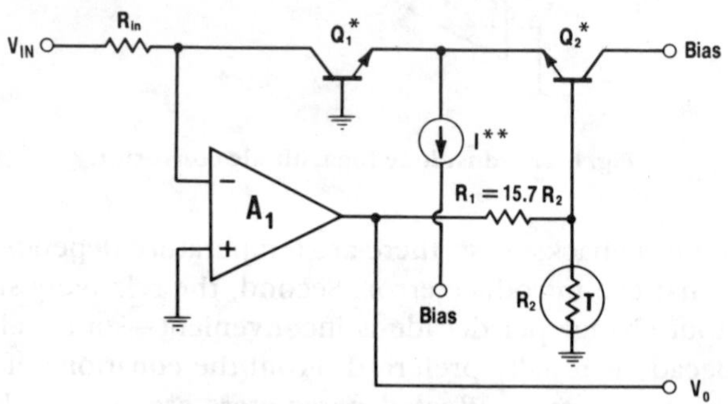


**Fig. 6-3. Log converter with temperature compensation of $\kappa T/q$ and $V_{BE}$.**

## 6.1.3 Other Logger Configurations

The transdiode connection is by no means the only transistor circuit useful in log converters. Two alternate feedback configurations are shown in Fig. 6-4. The feedback element can be connected either as a diode (Fig. 6-4A) or as a transistor (Fig. 6-4B).

The diode-connected transistor configuration of Fig. 6-4A is useful from the standpoint that it is a two-terminal device and is, therefore, easily reversible to accommodate input signals of negative polarity. It becomes limited in accuracy at lower input currents, however. The falloff in transistor gain ($\beta$ or $h_{FE}$) creates an increasing error since all of the input current does not flow through the collector as it does in the transdiode. This circuit is useful over about 5 decades of input current, from approximately 10 nA to 1 mA, using high-performance transistors as diodes. Examples are monolithics, such as the MAT-01, MAT-02, MAT-04, and the LM394 series, or discrete devices, such as the 2N5087 (pnp) or 2N5089 (npn).

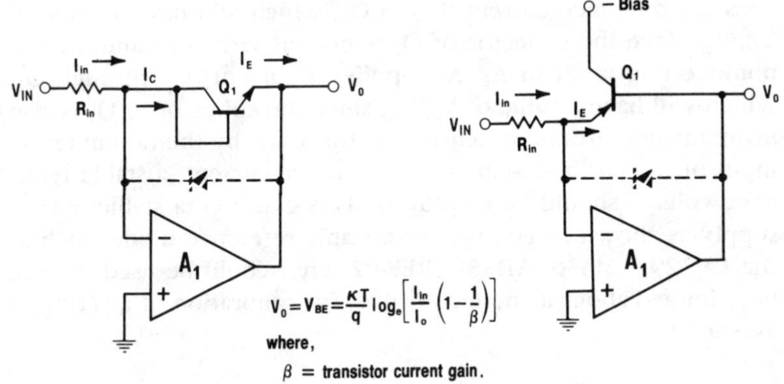


*(A) Diode feedback.* *(B) Transistor feedback.*

**Fig. 6-4. Alternate log configurations.**

The transistor connection of Fig. 6-4B is similar in performance to the diode connection. It does reduce loading on the op amp by virtue of the transistor current gain, but it also introduces some additional bias error due to the higher collector-base voltage. The dynamic range of this circuit is comparable to that of the diode connection, and it has similar errors due to gain falloff. In either instance, a reverse-connected low-leakage diode should be used for protection of the actual logging transistor. These types of logging elements will not be covered here in great detail; interested readers should consult the references cited at the end of the chapter for further information.

Log generators using transdiode connections are notoriously difficult to stabilize, because the feedback network can actually contribute gain rather than the loss which is usually the case. The gain is in excess of the open-loop gain of the op amp and must be considered in the overall loop. Further complicating matters is the fact that the gain of the transdiode changes with $I_C$. As a result, circuit stability requires either heavy frequency compensation of the amplifier or a gain reduction within the loop. Both techniques are commonly utilized, as will subsequently be seen.

### 6.1.4 Basic Log Generator

A circuit employing many of the design principles set down earlier is the log generator of Fig. 6-5. In this circuit, $A_2$ supplies a constant reference current ($I_{C_2}$) to $Q_2$, which will have a value of $V_{ref}/R_3$, since the collector of $Q_2$ is held at virtual ground by the noninverting input of $A_2$. $A_1$ supplies a current ($I_{C_1}$) through $Q_1$, which will have a value of $V_{in}/R_{in}$, since the collector of $Q_1$ is also maintained at ground potential—in this case, by the noninverting input of $A_1$. For best stability of reference current, a stable reference voltage should be employed. This could be a stable +15-V supply as shown in Fig. 6-5, or a stable reference diode, such as the LM329, LM336, AD589, REF-02, etc., could be used. $R_3$ can be trimmed about its nominal value, for calibration of $I_{C_2}$ (100 $\mu$A as shown).

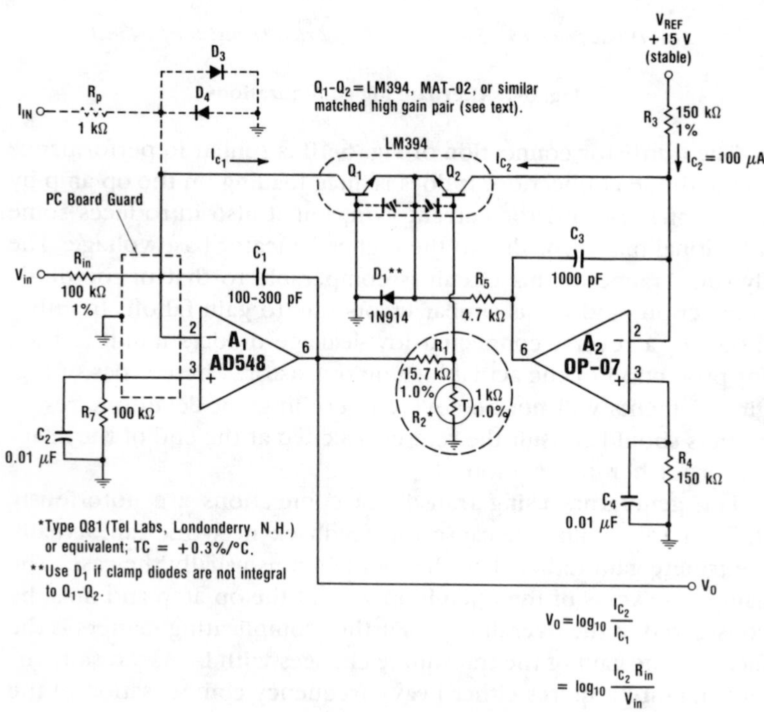


**Fig. 6-5. Basic log generator stage.**

$Q_1$ is being operated here as a transdiode connection, with $Q_2$ supplying temperature compensation of the offset voltage. The logged output of $A_1$ is temperature-compensated by divider $R_1$-$R_2$ and scaled to 1 V/decade, as explained in Section 6.1.2. Using the thermistor option specified for $R_2$, temperature compensation is within 1% over the range given. Feedback capacitor $C_1$ helps to stabilize the loop gain, as does the loss across divider $R_1$-$R_2$. Similarly, $C_3$ stabilizes $A_2$.

Clamping diode $D_1$ (in this circuit) provides reverse protection for the matched transistor pair. This will be necessary if clamping diodes (shown dotted) are not provided integral to $Q_1$-$Q_2$.

The transistor pair, $Q_1$-$Q_2$, is a well-matched, high-gain npn pair, preferably monolithic types for direct tracking/match. Example devices are the LM394, the MAT-04, MAT-02, and MAT-01 types, and the MP818. Other matched pairs include JEDEC types, such as the 2N2979, the 2N4044, and such IC-array types as the CA3096, etc. Discrete high-gain units, such as the 2N5089s, can also be selected for matching, and will work in these circuits. A special caveat goes for certain monolithic duals, such as the LM394 and the MAT-02/04. These units have a relatively high capacitance, which can lead to instability in some cases. Consult the manufacturer's application notes for specific compensation techniques for high-capacitance units.

These circuits are also configurable for pnp duals, which are more scarce than npn's. Types suitable for use are the MP352, IT130, 2N3810, and 2N3811 types. Paired discrete units, such as the 2N5087s, are also useful.

For selection of the two amplifiers, $A_2$ is relatively noncritical and can be an OP-07 or one of the JFET types. If bipolar types are used for $A_2$ (or $A_1$), the two source resistances should be matched, as shown. Amplifier $A_1$ should be the lowest offset-voltage and lowest bias-current unit consistent with the budget. This is simply because the input errors of this stage will limit the dynamic range, particularly when logging voltage inputs. In bipolar amplifier types, such low-bias-current units as the OP-07, the LM11, the 308A/312 types, or the LT1001 and LT1012 units are suggested. Either a very-low-trimmed offset selection should be used (preferable), or offset null should be employed to minimize the offset-voltage error of $A_1$. Among FET units, the OPA111 (or OPA121), the AD547, AD548, or AD549, the OP-41, or similar low-bias-current types will be very useful. In general, the low bias currents of an FET amplifier are very useful to the circuit, but low offset voltage and drift are also desirable for $A_1$, in particular, for voltage logging.

This circuit actually generates a voltage output which is the log of the ratio of two currents, $I_{C_1}$ and $I_{C_2}$. Normally, $I_{C_2}$ is fixed at some reference value, which, in this case, is 100 μA. This reference current determines the value of $I_{C_1}$, for which the log output will be zero, since $\log (I_{C_1}/I_{C_2}) = 0$ when $I_{C_1} = I_{C_2}$. The output is therefore simply

$$V_O = \log \frac{I_{C_2}}{I_{C_1}}$$

For the reference current of 100 μA as used here, the theoretical output for an input current of 100 pA will be 6 V; if the input were 1 mA, the output would be −1 V. $I_{C_1}$ can be equal to $I_{in}$ (for current sources), or equal to $V_{in}/R_{in}$ (for voltage sources).

In actual practice, both of these two range extremes are possible with this circuit, using the better logging transistors (LM394, etc.) and op amps with bias currents of 1–10 pA (see Chapter 2, and preceding discussion).

The AD548 shown is a low-power FET-input type with a 10-pA (max) input current. It is available with offset grades as low as 0.25 mV (max). This makes it a good all-around choice for this application.

Since the circuit actually responds to input current, true current sources will not require the precision input resistance ($R_{in}$). For these instances, a low-value, noncritical, input resistor (1 kΩ) should still be used for protection ($R_p$). Low-leakage clamp diodes should be added, if the amplifier used does not have internal protection (shown by dash lines in Fig. 6-5). Note also that the input offset voltage of $A_1$ is much less critical for current-source inputs.

For any low current (1 nA or less) logging use, *very* careful construction techniques should be used. This includes guarding around pins 2 and 3 of $A_1$ (and all connected common points), good PC layout (including guards on top and bottom), careful cleaning and preservation of the board, heat sinking of an FET device used for $A_1$, and low-leakage components ($C_1$ is a polystyrene type, for example). If these precautions are not taken, the ultimate performance of the better ICs will not be realized.

Fig. 6-6 illustrates a circuit which is of the same general type, but which is arranged to handle negative input currents, over a range of 10 pA to 1 mA. Here, a pair of LM11-type amplifiers are used, and the reference current is set up by the −15-V supply voltage. This implies that this supply be stable.

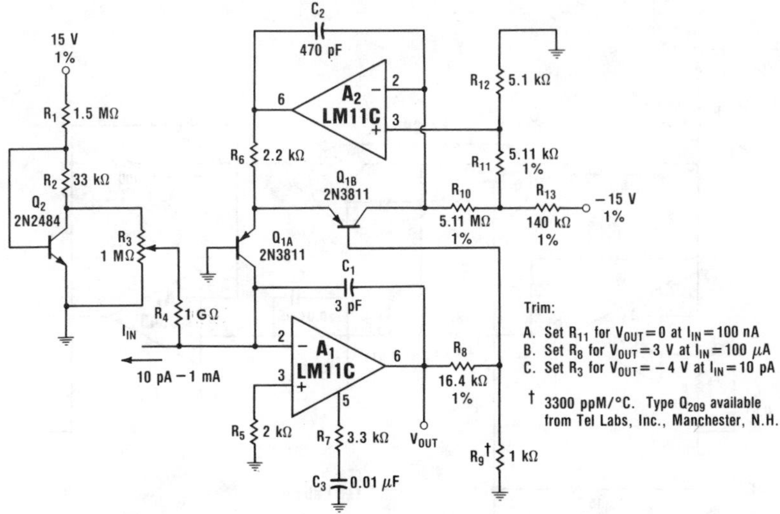


**Fig. 6-6. Log generator circuit for negative input currents.**
*(Courtesy National Semiconductor Corp.)*

For lowest errors due to the bias current of $A_1$, the bias current is (optionally) compensated by the $R_1$-$R_4$/$Q_2$ circuit. A matched pnp logging pair, a 2N3811, is used. Calibration of the circuit is accomplished by the trim sequence noted in Fig. 6-6 (using potentiometers for resistors $R_{11}$ and $R_8$).

### 6.1.5 Antilog (Exponential) Generator

The generation of an antilog response is a straightforward matter of rearranging the basic log circuitry to perform the inverse operation. An example is the antilog circuit of Fig. 6-7.

In this circuit, $Q_2$ generates the exponential current, and its output current is summed at the current-to-voltage converter, $A_2$. $Q_1$ provides voltage drive to $Q_2$ such that the collector current of $Q_2$ is exponentially related to the voltage at the base of $Q_1$. A temperature-compensating divider scales the input sensitivity to 1 V/decade as previously. The feedback of $A_1$ maintains the collector current of $Q_1$ constant at 100 μA. In this circuit, similar comments apply to the selection of $Q_1$-$Q_2$, $A_1$, and $A_2$, as was given for components in Fig. 6-5.

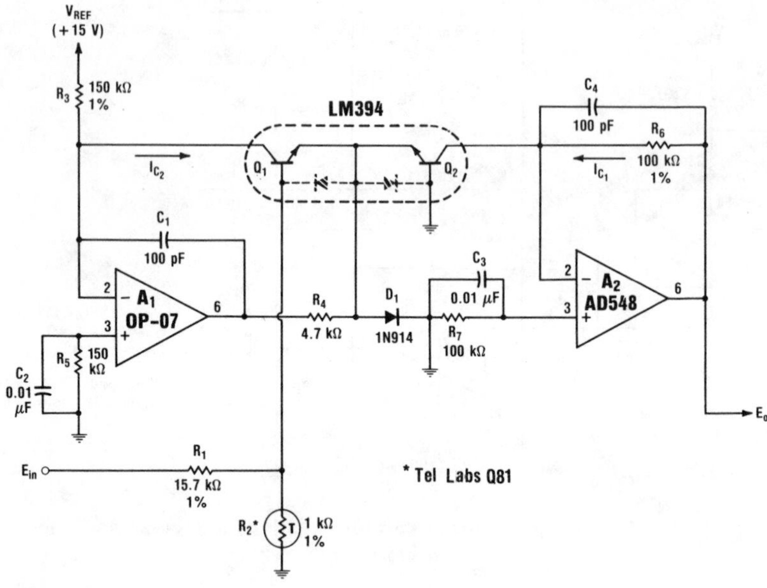


**Fig. 6-7. Antilog generator.**

## 6.1.6 Multifunction IC Log Applications

Performance limits aside, one aspect of log and antilog circuits is that they are complex in the number of separate precision components needed to realize a given function. Fortunately, multifunction ICs for this purpose are now available in monolithic form. This allows the designer to build circuits which are more compact and easily controlled, using just a single active device. An example is the Analog Devices AD538 multifunction IC, which will be illustrated using several log-type applications.

A log ratio circuit using the AD538 is shown in Fig. 6-8. This circuit responds to two input voltages, $V_x$ and $V_z$ (both positive). As can be noted from the internal elements of the 18-pin IC, the AD538 contains (shown at the top) two log stages and a difference amp (log ratio), as well as log and antilog stages (shown at the bottom), and a 2 V/10 V reference. In addition, pretrimmed input/output scaling resistors are included in the chip, for standard scaling. The user can also use external resistors for customized voltage scale(s), or can operate from current sources.

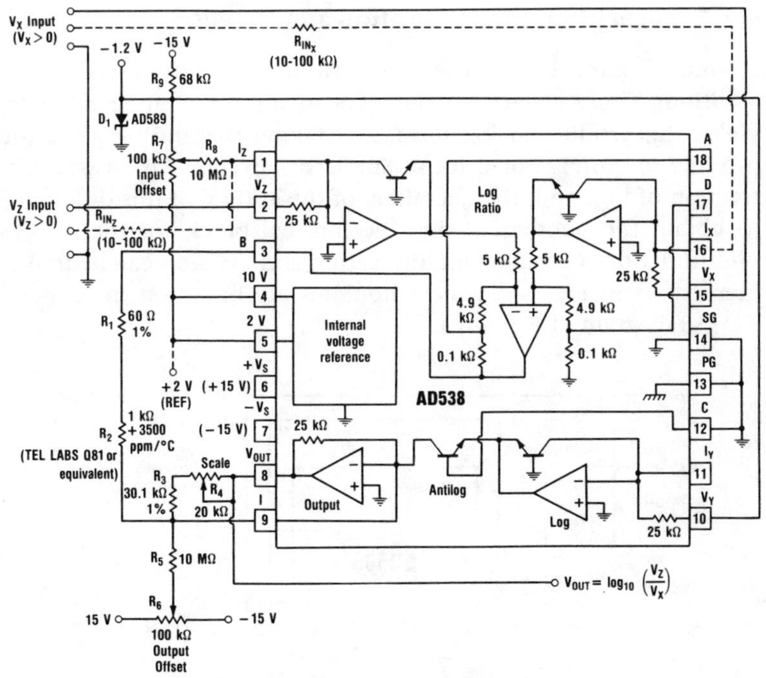


**Fig. 6-8. Log ratio circuit using multifunction IC.**

In this circuit, the log ratio circuit (top of sketch) generates a voltage equal to the difference of the logs of $V_x$ and $V_z$. This voltage is amplified and inverted in the output section (bottom of sketch), with temperature & compensation furnished by R2. $R_4$ sets the circuits scale factor to 1 volt per decade of input ratio. For this application, the log-antilog section is biased OFF by a negative voltage at the $V_y$ pin.

This circuit, as shown, responds to voltages in the +10-mV to +10-V range for $V_x$ and $V_z$. It can also be used with either input variable fixed, simply by tying that voltage input to the reference output, a +2 V here. Note also that virtually any two arbitrary voltage scales can be customized by the user, by adding a pair of resistors ($R_{in}$) at the $I_z$ and $I_x$ inputs. These resistors should be in the 10-kΩ to 100-kΩ range, so as to produce a 100-μA to 1-mA current, at the desired full-scale voltage. They should be stable types, preferably matched in ratio.

### 6.1.7 Logarithmic Multiplication and Division

Multiplication by the use of logarithms is a simple process of addition. Thus, given two numbers in log form, their addition yields the product in log form, and taking the antilog gives the product in conventional form. Similarly, division involves a subtraction of logs and, then, a taking of the antilog of this difference to obtain the quotient. Using the internal blocks of the AD538 multifunction IC, analog multiplication and division can be implemented, as in Fig. 6-9. (These hookups are based on the general AD538 diagram of Fig. 6-8.)

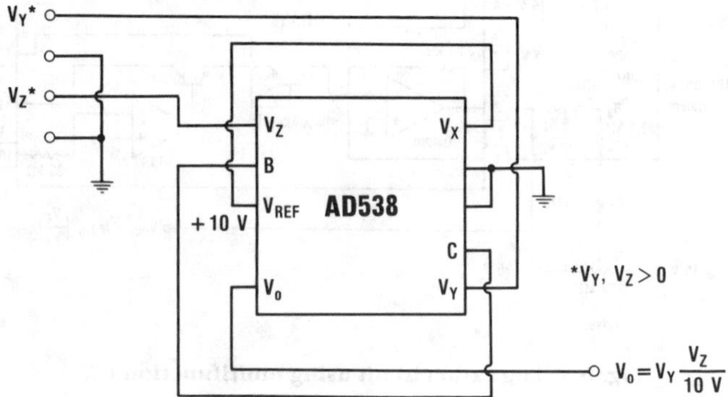


**Fig. 6-9. Multifunction IC analog multiplier (one quadrant).**

In Fig. 6-9, an analog one-quadrant multiplier can be implemented, simply by making the appropriate pin connections. This circuit generates the output

$$V_o = V_y \frac{V_z}{10\ V}$$

where

$V_y$ and $V_z$ are input variables (in the range of +10 mV to +10 V)

and 10-V scaling is accomplished by strapping the $V_x$ input to the +10-V reference pin. (If the reference is strapped for +2 V, 2-V scaling is achieved.) To implement the circuit, use the pins as shown (typically) in Fig. 6-8. Offset trim at $V_z$ is optional.

A divider, with 10-V scaling, is implemented as shown in Fig. 6-10. The major difference in terms of connection is the use of

$V_x$ here as the denominator, with $V_y$ strapped to the reference (to +10 V, for 10-V scaling).

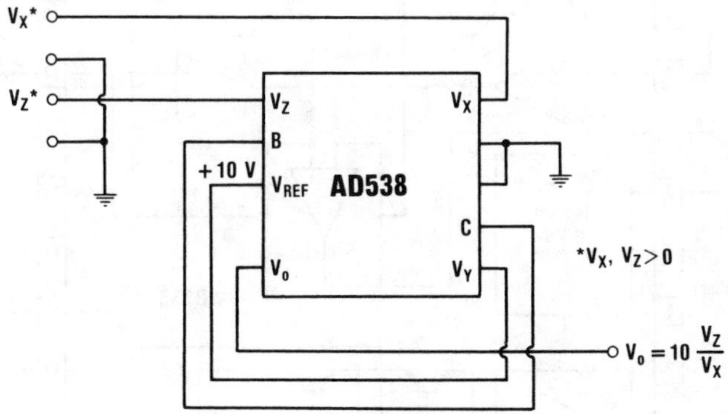


**Fig. 6-10. Multifunction IC analog divider (one quadrant).**

Both of these circuits are not only realized simply, but can actually outperform some dedicated analog function ICs.

### 6.1.8 Power Functions of an Input

Just as multiplication and division become processes of addition and subtraction when using logarithms, the generation of power functions becomes a process of scaling (multiplying) logarithms. Raising to powers greater than unity ($x^2$, $x^3$, etc.) is the scaling upward of the number in log form; taking a root ($x^{1/2}$, $x^{1/3}$, etc.) is the scaling downward of the number in log form. A general circuit that will raise the input to a power or take roots is shown in Fig. 6-11.

The circuit in Fig. 6-11 is a combination log converter and antilog generator using the AD538, with a variable scale adjustment between the log/antilog interface. The variable scaling modifies the exponent of the logged input to a value other than unity; therefore, a general-transfer expression of the circuit is:

$$V_O = V_Y \left( \frac{V_Z}{V_X} \right)^M$$

The above mentioned variable scaling provides the variable "M," while summing in log form provides the multiplication times $V_Y$.

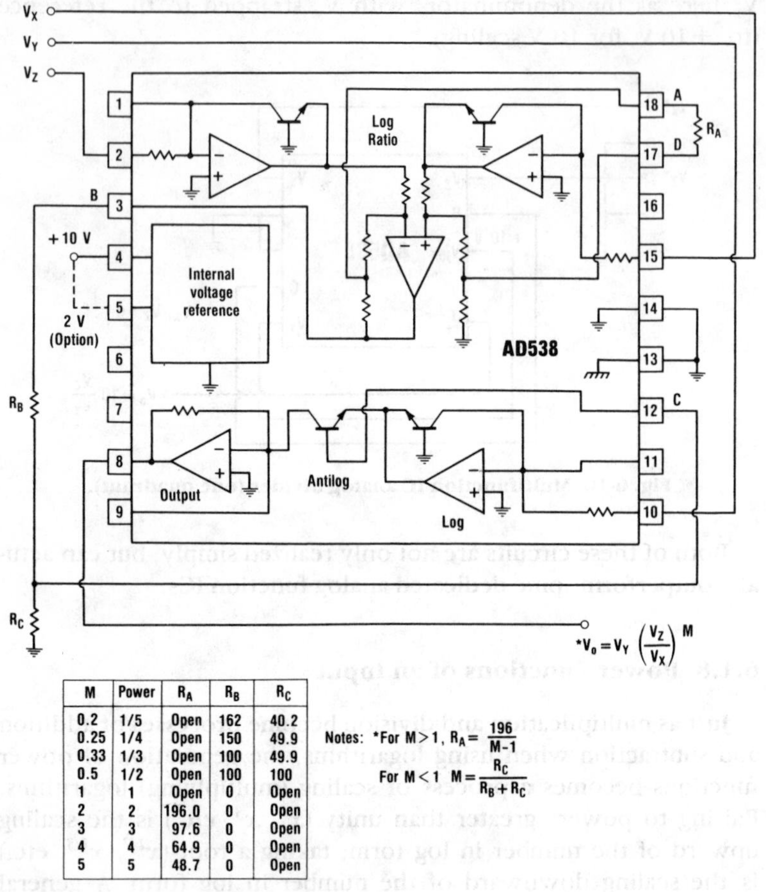


| M | Power | $R_A$ | $R_B$ | $R_C$ |
|---|---|---|---|---|
| 0.2 | 1/5 | Open | 162 | 40.2 |
| 0.25 | 1/4 | Open | 150 | 49.9 |
| 0.33 | 1/3 | Open | 100 | 49.9 |
| 0.5 | 1/2 | Open | 100 | 100 |
| 1 | 1 | Open | 0 | Open |
| 2 | 2 | 196.0 | 0 | Open |
| 3 | 3 | 97.6 | 0 | Open |
| 4 | 4 | 64.9 | 0 | Open |
| 5 | 5 | 48.7 | 0 | Open |

**Fig. 6-11. Multifunction IC general-transfer circuit (power function generator).**

As noted by the table in Fig. 6-11, either *attenuation* in the log domain is provided by $M > 1$ (powers) or *gain* in the log domain is provided by for $M < 1$ (roots). This can be done with $R_A$ for $M > 1$, and $R_B + R_C$ for $M < 1$. The table shown provides standard resistance values for values of M from ⅕ to 5 (the rated range). Resistance values should be 200 ohms or less for best stability.

Of course, all three inputs (X, Y, and Z) will not necessarily be used. As with the multiplier-divider circuits, constants can be assigned by tying a given input to the reference voltage (+2 V, or +10 V).

Variable or selectable power circuits are extremely useful in analog signal processing, for the generation of specific trigonometric functions, curve fitting, and linearization of transducer outputs.

## 6.2 ANALOG MULTIPLIERS

In Section 6.1, it was seen how the multiplication of two dc voltages (or currents) is accomplished using logarithmic modules. Generally, logarithmic multiplication is confined to unipolar dc signals rather than bipolar dc or ac signals. When ac signals are to be multiplied, either by a unipolar dc signal or by a bipolar signal, the need arises for a different type of multiplier. This type is what is generally termed as an *analog multiplier,* since it processes two analog signals to arrive at their product in analog form. Analog multipliers come in two basic types—two-quadrant and four-quadrant. The difference between these two types is in the response to different input-signal polarities.

### 6.2.1 Two-Quadrant Analog Multiplier

Fig. 6-12 depicts the output response of a two-quadrant multiplier to inputs $E_x$ and $E_y$. Several key features of operation may be noted from this graph. First, and perhaps most important, is the fact that if either $E_x$ or $E_y$ is zero, the output is zero. This satisfies the mathematical definition of multiplication by zero. Several curves of $E_o = E_xE_y$ are drawn for different values of $E_x$; $E_y$ ranges continuously from 0 V to +10 V. The output resulting for a given $E_x$ and $E_y$ can be determined graphically by locating the intersection of the $E_x$ and $E_y$ curves. For example, if $E_x = +5$ V and $E_y = +6$ V, then $E_o = +30$ V. A second example is $E_x = -10$ V and $E_y = +8$ V; in this case, $E_o = -80$ V. These two examples illustrate two-quadrant operation—that is, the output responds to both polarities of $E_x$ but to only one polarity of $E_y$. The output will follow the sign of $E_x$, but $E_y$ must always remain in the range of 0 V to +10 V. In circuit terms, this means that $E_x$ can be either a bipolar dc signal or ac signal, but $E_y$ must always be a single polarity of dc. This type of multiplier is often used as a linear gain control. Its usefulness can be appreciated by considering a fixed value of $E_x$ as $E_y$ ranges from zero to maximum. With $E_x$ a constant dc value (or steady ac level), the output will be directly and linearly controlled (multiplied) by $E_y$. Conversely, if operation is viewed from the standpoint of $E_y$ as a signal controlled by $E_x$, the output will be a varying level as $E_x$ varies, but with the additional capability of sign reversal as $E_x$ goes through zero.

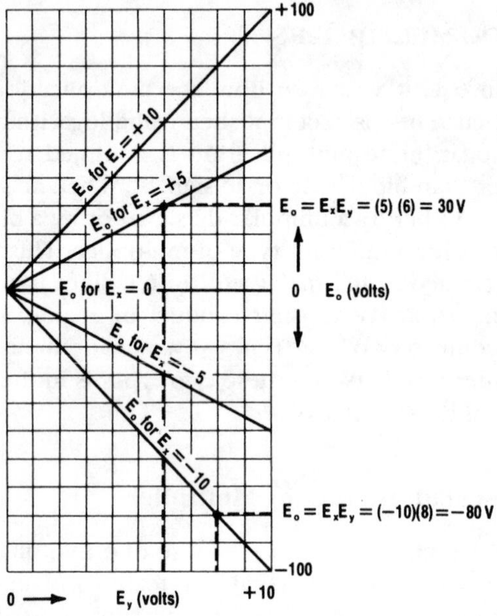


**Fig. 6-12. Two-quadrant multiplier characteristic.**

## 6.2.2 Four-Quadrant Analog Multiplier

Going a step further, the four-quadrant multiplier extends operation of the $E_y$ input into the opposite polarity (Fig. 6-13), allowing unrestricted operation with either $E_x$ or $E_y$ in the range of $\pm 10$ V. For operation with $E_y$ from 0 V to +10 V, operation of a four-quadrant multiplier is completely like that of a two-quadrant type. This is shown by the right half of Fig. 6-13. However, for operation with $E_y$ from 0 V to −10 V, the outputs from a four-quadrant type are as shown in the left half of Fig. 6-13.

To illustrate four-quadrant operation in $E_x$ and $E_y$, let $E_x = +5$ V and $E_y = -6$ V. The output is read from the curve of $E_o$ for $E_x = +5$ V in the lower left quadrant. For this example, $E_o = -30$ V. Note that any product obtained in the lower two quadrants will

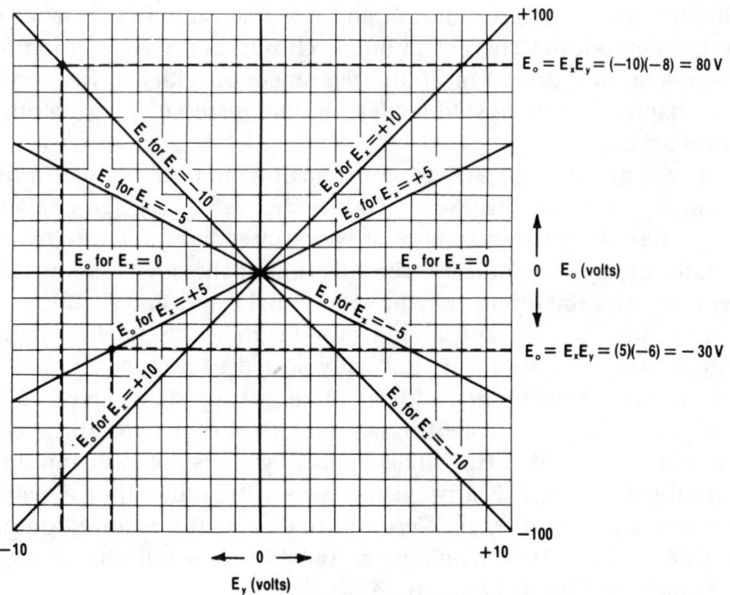


**Fig. 6-13. Four-quadrant multiplier characteristic.**

be negative. As a second example, let $E_x = -10$ V and $E_y = -8$ V. The output is in the upper left quadrant on the $E_x = -10$-V line; $E_o = 80$ V. This is the correct mathematical expression for the product of two negative numbers—a positive result. The output of a four-quadrant multiplier will be the product of the two inputs, always with the correct sign. Finally, as with the two-quadrant type, the output will always be zero if either $E_x$ or $E_y$ is zero. In operation, a four-quadrant multiplier may be considered fully symmetrical. Either input offers control of amplitude and sign, and the output is always correct for instantaneous sign(s) and magnitude(s) of $E_x$ and $E_y$.

So far we have spoken of multipliers in general terms only. To move into practical considerations requires some further definition of multiplier considerations and an understanding of the operation of typical devices. There are a number of circuit techniques for performing analog multiplication. Among these are the logarithmic technique mentioned earlier, the quarter-square technique (piece-wise approximation), pulse-width and/or pulse-height modulation techniques, and variable transconductance methods. Of these, the variable transconductance multiplier is the simplest and most economical, and it is realizable in a number of IC configurations. This technique is chosen as the foundation for discussion in this section. In the references provided at the end of the chapter, the interested reader can find material covering other approaches.

The symbolic representation for analog multipliers used in this book is shown in Fig. 6-14. The symbol is largely self-explanatory—like the op-amp symbol, it is not intended to denote circuit details, but simply indicates the function of the device. As will be seen shortly, the inputs may be differential for $E_x$ and $E_y$, allowing a great deal of sign and scale-factor flexibility. Also, the output signal from a multiplier is normally scaled to a range of ±10 V to provide compatibility with standard op-amp signal levels. Typically, the output is scaled by a factor of K = 0.1, to allow two full-scale inputs of ±10 V. This in no way alters the basic multiplication principles; it is a provision for setting the output dynamic range at a practical level. Were this not done, the required output with $E_x = E_y = 10$ V would be ±100 V, as shown in the theoretical graphs of Figs. 6-12 and 6-13.

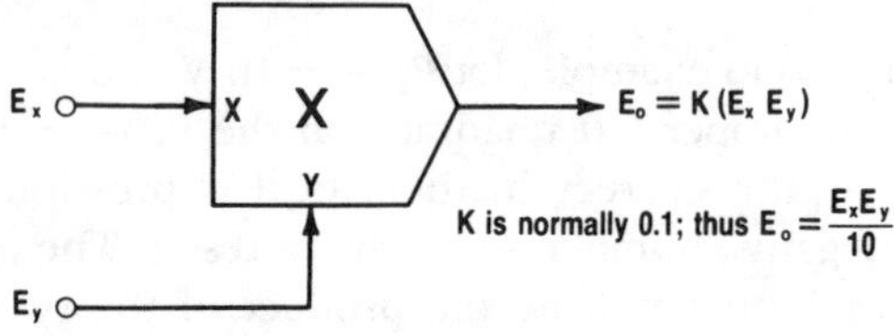


All voltages ($E_x$, $E_y$, and $E_o$) may be single-ended or differential.

**Fig. 6-14. Symbolic multiplier representation.**

### 6.2.3 Basic Arithmetic Functions Using Multipliers

#### *Division*

The process of multiplication is a function upon which many other operations are based. Division can be regarded as multiplication by the reciprocal of the divisor; thus, $E_o = E_x/E_y$ can also be written $E_o = (E_x)(1/E_y)$. Since we can generate the reciprocal of a transfer function by placing it within a feedback loop, division is accomplished as shown in Fig. 6-15.

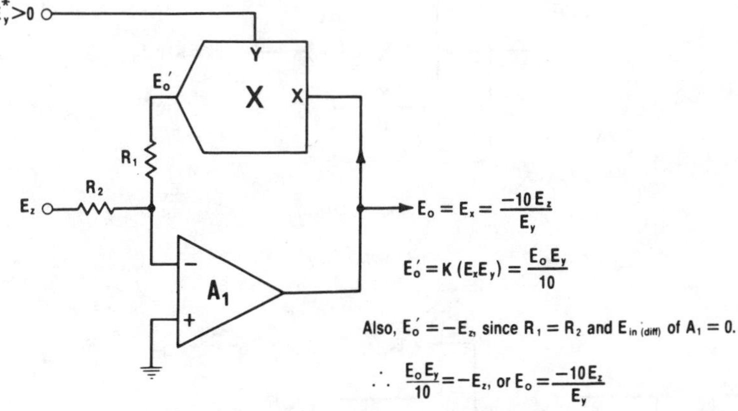


**Fig. 6-15. Division using a multiplier.**

The output of a multiplier is the scaled product of its two inputs ($E_x$ and $E_y$). In this case, $E_o' = K(E_xE_y) = 0.1\ (E_oE_y)$. Since the $E_o'$ voltage also feeds the summing point of $A_1$ through resistor $R_1$, it must also be true that $E_o' = -E_z$, since both these voltages must sum to zero at the summing point of $A_1$. Substituting,

$$E_o' = 0.1(E_oE_y) = -E_z$$

which yields for $E_o$

$$E_o = \frac{-10E_z}{E_y}$$

It should be noted that for this circuit to function correctly, $E_y$ must always be a positive voltage. If $E_y$ goes negative, the multiplier feedback becomes positive and a latch-up condition of $A_1$ occurs.

### *Squaring and Higher Powers*

An obvious use for a multiplier is the multiplication of a signal by itself, or the square function, illustrated in Fig. 6-16A. Since $E_o = E_xE_y/10$, and $E_x$ and $E_y$ are identical signals, the output is then $E_o = E_x^2/10$. In similar fashion, the output of a squaring circuit may be fed into a second multiplier (Fig. 6-16B) along with the original signal to obtain a cube function: $E_o = E_x^2E_x = E_x^3$. This concept can be extended to generate high-order functions to whatever degree necessary.

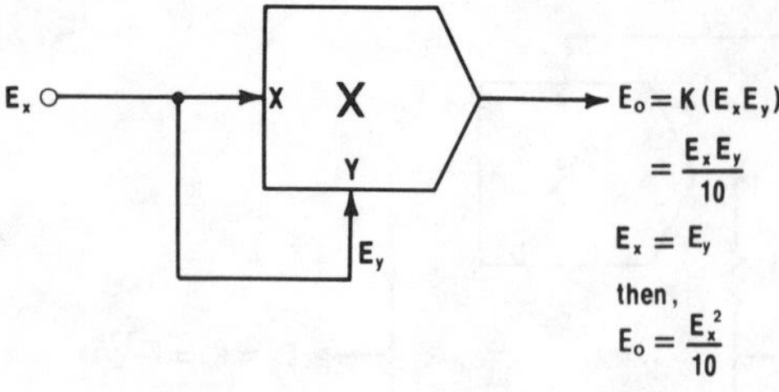


*(A) Squaring.*

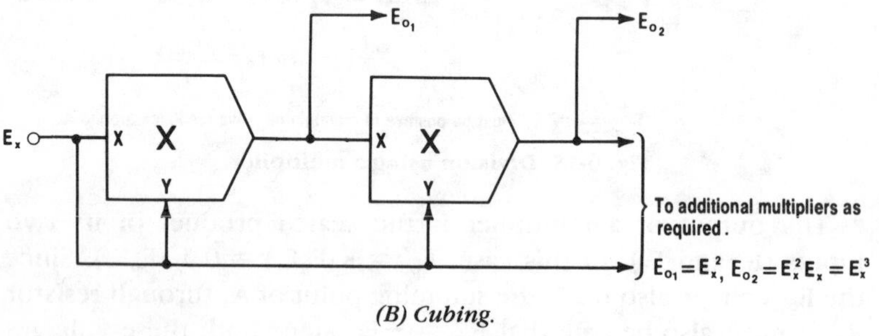


*(B) Cubing.*

**Fig. 6-16. Generating powers with a multiplier.**

### *Square Root*

A square-root function can be realized by placing a squarer within the feedback loop of an op amp, as in Fig. 6-17. Since the output of the multiplier is $E_o^2/10$, which will always be positive, the equal values of $R_1$ and $R_2$ will cause this value to be equal to $-E_z$. Thus, input $E_z$ must be *negative* to produce a positive output at the amplifier. Finally,

$$\frac{E_o^2}{10} = -E_z$$

and if $E_z < 0$,

$$E_o^2 = 10\ E_z$$

so,

$$E_o = \sqrt{|10E_z|}$$

These are the basic arithmetic functions that can be generated using multipliers. There are elaborate variations and combinations

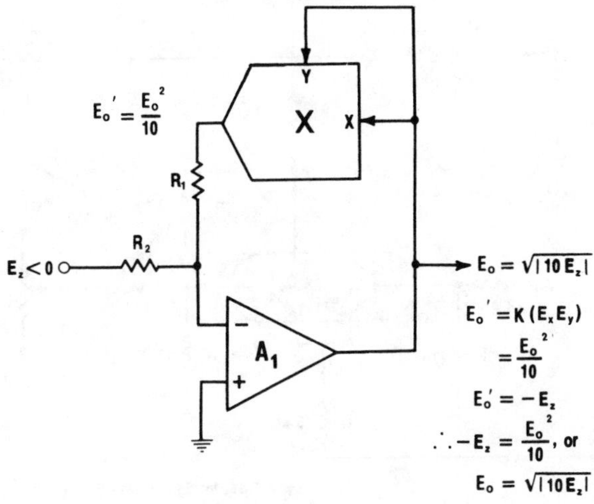


**Fig. 6-17. Square-root generator using a multiplier.**

of basic arithmetic functions, but they will not be detailed here. The interested reader should consult the references listed at the end of the chapter for further information.

### 6.2.4 Transconductance Multipliers

Virtually all popular IC multipliers operate on the principle of variable transconductance. A very rudimentary circuit that performs two-quadrant multiplication using this technique is shown in Fig. 6-18; it will be used as a developmental model for the more detailed structures to follow.

The circuit works by varying the emitter current of a matched pair of bipolar transistors. If the differential voltage, $E_x$, applied to $Q_1$-$Q_2$ is zero, the output is zero. As $E_x$ varies, the output signal ($I_o$) from $Q_1$-$Q_2$ will be

$$I_o = \frac{q}{2\kappa T} I_E E_x$$

This differential output is applied to $A_1$, a differential current-to-voltage converter. The output from $A_1$ will be

$$E_o = I_o R_o = \left(\frac{q}{2\kappa T} I_E E_x\right) 2R_{10}$$

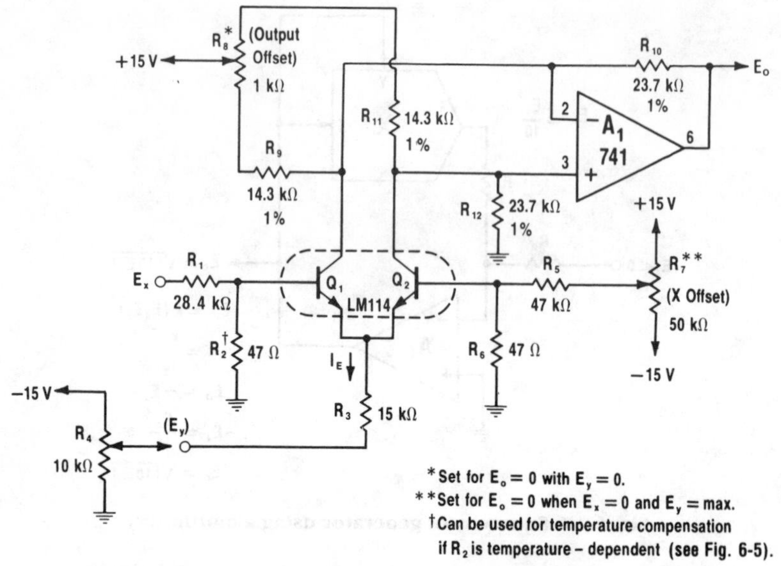


**Fig. 6-18. Basic two-quadrant transconductance multiplier.**

Finally, if we vary the emitter currents of $Q_1$ and $Q_2$ with a second input, $E_y$, the value of $I_E$ will be approximately $I_E = E_y/R_3$. This can be substituted into the equation for $E_o$ and, with some rearrangement,

$$E_o = E_x E_y \left(\frac{q}{2\kappa T}\right)\left(\frac{2R_{10}}{R_3}\right)\left(\frac{R_2}{R_1 + R_2}\right)$$

From this, it can be seen that the circuit accomplishes multiplication of $E_x$ and $E_y$ if the other terms of the equation are considered constant. Note that this expression includes the attenuation of the $R_1$-$R_2$ divider.

An examination of Fig. 6-18 will reveal that this circuit responds only to single polarities of $E_y$ (negative in this example), but to bipolar signals at $E_x$. This is (by definition) two-quadrant multiplication. As $E_y$ is varied to change $I_E$, the collector voltages of both $Q_1$ and $Q_2$ rise and fall with the change in common-mode signal. $A_1$ and the differential bridge resistors ($R_9$ through $R_{12}$) reject this common-mode signal while amplifying the desired differential signal ($I_o$) from $E_x$. Balance of the bridge is critical to good rejection of the $E_y$ signal at the output; therefore, the bridge should either use close-tolerance (0.1%) resistors or be nulled, as shown, with $R_8$. Resistor $R_7$ zeros the offset voltage of $Q_1$-$Q_2$.

An inherent property of this circuit is that linearity holds only for relatively small-signal amplitudes of $E_x$. Directly at $Q_1$, this limit is ±10 mV for 1% linearity; thus, an attenuator network such as $R_1$-$R_2$ is necessary to scale the input range up to the standard ±10-V levels. Also, to compensate the TC of gain due to the $\kappa T/q$ factor, $R_2$ can be made a thermistor with a TC of +0.3%/°C, as in Fig. 6-5.

### 6.2.5 Improving the Basic Two-Quadrant Multiplier

The circuit of Fig. 6-19 is an improvement on the basic multiplier circuit in two major respects. First, the nonlinearity displayed by $Q_1$-$Q_2$ at the $E_x$ input has been compensated by developing a predistorted voltage drive with the feedback loop around $A_2$ and $Q_3$-$Q_4$. $Q_3$ and $Q_4$ are a differential transistor pair, operated at a fixed emitter current. They are connected so that the current swing in $Q_3$ must equal the current swing in $R_1$. $A_2$ will force this current to be linear, even though the voltage drive at the input to $Q_4$ (and $Q_1$) will be nonlinear. This method ensures a linear current output from the differential pair, $Q_3$-$Q_4$; therefore, an identical pair ($Q_1$-$Q_2$) can also be driven in parallel, with a similar linear output. $R_1$ is selected for the desired full-scale input current, $I_x$; in this case, $I_x = \pm 250\ \mu A$ at ±10 V.

The other improvement is the method of current drive for the controlled pair, $Q_1$-$Q_2$. The simple resistive current drive ($R_3$) used in Fig. 6-18 is not entirely satisfactory in providing a linear variation of $I_E$ as $E_y$ approaches zero. A feedback configuration improves linearity, and this is accomplished by a bilateral VCCS (see Chapter 4), consisting of $A_3$ and associated components. The VCCS circuit supplies an output current that is linear at any level of $E_y$, thus extending linearity to cover very small $E_y$ voltages. $R_{1A}$ trims the overall circuit scale factor by setting the desired value of $I_x$ when $E_x$ is at a maximum. The other trim adjustments perform functions similar to those of Fig. 6-18. Optimum results with this circuit require that the four transistors be matched fairly well to track with temperature. The LM394 pairs provide for this; as will a (single) MAT-04. To a lesser extent, array types such as the CA3046 will be useful.

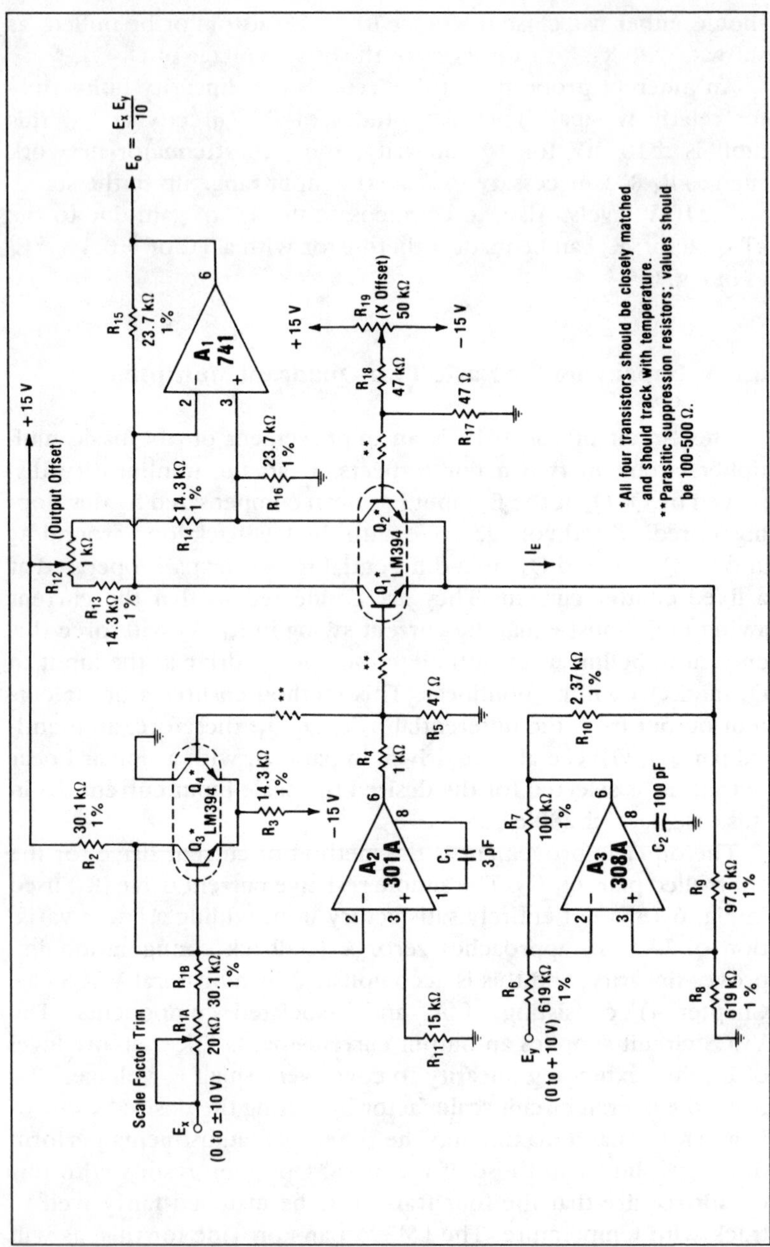


**Fig. 6-19. Improved two-quadrant multiplier.**

### 6.2.6 Monolithic IC Multipliers

The circuitry described so far, while useful, is intended primarily to demonstrate the general principles of variable-transconductance multiplication. This includes not only the basic premise of gain variation by control of emitter current, but also the requirements for linearization and conversion from differential to single-ended operation.

Commercial IC multipliers are typically four-quadrant extensions of the basic two-quadrant concept. These monolithic units often utilize a circuit structure similar to Fig. 6-20A.* The circuit can be viewed as a pair of cross-connected differential pairs ($Q_1$-$Q_2$ and $Q_3$-$Q_4$), fed by controlled emitter currents (from $Q_5$-$Q_6$), and having a common differential base drive ($E_x$). If the circuit is sectioned between $Q_2$-$Q_3$ and $Q_5$-$Q_6$, each half is a single differential pair similar to $Q_1$-$Q_2$ in Fig. 6-18. In fact, this circuit is actually a pair of two-quadrant multipliers interconnected to form one composite four-quadrant multiplier. Operation of the circuit is best understood by examining the output current(s) of transistors $Q_1$ and $Q_4$.

Current-source transistors $Q_5$ and $Q_6$ transform one input voltage ($E_y$) into complementary currents $I + I_y$ and $I - I_y$. Regarding output terminal $I_{o_1}$ first, we can see how multiplication occurs. If $E_x$ and $E_y$ have the polarities shown (call these positive), $Q_4$ will be on and receive the current $I + I_y$. When $E_x$ is negative, $Q_2$ will be on and the $I_{o_1}$ output will be $I - I_y$. This illustrates the control of polarity that $E_x$ exercises over the $I + I_y$ currents. With $E_x$ positive, $Q_4$ will be on and the $I_{o_1}$ will be the current from $Q_6$. If $E_y$ is also positive at the same instant, this value of $I_{o_1}$ will be $I + I_y$; if $E_y$ is negative instead, the value of $I_{o_1}$ is $I - I_y$. If both $E_x$ and $E_y$ are negative, $I_{o_1}$ becomes the current from $Q_5$, or $I + I_y$.

When the preceding four cases are summarized, it can be seen that operation is four-quadrant—always with the correct sign for any combination of $E_x$ and $E_y$. An additional feature of operation occurs when either $E_x$ or $E_y$ is zero. If $E_y = 0$, $I_y = 0$, and $I_{o_1} = I$. As $E_x$ alternates polarity, $Q_4$ turns on as $Q_2$ turns off, and vice versa. However, the currents in $Q_5$ and $Q_6$ remain the same and there is no net change in $I_{o_1}$. Likewise, if $E_x = 0$, no net change in $I_{o_1}$ will occur as $E_y$ is varied, because the output will contain equal

* This "gain cell" multiplier block was first described by B. Gilbert. See reference 14 at the end of the chapter.

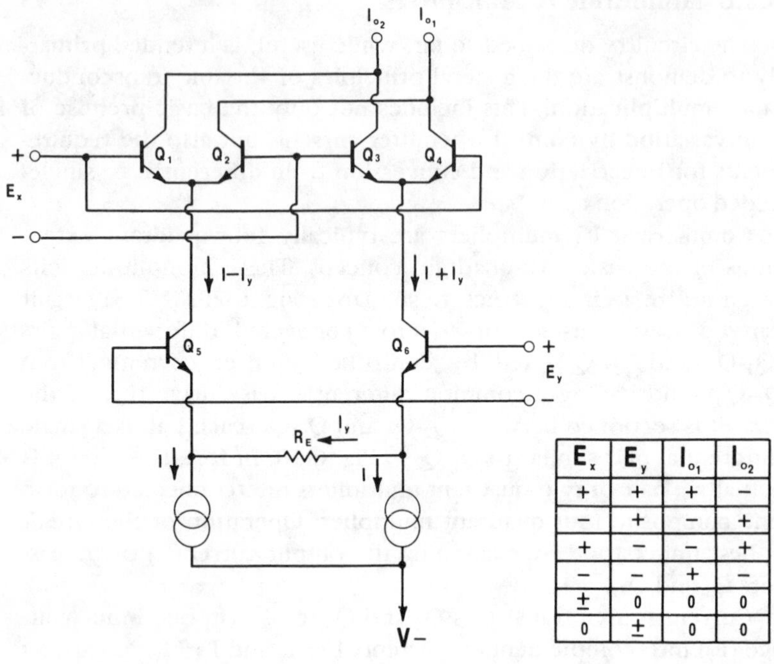


| $E_x$ | $E_y$ | $I_{o_1}$ | $I_{o_2}$ |
|---|---|---|---|
| + | + | + | − |
| − | + | − | + |
| + | − | − | + |
| − | − | + | − |
| ± | 0 | 0 | 0 |
| 0 | ± | 0 | 0 |

*(A) Circuit structure.* *(B) Input/output polarities.*

**Fig. 6-20. Basic operation of a monolithic four-quadrant multiplier.**

proportions from $Q_2$ and $Q_1$. Due to symmetry, the currents $I + I_y$ and $I - I_y$ are equal and opposite and will, therefore, cancel each other.

The operation of the circuit for different conditions of $E_x$ and $E_y$ is summarized in Fig. 6-20B. The complementary output ($I_{o_2}$) will always receive the negative of the current appearing at $I_{o_1}$. Thus, the output signals at $I_{o_1}$ and $I_{o_2}$ are differentially multiplied *currents.* To convert them back to a voltage requires a differential current-to-voltage converter such as that used in Fig. 6-18 or 6-19.

At this point, it should be noted that there is no fundamental guide to the designation of X and Y inputs. The multiplication principle is the same regardless of which input is termed "X" or "Y." This is mentioned only because there is no complete standardization of X and Y designations among manufacturers; however, this should not be a handicap if the fundamental principles are kept in mind.

To complete the multiplier, a diode linearity correction circuit is added to develop a nonlinear voltage drive for the bases of $Q_1$-$Q_4$. This linearizes the X input similar to the circuit used in Fig. 6-19. Darlington devices are used for the $Q_5$-$Q_6$ voltage-to-current converters to minimize error, with similar devices added to the X input.

### *The MC1495*

A complete circuit that accomplishes linear four-quadrant multiplication using the Motorola MC1495 is shown in Fig. 6-21. Both the X and Y inputs are fully differential and are scaled in sensitivity with an external emitter resistor. The differential- and common-mode input range is ±10 V. The flexibility of this multiplier block is maximized since virtually all circuit nodes are externally accessible, allowing operation in a variety of modes.

Due to the additional stage in the $E_x$ input of this multiplier, a redefinition of polarities is necessary. This is shown in the functional block diagram of Fig. 6-22. Pins 4 and 8 are the differential $E_x$ inputs, and pins 9 and 12 are the differential $E_y$ inputs. Current scaling in both inputs is set by resistors $R_x$ and $R_y$, which set currents $I_x$ and $I_y$ according to $I_x = E_x/R_x$ and $I_y = E_y/R_y$. In using the device, either sign may be realized by applying an input signal to the inverting or noninverting input as desired, with scaling set by $R_x$ and $R_y$. A summary of input/output relations is also given in Fig. 6-22.

### *The AD532 and AD534*

As can be appreciated from the preceding discussion, the MC1495 is in actuality a "gain-cell" chip, and requires much in the way of support circuitry to implement complete multipliers and dividers. Ready-to-use complete multiplier ICs are also available from such firms as Analog Devices and Burr-Brown. In this section, we will examine the basic applications of some of the representative units of this type.

The Analog Devices AD532 and AD534 are complete multiplying ICs, with pretrimmed adjustments for scale factor and offset. Both devices perform the basic multiplying function of $E_o = E_xE_y/10$, but they do so with different degrees of accuracy and flexibility. Functional block diagrams of both devices are shown in Fig. 6-23.

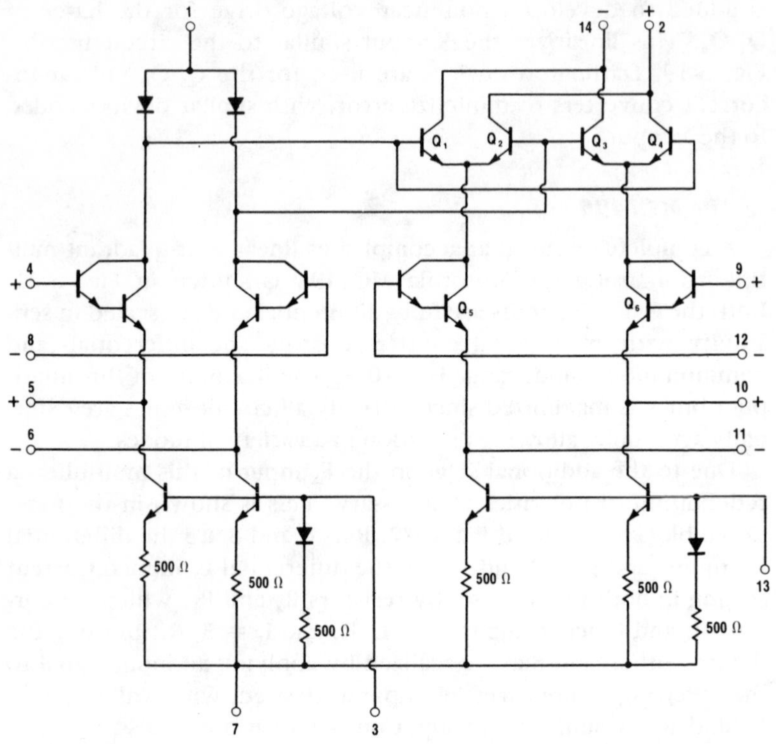


**Fig. 6-21. Motorola MC1495 four-quadrant multiplier.**

The AD532, shown in Fig. 6-23A, consists of differential input amplifiers for the X and Y channels a multiplier cell, and an output amplifier. This device is capable of a total error of 2% or less for the AD532J version, or 1% for the AD532K version. The dynamic range of the inputs and the output is ±10 V, and the device uses ±15-V power supplies.

The AD534, shown in Fig. 6-23B, consists of differential input amplifiers for the X and Y channels, a multiplier cell, a reference and bias circuit, an output amplifier, and differential Z inputs. This device is very accurate, with total errors of 1%, 0.5%, and 0.25% for the AD534J, K, and L grades, respectively.

The relative brevity of this section precludes going into all the details of the operation of these devices, and the interested reader should consult the references at the end of the chapter.

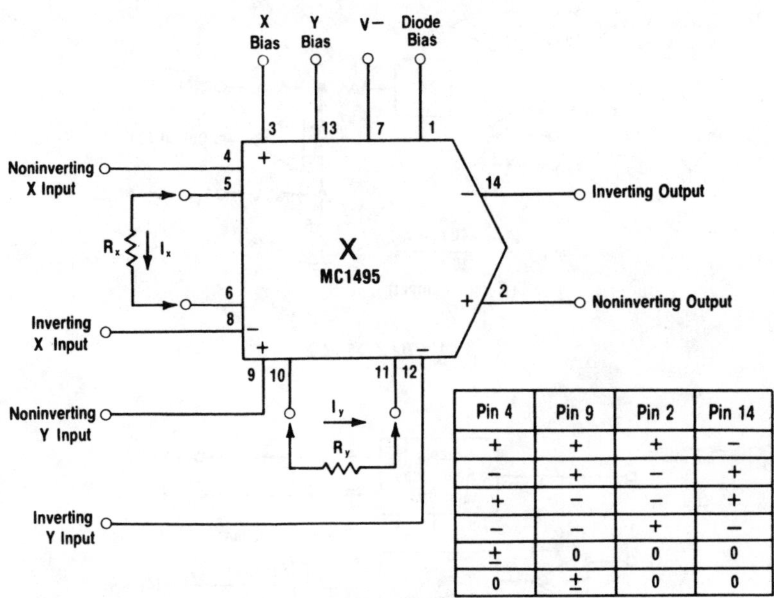


| Pin 4 | Pin 9 | Pin 2 | Pin 14 |
|---|---|---|---|
| + | + | + | − |
| − | + | − | + |
| + | − | − | + |
| − | − | + | − |
| ± | 0 | 0 | 0 |
| 0 | ± | 0 | 0 |

**Fig. 6-22. Functional block diagram of the Motorola MC1495 multiplier.**

## 6.2.7 Arithmetic Functions Using IC Multipliers

In this section, the basic arithmetic functions are illustrated, as they can be realized with pretrimmed IC multipliers.

### *Multipliers*

Two simple, yet sufficiently complete circuits for multiplying two voltages, X and Y, using the AD532 and AD534, are shown in Fig. 6-24.

In the AD532 circuit (Fig. 6-24A), connecting the Z input to the output sets the device up for multiplication. Although it is not basically necessary, overall dc accuracy can be optimized by the optional output offset trim, using the 20-kΩ potentiometer. The input voltages, X and Y, can be either differential or single ended, as appropriate. The fully differential inputs optimize input flexibility.

The AD534 circuit (Fig. 6-24B) uses a similar differential input scheme, but also adds the flexibility of differential Z inputs. For normal uses, $Z_1$ is tied to the output and $Z_2$ is grounded. If desired,

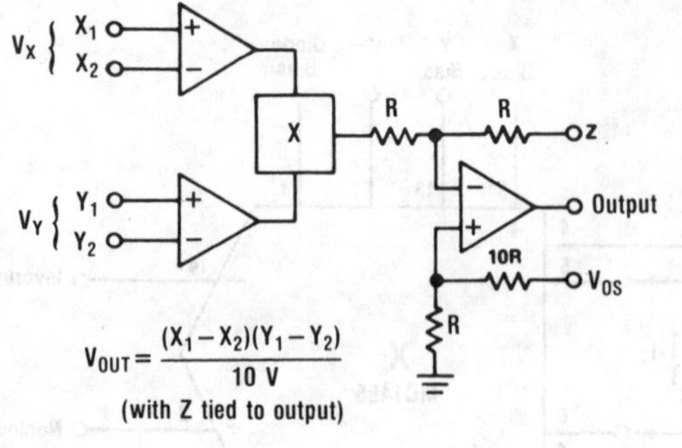


*(A) The AD532.*

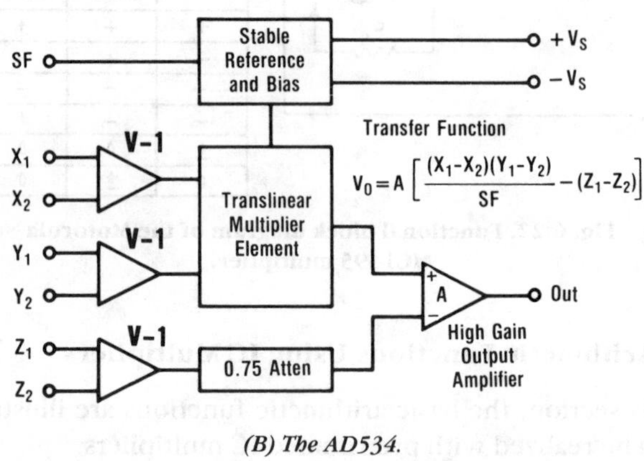


*(B) The AD534.*

**Fig. 6-23. Functional block diagrams of pretrimmed IC multipliers.**
*(Courtesy Analog Devices, Inc.)*

the output can be referred to a reference voltage by applying the reference to the $Z_2$ input (up to ±10 V).

Note that both devices are scaled by a 1/10 factor, to allow a ±10-V output for ±10-V inputs (as discussed in the introduction).

### *Dividers*

The arithmetic function of division can be accomplished as illustrated in Fig. 6-25. As can be noted, this function can be somewhat more difficult to implement, depending on which device is actually to be used.

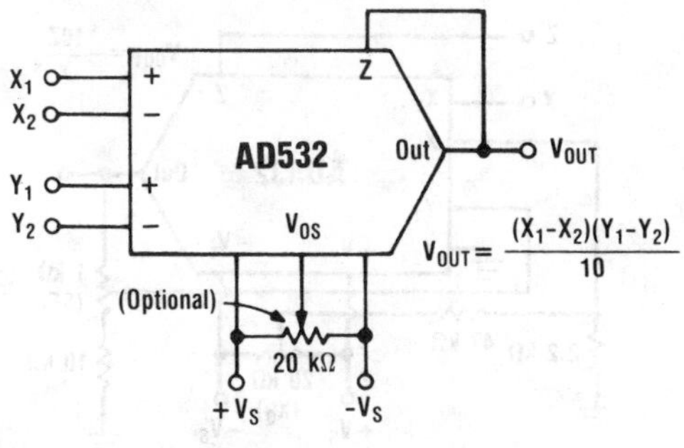


*(A) Using the AD532.*

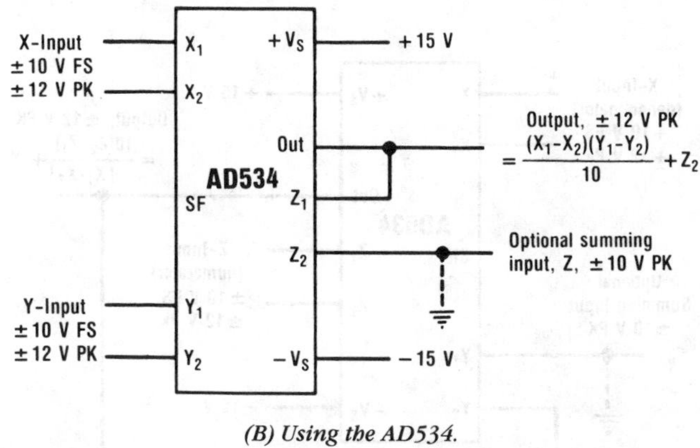


*(B) Using the AD534.*

**Fig. 6-24. General-purpose four-quadrant multiplier circuits using pretrimmed devices.** *(Courtesy Analog Devices, Inc.)*

The AD532 is set up for division as shown in Fig. 6-25A. This circuit divides an input voltage, Z, by a second voltage, X, with a scaling constant of 10. The 1-kΩ potentiometer trims the scale factor with the X input at −10 V, and the $X_o$ offset potentiometer trims the offset with Z at +10 V; X at −1 V. Note that the X input is restricted to negative-range voltages only (to avoid latch-up), and the most useful dynamic range is between −0.5 V and −10 V.

The AD534 is used as a divider as shown in Fig. 6-25B. This circuit divides a differential voltage, Z, by a second differential voltage, X, with a scaling constant of 10. If the (optional) Y input is used, a constant proportional to the Y voltage is added to the output.

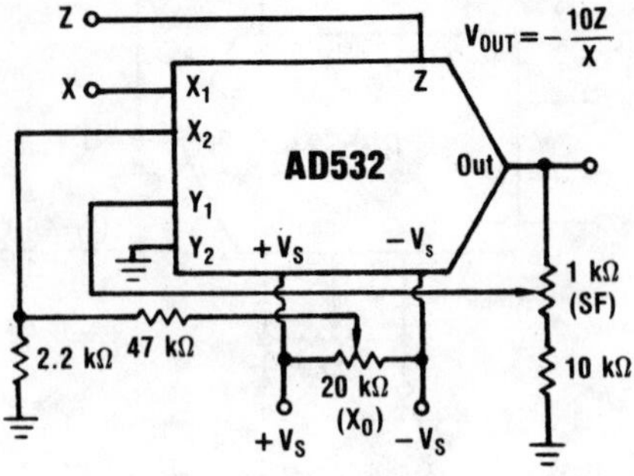


*(A) Using the AD532.*

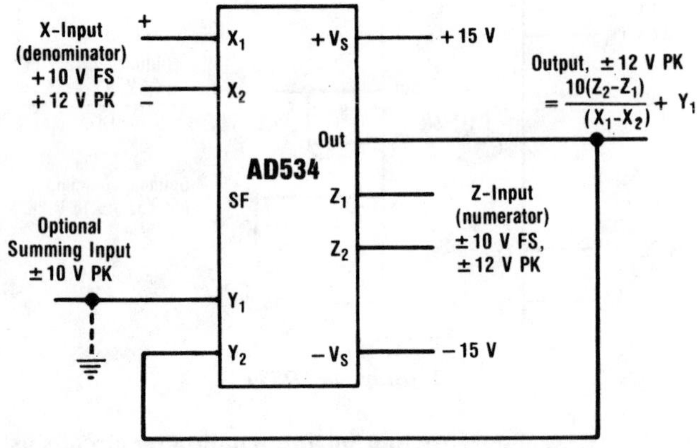


*(B) Using the AD534.*

**Fig. 6-25. Divider circuits using pretrimmed devices.**
*(Courtesy Analog Devices, Inc.)*

Using the AD534 (K or L) with a +10-V to +1-V X-input range, the error will be no more than 1% with this circuit. Greater output gain (increased scale factor) can be obtained simply by dividing down the feedback applied to the $Y_2$ input (a 2/1 divider will, for example, increase the scale factor by 2/1).

### *Square Rooters*

Circuits which provide the mathematical function of square root are shown in Fig. 6-26.

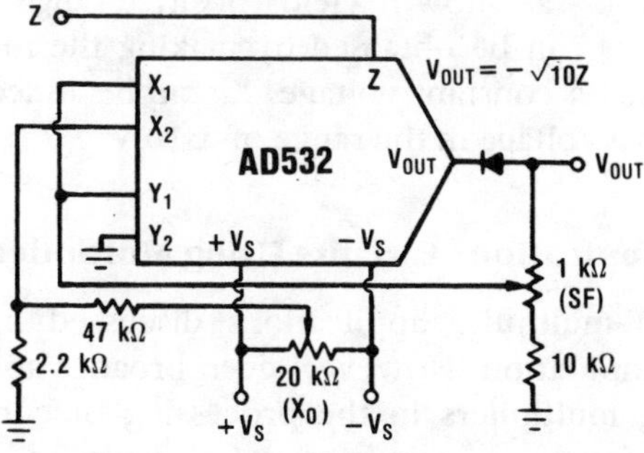


*(A) Using the AD532.*

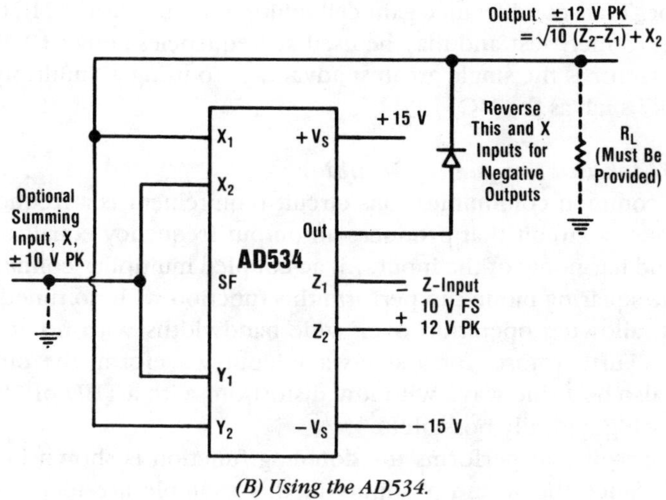


*(B) Using the AD534.*

**Fig. 6-26. Square-root circuits using pretrimmed devices.**
*(Courtesy Analog Devices, Inc.)*

The AD532 circuit, shown in Fig. 6-26A, solves for the square root of the product of the input voltage, Z, times the constant, 10, with a negative range of output. The input voltage is confined to positive ranges only, and a diode must be used (a 1N914 type is adequate), as shown, to prevent latch-up. To optimize accuracy, trim the offset terminal (see Fig. 6-24A) for −1 V dc out with +0.1 V dc in, and also trim SF and $X_0$.

The AD534 is used in a square-root mode as shown in Fig. 6-26B. The Z input has a polarity as indicated, and a diode must be used. The output (as shown) yields positive-range voltages, but negative outputs can be obtained by making the indicated alterations. Note that a constant voltage, X, can be added (optionally) by summing in a voltage in the range of ±10 V.

### 6.2.8 Communications Circuits Using Multipliers

So far, the multiplier applications discussed have involved arithmetic computation. However, even broader applications are possible using multipliers in the processing of communications signals. These functions include modulation (am and dsb), demodulation, electronic gain control, and many others. When operating at high frequencies, the output op amp becomes a limiting factor. Transconductance gain-cell multipliers such as the MC1495 are extremely fast and may be used at frequencies above 10 MHz. This factor is the single greatest advantage to using a "multiplying block" such as the MC1495.

#### *Wideband Frequency Doubler*

A common communications circuit requirement is a frequency doubler, a circuit that produces an output frequency equal to the second harmonic of the input. An ac-coupled multiplier connected in the squaring mode can perform this function with no tuned circuits, allowing operation over wide bandwidths with no adjustment. Furthermore, for a sine-wave-input waveform, the output will also be a sine wave with low distortion, with a THD of 1% or less being typically possible.

A circuit that performs the doubling function is shown in Fig. 6-27. Since the X and Y inputs in this example are identical, a common bias-setting resistor ($R_8$) is used. Outputs can be taken from either pin 2 or pin 14, depending on the desired phase. $C_{O_1}$-$C_{O_2}$ should be chosen for negligible reactance with respect to $R_{L_1}$ and $R_{L_2}$, the load impedance of the following stage. The circuit will work as shown with $R_7$ grounded, but an offset adjustment can be utilized for minimum distortion if desired.

The high-frequency performance of this circuit is dependent on the RC time constant at several circuit points. $R_{10}$-$R_{11}$ and/or $R_L$ may be reduced to decrease the effects of stray capacitance at the output, and buffering with an emitter follower will also help. Similarly, reducing $R_3$ and $R_6$ will reduce the effects of stray capacitance at the input. With circuit impedances optimized for high-

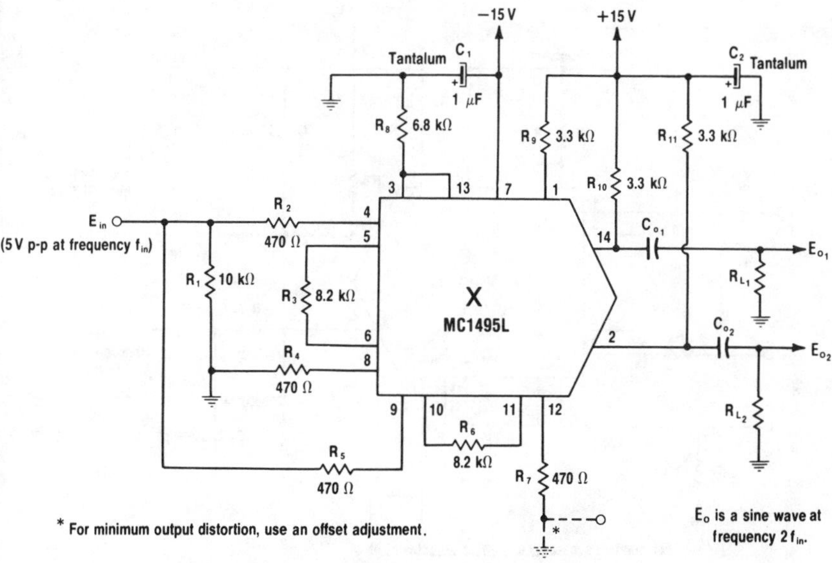


**Fig. 6-27. Wideband frequency doubler.**

frequency performance, this basic configuration can operate at frequencies as high as several megahertz.

### *Balanced Modulation/Demodulation*

A circuit very similar to the previous one is the balanced modulator/demodulator shown in Fig. 6-28. This is virtually the same circuit as Fig. 6-27, but with the X and Y inputs separated. If two separate frequencies, $f_c$ and $f_m$, are applied to the two inputs, the output will contain neither of the two original inputs, only their products, in the form of $f_c \pm f_m$. There is no restriction on which input is used for the carrier or modulation, since they are symmetrical. There is also no limitation on either of the two frequencies; they may be widely separated (as in conventional modulation systems) or nearly identical, because of the untuned wide-bandwidth method of modulation.

The circuit will work with $R_4$ and $R_7$ grounded, as shown, but for optimum suppression of the input signals, both channels should be nulled. This is accomplished by applying either input signal and then adjusting the offset control on the opposite channel for minimum output. With nulling, carrier suppression can

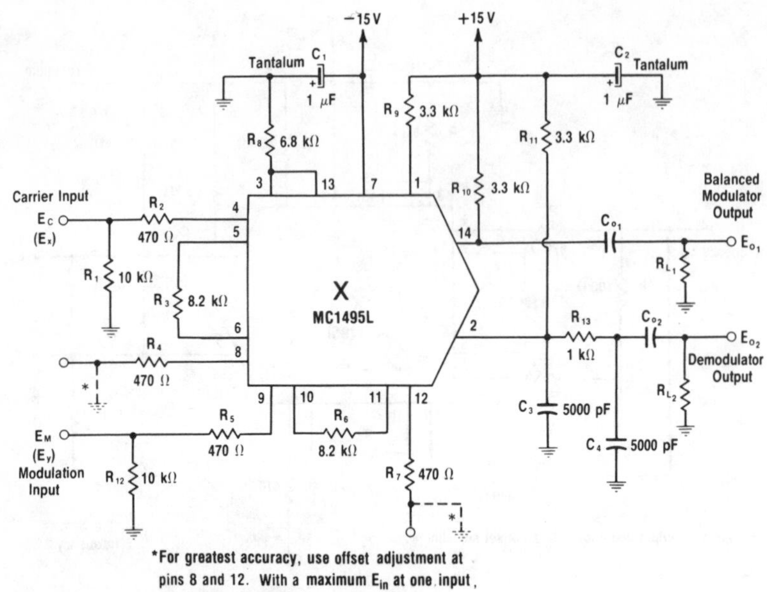


**Fig. 6-28. Balanced modulator/demodulator.**

approach 60 dB. If very wide bandwidth is desired, the time-constant considerations discussed for Fig. 6-27 should be applied.

The modulator/demodulator circuit can be used for am or ssb demodulation by applying a modulated carrier to one input and an unmodulated carrier of identical frequency and phase to the second input, and by adding a low-pass filter to the output. The multiplying action of the constant-level carrier and the varying-level modulated wave results in a series of rectified output pulses corresponding to the modulation envelope. The low-pass filter then removes the residual carrier components, leaving only the modulation. This form of demodulation is very linear and has a wide dynamic range, since it does not have the "diode threshold" of standard am demodulators. The low-pass filter used to recover the modulation is formed by $R_1$ (or $R_{10}$ if taken from pin 14) and $C_3$ plus $R_{13}$-$C_4$. $C_{o_2}$ is a coupling capacitor chosen for low reactance in relation to $R_{L_2}$. The demodulator circuit, like the modulator, may be used on either output terminal. Offset adjustments are generally not necessary with the demodulator, however, since the output filter rejects any residual carrier which might leak through.

### *AM Modulator*

Conventional amplitude modulation can also be accomplished with a multiplier as shown in Fig. 6-29. In reality, am is a variation in level of a constant carrier. In this circuit, the carrier is applied to one input of the modulator, and a fixed dc level (bias) is applied to the second input. This results in a constant-level carrier output. If modulation is then applied in addition to the bias voltage, with a p-p amplitude of twice the bias level, 100% am will result.

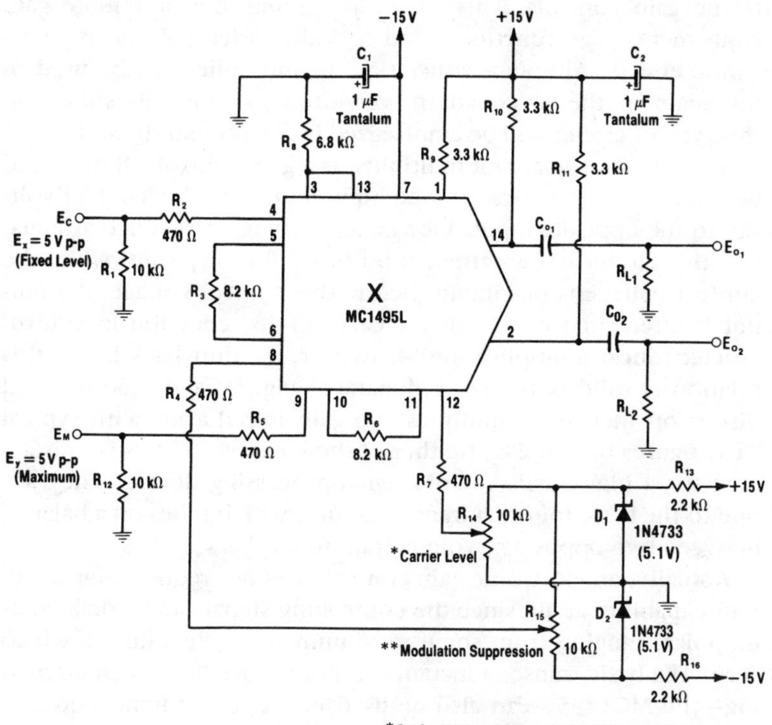


**Fig. 6-29. Am modulator.**

In Fig. 6-29, $R_{14}$ inserts a fixed 2.5-V bias at the inverting Y input, producing a constant-level carrier at the output. A 5-V p-p signal at the modulation input will alternately double and cancel the carrier level at the output by summing with this bias. With the values shown, a 5-V p-p signal at the modulation input will result in 100% modulation. The offset adjustment for the Y channel, $R_{15}$, is used to suppress any direct feedthrough of the modulation signal. The circuit can be optimized for high-frequency carrier operation by reducing the values of $R_{10}$ and $R_{11}$ at the output, and $R_3$ at the input.

### *Electronic Gain Control*

As was mentioned briefly in the introductory comments on multipliers, both two- and four-quadrant types can be used as electronic gain controls. This allows programmable or remote gain adjustments, agc functions, and a wide variety of special electronic effects. Although either type of multiplier can be used in this manner, the two-quadrant version is more ideally suited, so this type of circuit will be emphasized in the present discussion.

To use a four-quadrant multiplier as a gain control, all that need be done is apply a signal to one input and a variable control voltage to the opposite input. Viewed as a voltage-controlled attenuator, this circuit has an attenuation of 0 dB (unity gain) when the control voltage is maximum. Below this control voltage, the output is attenuated by 20 dB for each 20-dB reduction in control voltage (linear multiplication). However, the drawback is that this relation is valid only over a dynamic range of about 60 dB, and distortion increases rapidly as the gain is reduced, with typical THD figures of 1 to 2%. Neither of these characteristics is satisfactory for a high-quality, audio-signal-processing device. They are due to the basic four-quadrant structure, which relies on a balance between two opposing two-quadrant multipliers.

Actually, an electronic gain control does not require four-quadrant capability at all, since the controlling signal can be defined as unipolar, ranging from some maximum full-scale value down to zero. The basic transconductance multiplier we have been discussing—the MC1495—can also be used in a two-quadrant mode, as illustrated in Fig. 6-30.

Conversion of the MC1495 to two-quadrant operation involves only the disabling of one of the two-quadrant multipliers shown in Fig. 6-21. In the version shown here (Fig. 6-30), $Q_5$ is disabled by grounding the base and emitter (pins 11 and 12), and no bias is supplied to pin 13. The remaining two-quadrant multiplier section using $Q_6$ is biased externally by op-amp $A_2$, which acts as a positive-input VCCS (see Chapter 4). Operated as shown, the circuit resembles the improved two-quadrant multiplier of Fig. 6-19.

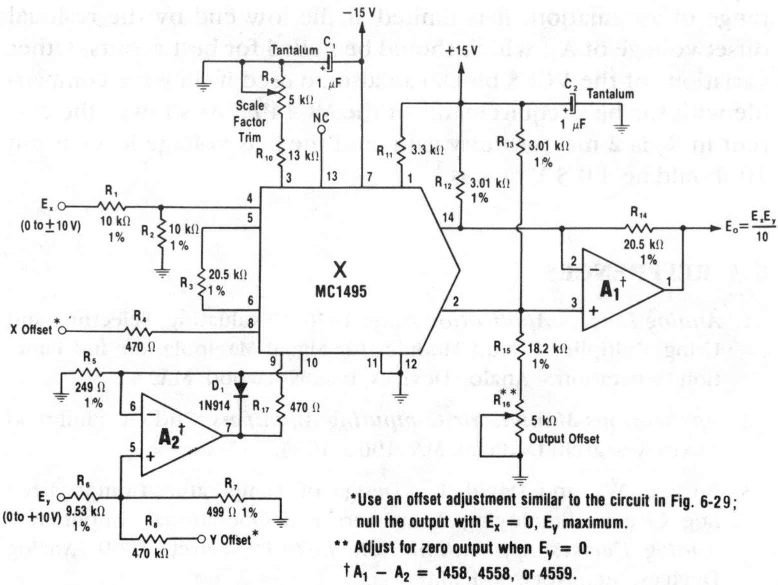


**Fig. 6-30. Two-quadrant multiplier using a four-quadrant device.**

Two-quadrant multiplier operation of a differential pair is precisely controllable over an extremely wide dynamic range—100 dB or better of gain control is readily achievable. In addition, the only significant distortion mechanism is "linearization," which comes from the diodes that process the $E_x$ signal internally. This distortion is mostly from the second harmonic; it can be controlled by scaling $R_3$ ($R_x$) appropriately, thus limiting the $I_x$ current. For example, in the circuit shown, the THD is below 0.5% at a full-scale input of ±10 V, and it reduces to negligible values at lower input levels, where it would typically be operated in an audio circuit.

A further asset of the two-quadrant multiplier over the four-quadrant is in terms of noise. The two-quadrant transconductance multiplier reduces the output noise level along with the signal, whereas a four-quadrant type does not. This factor is important where good signal-to-noise ratios and wide dynamic range are important.

The circuit is otherwise similar in many regards to previous ones, with the exception of the VCCS, $A_2$. The accuracy of this voltage-to-current converter will determine the overall dynamic range of attenuation; it is limited at the low end by the residual offset voltage of $A_2$, which should be nulled for best results. Other variations of the VCCS block can also be used if they are compatible with the bias requirements of the MC1495. As shown, the current in $R_5$ is 2 mA for unity gain, and the bias voltage level at pin 10 should be +0.5 V or less.

## 6.3 REFERENCES


1. *Analog Devices Application Note,* 1970. "Evaluating, Selecting, and Using Multiplier Circuit Modules for Signal Manipulation and Function Generation." Analog Devices, Inc., Norwood, MA.

2. *Applications Manual for Computing Amplifiers,* 2nd ed. Philbrick/Nexus Research, Dedham, MA, 1966, 1985.

3. Borlase, W., and David, E. "Design of Temperature-Compensated Log Circuits Employing Transistors and Operational Amplifiers." *Analog Devices Application Note E020-10,* March 1969. Analog Devices, Inc., Norwood, MA.

4. Burwen, R. "A Complete Multiplier/Divider on a Single Chip." *Analog Dialogue,* Vol. 5, No. 1, January 1971.

5. __________, and Sullivan, D. *AD530 Complete Monolithic MDSSR Technical Bulletin,* July 1971. Analog Devices, Inc., Norwood, MA.

6. Counts, L., Kitchin, C., and Sherman, S. "One-Chip 'Slide Rule' Works with Logs, Antilogs for Real-Time Processing." *Electronic Design,* May 2, 1985.

7. __________. "Complete Monolithic Analog Multifunction Chip." *Analog Dialogue,* Vol. 19, No. 1, 1985.

8. Dobkin, R. C. "Logarithmic Converters." *National Semiconductor Application Note AN-30,* November 1969. National Semiconductor Corp., Santa Clara, CA.

9. __________. "Op Amp Circuit Collection." *National Semiconductor Application Note AN-31,* February 1970. National Semiconductor Corp., Santa Clara, CA.

10. __________. "Precision AC/DC Converters." *National Semiconductor Linear Brief LB-8,* August 1969. National Semiconductor Corp., Santa Clara, CA.

11. Fullagar, D. "The 8007—A High-Performance FET-Input Operational Amplifier." *Intersil Application Bulletin A005,* March 1972. Intersil, Inc., Cupertino, CA.

12. Gilbert, B. "Accurate, Low-Cost, Easy-to-Use IC Multiplier." *Analog Dialogue,* Vol. 11, No. 1, 1977.

13. ________. "A High-Performance Monolithic Multiplier Using Active Feedback." *IEEE Journal of Solid State Circuits,* December 1974.

14. ________. "A Precise Four-Quadrant Multiplier With Sub-Nanosecond Response." *IEEE Journal of Solid State Circuits,* December 1968.

15. Giles, J.N. *Fairchild Semiconductor LIC Handbook.* Fairchild Semiconductor, Mountain View, CA, 1967.

16. Gurski, R. J. "Logarithmic Devices." *The Lightning Empiricist,* Vol. 17, No. 1, March 1969. Philbrick/Nexus Research, Dedham, MA.

17. Huehne, K. "The MC1556 Operational Amplifier and Its Applications." *Motorola Application Note AN-522,* 1970. Motorola Semiconductor Products, Inc., Phoenix, AZ.

18. ________. "Transistor Logarithmic Conversion Using an Integrated Circuit Operational Amplifier." *Motorola Application Note AN-261A,* 1971. Motorola Semiconductor Products, Inc., Phoenix, AZ.

19. Korn, G. A., and Korn, T. M. *Electronic Analog and Hybrid Computers,* 2nd ed., McGraw-Hill Book Co., New York, 1972.

20. "FET Circuit Applications." *National Semiconductor Application Note AN-32,* February 1970. National Semiconductor Corp., Santa Clara, CA.

21. Nelson, C. T. "Super Matched Bipolar Transistor Pair Sets New Standards for Drift and Noise." *National Semiconductor Application Note AN-222,* July 1979. National Semiconductor Corp., Santa Clara, CA.

22. Paterson, W. L. "Multiplication and Logarithmic Conversion by Operational Amplifier-Transistor Circuits." *The Review of Scientific Instruments,* Vol. 34, No. 12, December 1963.

23. Renschler, E. "Analysis and Basic Operation of the MC1595." *Motorola Application Note AN-489,* March 1970. Motorola Semiconductor Products, Inc., Phoenix, AZ.

24. ________. "The Monolithic Multiplier." *Motorola Monitor,* Vol. 7, No. 3, 1969.

25. Rudin, M. "Applying Microvolt/Picoamp Instrumentation Operational Amplifiers." *Precision Monolithics Application Note.* Precision Monolithics, Inc., Santa Clara, CA.

26. Sheingold, D. H. (ed.). *Analog-Digital Conversion Handbook,* Third Edition. Analog Devices, Inc., Norwood, MA, 1986.

27. ________. *Multiplier Application Guide.* Analog Devices, Inc., Norwood, MA, 1978.

28. ________. *Nonlinear Circuits Handbook.* Analog Devices, Inc., Norwood, MA, 1974.

29. "Logarithmic and Exponential Amplifiers With SN72709." *Texas Instruments Application Bulletin CA-149,* March 1970. Texas Instruments Incorporated, Dallas, TX.

30. Tobey, G. E., Graeme, J. G., and Huelsman, L. P. (eds.). *Operational Amplifiers—Design and Applications.* McGraw-Hill Book Co., New York, 1971.

31. Welling, B., and Kinsey, L. "Using the MC1595 Multiplier in Arithmetic Operations." *Motorola Application Note AN-490,* October 1970. Motorola Semiconductor Products, Inc., Phoenix, AZ.

32. Widlar, R. J. "Designing With Super-Beta Transistor Op-Amp ICs." *EEE,* December 1969, February and March 1970.

33. ________. "IC Op Amp Beats FETs on Input Current." *National Semiconductor Application Note AN-29,* December 1969. National Semiconductor Corp., Santa Clara, CA.

34. ________. "Monolithic Operational Amplifiers—The Universal Linear Component." *National Semiconductor Application Note AN-4,* April 1968. National Semiconductor Corp., Santa Clara, CA.

35. Manufacturers' Data Sheets:

|  |  |  |
|---|---|---|
| Analog Devices | AD532 | Multiplier |
| | AD534 | Multiplier |
| | AD536 | RMS Converter |
| | AD538 | Multifunction IC |
| | AD542 | Operational Amplifier |
| | AD545 | Operational Amplifier |
| | AD548 | Operational Amplifier |
| | AD549 | Operational Amplifier |
| National Semiconductor | LF356 | Operational Amplifier |
| | LM11 | Operational Amplifier |
| | LM394 | Matched Transistor Pair |
| Precision Monolithics | MAT-01 | Matched Transistor Pair |
| | MAT-02 | Matched Transistor Pair |
| | MAT-04 | Matched Transistor Pair |

SEVEN

# *Amplifier Techniques*

This chapter addresses the wide range of considerations necessary for the optimum design of amplifiers used for a wide variety of tasks. This includes instrumentation uses as well as general-purpose signal processing. Also included are a variety of amplifier techniques useful in augmenting the performance characteristics of individual ICs.

## 7.1 INSTRUMENTATION AMPLIFIERS

One of the most useful applications of an IC op amp is the differential-input dc-amplifier configuration, illustrated in Fig. 7-1. This configuration is generally called an ''instrumentation amplifier,'' or often, simply an ''IA.'' Due to the inherent rejection of common-mode noise ($E_{cm}$) and the linear amplification of desired signals ($E_s$) that it provides, this circuit and variations of it find widespread use in signal processing. Dc and low-frequency signals are commonly received from a transducer, amplified, and then transmitted in a single-ended mode. Since the desired signal may have a full-scale amplitude of typically only a few millivolts, while common-mode noise may be several volts, such factors as amplifier input drift and noise are critical to overall accuracy. Equally important are the bias current(s) and the common-mode rejection of the amplifier configuration. Several IA circuits are discussed in this section, with the unique and distinguishing features of each highlighted.

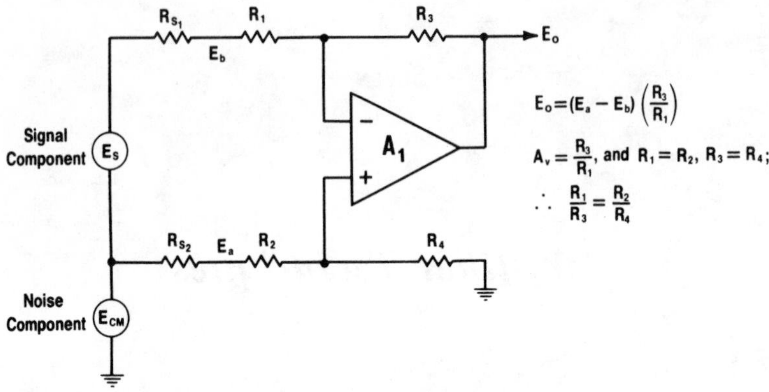


**Fig. 7-1. Basic differential-input instrumentation amplifier.**

### 7.1.1 Buffered Differential-Input IA

One obvious disadvantage of the basic differential circuit of Fig. 7-1 is the loading effect of $R_1$ and $R_2$ on the signal source(s). If a high gain is to be realized from the stage, $R_1$ and $R_2$ must necessarily be low. However, this tends to impair common-mode rejection, as there is usually little control over the source resistances of the two inputs. Consequently, even a slight mismatch in these resistances appears as a mismatch between $R_1 + R_{S_1}$ and $R_2 + R_{S_2}$ which degrades the CMRR because the gains for both signals are no longer equal.

The positive cure for impedance mismatch between the differential inputs is a very high impedance at both inputs. This can be accomplished by a voltage-follower (input-buffer) arrangement as shown in Fig. 7-2. This circuit raises the input impedance to $10^{12}$ Ω, has a 2-nA input current, and the ability to amplify differential signals superimposed on ±11 V of common-mode noise.

The second stage of the circuit, $A_3$, provides a differential voltage gain of 40 dB while further suppressing the common-mode component. The common-mode rejection properties of this configuration can be realized only when the arms of the differential "bridge" are matched. This requires that $R_2/R_4 = R_3/R_5$, as shown in the figure. The close tolerances indicated are necessary to guar-

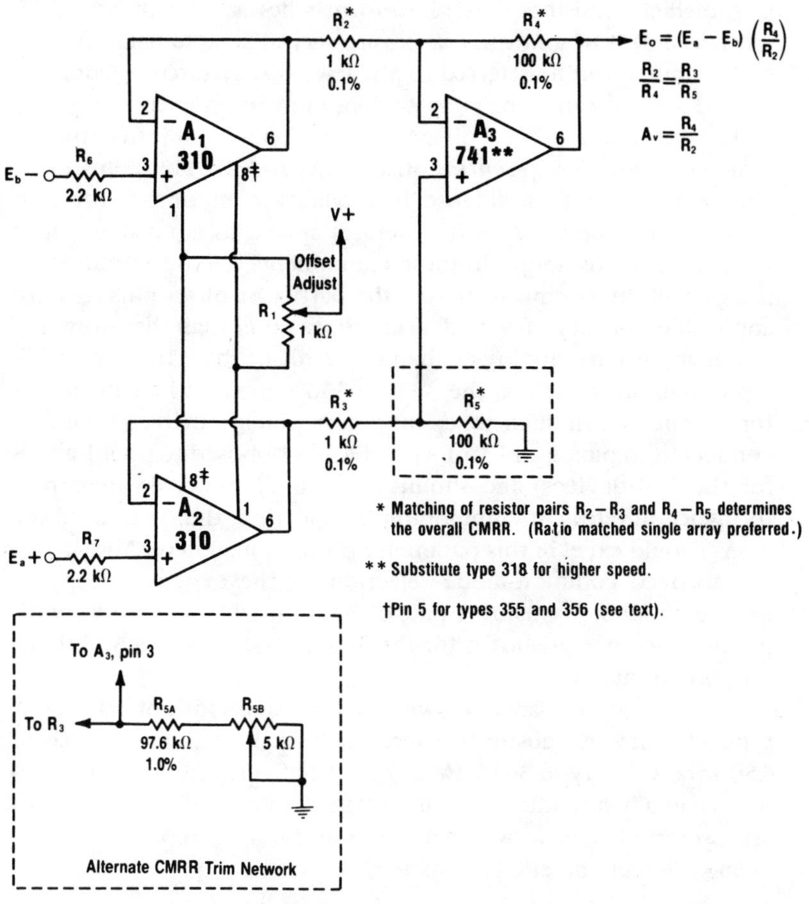


**Fig. 7-2. Buffered differential-input instrumentation amplifier.**

antee this degree of match without "tweaking" of the circuit. If an adjustment is desired, either ratio may be trimmed for a match. It is usually preferable to trim the noninverting side ($R_3$-$R_5$) to avoid upsetting the gain relation of $R_4$ to $R_2$.

An alternate network is shown which may replace $R_5$ to allow the CMRR to be trimmed to 100 dB or better, the same as with the 0.1% network. Perhaps the best *practical* solution to the resistance tolerance problem is to use a very tightly *ratio*-matched and TC-tracking film network for these resistances. This way, a single part is a solution—there is no need to pay a cost penalty for overly tight *abso-*

*lute*-tolerance resistors (0.01% *ratio* matches are readily available, whereas 0.01% *absolute* tolerances are very difficult to obtain!).

The output can be referred to a voltage ($E_R$) by disconnecting the ground at $R_5$ and applying $E_R$ to $R_5$ from a low-impedance source.

The net input offset voltage of the circuit is adjusted by $R_1$, which provides a *differential* offset to $A_1$ and $A_2$. This adjustment (and $R_{5B}$, if used) should use high-quality components for best results, since the voltages it generates appear as equivalent input voltages. Any instability in these components will be amplified by the gain of the circuit. Note that the particular offset pins that are applicable can vary, if $A_1$ and $A_2$ are types other than that shown.

For input currents lower than the 2 nA of the 310 units, FET-input amplifiers such as the 355 or 356 types can be substituted for $A_1$ and $A_2$. In such case, the offset potentiometer should be connected to pins 1 and 5 of $A_1$ and $A_2$ (as opposed to pins 1 and 8 for the 310 devices) and should be a 10-kΩ value. For best performance in terms of common-mode rejection, the types used for $A_1$-$A_2$ should excel in this parameter (or have matched CMRR).

Improved common-mode rejection (at the expense of speed) can be obtained by using a pair of 312 units for $A_1$ and $A_2$, with pinouts the same as shown for the 310 units, but with a 50-kΩ offset potentiometer.

A 741 or 307 op amp ($A_3$) will yield a bandwidth of 10 kHz. A type-318 may be substituted here, if desired, for a bandwidth of 150 kHz. Or, a type-301A (with a reduced compensation of 5 pF) will yield a bandwidth of about 100 kHz. In general, bipolar types are preferred for the $A_3$ stage for best common-mode rejection, even with well-matched resistances.

### 7.1.2 Differential-Input Amplifier With Variable Gain

While the circuit of Fig. 7-2 is generally very useful, it has one problem inherent to many IAs—an inability to adjust gain without upsetting the bridge balance and, thus, the CMRR. If $R_4$ of Fig. 7-2 is varied, for example, it will indeed adjust gain, but it will simultaneously ruin the CMRR because of the resulting bridge imbalance. What is needed is a method of varying the differential gain of the circuit with a single adjustment, while retaining a high and constant CMRR.

A modification of Fig. 7-2 allows this and is shown in Fig. 7-3. This circuit varies the gain of stage $A_3$ by modifying the feedback returned to $R_4$; this, in essence, places a calibrated attenuator within the feedback loop. Since feedback attenuation results in higher gain, this achieves the desired gain variation. An active attenuator ($A_4$) is used rather than a simple divider because $A_4$ presents a constant zero-impedance source to $R_4$, exactly the condition needed to maintain good balance and a high CMRR. The output impedance of $A_4$ remains low for any setting of $R_6$; therefore, gain adjustments do not alter the CMRR in this circuit.

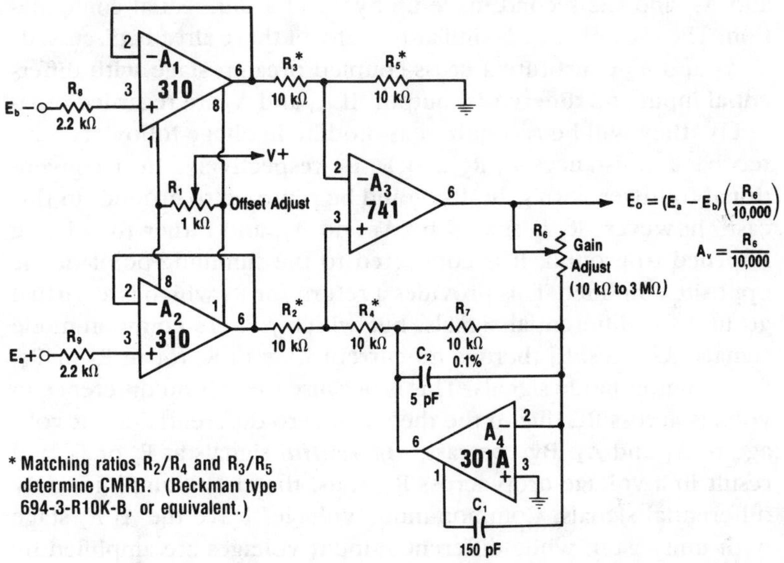


**Fig. 7-3. Variable-gain instrumentation amplifier.**

An interesting feature of this method of gain adjustment is that the gain is linearly proportional to $R_6$, which allows $R_6$ to be a calibrated rheostat for repeatable gain settings. With the values shown, gain can be varied from unity to 300. Offset-adjustment $R_1$ operates here in the same fashion as described for Fig. 7-2, and the same constraints apply. For both these circuits, good layout techniques are recommended—rf bypassing for amplifier supply leads (particularly at $A_4$) and short, direct, signal leads. It is also recommended that the bridge resistances be low-temperature-coefficient types for best dc stability, preferably in a single array. The type specified in the drawing will provide a matching of 0.05%.

### 7.1.3 IAs With Differential-Input Preamplifier Stages

For situations where high-input impedance and the lowest-offset-voltage drift are required, the configurations of Fig. 7-4 can be used. None of the circuits described previously have been optimized from the standpoint of minimum offset drift. The configuration here is unique in this regard, and a closer examination of how this is accomplished will reveal some attractive properties of the circuit. The circuit consists of two stages—the first consisting of $A_1$ and $A_2$, and the second made up by $A_3$ in a differential configuration. The second stage is similar to many of those already discussed.

$A_1$ and $A_2$ constitute a cross-coupled preamp stage with differential input and differential output. If $A_1$ and $A_2$ are regarded separately, they will be recognized as modified voltage followers with feedback resistances $R_1$-$R_3$ and $R_4$-$R_3$ respectively. In a conventional follower with gain, $R_3$ would be returned to ground. In this case, however, $R_3$ is shared by $A_1$ and $A_2$, and rather than being returned to ground, it is connected to the summing point of the opposite amplifier. This provides a return for $R_3$ which is at virtual ground for differential signals, but which *follows* common-mode signals. As a result, there is no current flow in $R_3$ (or in $R_1$ or $R_4$) for common-mode signals. This is because there is no difference in voltage across $R_3$, due to the theory of zero differential-input voltage to $A_1$ and $A_2$. By contrast, ***differential*** signals at $E_a$ or $E_b$ will result in a voltage drop across $R_3$; thus, the circuit amplifies only differential signals. Common-mode voltages leave the $A_1$-$A_2$ stage with unity gain, while differential-input voltages are amplified by a factor of $(R_3 + 2R_1)/R_3$ [alternately expressed as $1 + (2R_1/R_3)$].

Several advantages may be inferred from this fact. One, the preamp stage provides an immediate gain in signal/noise ratio, because the differential signal is raised above the common-mode (noise) component by the ratio of the differential gain of the stage. Two, the gain-determining resistances ($R_1$, $R_3$, and $R_4$) ***do not*** affect the CMRR and are therefore much less critical as to tolerance. Any mismatch in $R_1$ and $R_4$ will show up as a mismatch in differential gain between the two outputs—a far less serious condition than a CMRR loss.

Another feature of this circuit is the minimum amount of gain for common-mode input signals. This factor has a bearing not only on actual common-mode inputs, but also on the offset voltages and drifts of $A_1$ and $A_2$. If $A_1$ and $A_2$ drift at an equal rate and in the same

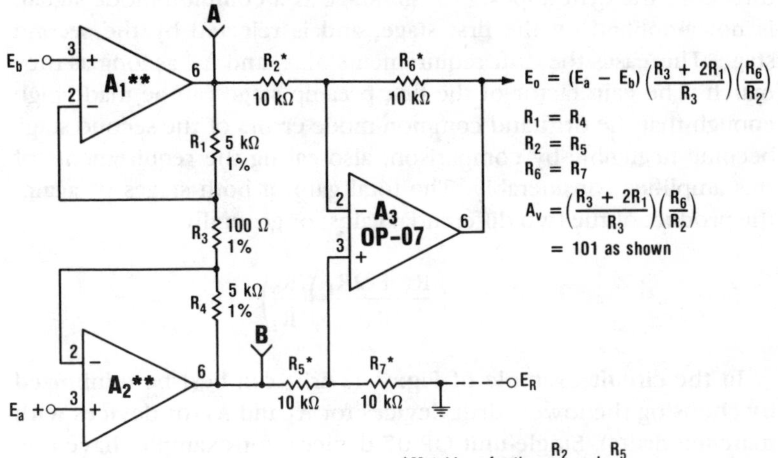


*(A) General-purpose circuit.*

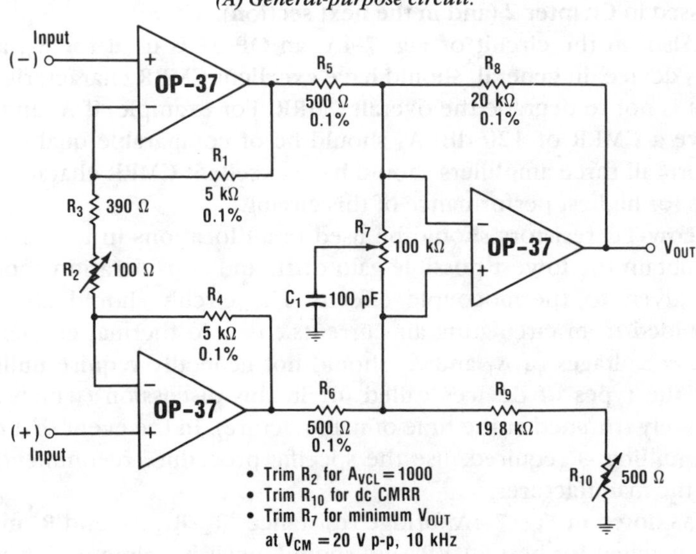


*(B) Low-noise circuit (gain = 1000).*

*(Printed with permission from Precision Monolithics Incorporated.)*

**Fig. 7-4. IAs with differential preamp.**

direction, the drift appears to this stage as a common-mode signal, is not amplified by the first stage, and is rejected by the second stage. This eases the drift requirements of $A_1$ and $A_2$, as long as they match. The gain factor of the first preamp stage can be made high enough that the drift and common-mode errors of the second stage become negligible by comparison, also easing the requirements of this amplifier considerably. The total gain of both stages is, again, the product of the two differential gains, or generally,

$$A_v = \left(\frac{R_3 + 2R_1}{R_3}\right)\left(\frac{R_6}{R_2}\right)$$

In the circuit example of Fig. 7-4, drift can best be minimized by choosing the lowest drift devices for $A_1$ and $A_2$ (or devices with matched drifts). Single-unit OP-07 devices, for example, have low drifts, low offset voltages, and low bias currents (as low as 1 μV/°C, 100 μV, and 2 nA, respectively, depending on the grade of device used). Alternatively, matched-parameter dual devices, such as the OP-10, can also be used. Other low-offset, low-drift devices suitable for $A_1$ and $A_2$ are the instrumentation-grade devices discussed in Chapter 2 (and in the next section).

Also, in the circuit of Fig. 7-4A, an OP-07 is used for $A_3$, and this device, in general, should have excellent CMRR characteristics if it is not to degrade the overall CMRR. For example, if $A_1$ and $A_2$ have a CMRR of 120 dB, $A_3$ should be of comparable quality. In short, all three amplifiers should have excellent CMRR characteristics for highest performance of this circuit.

Low-TC resistors should be used in all locations in this circuit to obtain the lowest possible gain drift, and consideration should be given to thermocouple effects. The circuit should also be shielded from circulating air currents to avoid thermal gradients. Offset voltages of $A_1$ and $A_2$ should not generally require nulling for the types of devices called for in this discussion (which are actively trimmed at the time of manufacture). In the event that offset nulling is required, use the specific procedure recommended by the manufacturer.

As noted in Fig. 7-4A, bridge resistances $R_2$, $R_5$, $R_6$, and $R_7$ must be matched for best CMRR, and should never be trimmed for gain adjustment. They should be ratio-matched and part of a common array, with low-tracking TC (such as the precision array specified in the notes). If gain trim or programming is desired, resistance $R_3$

should be used for this purpose. Although the gain relationship is unwieldy, it nevertheless does not compromise the CMRR.

An alternate form of this basic circuit is shown in Fig. 7-4B, and uses the OP-37-type op amp for both low noise and wide bandwidth. Gain of this circuit is 1000, and the bandwidth is 800 kHz. Three trimming resistors are provided to optimize performance.

With a very high, second-stage, gain ratio, a single-array bridge-resistor network is more difficult. This example uses precision, discrete, 0.1% film units, with the trim for optimum dc CMRR accomplished with $R_{10}$. Gain is trimmed via $R_2$, similar to the previous circuit. This particular circuit also uses an ac trim, $R_7$. This control balances the high-frequency impedances, to provide more than 100 dB of CMRR at 20 kHz, even for source resistances up to 1 kΩ.

The equivalent input noise of this circuit is less than 5 nV/$\sqrt{\text{Hz}}$, which is about three times better than if OP-07s were to be employed. Because the OP-37 devices are not stable for gains of less than 5, very low gains will not be possible with this circuit. However, the OP-27 can be readily substituted for operation at low gains, with similar low-noise performance.

Other examples of this popular type of instrumentation-amplifier stage are covered in the discussions that follow. Note that the performance will, in most cases, be dependent upon the ratio-matching of the bridge resistances.

### *FET-Input IA*

Fig. 7-5 illustrates an amplifier of this general type that uses FET-input amplifiers for the $A_1$ and $A_2$ stages. The big advantage that a circuit of this type has over those discussed previously is in the dramatic reduction in room-temperature bias currents. With high-performance, low-drift, FET input devices, the drift can approach that of the best bipolar amplifiers, with the advantage of drastically lowered bias currents. CMRR will be good, but somewhat dependent upon the specific devices used for $A_1$-$A_2$.

This circuit uses a pair of AD548C, low-power, low-drift, FET-input op amps for $A_1$ and $A_2$, which results in a typical input current of 10 pA (either input), an offset of 500 μV, and a 4 μV/°C drift. The second stage uses an OP-07 type for minimum additional drift, and the overall gain is 197. Gain can be trimmed via $R_3$, but the bridge resistors should be stable, well-matched types (such as those recommended for use in Fig. 7-4B). The (optional) offset null used with $A_2$ adds minimal drift.

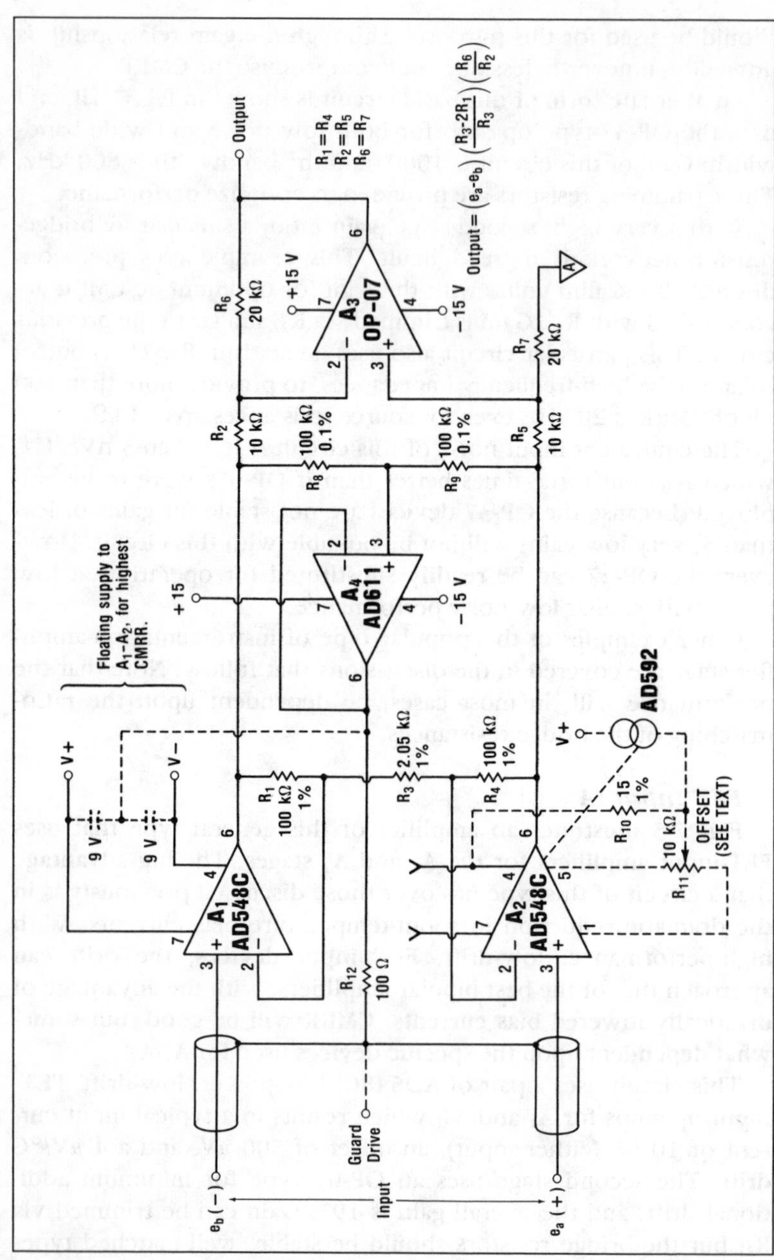


Fig. 7-5. FET-input IA with differential preamp.

This circuit adds a common-mode buffer ($A_4$), which can be used to drive the shields of input cables (for lowest capacitance). If the input op-amp pair (only) is optionally powered with a pair of 9-V batteries, their op-amp output can drive the common point as shown, for lowest CMRR. Total current drain is less than 0.5 mA in $A_1$-$A_2$.

Alternate high-performance types for $A_1$-$A_2$ include the AD648 (dual AD548). $A_3$-$A_4$ could also be a second AD648, for a total overall drain of less than 1 mA.

### *Guarded FET-Input Femtoampere IA*

The circuit of Fig. 7-6 is one that is optimized from the standpoint of the absolute lowest input current(s). Using two AD549 op amps, and appropriate construction methods (see Chapter 3 on guarding), this circuit is capable of input currents as low as 75 fA (0.075 pA), depending on the specific grade of op amp used for $A_1$-$A_2$.

As noted in the drawing (Fig. 7-6), complete guarding is used at the input terminals of the two input amplifiers; this includes driving the two cases of the devices (pin 8), as well as two Teflon standoffs, with printed-circuit guards around terminals 2 and 3. Buffer amp $A_4$, an AD611 device, drives the shields of the input cables at the common-mode voltage, which is picked off from the input stage by 0.1% ratio-matched resistors, $R_8$-$R_9$. This reduces both the leakage and the effective capacitance in the cables (which should be of the highest-quality dielectric, such as Teflon). $A_3$ is an OP-07-type bipolar instrumentation-grade op amp, for lowest CMRR error in this stage.

Using AD549B devices in the input stage, the maximum input offset voltage will be 2 mV, while drift is a maximum of 20 $\mu$V/°C, and the minimum CMRR will be 90 dB. The AD549B input current will be 150 fA (max); substitution of the AD549C will lower this to 75 fA (max). Gain of this IA can be made jumper programmable, for gains of 2, 4, or 8, as shown in the table in Fig. 7-6. A single, common, ratio-matched ($7 \times 10$ k$\Omega$) resistor array is used to set the gain of $A_1$-$A_2$, and a second ($4 \times 10$ k$\Omega$) array sets the gain of $A_3$ (which can, of course, be higher, if preferable, or it *also* can be made programmable).

### *Picoampere IA*

The circuit of Fig. 7-7 illustrates an IA that has both a picoampere-level of input current(s) and, also, a CMRR of 106 dB.

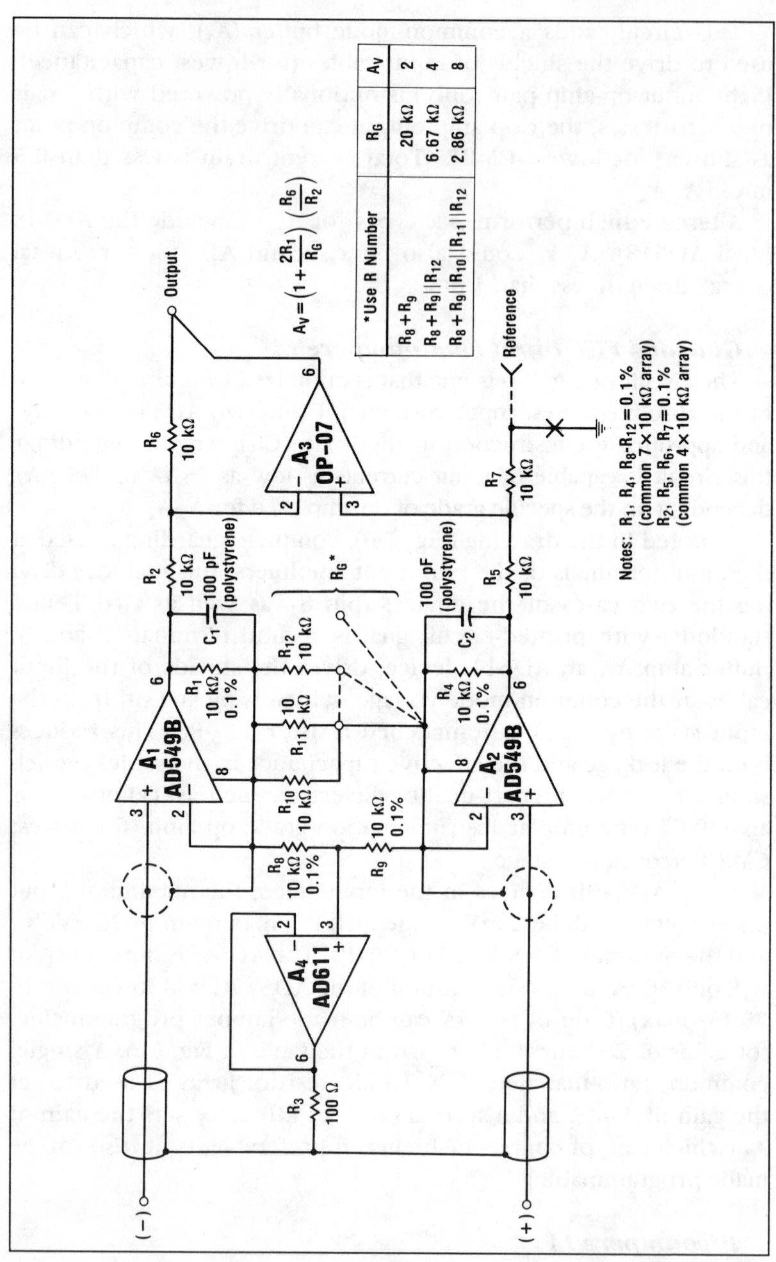


**Fig. 7-6. Femtoampere IA.**

Overall simplicity in this circuit is achieved by the use of the OPA2111 dual-FET input device in the input stage, plus the use of an integral, fixed-gain, differential amp, the 3627. The use of the 3627 provides the user with an op amp plus four bridge resistors, pretrimmed to the required accuracy; this considerably simplifies the task. The gain resistances, $R_F$-$R_G$, can be conventional film types, appropriate to the gain, as needed. Alternately, the INA105 differential amp can be used, with slightly reduced performance.

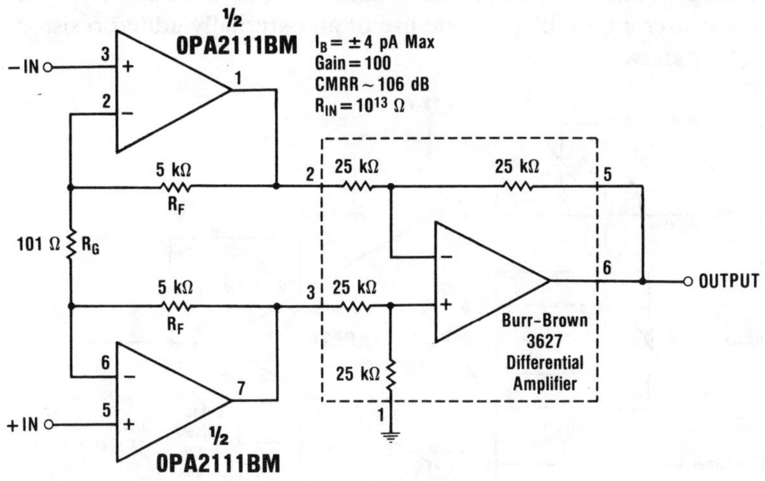


**Fig. 7-7. Picoampere IA.**

*(© 1984 Burr-Brown Corp. Reproduced with permission.)*

Input bias current is 4 pA at either input for the OPA2111 device, and offset voltage will be 1 mV (max). If desired, these figures can be lowered to 1 pA and 0.5 mV, respectively, by using a pair of OPA111Bs.

### *Integral Monolithic IA*

Thus far, a variety of different IA circuits have been discussed, optimized towards various performance aspects. While they are all useful to their intended purposes, for general applications, some may be considered too complex.

Inasmuch as the IA is an extremely useful circuit building block, monolithic single-chip versions have appeared. These offer virtually all of the performance of the circuits thus far discussed, but generally without the trimming and complexity problems.

One example in the Analog Devices AD524, illustrated in Fig. 7-8. This device offers pin-selectable differential gains of 10, 100, or 1000, for analog inputs with a CM range of ±10 V. Gains are internally trimmed to these levels (to tolerances as low as 0.02%), and CMRR ranges from 90–120dB (dependent upon gain). The output signal has a range of ±10 V, and can be referred to either ground (as shown), or optionally offset by applying a dc bias. Gain is also user adjustable, via the use of an externally added resistor, ($R_G$), as shown.

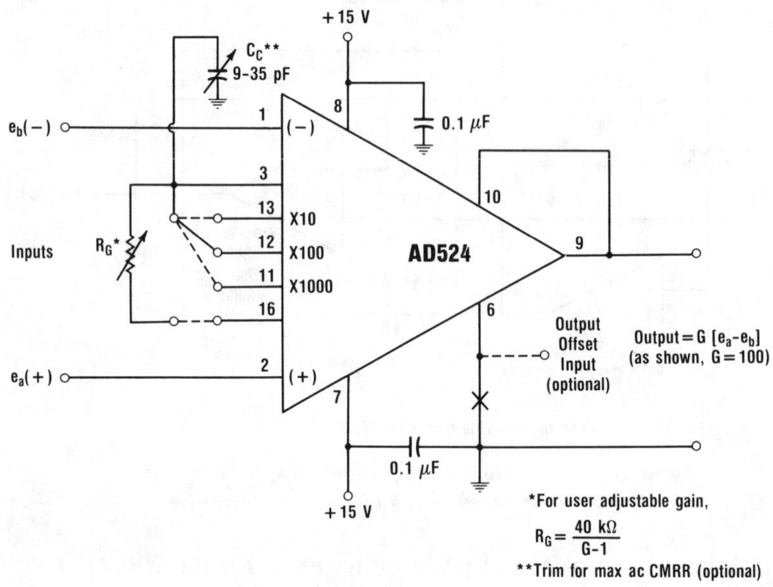


**Fig. 7-8. Integral monolithic IA.**

The AD524 is easy to apply and, in most instances, no additional components are necessary in order to use it. The ac CMRR can be optimized, when necessary, by the use of a trimmer cap, $C_c$.

The AD524 has a bandwidth of 150 kHz, operating at a gain of 100 (as shown here), and the input bias current is as low as 15 nA (max), while the offset voltage is as low as 50 $\mu$V (max). A similar (but lower-noise device) is the AD624, offering slightly different pretrimmed gains. A related device, the AD625, offers a user-programmable gain feature.

Integral monolithic IAs of this type can be effectively applied to a wide range of signal-processing tasks, with high precision.

### *FET Buffered Monolithic IA*

While monolithic instrumentation amplifiers are excellent all-around devices from the standpoint of precision and utility, one performance area still open for improvement is bias current. The buffer circuit of Fig. 7-9 can reduce the two bias currents of a typical bipolar-input IA to the low picoampere range.

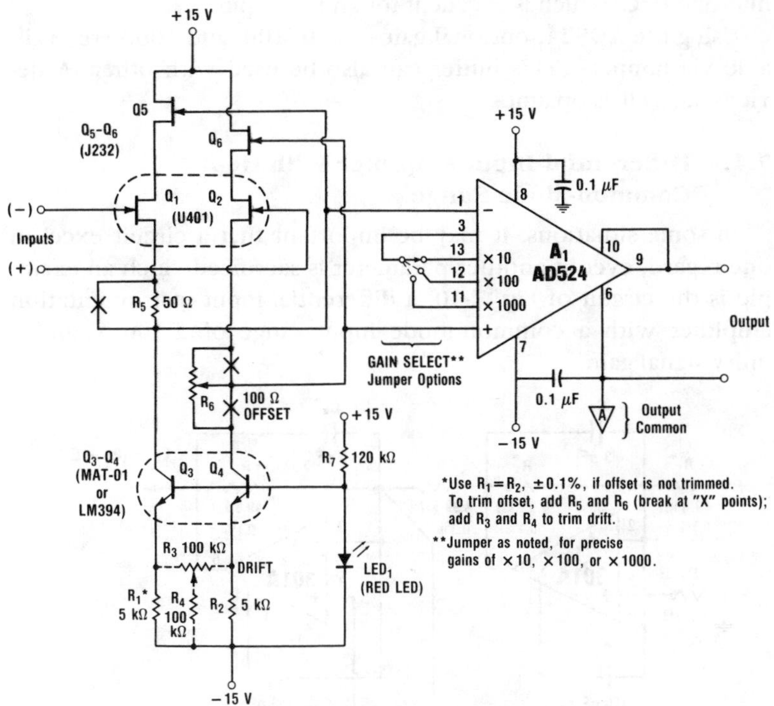


**Fig. 7-9. FET-buffered monolithic IA.**

This circuit is a dual, precision, source follower, using a high-CMRR FET pair, the U401. This device series (U401 through U406) is available with offsets as low as 5 mV and drift as low as 10 μV/°C. While the CMRR is 95 dB, well below that of a good IA, such as the AD524, bootstrapping can raise this to 120 dB or more. Here, the bootstrap devices are $Q_5$-$Q_6$, a pair of J232 JFETs.

In addition, the use of a high-CMRR dual-current source is necessary; this is provided by the $Q_3$-$Q_4$ pair.

While this circuit essentially preserves the basic CMRR of the instrumentation amplifier and lowers the input current, it does degrade the overall offset and drift, due to the FETs. If the optional resistances $R_3$-$R_4$ and $R_5$-$R_6$ are not included, the best performance achievable is that of the basic (untrimmed) U401—5 mV and 10 μV/°C. However, if these two trims are used, $R_6$ can be used to trim the OFFSET, and $R_3$, the DRIFT. Careful sequential adjustment of these trimmers will allow net offsets of less than 1 mV, and drifts of a few microvolt/°C, which is excellent for an FET input.

Using the AD524, optional gains of 10, 100, and 1000 are available via jumpers. This buffer can also be used with other IA devices, as well as op amps.

### 7.1.4 Differential-Input Amplifier With High Common-Mode Range

In some situations, it may be important that a circuit excel in one regard, even if another parameter is sacrificed. Such an example is the circuit of Fig. 7-10, a differential-input instrumentation amplifier with a common-mode input range of ±100 V and a unity signal gain.

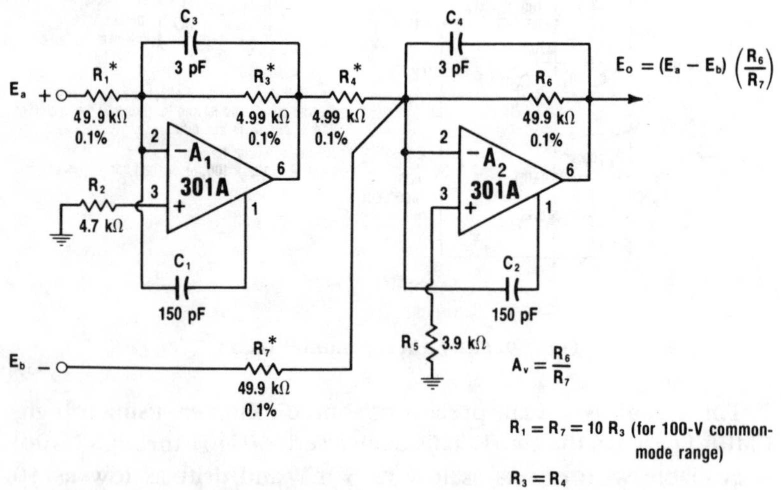


**Fig. 7-10. Instrumentation amplifier with common-mode range of ±100 V.**

This circuit trades off input loading in return for a higher common-mode range and reduced common-mode errors. Both $A_1$ and $A_2$ are used as inverting stages; thus, both have a characteristically low input impedance—in this case, the values of $R_1$ and $R_7$, or 49.9 kΩ. A virtue of the inverting configuration is the fact that there is no common-mode voltage applied to the amplifier input terminals; therefore, no error can exist due to variations in this parameter.

The circuit is a combination inverting attenuator and scaling adder which rejects common-mode input components while amplifying differential ones. $A_1$ amplifies signal $E_a$ by a factor of $-R_3/R_1$ or, in this case, $-1/10$. This signal is combined with $E_b$ at summer $A_2$. $E_b$ is passed to the output with a gain of $-R_6/R_7$, or $-1$, while $E_a$ is amplified a second time by $-R_6/R_4$, or $-10$. The total gain for the $E_a$ signal is then $(-1/10)(-10) = +1$. The resultant gain from both inputs to the output is therefore unity, with signs as shown. A common-mode signal, however, becomes cancelled in stage $A_2$; it will appear at $A_2$ as equal and opposite signals, summing to zero. Differential inputs are amplified by the ratio $R_6/R_7$.

For this circuit to work, the matching of resistor ratios is again critical and should be maintained as shown by the tolerances indicated, with resistor pairs matched as shown. Note that $A_1$ must scale down the common-mode input voltage linearly; therefore, the ratio of $R_3/R_1$ must be selected to attenuate the common-mode input voltage to $\pm 10$ V or less at the output of $A_1$.

Gain can be adjusted by varying $R_6$, since the CMRR will not be affected as long as $R_3/R_4 = R_1/R_7$. Low TC resistors should be used for all resistances (except possibly $R_2$ and $R_5$) to maintain good thermal stability of gain and CMRR. Due to the wide bandwidth (approximately 1 MHz), rf bypassing of the power supply is also recommended (if a 301A is used, as shown). If wide bandwidth is not required, conventional inverters, such as 741s, may be used for $A_1$-$A_2$. Many duals will also work well here, and FET units will allow large values for $R_1$-$R_7$. Single-supply types, such as the 358, 1013, or 3240, etc., can also be useful.

### 7.1.5 IA With High Input Impedance

A circuit with high input impedance and using only two amplifiers is illustrated in Fig. 7-11. Like the previous circuit, it uses a combination of inversion and a summation to cancel the common-mode component at the two inputs. In this case, however, a combination of follower configurations also provides an extremely high input impedance.

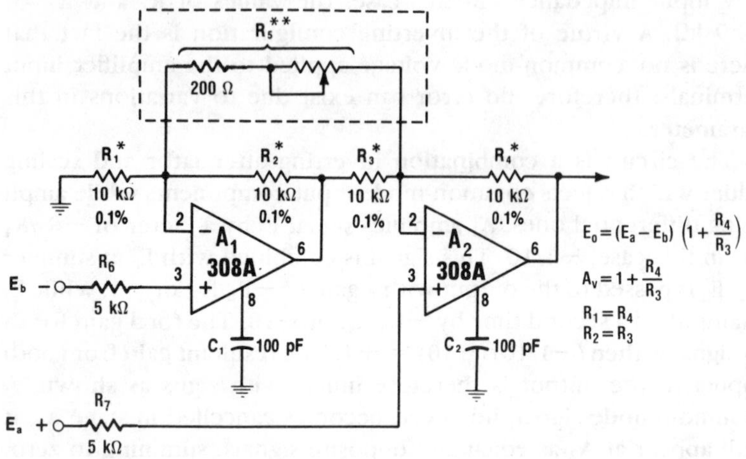


**Fig. 7-11. Instrumentation amplifier with high input impedance.**

$A_1$ amplifies $E_b$ by a factor of $(R_1 + R_2)/R_1$ (in this case, 20/10). The signal is further amplified by $A_2$, which operates as an inverter for signals from $A_1$. The stage gain of $A_2$ is $-R_4/R_3$ (in this case, $-10/10 = -1$). The overall gain to the $E_b$ signal is, therefore, the product of both stage gains, or $(2)(-1) = -2$.

For signals at $E_a$, $A_2$ operates as a follower with a gain of $(R_3 + R_4)/R_3 = 2$. Therefore, the overall gain of $E_a$ is $+2$, exactly equal and opposite that for $E_b$. If a common-mode signal appears at both inputs, it will be amplified by equal and opposite amounts and cancelled at the output of $A_2$.

Note that because of the gain relationships of this circuit, high differential gains provide the most effective common-mode rejection due to the relative boost in signal/noise ratio. The circuit as shown can accommodate ±5-V common-mode inputs while amplifying a ±100-mV signal. The optional gain trim potentiometer,

$R_5$, can be used to adjust gain, since it does not affect the CMRR. In this example, a 10 kΩ network is used, so the gain is 2 without the use of $R_5$. With a minimum $R_5$ of 200 ohms, the gain will be $2 + 100 = 102$.

As before, resistor ratio-matching is critical to good common-mode rejection. Common-mode rejection is also dependent on the op amp used; therefore, devices with a good CMRR characteristic, such as the 308A (typically 110 dB), are desirable. The 308-type amplifiers use shunt compensation ($C_1$ and $C_2$) to reduce their susceptibility to power-supply-induced noise. The circuit bandwidth is approximately 5 kHz, and can be extended by reducing the value of $C_2$ for stage $A_2$. This should be done, if possible, as it greatly improves the ac CMRR.

Note that the offset voltage *match* between the two amplifiers is most critical to this circuit. The dc CMRR match is also critical. Amplifiers with matched input characteristics, such as the many dual bipolar-input types, will thus work very well in this type of circuit. Some examples are the OP-10, OP-207, OP227, and the general-purpose duals, such as the 4558, 4559, and 358.

If suitable high-CMRR FET-input op amps are used, this circuit can also be an effective, very-high-input-impedance configuration, with minimal source loading. Dual devices, such as the AD642, A647, OPA2111 can achieve picoampere input currents, with low offsets. For single-supply use, a 3240 can handle 0-V to +10-V ranges with a few picoamperes of input current, and a pair of 3420s will allow sub-picoampere bias currents. Micropower single-supply dual types include the OP-221.

### 7.1.6 High-Gain, Wide-Bandwidth IA

There are alternate methods of achieving high common-mode rejection that are worth noting. Although not based entirely on op-amp theory, the circuit of Fig. 7-12 is one such example. $Q_1$ and $Q_2$ in this circuit are matched monolithic transistors, used as an emitter-coupled pair biased by constant-current sources $Q_3$ and $Q_4$. Since the output impedance of a current source, such as $Q_3$ or $Q_4$, is extremely high, it raises the common-emitter impedance of $Q_1$ and $Q_2$. As a result, neither $Q_1$ nor $Q_2$ will have significant gain for a signal applied to both bases. Therefore, this stage by itself possesses a fairly high common-mode rejection.

*Differential* signals applied to $Q_1$ and $Q_2$ appear across $R_g$ and

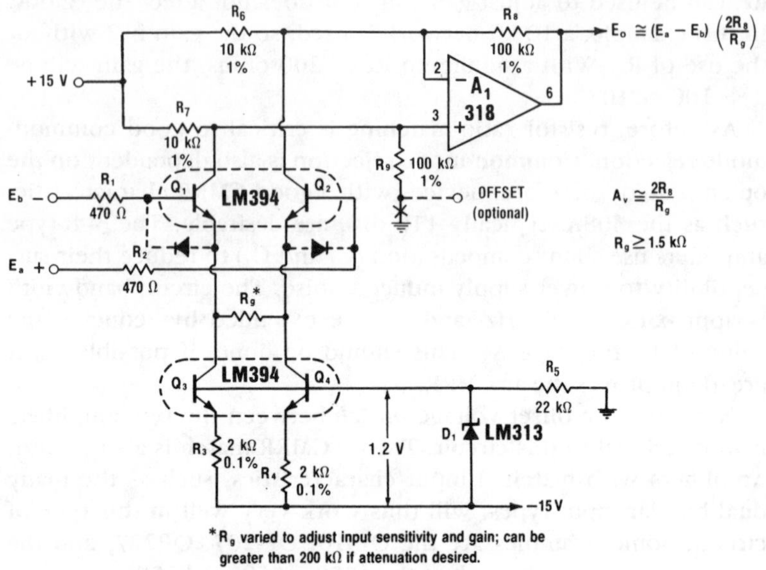


**Fig. 7-12. High-gain, wide-bandwidth instrumentation amplifier.**

are amplified by a factor inversely proportional to the value of $R_g$, since the signal current flowing in $R_g$ appears doubled in $R_8$. Common-mode signals are attenuated by the $Q_1$-$Q_2$ stage by a factor of 80 dB or more, and appear at the input to $A_1$ as a common-mode signal from $Q_1$-$Q_2$. Since this signal has already undergone a CM rejection in $Q_1$-$Q_2$, the balance of this bridge is not as critical—allowing 1% resistor tolerances. To null the output of $A_1$ when no input is present, an optional OFFSET voltage can be applied to the bottom of $R_9$.

A unique feature is the stable temperature-independent gain, even though $Q_1$ and $Q_2$ are not within a feedback loop. This is provided for by $D_1$ and $Q_3$-$Q_4$, with an output current that varies with temperature, complementing the change in transconductance of $Q_1$-$Q_2$ with temperature.

The circuit bandwidth is limited to approximately 1 MHz by the response of $A_1$ if the 318 is used. The circuit can handle common-mode input voltages up to ±10 V, but the differential-input range (or sensitivity) is proportional to $R_g$. As $R_g$ is lowered, gain is increased and less input voltage is required for full-scale output. The range of linear gain is slightly over 40 dB—at lower values of

$R_g$, the effect of emitter resistance in $Q_1$ and $Q_2$ begins to produce nonlinearities in the full-scale input.

*A final, practical note*—if any input transistor other than the 394 is used for $Q_1$-$Q_2$, protection diodes should be added (if they are not integral to the device itself).

### 7.1.7 Single-Supply "+5-V" Instrumentation Amplifier

With the increasing application of analog circuits from low-voltage single-polarity supplies, such as +5 V, there is a real need for precision analog signal processing which can operate in such an environment without compromise.

The circuit of Fig. 7-13 is a single IA which operates from a supply as low as +5 V and handles CM inputs of 0 to +3.5 V. CMRR and PSRR of the circuit are 100 dB, while the offset is 500 $\mu$V and drift is 4 $\mu$V/°C. The two input resistors provide both overvoltage protection and low-pass filtering, in concert with the capacitance across the inputs. As with the previous IAs, well-matched (in ratio) array resistors should be used in the 10-kΩ bridge. $R_g$ is used to manipulate the gain, and is typically 2 kΩ, for a gain of 201. The output range of the circuit is 4 V or more, typically, and the no-load supply current is slightly more than 1 mA.

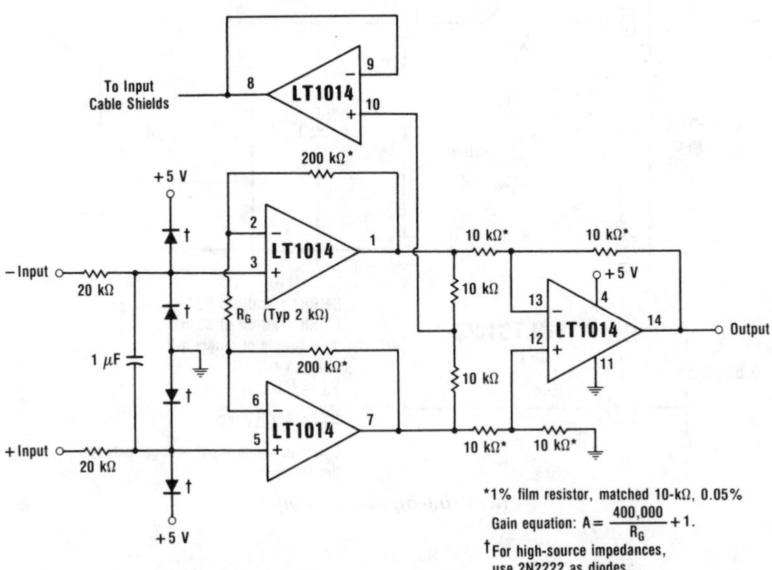


**Fig. 7-13. Single-supply "+5-V" IA.** *(Courtesy Linear Technology Corp.)*

### 7.1.8 "Flying Capacitor" IAs

The two circuits of Fig. 7-14 utilize an input-commutation principle and an IA in what is commonly referred to as the "flying capacitor" technique. While this is not a true real-time signal-amplification method, it can excel when used with low-frequency signals, and it offers superior CM rejection.

In Fig. 7-14A, the differential-signal input is applied to pins 7 and 13 of an LTC1043-type analog switch, which self-clocks at 500 Hz. When $C_S$ is connected to the input (the cycle shown), it stores the input differential voltage. On the alternate cycle, $C_S$ transfers charge to $C_H$, which drives the amplifier. As noted, the configuration of this amplifier is single-ended, yet it effectively amplifies differential signals, by the standard gain of $1 + R_2/R_1$.

In this circuit, an LT1013 section is the amplifier, and the circuit features 120 dB or more of CMRR, up to 10 kHz or more. The input CM range is ±5 V, as determined by the supplies to the LT1043.

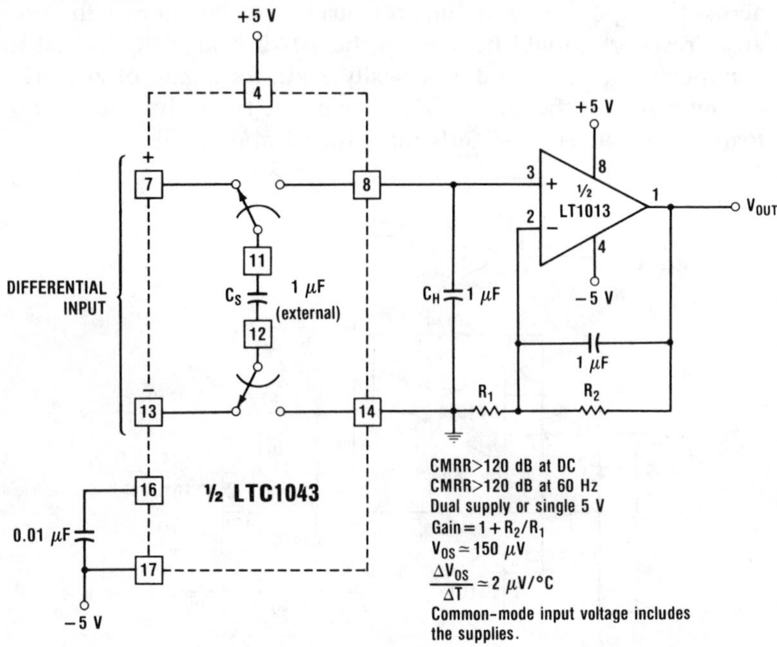


*(A) Using bipolar op amp.*

**Fig. 7-14. Flying capacitor IAs.**

The companion circuit of Fig. 7-14B is very similar, but it uses an LT1052 chopper-stabilized op amp to achieve an offset of only 3 μV along with a gain of 1001. The amplification (top) portion of the sketch is similar to that of Fig. 7-14A, except that in this case, the LT1052 is operated from a single +5-V supply. The optional circuit at the bottom of the drawing constitutes a chopper to produce a −0.5-V supply, which allows the LT1052 to swing to

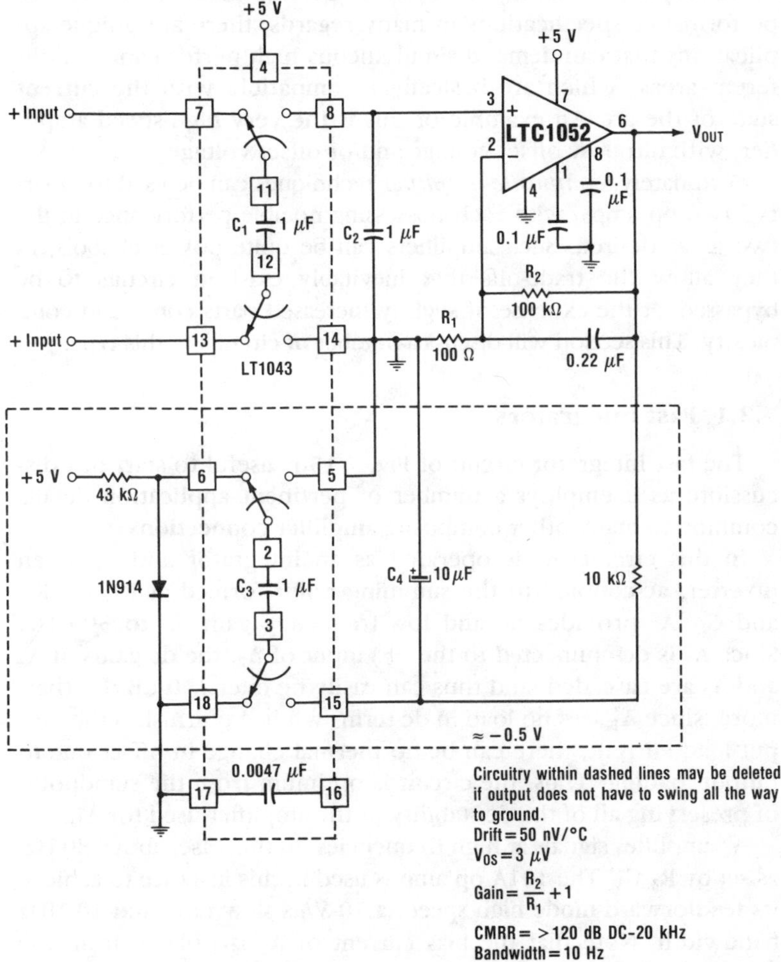


*(B) Using chopper-stabilized op amp.*

*(Courtesy Linear Technology Corp.)*

ground. The CM-input range of this circuit is 0 to +5 V, as determined by the switch supply voltage.

In both of these circuits, the signal-path capacitors ($C_S$, $C_H$, etc.) should be low-dielectric absorption types, such as polypropylene.

## 7.2 COMPOSITE AMPLIFIERS

Although present-day IC op-amp devices achieve superlative performance specifications in many regards, there are unique applications that can demand simultaneous high performance in different areas, which are basically incompatible with the current state of the art. An example of this is the very high-speed amplifier, with ultralow offset voltage and/or offset voltage drift.

Fortunately, ***composite-amplifier*** techniques can be used to "marry" two op amps, with each possessing notable performance in the two areas desired. Such amplifiers can be quite powerful tools, as they allow the trade-offs that inevitably exist in circuits to be bypassed, at the expense of slightly increased parts count and complexity. This section will discuss a number of circuits of this type.

### 7.2.1 Fast Integrators

The fast integrator circuit of Fig. 7-15 is useful to start this discussion, as it employs a number of pertinent application details common to many other composite amplifier connections.

In this circuit, $A_1$ is operated as an integrator and $A_2$ as an inverter, ac-coupled to the summing point formed at $R_1$, $R_4$, $R_S$, and $C_L$. $A_1$ provides dc and low-frequency gain up to 500 Hz. Since $A_1$ is dc-connected to the (+) input of $A_2$, the dc gains of $A_1$ and $A_2$ are cascaded, and thus can easily be over 150 dB. Furthermore, since $A_1$ sees no load in dc terms while $A_2$ furnishes the output load current, there can be no thermal change in offset due to output loading. Thus, the circuit is optimum from the standpoint of preserving all of the dc stability of the amplifier used for $A_1$.

$A_2$ amplifies signals at high frequencies; in this case, above 80 Hz, as set by $R_3$-$C_3$. The 101A op amp is used in this instance to achieve its feedforward-mode high speed, a 10-V/μs slew rate, and 10-MHz bandwidth. Note that the bias current of $A_2$ is isolated from the summing point, thus the net bias current is determined by $A_1$.

As a composite amplifier, this circuit has the input offset voltage, bias current, and drift of $A_1$, but the speed and drive capabil-

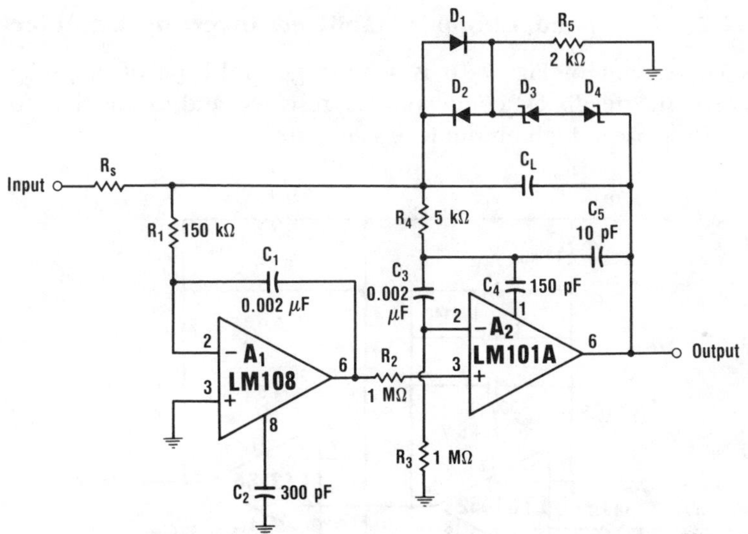


**Fig. 7-15. Fast integrator using composite amplifier.**
*(Courtesy National Semiconductor Corp.)*

ity of $A_2$. Note that the bias current, offset voltage, and drift of $A_2$ are not material to overall operation. Offset voltage adjustment is not shown in this example but can be accommodated by a small voltage at the (+) input of $A_1$, or by substituting a comparable offset-nullable device for $A_1$, such as the 112 (or 312). Also, a very-low-offset device, such as the OP-07, could be used.

Using the devices shown in Fig. 7-15, the bias current is 2 nA or less. This can be reduced by employing an LM11 for $A_1$, which will lower the bias current to 50 pA maximum, with a 300-μV maximum offset voltage (adjustable to zero). The 308 (308A) and 301A commercial temperature range (0°C to +70°C) amplifiers can also be used in the $A_1$-$A_2$ positions of this circuit.

The $D_1$-$D_4$ "T" diode-clamping network shown in Fig. 7-15 is not necessary for basic operation of this composite amplifier, but should be seriously considered, particularly for an integrator (see Chapters 5 and 9). Because of the very high dc gain of the loop, overload and subsequent recovery problems should be borne in mind. This is generally true of most composite amplifiers (not just this example), and serious attention should be given to preventing overload, or minimizing its consequences. Integrators, such as $A_1$, might best be clamped with reverse-connected low-leakage diodes across $C_1$, for example, to prevent possible latch-up modes.

### 7.2.2 High-Speed, Chopper-Stabilized Inverting Amplifier

The circuit of Fig. 7-16 is a more general type of amplifier, where the feedback connection is resistive and is suitable for inverting use at high output levels and rates.

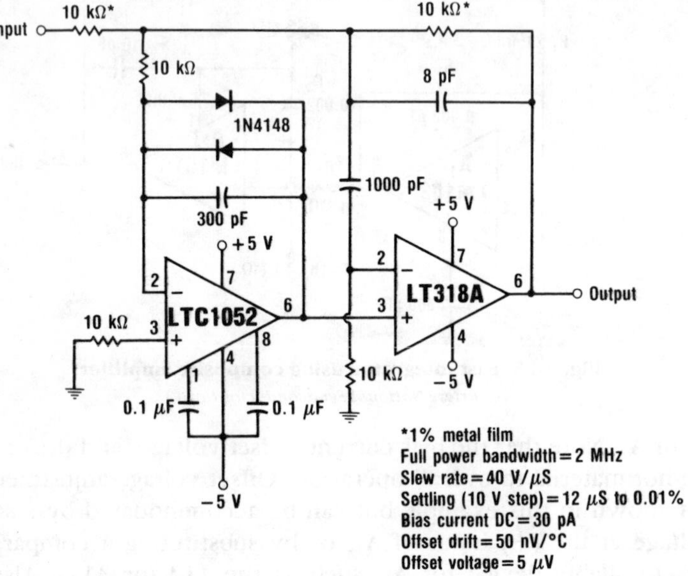


**Fig. 7-16. High-speed, chopper-stabilized inverter.**
*(Courtesy Linear Technology Corp.)*

This circuit uses the LT1052 chopper-stabilized op amp as the input integrator, driving a 318-type output amplifier. A pair of 1N4148 diodes serve to clamp the integrator in case of overload, thus enabling faster recovery. Both amplifiers are operated from ±5 V in this circuit (although the 318 could use ±15-V supplies, for more output).

The circuit is configured for unity gain, has a slew rate of 40 V/μs, and settles to 0.01% in 12 μs. This is achieved with an offset of only 5 μV, due to the chopper-amp input stage.

### 7.2.3 High-Speed FET Composite Inverter

The circuit of Fig. 7-17 operates from ±15-V supplies, and achieves a 100 V/μs slew rate by operating the 318 amplifier in an unusual mode. Here, the 318's internal npn input stage is turned OFF, by tying pins 2 and 3 to V−. The "outboard" JFET pair of 2N4393s is used as an offset-corrected differential pair, supplying a relatively high bias current to the 318 at pins 1 and 5.

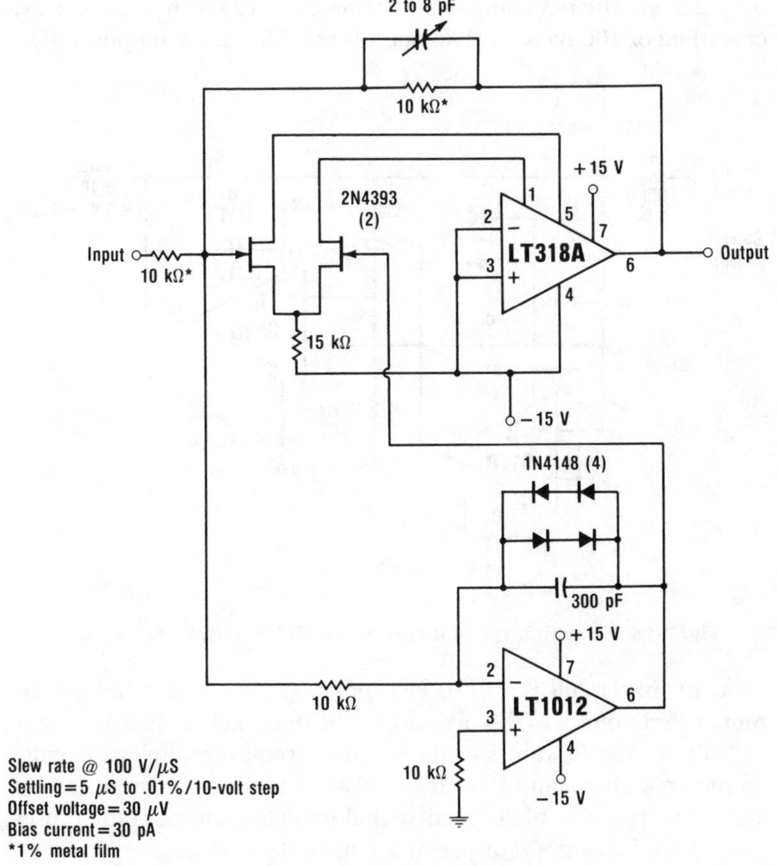


**Fig. 7-17. High-speed FET composite amplifier.**
*(Courtesy Linear Technology Corp.)*

Normally, discrete random JFETs would be impossible to use as a differential pair. However, in this case, they are offset-zeroed by the 1012 integrator. As a result, the composite amplifier has 100-pA typical bias currents and the offset voltage of the 1012, typically only 30 μV. Settling time of the circuit is 5 μs to 0.01%. (Note—for simplicity, a 2N5566 differential pair can also be used.)

### 7.2.4 Very-High-Speed Inverting Amplifier

The circuit of Fig. 7-18 is useful at even higher speeds, as it employs a 2540 in the output. This device has a 400 V/μs slew rate, a GBP of 400 MHz, and can settle to a 0.1% level in 200 ns for a 10-V output step. The pin-compatible alternate (a 5195) can achieve a faster settling of 100 nS to 0.01%, but it is rated for ±5-V outputs only.

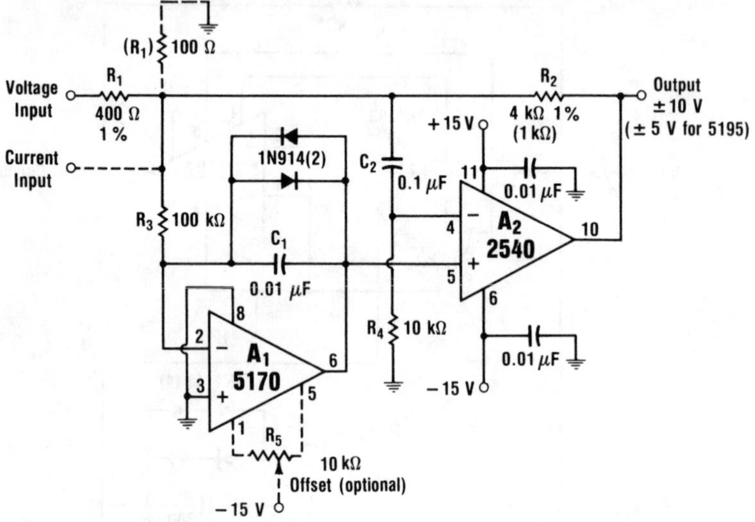


**Fig. 7-18. Very-high-speed inverting amplifier with low dc errors.**

$A_1$ of this circuit is a 5170 FET-input op amp, which has a maximum offset voltage of 0.3 mV and can be optionally nulled as shown.

With $R_1$ and $R_2$ values as shown, the circuit is an inverting video amplifier with a gain of 10. It can also be used as a current-to-voltage converter for high-speed digital-to-analog output-buffer uses, with minor changes. For use as a current-to-voltage converter, the current source drive is connected directly to the summing point, $R_1$ is connected to ground, and the value of $R_2$ becomes 1 kΩ. With a 0- to −5-mA drive from a high-speed digital-to-analog converter, the output-voltage range becomes 0 to +5 V.

### 7.2.5 High-Accuracy, Low-Bias-Current Follower

The circuit of Fig. 7-19 operates similarly to the previous ones, but adds the correcting dc offset to the balance pins of the second stage. This circuit is arranged as a follower, and will have the low bias current and offset voltage of $A_1$, and, in addition, the low-frequency CMRR. An option shown for this circuit is the dc offset trim method used for the LM11-type op amp ($A_1$).

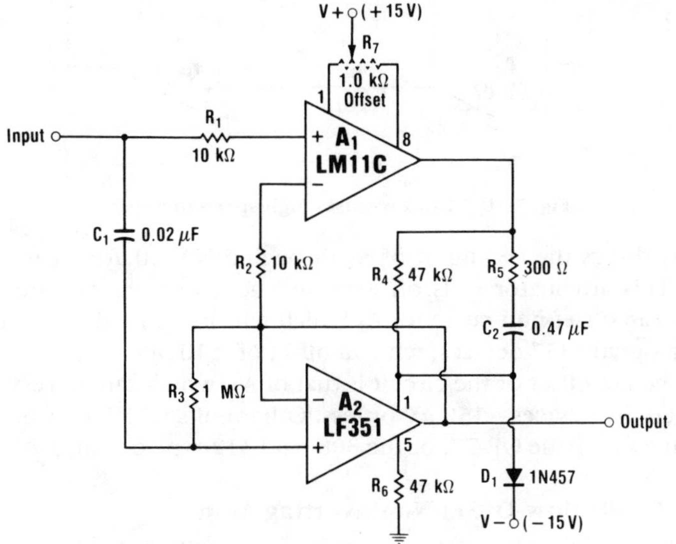


**Fig. 7-19. High-accuracy, low-bias-current follower.**

This circuit also ac-couples the $A_2$ stage to minimize bias currents, via $C_1$. As such, it also exhibits a relatively long overload recovery time and should not be driven outside its common-mode range of $\pm 10$ V. Direct coupling of $A_2$ will eliminate this at the expense of an increased high-temperature bias current. Settling time of the circuit is 10 $\mu$s to a 0.01% accuracy level for a 10-V step input.

### 7.2.6 Ultralow-Offset High-Speed Inverter

The circuit of Fig. 7-20 is a much simplified form of a composite inverting amplifier, optimized for very low offset and high speed.

As can be noted in the circuit, both $A_1$ and $A_2$ are directly coupled to the summing point. $A_1$ is a low-speed device, selected for its very low offset voltage—75 $\mu$V maximum for the OP-07E. $A_2$ is a high-speed FET-input unit, selected for its high slew rate and GBP—60 V/$\mu$s and 26 MHz, respectively, for the OP-17; slightly less if the similar 357 device is used.

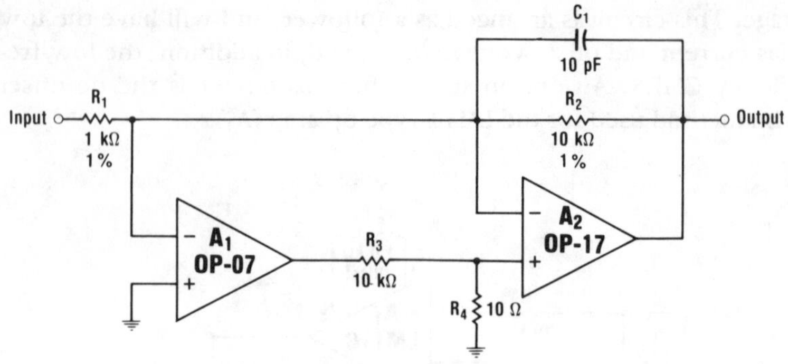


**Fig. 7-20. Ultralow-offset high-speed inverter.**

$A_1$ drives the (+) input of $A_2$ through a ×1000 attenuator, $R_3$-$R_4$. This attenuator rolls off excessive ac gain from $A_1$, and provides an offset-null range for $A_2$ which will accommodate even the worst-grade 357 devices, with an offset of ±10 mV.

The net offset of the circuit is that of $A_1$, which can be very low for OP-07 devices—150 μV or less. Other units which may be useful for $A_1$ are the OP-77, or the 308- and 312-type op amps.

### 7.2.7 Ultralow-Offset Noninverting Amp

The circuit of Fig. 7-21 is a composite amplifier that is arranged to offset-zero a low-noise IC preamp, allowing microvolt-level signals to be amplified without ac coupling.

As shown, the 1007 low-noise amplifier is used as a gain-of-1001 noninverting stage with a 100-ohm feedback source resistance (for lowest noise). The 1052 chopper amp integrates the residual input offset error that is inherent to the 1007, and drives the device's null pins to zero the overall circuit to its own 3-μV (typical) offset voltage, along with the 50-nV/°C drift of the chopper.

Bandwidth, as shown, is 8 kHz, but this can be extended to 60 kHz by the use of a 1037 unit. The 1037 will, however, be limited to closed-loop gains of 5 (or more). OP-27 and OP-37 units will also work in this circuit, but with slightly greater noise, vis-a-vis the 1007 or 1037 units.

### 7.2.8 General-Purpose Composite Amplifier

The composite amplifier of Fig. 7-22 is a more general-purpose one, and can be adapted for either inverting or noninverting modes of use over a wide range of gains.

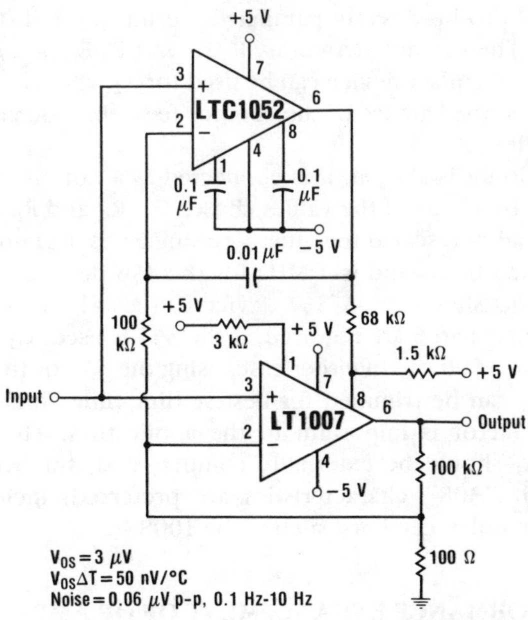


**Fig. 7-21. Ultralow-offset noninverting amp.**
*(Courtesy Linear Technology Corp.)*

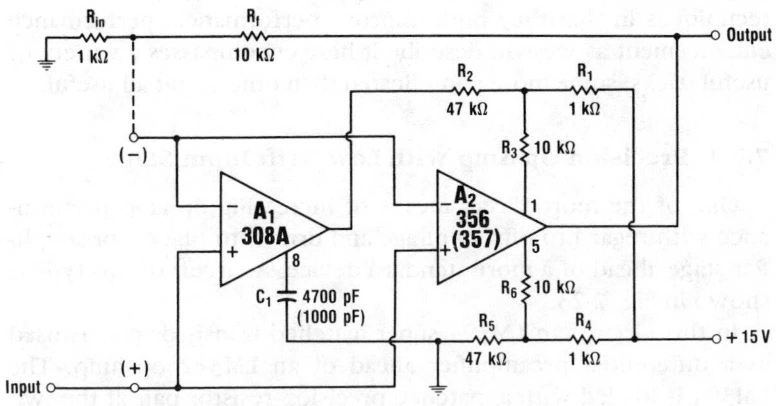


**Fig. 7-22. General-purpose composite amplifier.**

This circuit gets its general-purpose capability by virtue of the drive method used to offset $A_2$. Rather than driving the (+) input of $A_2$, the balance input (pin 1) is driven. This allows *both* inputs of $A_1$ and $A_2$ to be directly paralleled, permitting full differential operation. The balance networks, $R_1$-$R_3$ and $R_4$-$R_6$, are chosen so that a ±10-mV offset device can be used for $A_2$, and $A_1$ will still be able to adjust the balance of the composite so that the net offset is equal to that of $A_1$.

As the circuit is shown, it is connected as a noninverting gain-of-11 stage by virtue of the values chosen for $R_{in}$ and $R_f$. However, it can be readily rescaled for other working gains. Gain-bandwidth is determined by $A_2$ and is 4 MHz for the 356 device, which also has a 15-V/μs slew rate. A 357 device can also be used for $A_2$ if gains of more than 5 are required. If the 357 is used, $C_1$ should be decreased to 1000 pF. In either case, using the 356 or the 357, the value of $C_1$ can be trimmed for best settling-time characteristics, when this factor is important to the application. The amplifier used for $A_1$ should be externally compensated, for best results. Types with "308" characteristics are preferred, including the more recent enhancements, such as the 1008.

## 7.3 PERFORMANCE ENHANCEMENT OF OP AMPS

Closely allied to composite amplifiers are methods to augment or enhance certain individual performance characteristics of op amps. Although there are not many differences between the two techniques in that they both improve performance, performance enhancement as we will describe it here encompasses a variety of useful tricks, some more complicated than others, but all useful.

### 7.3.1 Precision Op Amp With Low-Drift Input Stage

One of the more direct means of increasing op-amp performance with regard to offset voltage and drift is to place a preamplifier stage ahead of a more standard device. A circuit of this type is shown in Fig. 7-23.

In this circuit, an LM394 super-matched transistor pair is used as a differential preamplifier ahead of an LM312 op amp. The LM394 is loaded with a matched precision-resistor pair at the two collectors; frequency compensation is provided by $C_1$ and $C_2$ to shape the overall frequency response of the two-stage amplifier, rolling off the gain at 6 dB/octave.

As might be expected from cascading two stages, the dc gain of this op amp is very high—128 dB or more. The output stage buff-

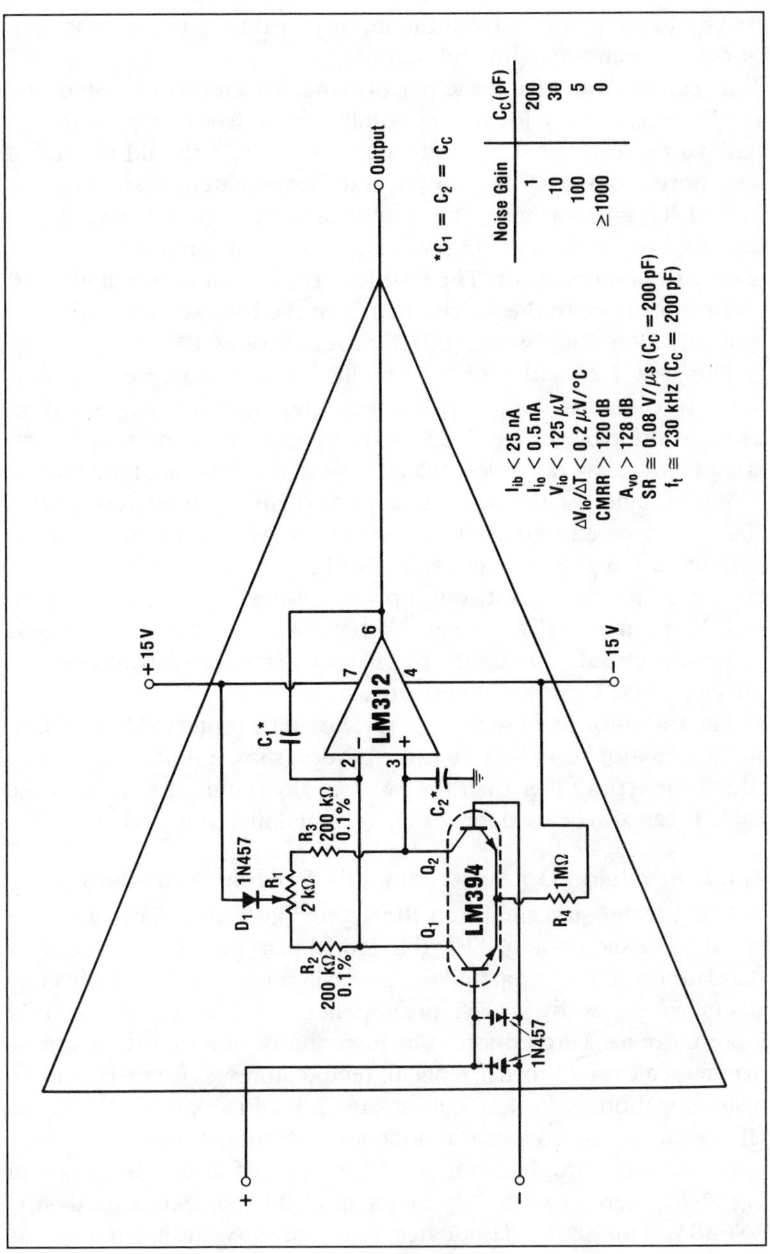


**Fig. 7-23. Precision op amp with low-drift input stage.**

ers any loading effects from the input pair due to the thermal isolation between the input and output.

As can be noted from the performance specifications listed, the amplifier has a very low offset voltage and offset voltage drift. To achieve the drift of 0.2 $\mu$V/°C, the LM394 pair should be nulled with potentiometer $R_1$ as shown, and a ratio-matched pair used for $R_2$ and $R_3$. Furthermore, the output amplifier should have a reasonably low offset drift (15 $\mu$V/°C or less), or it can indirectly contribute to nonideal drift. The CMRR of the circuit is very high (120 dB or more) due to the precise match of the LM394. Bias current is typically 25 nA or less, and offset current is only 0.5 nA.

The slew rate and $f_t$ of the amplifier will be inversely proportional to $C_1$ and $C_2$ if $C_C$ is varied for different closed-loop gains, as listed in the table in Fig. 7-23. At unity-gain compensation ($C_C$ = 200 pF), the slew rate is less than 0.1 V/$\mu$s, and $f_t$ is about 0.25 MHz.

While many of the individual performance parameters of this circuit can be exceeded by several of the higher-performance instrumentation-grade op amps, it is still useful. One reason is that it is an example of several sound precision-design practices, in general. More specifically, for very high gains where bandwidth needs to be optimized, the ability to custom "frequency compensate" the circuit via $C_C$ makes it very attractive.

The transistor-pair used for $Q_1$-$Q_2$ should be protected from differential breakdown, so therefore the diodes (shown dotted) should be added for types other than the LM394. Monolithic ICs, such as the LM321, can also be used for the $Q_1$-$Q_2$ function in this circuit.

### 7.3.2 Precision Op Amp Using FET Preamp With Gain

Using techniques similar to those previously described, it is also possible to incorporate FET differential-pair gain stages ahead of standard op amps. Compared to bipolar-input preamplifiers, the large advantage gained by an FET preamplifier is a dramatic reduction in input currents. Furthermore, this lowering of input currents can be maintained over a broader range of temperatures, since there is minimal dissipation in the FET devices. In an IC op amp, the dissipation effects can be serious, even at moderate ambient temperatures.

A relatively simple example of this type of circuit is shown in Fig. 7-24. Here, a monolithic n-channel JFET device is used—the 2N6483. This particular device is specifically designed for this type of application and features controlled offset, drift, and CMRR specifications.

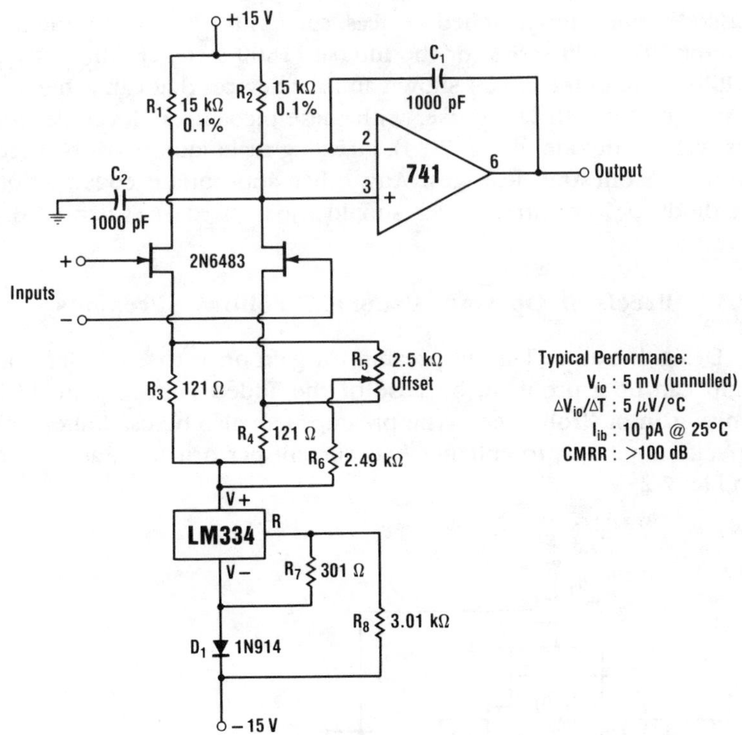


**Fig. 7-24. Precision op amp using FET preamplifier.**

In this circuit, the 2N6483 pair is loaded by a matched pair of 15-kΩ resistors and drives a standard 741 op amp as the output stage. Compensation is provided by $C_1$ and $C_2$, which results in an $f_t$ on the order of 300 kHz with the values shown.

The 2N6483 device has the tightest specifications of its series. It has a maximum offset voltage of 5 mV, and an offset voltage drift of 5 $\mu$V/°C. The offset voltage can be adjusted to zero with the offset potentiometer, as shown. A constant drain current of 400 $\mu$A is provided by the LM334 current source, and $R_7$, $R_8$, and $D_1$.

The input bias current of the 2N6483 is dependent upon the drain-to-gate potential or, as used here, the common-mode voltage. Over a common-mode range to −10 V, it typically will be 10 pA or less (at room temperature). The input currents will double with each 10°C increase in temperature, as in any JFET.

Increased CMRR in an FET preamplifier can be realized by using cascode-connected matched devices, such as the National Semiconductor NDF9406 series, or the Intersil IT500 series. Both of these will operate in the circuit shown and are devices that can achieve a CMRR of 120 dB. Of course, with these picoampere-level circuits (as well as the one of Fig. 7-24), guarding techniques must be used to screen off stray leakages. Any other appropriate construction methods for low-current levels should also be used (see Chapter 3).

### 7.3.3 Precision Op Amps Using FET-Follower Preamps

Designing an FET-input preamp for gain prior to a standard op amp can be a problem, because of the added voltage gain. FET unity-gain or "follower" type preamps can also be used ahead of specific op amps, to enhance bias-current performance, as shown in Fig. 7-25.

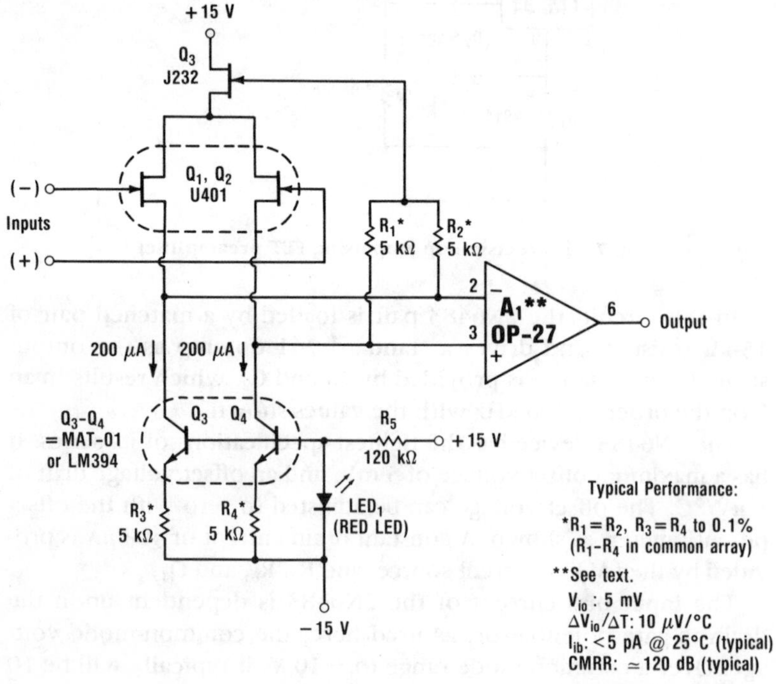


*(A) Medium speed, high CMRR.*

**Fig. 7-25. FET-**

Although many FET IC op amps exist, high CMRR performance is not generally a strong point with them (except for a notable few). The FET preamp of Fig. 7-25A features a 120-dB CMRR performance, a bias current of only a few picoamperes, and low offset. It uses a U401 matched JFET pair with bootstrapping by $Q_3$, and a high-CMRR current source. While this dual-follower circuit is capable of high CMRR, the performance would be lost if it were not also used with a very-high-performance op amp. Very simply, this means a device with a comparable (or higher) CMRR, such as an OP-07, OP-27, or OP-37.

The speed of this circuit will be that of the op amp used, as will the output drive and open-loop gain. The bias current is that of the preamp, and the noise will be the RSS of the preamp, plus that of the op amp.

The two key parameters where this circuit is weak, vis-a-vis the

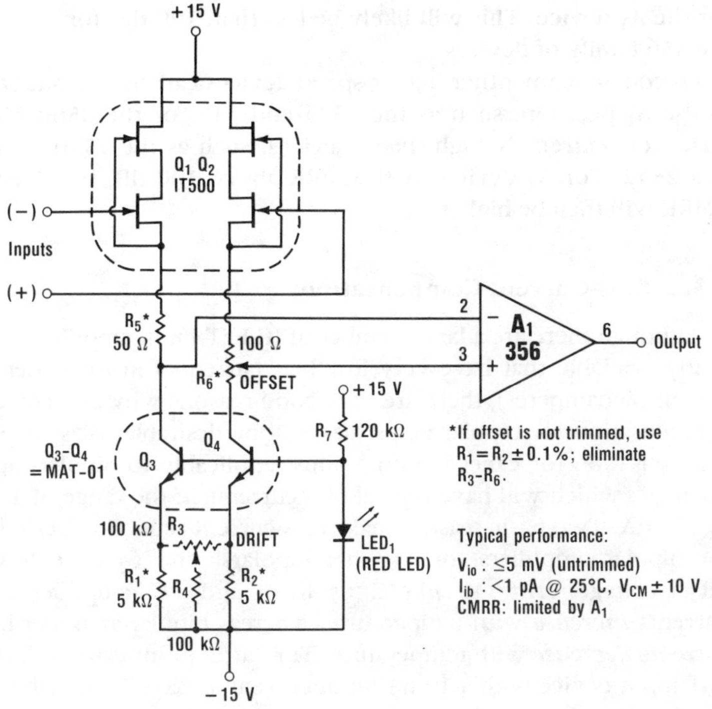


*(B) High speed and low $I_B$.*

**follower preamps.**

OP-27 (or OP-07), are the offset voltage and drift. These can be carefully trimmed (as in the next circuit) for instances where this is critical. Because of the circuit's high CMRR, however, bias current tends to stay very constant with CM voltage, unlike many general-purpose FET op amps. Note that the U401-J232 combination is functionally equivalent to a cascode JFET dual, such as the ID500 (refer to the next circuit).

The FET-follower preamp of Fig. 7-25B uses a dual-JFET cascode pair, the ID500. This device features a bias current of less than 1 pA, which stays constant for CM voltages of ±10 V. Otherwise, this circuit is similar to the one in Fig. 7-25A as far as the preamp portion is concerned, and it illustrates both optional nulling and drift trim.

This example uses a 356-type output amplifier, which allows the high slew rate of this IC to be realized. Overall circuit CMRR will be limited to whatever the particular CMRR performance is for the $A_1$ device. This will likely be less than 100 dB, for most of the 356 family of devices.

Of course many other higher-speed devices can also be buffered in the $A_1$ position, such as the 5534 and 318, or the 2500/2525 series, or, extremely high-speed devices, such as the 2540, 2541, and 2542. For $A_1$ devices with CMRR above 100 dB, the overall CMRR will then be higher.

### 7.3.4 Bias-Current Compensation

Although there are a large number of IC FET-input amplifiers currently available that have very low bias currents (on the order of tens of picoamperes), there are very good reasons why bias-current compensation for bipolar amps may still be desirable. Bias-current compensation (or cancellation) is most applicable to bipolar-input op amps, which will have typical bias currents in the range of 1 nA to 100 nA. Two basic reasons make bias-current reduction desirable for bipolar amplifiers: one is that bipolars tend to have lower offset-voltage drift (10 $\mu$V/°C or less); and FET-amplifier bias currents *increase* with temperature, whereas bipolar-amplifier bias currents *decrease* with temperature. This latter point can result in a FET-input device with a 10-nA input current at (say) 70°C, where a bipolar device can be held to below 1 nA. It is not at all difficult to reach 70°C *junction* temperature in ICs with moderate dissipation, even in relatively modest environments.

Some simple and effective schemes for compensating bias currents have been described by Widlar,* and their usefulness is illustrated in Figs. 7-26 and 7-27.

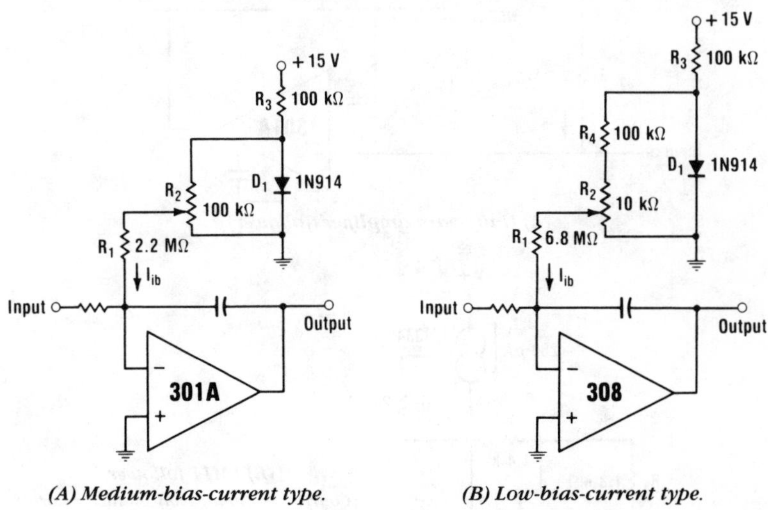


*(A) Medium-bias-current type.*

*(B) Low-bias-current type.*

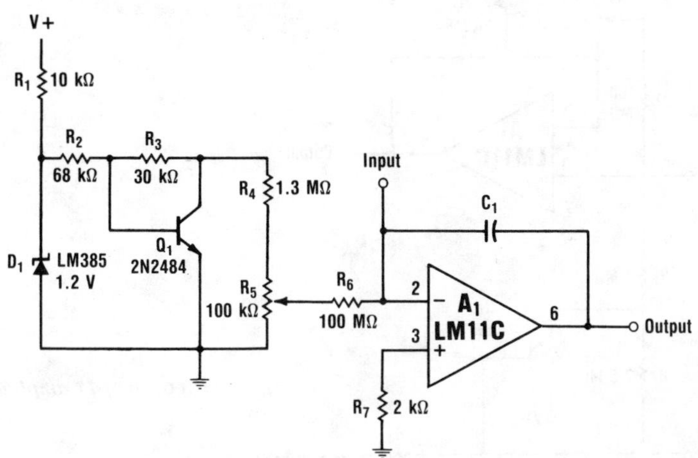


*(C) Very-low-bias-current (LM11) device.*

*(Courtesy National Semiconductor Corp.)*

**Fig. 7-26. Bias-current compensation—inverting amplifiers (integrators).**

* See references 34 and 35 at the end of the chapter.

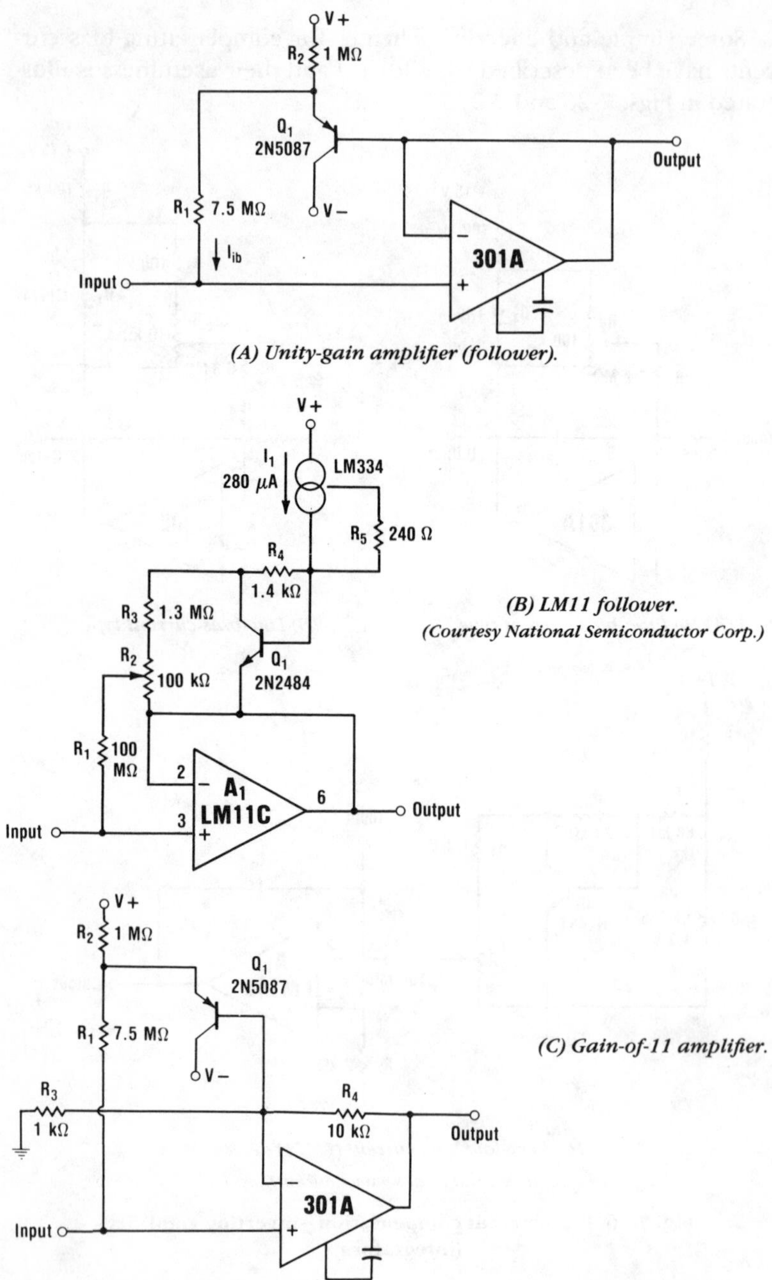


*(A) Unity-gain amplifier (follower).*

*(B) LM11 follower.*
*(Courtesy National Semiconductor Corp.)*

*(C) Gain-of-11 amplifier.*

**Fig. 7-27. Bias-current compensation—noninverting amplifiers.**

In inverting amplifiers, the bias compensation is most simple, as there is no common-mode voltage and the compensating voltage can be referred to ground. Three variations of a scheme for compensating integrators for lowest bias current (and, thus, lowest drift) are shown in Fig. 7-26.

In Fig. 7-26A, a medium-bias-current device, such as the 301A (80 nA), is compensated by $R_1$, which is returned to a variable voltage from $R_2$. This voltage is derived from a forward-biased diode; thus it will tend to track the bias current (which decreases as the temperature rises). $R_1$ is selected to provide the worst-case bias current for the IC in use (if different), and either $R_2$ can be adjusted, or the value of $R_1$ can be trimmed. For pnp-input amplifiers, the diode and supply voltage should be reversed. Trimming is a necessary evil with these circuits, since the IC bias currents are not precisely known. Compensation can be expected to provide a 5-to-10-times reduction in bias current over a limited temperature range.

With amplifiers that have very low initial-bias currents, such as the 308 types, the resistance values can easily become troublesome in the circuit of Fig. 7-26A. The variation shown in Fig. 7-26B divides down the diode voltage to 50 mV, which allows a more reasonable value for $R_1$ to be used. This division scheme effectively multiplies the value of $R_1$; and it can be adapted to other circuits as well.

The circuit variation shown in Fig. 7-26C is applicable to the LM11, an amplifier which has a very linear and predictable bias current. The circuit also includes a preregulator ($R_1$-$D_1$), which provides immunity from supply changes. This circuit is capable of bias currents that are lower than 100 pA over a 0 to +100°C range, after compensation by $R_5$ (at 25°C).*

A bias-current-compensated follower is shown in Fig. 7-27A, which uses a bootstrapped pnp transistor to provide a floating-current source via $R_1$. $R_1$ can be trimmed for compensation, or a potentiometer can be used as in Fig. 7-26A. For simplicity, a diode can be used for $Q_1$ also, but the pnp transistor minimizes loading on the op amp. For pnp-input op amps, use an npn transistor returned to V−.

The LM11 circuit of Fig. 7-27B allows picoampere-level bias currents, in a follower. Regulation for the compensation is provided by the LM334.

---

* See reference 37 at the end of the chapter.

In a noninverting amplifier, the transistor method of deriving the compensating current is very useful, as shown in Fig. 7-27C. The base of $Q_1$ is returned to the (−) input, which bootstraps $R_1$ as in Fig. 7-27A. The low base current of the pnp will not introduce large errors if the values of $R_3$ and $R_4$ are maintained relatively low. $R_1$ is trimmed for compensation in this circuit to avoid error currents in $R_3$ and $R_4$.

Two alternate methods of bias current compensating a follower for high common-mode inputs are shown in Fig. 7-28. In Fig. 7-28A, a scheme for providing a bias current for npn inputs is shown using a transistor array—the CA3096.

Here, the npn transistor of the array ($Q_3$) is forced to conduct a base current that will be nominally equal to the bias current to be compensated. The pnp current mirror returns this current to the (+) input of the op amp, resulting in the desired compensation. A drawback of this scheme is that it is not regulated against the supply voltage. (Not as shown, but it can be so adapted by using a zener connection for one of the remaining CA3096 devices.) $R_1$ is trimmed for compensation.

A version of this circuit that is useful for pnp inputs is shown in Fig. 7-28B. It works similarly, but with an npn current mirror.

Both of these circuits are really general-purpose circuits, and can accommodate followers or inverters. The circuit of Fig. 7-28B can also accommodate *differential* op-amp inputs (with two compensation currents) by adding a second npn current-source transistor to the current mirror, with an additional 1-kΩ emitter resistor.

### 7.3.5 Bias-Current Cancellation Using Matched Amplifiers

Some of the more effective methods of minimizing bias currents come about when *matched* amplifiers are used. Basically, such methods take advantage of the fact that all input bias currents in a multiple IC op amp will be matched within a few percent. The bias current of one amplifier of a dual or a quad device can then be used as a reference to develop a compensating current which effectively cancels the bias current of another amplifier on the same chip. Although the degree of compensation might be somewhat better than the more general methods described previously, the real attractiveness of this method is in the fact that it does not require trimming for good results (whereas the others do).

The scheme of Fig. 7-29 is a general-purpose one, previously

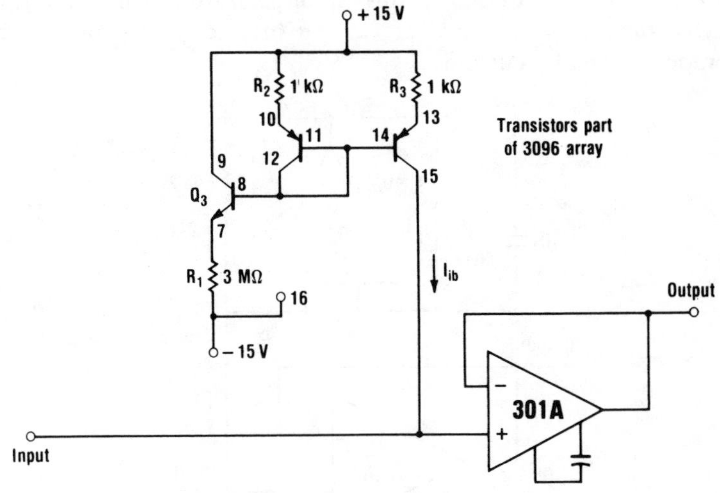


*(A) Npn inputs.*

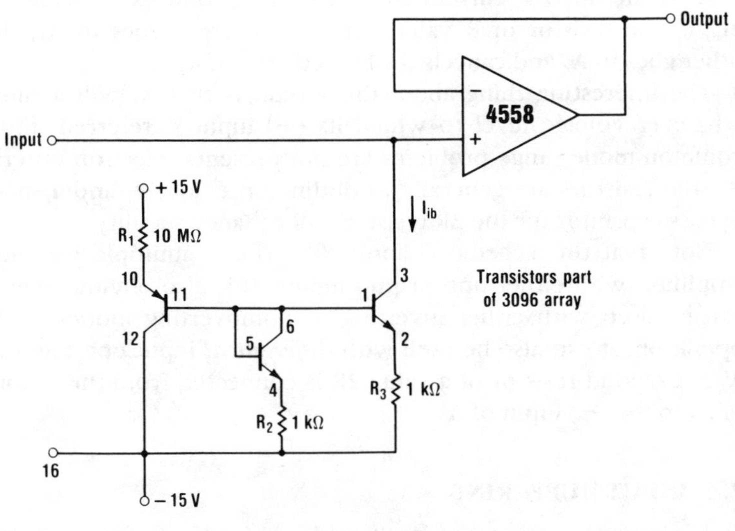


*(B) Pnp inputs.*

**Fig. 7-28. Alternate bias-current compensation methods for high common-mode inputs.**

described by the author.* In this circuit, a section from a multiple device functions as a floating current inverter and provides the compensating bias current.

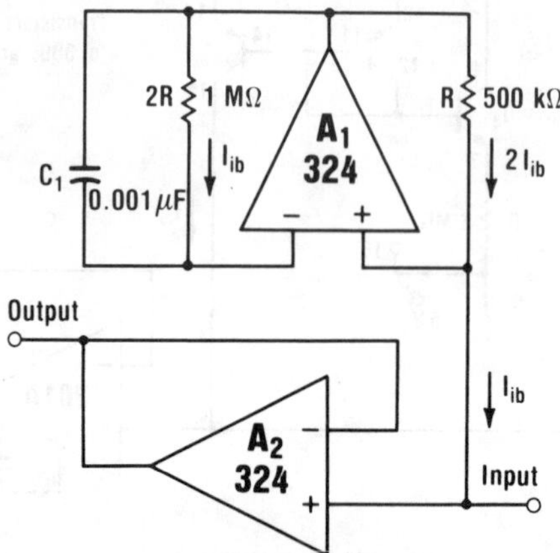


**Fig. 7-29. General method of bias-current cancellation using matched amplifiers.**

$A_1$ is the floating current source, which produces a current of $2I_{ib}$ in the resistor of R value. One bias current goes to $A_1$, the other goes to $A_2$ and cancels the bias current of $A_2$.

The interesting thing about this circuit is that $A_1$ will assume whatever voltage level to which its (+) input is referred; thus, common-mode range problems are not present. Selection criteria for the resistors are generally as outlined previously, and a small bypass capacitor for the 2R resistor will enhance stability.

Note that this scheme is applicable to *any* multiple matched amplifier, with either npn or pnp inputs. It is also obvious that it can be used with either inverting or noninverting modes of $A_2$ operation. It can also be used with *differential* input operation of $A_2$ if a second resistor of a value 2R is connected from the output of $A_1$ to the (−) input of $A_2$.

## 7.4 INPUT BUFFERING

A previous section has dealt with methods of designing op-amp preamplifier stages, including FET-input types. FETs are, of

* See reference 20 at the end of the chapter.

course, the most direct and straightforward way by which amplifier input currents can be reduced, but, often, other rather simple circuitry can be used to provide effective input buffering, with minimal trade-offs.

### 7.4.1 Zero-Offset FET Buffer/Follower

The circuit of Fig. 7-30 is not a general-purpose op amp, but a specialized FET-input unity-gain follower, configured for low offset and high accuracy.

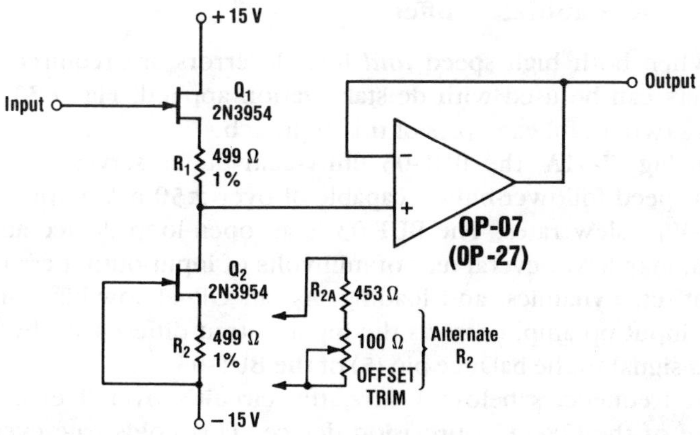


**Fig. 7-30. Zero-offset FET buffer/follower.**

In this circuit, $Q_1$ and $Q_2$ are a well-matched FET differential pair arranged in a totem-pole follower circuit. $Q_2$ operates as a current source, biasing $Q_1$ at a constant current. With $R_1$ equal to $R_2$, and $Q_1$ and $Q_2$ matched, this results in nominally zero offset between the input and the bottom of $R_1$. Offset can be made adjustable to "zero," by using the alternate $R_2$ configuration. The op amp preserves the linearity of the configuration by reducing the effective loading of the follower to essentially zero.

As with the previous circuit, the offsets and drifts of the two stages of this circuit add directly. As noted, $R_2$ can be trimmed to adjust the offset if desired. An OP-07 or OP-27 low-drift op amp is shown as the output buffer to minimize the offset and drift contribution of this stage.

As shown, the circuit is not a particularly fast one, mainly because of the OP-07. The FET follower *is* itself, however, a fast circuit and more of this speed can be made available to the output by employing a fast op-amp buffer, such as the 318 or the 2500 series of bipolar op amps. Even faster FET followers are possible, using this basic configuration, with $Q_1$-$Q_2$ using a 2N5911 or U440 FET pair. In such case(s), the op amp would preferably be a very-high-SR device, such as a 318 or one of the 2500 series.

The overall CM error of this follower will be essentially that of the JFET pair. This can be optimized by careful selection, or by the use of a cascode pair for $Q_1$-$Q_2$.

### 7.4.2 DC Stabilized Buffers

When both high speed *and* low dc errors are required, FET buffers can be used with dc stabilization applied. Fig. 7-31 illustrates two useful examples of this approach.

In Fig. 7-31A, the BUF-03 unity-gain buffer serves as the $A_1$ high-speed follower-buffer, capable of over ±50-mA outputs with 200-V/μs slew rates. The BUF-03 is an open-loop device and, as such, may have several tens of millivolts of input-output error due to offset, dynamics, and loading. $A_2$, an OP-41 low-bias current FET-input op amp, corrects the input-output differences, by feeding a signal to the balance pin (5) of the BUF-03.

At frequencies below 1 kHz, the circuit's overall errors are those of the OP-41, a precision device. This holds true even for heavy loads, since the BUF-03 furnishes the load drive. At higher frequencies, the response is that of the BUF-03. This circuit makes a good input buffer for high-speed high-resolution a/d's. A clip-on heat sink for $A_1$ is recommended, to minimize its temperature (and, thus, the input bias current).

The version of Fig. 7-31B is somewhat similar conceptually, but it features the very popular "0033" type of high-speed hybrid FET-input buffer. This device is capable of slew rates of over 1000 V/μs, with input-output delays of less than 5 ns. With up to 100 mA of peak output available, it makes an excellent dc-response video-bandwidth buffer, with ample output.

The LH0033-type buffer is also an open-loop type of circuit, with only the intrinsic match of the FET devices used and the initial trimming at manufacture used to control offset. As a result, the dc stability of the circuit is relatively poor (vis-a-vis a precision op amp).

In this circuit, the $A_2$ op amp corrects the input-output offset by dynamic control of the $R_{DS(on)}$ of $Q_1$, a MOSFET device. This tech-

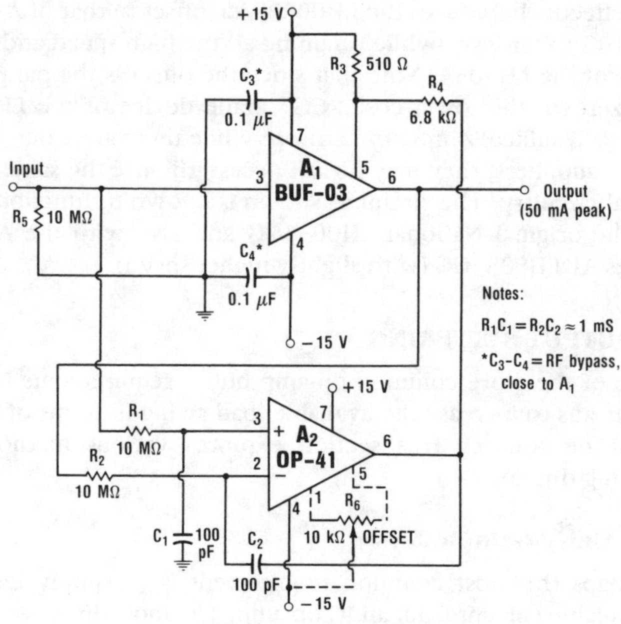


*(A) High-speed circuit.*

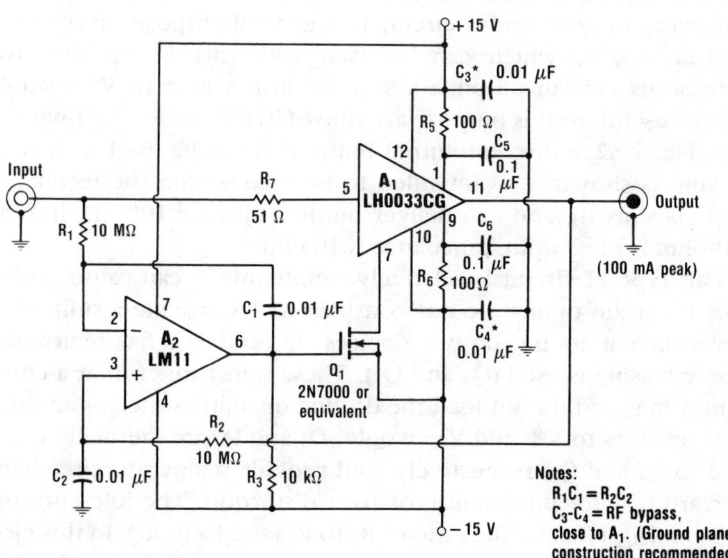


*(B) Video circuit.*

**Fig. 7-31. Dc stabilized buffers.**

nique effectively reduces the LH0033's dc offset to that of $A_2$ (typically 100 $\mu$V or less), while retaining all the high-speed and drive virtues of the LH0033. Note that since the offset is the parameter being zeroed, the lower-cost "CG"-grade device of the LH0033 device will suffice. *A note of caution:* while there are a number of "0033" suppliers, they may not all necessarily use the same exact internal circuitry. The circuit as shown is known to function with both the original National LH0033CG and also with the Analog Devices ADLH0033CG (with slightly higher speed).

## 7.5 OUTPUT BUFFERING

One of the more common op-amp buffer requirements is finding a means to increase the available load swing in terms of either current or voltage. This section explores various methods of achieving this goal.

### 7.5.1 Unity-Gain Buffers

Perhaps the most common requirement is to simply increase the available current from an IC op amp. The most direct way is to use a higher-current device, such as the 759, OP-50, 2541, etc. However, another useful circuit is a general-purpose unity-voltage-gain *buffer,* which can be used with *any* IC op amp to increase its current output to, say, ±100 mA at ±10 V. Several circuits useful for this purpose are shown in the section following.

In Fig. 7-32, a discrete output buffer that can be used with any op amp is shown. It is intended to be used within the feedback loop (as shown), and can deliver outputs up to ±100 mA into a 100-ohm load. Output impedance is 10 ohms.

This type of circuit is relatively simple, but it can cause problems if certain points are not considered. To provide a sufficient drive current to the output devices, $Q_1$ and $Q_2$, fixed-current-source biasing is used ($Q_3$ and $Q_4$). These transistors run at a constant 4 mA, and do not load the driving op amp as the commonly used resistors to V+ and V− would. $D_1$ and $D_2$ are thermally coupled to $Q_1$ and $Q_2$, respectively, and provide temperature-tracked forward bias for minimum crossover distortion. The idle current in $Q_1$ and $Q_2$ is set by the ratio of $R_4$ to $R_7$ (and $R_5$ to $R_8$). In this circuit, $Q_1$ and $Q_2$ idle at just over twice the current in $D_1$ and $D_2$. $R_7$ and $R_8$ also provide short-circuit protection for $Q_1$ and $Q_2$, which are 1-A high-speed units.

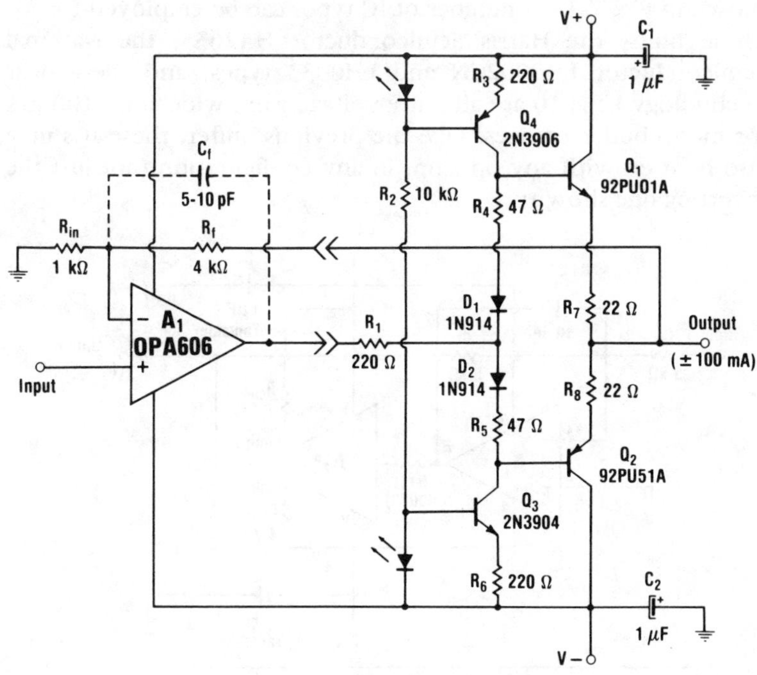


**Fig. 7-32. Discrete medium-current output buffer.**

The design can either be used as shown, or tailored for individual uses. The function of $R_1$ may not be obvious; it is simply to prevent parasitic oscillation in $Q_1$ and $Q_2$. Good bypassing of the supplies is mandatory for this circuit, as it will deliver heavy current spikes to the supply buses. $C_1$ and $C_2$ should be solid tantalum (or other low-impedance at high-frequency) capacitor types.

Because of the bias-current regulation for $Q_1$ and $Q_2$, this circuit can be used over a ±10-V to ±20-V supply range with little deterioration. Op amp $A_1$ is selected to optimize the performance parameter of interest (i.e., speed, dc accuracy, etc.). In Fig. 7-32, an OPA606, a high-performance 356 type, is used. Other FET types with similar speed are the AD711, the 1022 or 1056, and the 34081 (series). To optimize for low noise and drift, the op amp can be an OP-07, OP-27, or OP-37 bipolar type. A gain-of-5 noninverting stage is shown, but lower (or higher) gains are possible, as well as inverting-mode operation.

A more general circuit, using an IC output-buffer stage, is

shown in Fig. 7-33; a number of IC types can be employed for $A_2$. These units, the Harris Semiconductor HA2635, the National Semiconductor LH0002CN and LH0033 types, and the Linear Technology LT1010 are all unity-voltage-gain, wideband, 100-mA (or more) buffer devices. Like the previous buffer, these ICs may also be used with any op amp, in any configuration (not just the inverting one shown).

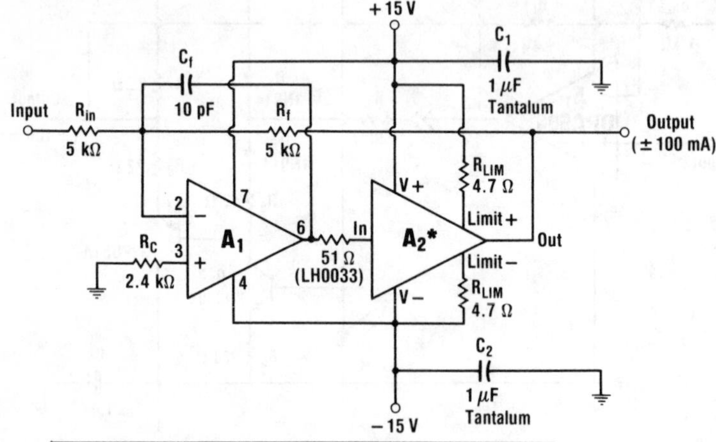


***$A_2$ Pin Numbering**

| | $A_2$ Device | | | |
|---|---|---|---|---|
| Pin | HA2635 | LH0002CN | LH0033 | LT1010 |
| In | 3 | 8 | 1 | 8 |
| Out | 9 | 4 | 4 | 4 |
| V+ | 7 | 1 | 8 | 2 |
| V− | 2 | 7 | 5 | 6 |
| Limit + | 8 | 2 | 7 | Internal |
| Limit − | 1 | 6 | 6 | Internal |
| | (12 pin, TO-8) | (8 pin, TO-5) | (8 pin, minidip) | (4 pin, TO-39) |

Notes:
1. Use R = 51 Ω in series with LH0033 input.
2. Offset is preset to zero on LH0033 with jumper (pins 2-3 on 8-pin device, pins 6-7 on TO-8 case).
3. LT1010 TO-39 package is the 4-pin equivalent to LH0002, with internal current limiting.

**Fig. 7-33. Typical IC medium-current output-buffer devices.**

$C_f$ is a feedback capacitor that should be adjusted for the desired bandwidth, in concert with $R_f$ and the IC to be used for $A_1$. $R_C$ is a bias-compensation resistor, usually necessary only with bipolar amplifiers used for $A_1$. Some specific examples will illustrate the representative use of these ICs. Fig. 7-33 is a *functional* diagram with regard to $A_2$; pin numbering details are in the table.

Fig. 7-34 illustrates how the 1010 buffer is used with a 1007 (or OP-27) type of op amp to provide over 100 mA of current drive to highly capacitive loads. In this circuit, the buffer's peak output current ($I_P$) and the load capacitance ($C_L$) set the loaded SR of the circuit. For sine-wave signals, the full power bandwidth ($f_p$) is set by the relation shown; for the values of $I_P = 0.25$ A (typical for the 1010) and $C_L = 0.33\ \mu F$, the power bandwidth is 12 kHz.

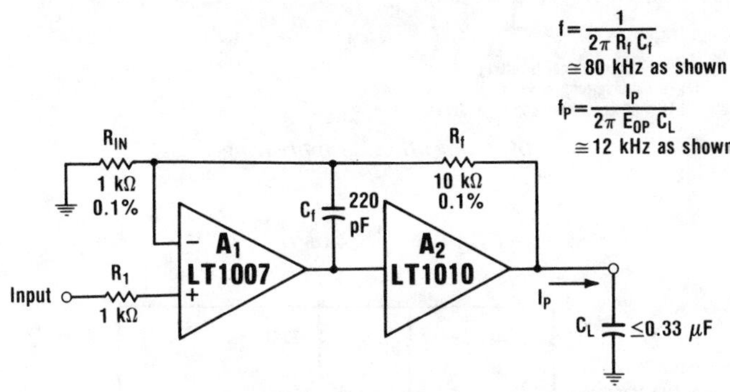


**Fig. 7-34. LT1010 100-mA noninverting buffer.**

Small-signal bandwidth is set by $R_f$-$C_f$ in a more or less conventional fashion. $C_f$ can be optimized for a given load capacitance, for minimum overshoot. Gain of this noninverting circuit is 11, as shown, and should be adjusted via $R_{IN}$ as necessary. Gain for dc or very low frequencies follows the standard relationship.

## 7.5.2 Buffer Stages With Voltage Gain

Current buffering output stages can also be set up to provide voltage gain as well as current gain. This approach is useful to increase the output of lower-voltage ICs to conventional levels, or to increase the output swings of standard ICs to beyond ±10 V.

The circuit of Fig. 7-35 illustrates how the 1052 (or 7652) low-voltage chopper op amp can be operated from ±15-V supplies, and its output raised to ±10-V swings. Here, the two 2N3904 transistors act as zener diodes, reducing the ±15-V supplies to ±7 volts to power the 1052. The 1052 drives the 318 amplifier, which is (locally) set up for a gain of 4, by the 3-kΩ/1-kΩ feedback.

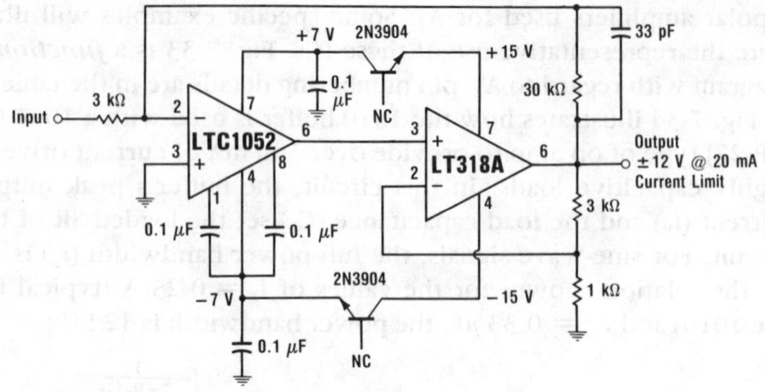


*(A) Conventional supply range.*

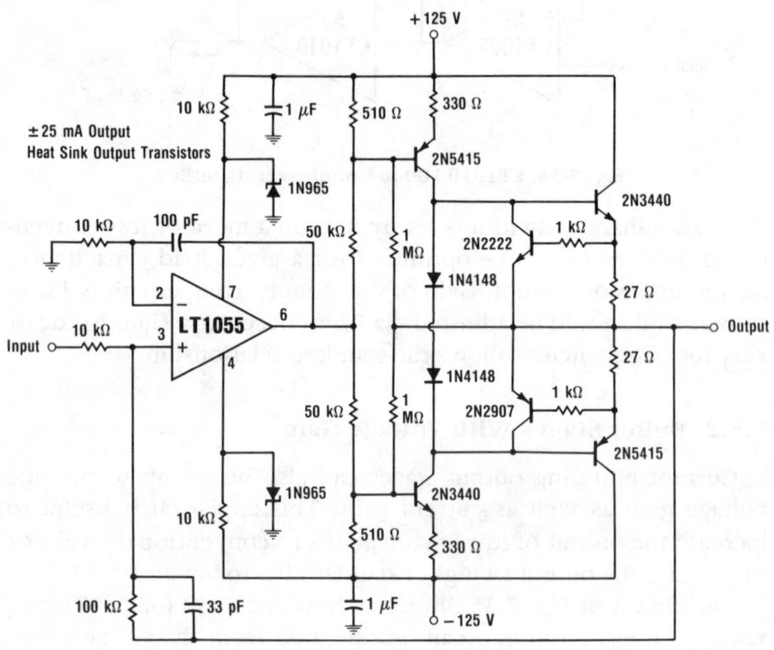


*(B) ±125-V supply range.*

**Fig. 7-35. Booster stage for low-voltage op amps.**

*(Courtesy Linear Technology Corp.)*

This raises an output of ±3 V from the 1052 to ±12-V levels at the main output. Overall loop feedback is provided by the 30-kΩ/3-kΩ resistors, back to the (−) input of the 1052, with stabilization set by the 33-pF capacitor. Although this circuit illustrates an inverting gain of 10, other gains plus noninverting are also possible. The input range limits of the 1052 amplifier (or any other similarly operated stage) must be observed, in practice.

The variation in Fig. 7-35B shows how similar principles can be applied to boost conventional-supply-range (±15 to ±20 V) op amps to ±120-V levels.

This circuit uses an IC input stage, with lower voltage supplies derived by zeners from the ±125 volts. The output from the 1055 op amp drives a complementary high-voltage push-pull buffer. This stage is buffered by a pair of complementary emitter-followers, at which the ±100-V output at 25 mA appears. While active current limitation is provided by the 2222/2907 transistor pair, heat sinking of the outputs is still necessary.

Gain is −10 as shown, but is adjustable via the 100-kΩ/10-kΩ resistor pair.

### 7.5.3 Multiple Amplifier Buffers

Since so many multiple IC op amps are readily available, a logical solution to moderate increases in output current is to parallel two or more sections. However, this cannot be done indiscriminately or disaster will result. When appropriate precautions are taken with multiple amplifiers, quite useful (and economical) circuits can result. These buffer types can be designed to be applicable to either unity- or higher-gain stages.

In Fig. 7-36, a configuration by Pease* is shown which is actually a form of bootstrapping. It operates by combining the outputs of two like amplifiers, to enable impedance loads down to one half that of the rating to be driven.

As can be noted, $A_2$ is a noninverting stage, set up for a gain of 1.1; thus, a voltage that is 1.1 times the output appears at the output of $A_2$. $R_2$ is set in value to be one fifth the load resistance; here, it is 120 ohms. Since $V_{out}$ appears at one end of $R_2$, and 1.1 $V_{out}$ appears at the other end, there exists 0.1 $V_{out}$ across $R_2$. Since $R_2$ is equal to $\frac{1}{5}$ $R_1$, $R_2$ must then conduct one half the load current, taking half the load burden from $A_1$.

* See reference 26 at the end of the chapter.

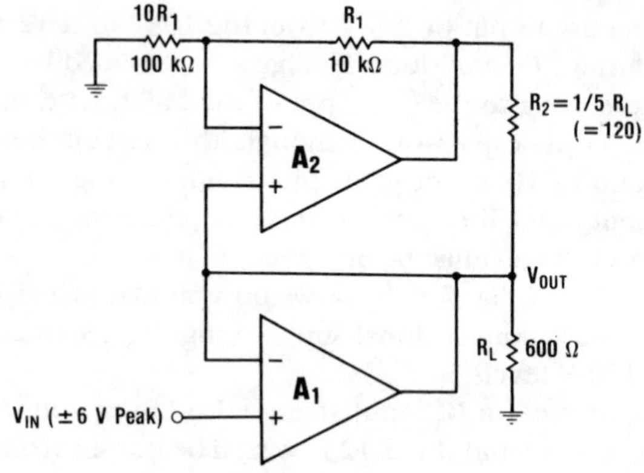


**Fig. 7-36. Bootstrapping multiple amplifiers.**
*(Courtesy National Semiconductor Corp.)*

Note that the conditions just described will exist for various other specific circuit values, as long as the key ratios are satisfied. $A_1$ can also be used in other configurations, such as inverting-gain stages. The key is that the (+) input of $A_2$ and the bottom of $R_2$ should be returned to the output of $A_1$ (and $R_L$).

A more general technique for combining the outputs of two or more amplifiers is shown in Fig. 7-37. Basically, it consists of summing the outputs of a number of op amps through like, small-value resistors (to force equal current distribution).

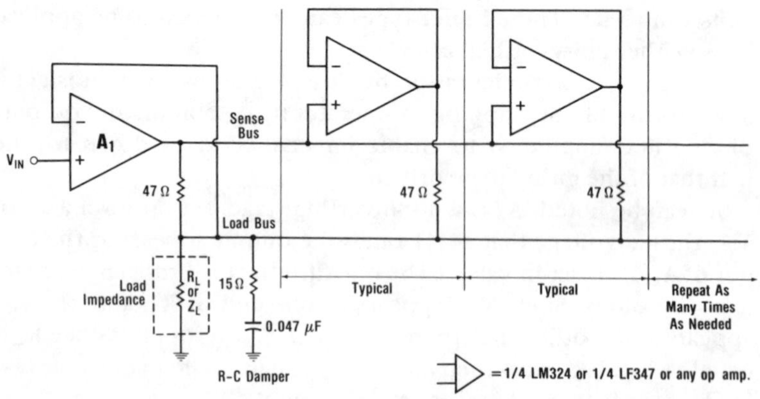


**Fig. 7-37. General technique for paralleling two or more amplifiers.**
*(Courtesy National Semiconductor Corp.)*

In Fig. 7-37, $A_1$ is a voltage-amplifier stage, and the follower-connected stages are fed from the output of $A_1$ and summed into a common-load bus. This technique is expandable to as many sections as may be necessary. With individual amplifier ratios of ±5 mA, this will allow a ±20-mA worst-case load current from a quad IC. Typically, much more than this figure can be obtained.

For stability in the presence of difficult loads, an RC damper, such as the 15-ohm/0.047-μF combination, may be necessary for wideband op amps; low-frequency devices, such as the 324 or 348, may not need it.

An example of this technique when applied to a noninverting dc amplifier is shown in Fig. 7-38. This circuit can drive loads of 500 ohms or less to levels of ±10 V with good linearity. $C_1$ will help with stability, and can be use to isolate and stabilize capacitive loads. Many quad ICs can be useful in this circuit.

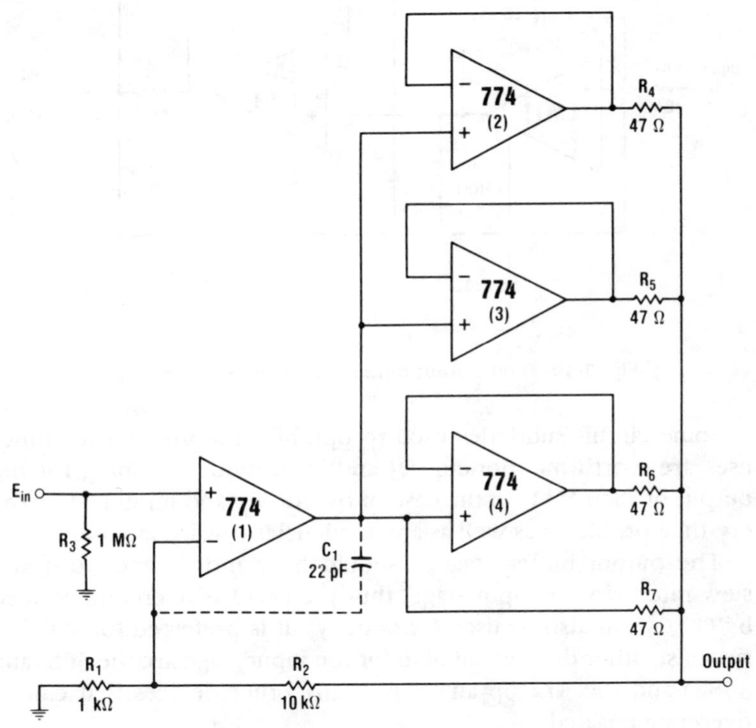


**Fig. 7-38. Multiple noninverting dc amplifier.**

The multiple-amplifier buffer technique can also be useful to increase current capacity and provide thermal isolation between a voltage amplifier and a power-dissipating output stage.

In Fig. 7-39, this idea is shown using the LM11, a precision op amp, as a voltage follower. The LM11 is buffered by a low-cost dual device—the 1458. Overall feedback is connected around $A_1$-$A_2$ and the LM11 so that the basic dc accuracy is that of the LM11. The 1458 sections serve only as low-cost current drivers at a cost and simplicity that is more attractive than most discrete circuits.

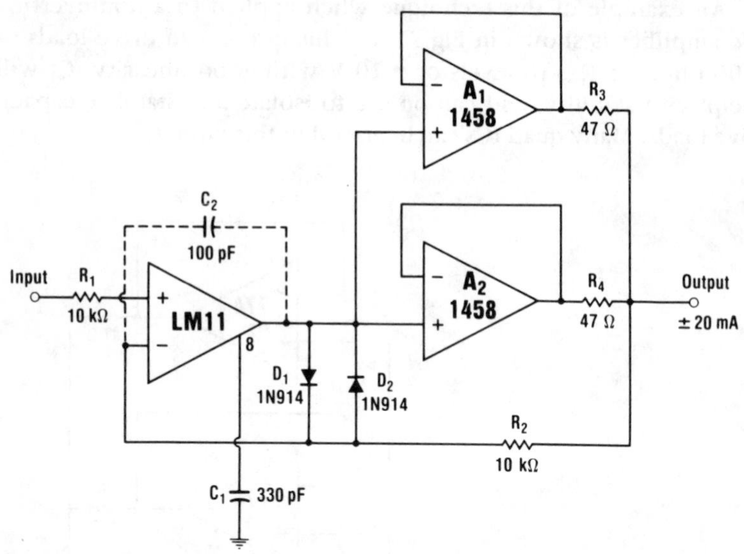


**Fig. 7-39. Dual-output buffer for a precision op amp.**

Some circuit subtleties used to optimize the hookup for other uses are worth mentioning. $D_1$ and $D_2$ provide clamping for the output of the LM11, in the case of overdrive, and minimize recovery-time problems as well as any tendencies toward latch-up.

The output buffer stage(s) should have more bandwidth and slew rate than the input stage, thus the LM11 is overcompensated by $C_1$. $C_2$ can also be used (optionally); it is preferred for stability reasons. Other devices suitable for the input stage are the 308 (and 308A) and the 312 op amps, plus any other devices that can be overcompensated.

A technique that increases load voltage is to drive the load *dif-*

*ferentially,* as shown in Fig. 7-40. Here, $A_1$ is a noninverting gain-of-5 stage, and $A_2$ inverts the output of $A_1$. The result is two equal but out-of-phase voltages across the output, as noted.

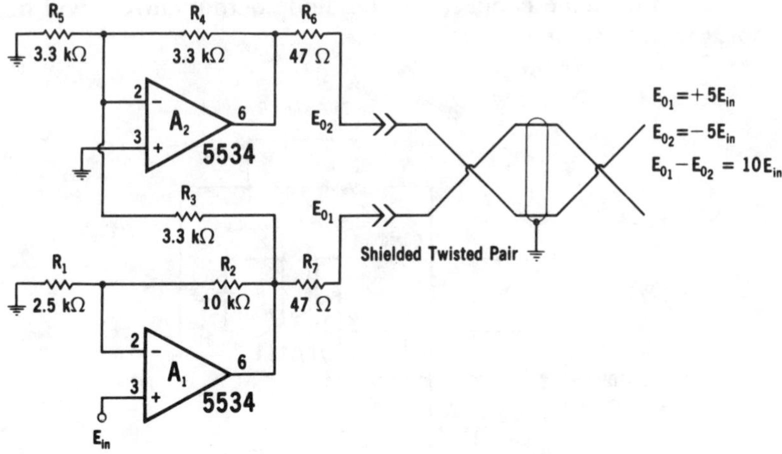


**Fig. 7-40. Differential line driver.**

In this particular example, the 5534 op amp is used, which can supply up to ±7 V across a differentially connected 300-ohm load, or ±20 V across a 1-kΩ differential load. In general, however, the technique is applicable to any pair of medium-level (or higher) output amplifiers.

With higher-voltage op amps, such as the 2645 or 343, this circuit can deliver up to 100-V p-p to a floating load using ±30-V supplies. Although it is not shown, additional current output can be realized by using parallel connections for $A_1$ and $A_2$.

## 7.6 CURRENT-TO-VOLTAGE CONVERTERS

Op amps used in the following current-to-voltage configuration (transimpedance amplifiers) are useful in medical instrumentation, for light sensors, etc., when used in conjunction with photodiode transducers. In such applications, the diode performs as a current source with an output I, and the op-amp output is $-(I)(R_F)$. With modern high-gain, low-offset, low-noise, picoampere-input op amps, $R_F$ can range up to 1000 megohms ($10^9$ ohms), allowing very high sensitivities.

Fig. 7-41 illustrates such an application, using an OPA111 op amp and a feedback resistor of $10^9$ ohms. This configuration allows an output scale factor of 1 V/nA of input. Here, the amplifier is used with a PIN photodiode, operated in a photovoltaic mode. With such a connection, the diode output current will be linear with light input.

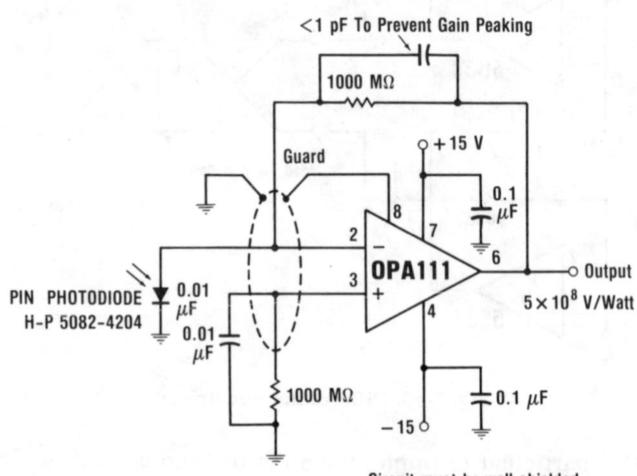


**Fig. 7-41. Photodiode current-to-voltage converter.**
*(© 1984 Burr-Brown Corp. Reproduced with permission.)*

For lowest output offset, a equal-value bias-compensation resistance is shown in the (+) input; however, this lead can also be grounded directly for simplicity. The circuit should be well constructed, with guarding used as noted.

This general principle can be used with a wide range of sensors, with the scale factor adjusted by $R_F$ over a range of $10^7$ to $10^9$ ohms. For best results, the op amp used should have a low-offset voltage, low noise, and a bias current of 1 pA or less.

## 7.7 REFERENCES


1. *Applications Manual for Computing Amplifiers,* 2nd ed. Philbrick/Nexus Research, Dedham, MA, 1966, 1984.
2. Choice, L. "Drift-Matched IC Op Amps Can Save Money." *Electronics,* May 22, 1972.

3. _________. "Series-Connected Op Amps Null Offset Voltage." *Electronics,* March 27, 1972.

4. Cotreau, G., and Creed, S. "Pair of One-Chip Op Amps Streamline the Design of Fast Analog Circuits." *Electronic Design,* February 21, 1985.

5. Counts, L., and Wurcer, S. "Instrumentation Amplifier Nears Input Noise Floor." *Electronic Design,* June 10, 1982.

6. Demrow, R. "Evolution from Operational Amplifier to Data Amplifier." *Analog Devices Application Note,* September 1968. Analog Devices, Inc., Norwood, MA.

7. _________. "Op Amps as Electrometers." *Analog Dialogue,* Vol. 5, No. 2, 1971.

8. Dobkin, R. C. "Feedforward Compensation Speeds Op Amp." *National Semiconductor Linear Brief LB-2,* March 1969. National Semiconductor Corp., Santa Clara, CA.

9. _________. "IC Preamp Challenges Choppers on Drift." *National Semiconductor Application Note AN-79,* February 1973. National Semiconductor Corp., Santa Clara, CA.

10. _________. "Instrumentation Amplifier." *National Semiconductor Linear Brief LB-1,* March 1969. National Semiconductor Corp., Santa Clara, CA.

11. _________. "Op Amp Circuit Collection." *National Semiconductor Application Note AN-31,* February 1970. National Semiconductor Corp., Santa Clara, CA.

12. Erdi, G. "A Low Drift, Low Noise Monolithic Operational Amplifier for Low Level Signal Processing." *Fairchild Semiconductor Application Bulletin APP-136,* July 1969. Fairchild Semiconductor, Mountain View, CA.

13. _________. "Minimizing Offset Voltage With Temperature in Monolithic Operational Amplifiers." *Proceedings of the NEC,* Vol. 25, 1969.

14. _________. "A 300 V/μs Monolithic Voltage Follower." *IEEE JSSC,* Vol. SC-14, #6, December 1979.

15. _________, Schwartz, T., Bernardi, S., and Jung, W. "Op Amps Tackle Noise—and for Once, Noise Loses." *Electronic Design,* December 20, 1980.

16. _________, and Williams, J. "Precision Op Amp Serves Host of Design Needs." *Electronic Design,* September 1, 1983.

17. *Fairchild Semiconductor Application Bulletin APP-138,* July 1969. "μA725 Instrumentation Applications." Fairchild Semiconductor, Mountain View, CA.

18. Frederiksen, T. M. "Multiple-Function LICs." *Proceedings of Electronic Products Magazine LIC Seminar,* March 1973.

19. Fullagar, D. "The 8007—A High-Performance FET-Input Operational Amplifier." *Intersil Application Bulletin A005,* March 1972. Intersil, Inc., Cupertino, CA.

20. Jung, W. G. "Bias-Current Cancellation Easily Implemented With Matched Op Amps." *Electronic Design,* June 21, 1977.

21. __________. "Stable FET-Input Op Amps Achieve Precision Performance." *EDN,* November 10, 1982.

22. *National Semiconductor Application Note AN-13,* September 1968. "Application of the LH0002 Current Amplifier." National Semiconductor Corp., Santa Clara, CA.

23. Nelson, C. T. "Super Matched Bipolar Transistor Pair Sets New Standards for Drift and Noise." *National Semiconductor Application Note AN-222,* July 1979. National Semiconductor Corp., Santa Clara, CA.

24. Ott, W. E. "Combine Two Op Amps to Avoid the Speed-Accuracy Compromise." *Burr-Brown Application Note AN-64,* July 1973. Burr-Brown Research Corp., Tucson, AZ.

25. Patstone, W. "Designing Femtoampere Circuits Requires Special Considerations." *EDN,* July 1, 1972.

26. Pease, R. A. "Get More Power Out of Dual or Quad Op Amps." *National Semiconductor Linear Brief, LB-44,* April 1979. National Semiconductor Corp., Santa Clara, CA.

27. Rudin, M. "Applying Microvolt/Picoamp Instrumentation Operational Amplifiers." *Precision Monolithics Application Note.* Precision Monolithics, Inc., Santa Clara, CA.

28. Sheingold, D. H. (ed.). *Analog-Digital Conversion Handbook,* 3rd ed. Analog Devices, Inc., Norwood, MA, 1986.

29. Siegel, B., and Van Der Gaag, L. "Applications for a New Ultra-High Speed Buffer." *National Semiconductor AN-48,* August 1971.

30. Soderquist, D., and Erdi, G. "The OP-07 Ultra-Low Offset Voltage Op Amp." *Precision Monolithics Application Note AN-13,* December 1975. Precision Monolithics, Inc., Santa Clara, CA.

31. Tobey, G. E., Graeme, J. G., and Huelsman, L. P. (eds.). *Operational Amplifiers—Design and Applications.* McGraw-Hill Book Co., New York, 1971.

32. Underwood, R. K. "New Design Techniques for FET Op Amps." *National Semiconductor Application Note AN-63,* March 1972. National Semiconductor Corp., Santa Clara, CA.

33. Widlar, R. J. "Designing With Super-Beta Transistor Op-Amp ICs." *EEE,* December 1969, February and March 1970.

34. ________. "Drift Compensation Techniques for Integrated DC Amplifiers." *National Semiconductor Application Note AN-3,* November 1967. National Semiconductor Corp., Santa Clara, CA.

35. ________. "IC Op Amp Beats FETs on Input Current." *National Semiconductor Application Note AN-29,* December 1969. National Semiconductor Corp., Santa Clara, CA.

36. ________. "Monolithic Operational Amplifiers—The Universal Linear Component." *National Semiconductor Application Note AN-4,* April 1968. National Semiconductor Corp., Santa Clara, CA.

37. ________, Pease, B., and Yamatake, M. "Applying a New Precision Op Amp." *National Semiconductor APP242,* April 1980.

38. Williams, J. "Applications for a New Power Buffer." *Linear Technology AN4,* September, 1984.

39. ________. "Application Considerations and Circuits for a New Chopper-Stabilized Op Amp." *Linear Technology AN9,* March 1985.

40. Witten, S. "Designing Junction FET Input Op Amps." *Siliconix Application Note AN74-3,* August 1974. Siliconix, Inc., Santa Clara, CA.

41. Wong, J., and Sherwin, J. "Applications of Wide-Band Buffer Amplifiers." *National Semiconductor AN-227,* October 1979.

42. Wurcer, S., and Counts, L. W. "A Programmable Instrumentation Amplifier for 12-Bit Resolution Systems." *IEEE JSSC,* Vol. SC-17, #6, December 1982.

43. Wurcer, S., and Jung, W. G. "Instrumentation Amplifiers Solve Unusual Design Problems." *EDN,* August 4, 1983.

44. Manufacturers' Data Sheets:

| Manufacturer | Part | Description |
|---|---|---|
| Analog Devices | AD542 | Operational Amplifier |
| | AD544 | Operational Amplifier |
| | AD545 | Operational Amplifier |
| | AD547 | Operational Amplifier |
| | AD548 | Operational Amplifier |
| | AD549 | Operational Amplifier |
| | AD611 | Operational Amplifier |
| | AD642 | Operational Amplifier |
| | AD647 | Operational Amplifier |
| | AD648 | Operational Amplifier |
| | AD711 | Operational Amplifier |
| | AD524 | Instrumentation Amplifier |
| | ADLH0033 | Buffer |
| Beckman | Thin-Film Resistor Catalog (Bulletin 8001) | |
| Burr-Brown | OPA111 | Operational Amplifier |
| | OPA606 | Operational Amplifier |
| | OPA2111 | Operational Amplifier |
| | 3627 | Instrumentation Amplifier |
| | INA105 | Instrumentation Amplifier |
| Harris Semiconductor | HA2635 | Buffer |
| | HA5105 | Operational Amplifier |
| | HA5170 | Operational Amplifier |
| | HA5195 | Operational Amplifier |
| | HA2500/2525 (series) | Operational Amplifier(s) |
| | HA2540 | Operational Amplifier |
| | HA2541, 2542 | Operational Amplifier |
| Intersil | ID500 | JFET Transistor Pair |
| | ICL7652 | Operational Amplifier |

| | | |
|---|---|---|
| Linear Technology Corporation | LT1007 | Operational Amplifier |
| | LT1010 | Buffer |
| | LT1012 | Operational Amplifier |
| | LT1013/14 | Operational Amplifier |
| | LT1022 | Operational Amplifier |
| | LT1043 | Switched Capacitor Building Block |
| | LT1052 | Operational Amplifier |
| | LT1055/56 | Operational Amplifier |
| Motorola Semiconductor Products, Inc. | MC34081 Series | Operational Amplifiers |
| National Semiconductor | LF351 | Operational Amplifier |
| | LF355/356/357 | Operational Amplifiers |
| | LH0002 | Buffer |
| | LH0033 | Buffer |
| | LM11 | Operational Amplifier |
| | LM121 | Preamplifier |
| | LM308/308A | Operational Amplifiers |
| | LM324/324A | Operational Amplifiers |
| | LM394 | Supermatched Transistor Pair |
| Precision Monolithics | OP-07 | Operational Amplifier |
| | OP-08 | Operational Amplifier |
| | OP-10 | Operational Amplifier |
| | OP-16 | Operational Amplifier |
| | OP-17 | Operational Amplifier |
| | OP-27 | Operational Amplifier |
| | OP-37 | Operational Amplifier |
| | OP-41 | Operational Amplifier |
| | OP-50 | Operational Amplifier |
| | BUF-03 | Buffer |
| | MAT-01 | Matched Transistor Pair |
| | MAT-02 | Matched Transistor Pair |
| | MAT-04 | Matched Transistor Quad |
| Siliconix | U401 | JFET Transistor Pair |
| | U440 | JFET Transistor Pair |
| | J232 | JFET |

EIGHT

# *Comparators*

A comparator is a special class of op amp, where the device is used to compare voltages between its two inputs. Used in this manner, the op amp operates open-loop; that is, there is no negative feedback applied to the (−) input as there is in the classical feedback circuits.

Circuits optimized for comparator use, therefore, need none of the phase/frequency compensation required for op-amp stabilization with feedback. In fact, these compensation components are detrimental in that they slow the response time of the comparator. Although in principle any op amp can be used as a comparator, using a compensated device, such as a 741, for example, will result in a response time in the tens of microseconds, whereas a device optimized for comparator use, such as the 311, will respond in several hundred nanoseconds.

Comparators usually interface at their outputs with digital logic; that is, with a standard family, such as TTL, CMOS, etc. Therefore, using standard op-amp outputs designed for linear ±10-V swings will require some level shifting and/or clamping in order to drive logic.

This chapter will treat the basic application considerations of comparators, how standard devices can be applied as comparators, and give representative application examples.

## 8.1 BASIC DIFFERENTIAL-INPUT COMPARATOR

A comparator circuit is one that provides an indication of the relative state of two input potentials. If one input is a reference potential and the other an unknown, the comparator output will indicate whether the unknown signal is above or below the reference potential. A basic op-amp comparator circuit is diagrammed in Fig. 8-1A.

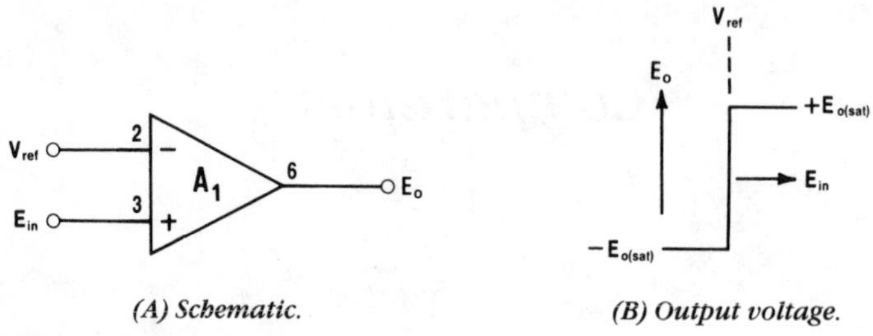


*(A) Schematic.* *(B) Output voltage.*

**Fig. 8-1. Basic comparator circuit.**

In this example, $V_{ref}$ is a positive voltage applied to the inverting input of op amp $A_1$. $E_{in}$ is the unknown potential, applied to the noninverting input. When $E_{in}$ is lower than reference voltage $V_{ref}$, the output of $A_1$ will be at the negative saturation limit, $-E_{o(sat)}$. As soon as $E_{in}$ rises higher than $V_{ref}$, the amplifier output flips to the positive saturation limit, $+E_{o(sat)}$. A graph of this transfer characteristic is shown in Fig. 8-1B.

In this circuit, the amplifier is operated in an open-loop condition; therefore, the voltage difference required to change the output from one state to the other is quite small, essentially $[+E_{o(sat)} - (-E_{o(sat)})]/A_{vo}$. Since the voltage is often but a few hundred microvolts (or less), the dominating factor that determines the exact threshold is the offset voltage of the amplifier, which may be as great as $\pm 10$ mV in some cases. For this reason, precision comparators should be nulled so that the input-differential voltage is as close to zero as practical when the output is zero. Furthermore, any source resistances in the input path should be selected so as to minimize the offset voltage. Alternatively, a precision low-offset-voltage device with low offset voltage and low currents can be employed. Many devices can today achieve offsets of 100 $\mu$V or less, i.e., the OP-07.

It is generally desired that a comparator be capable of changing output states as rapidly as possible. Since comparator circuits normally operate without negative feedback, there is no necessity for frequency compensation, since there is no closed loop to stabilize. Frequency compensation in a comparator is usually detrimental, because it slows down the open-loop speed of the amplifier. A type-741 op amp, for example, with a slew rate of 0.5 V/μs will require 40 μs to swing from −10 V to +10 V, due to its slew-rate limitation. By contrast, a 101-type device, which has the same (0.5 V/μs) slew rate when compensated, will have a slew rate of 20 V/μs or greater with the compensation removed, so that it can accomplish the −10-V to +10-V swing in 1 μs or less.

Removal of frequency compensation also increases the high-frequency gain of the op amp as discussed in Chapter 1. This increases the frequency range over which the comparator will have maximum sensitivity, since a high open-loop gain implies that a small input voltage is required to initiate an output transition.

Comparators are subject to wide differential-input voltage swings, since both $V_{ref}$ and $E_{in}$ may be anywhere within the common-mode input range of ±10 V. This requires a differential-input voltage rating of at least ±20 V to accommodate the worst-case limits in a circuit such as Fig. 8-1A. Other comparator configurations can ease this wide voltage-swing requirement as will be seen shortly. Examples of wide differential-input capability in op amps are the 741 and the 301A (±30 V), and many of the newer FET-input devices, such as the 356 and 357 types (±30 V).

Beyond the basic requirement of input-voltage range, speed is the next criterion in selecting op amps suitable for comparators. Speed has two meanings for comparators: response time and slew rate. Response time is simply the total time required for an output transition to occur once an input change of state has taken place. It is specified both for a positive step input (minus to plus) and a negative step input (plus to minus). Slew rate indicates how rapidly the amplifier output can change during a transition. It can also be different for positive- and negative-going slopes.

Fig. 8-2 is a simple test circuit which can be used to evaluate different op amps, showing response times in a standard condition. It is worth noting that devices having the shortest response times and fastest slew rates are those which normally use the *most* frequency compensation—removing this compensation allows greater inherent bandwidth, or the highest GBP.

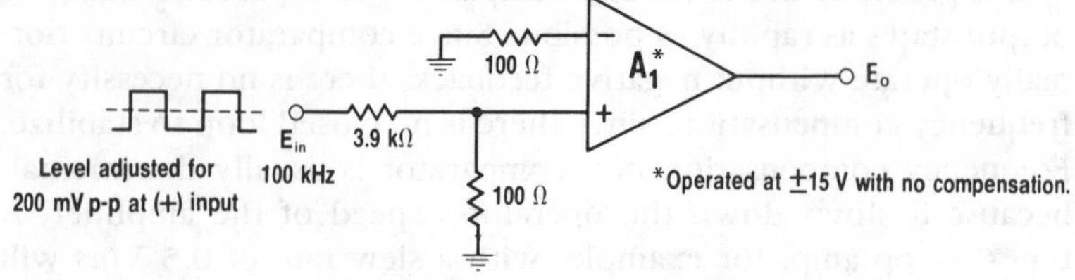


**Fig. 8-2. Test circuit for using op amps as comparators.**

Figs. 8-1 and 8-2 are both examples of noninverting comparators; that is, the output has the same direction of polarity as the input. An inverting comparator may be required in some cases, and this configuration is obtained simply by reversing the inputs so that $E_{in}$ is connected to the inverting (−) op-amp terminal. In either case, the op amp presents a high impedance to both the $V_{ref}$ and $E_{in}$ sources.

## 8.2 BASIC SINGLE-ENDED-INPUT COMPARATOR

The differential-input comparator described in the previous section is the most general configuration and the one most often used. Single-ended-input configurations are also possible, however, as illustrated in Fig. 8-3. In this example, both the signal and reference voltages are applied to a common input with the opposite input grounded. $R_1$ and $R_2$ form a voltage divider between $E_{in}$ and $V_{ref}$, and $V_{ref}$ is a voltage of polarity opposite to that of $E_{in}$. $V_{ref}$ and $R_1$-$R_2$ define the threshold value of $E_{in}$ that will cause the voltage at the junction of $R_1$ and $R_2$ to cross ground, which changes the state of the comparator. This threshold voltage is $E_{in(th)} = V_{ref}(R_1/R_2)$.

In practice, voltage $V_{ref}$ and resistance $R_2$ will define a current in $R_2$ for $E_a = 0$ (trip point); thereafter, $R_1$ can be selected for the desired trip level of $E_{in}$. Although noninverting operation of the comparator is shown, inverting operation is also possible by reversing the inputs, as in the previous example.

This comparator has the advantage of being noncritical as to the specific device used for $A_1$ in regard to differential-input rating. The voltage between the inputs of $A_1$ will be smaller than $E_{in}$ and $V_{ref}$ by the division of $R_1$-$R_2$, and, therefore, a smaller input rating is tolerable for $A_1$. This allows use of such op amps as the 709, which is an unprotected type. $E_{in}$ can be unrestricted in magnitude

(A) Schematic.

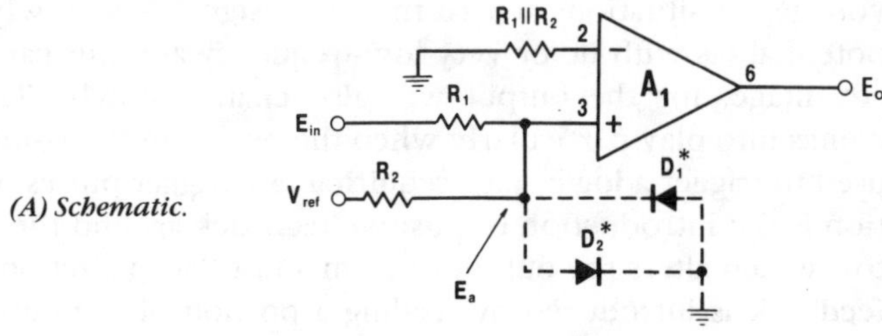


*May be necessary for input overvoltage protection.

(B) Output voltage.

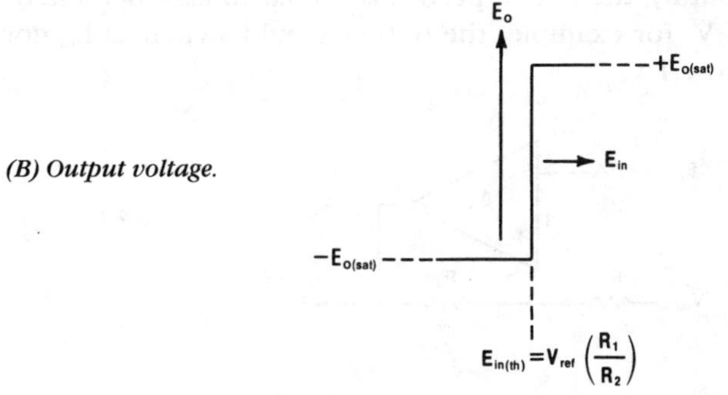


**Fig. 8-3. Single-ended comparator.**

since $R_1$ can always be selected to scale down this voltage. It is also good practice to add a pair of back-to-back clamp diodes ($D_1$ and $D_2$) to constrain the $E_a$ voltage. They add no error in the threshold point, because when $E_a = 0$ at the trip point, both diodes are nonconducting. The diodes can be connected as diagrammed in Fig. 8-3 or applied directly across the inputs for differential clamping. Note that many ICs, such as the OP-27 and OP-37, for example, have such diodes built into their input stages. When using such devices, the input resistor(s) *must* be used in order to limit the maximum current when there are large voltage differentials (see the discussions of Chapter 3).

## 8.3 COMPARATORS WITH HYSTERESIS

The comparators described thus far have been basic open-loop types, which amplify an input signal by comparison to a reference voltage. In situations where the input signal is a slowly varying potential (as with dc or very-low-frequency ac), this can be a disadvantage, for the output will also change slowly. This factor comes into play particularly when the output of the comparator is used to trigger a logic stage requiring fast trigger pulses. One solution is the introduction of positive feedback around the comparator, which alters the output to form a fast "snap" action. Positive feedback is introduced by feeding a portion of the output signal back to the noninverting input.

Fig. 8-4A is an inverting differential-input comparator with the reference voltage applied to the noninverting input through $R_1$. Without $R_2$, the circuit performs similar to Figs. 8-1 and 8-2. If $V_{ref}$ is +1 V, for example, the output would switch as $E_{in}$ goes above or below 1 V.

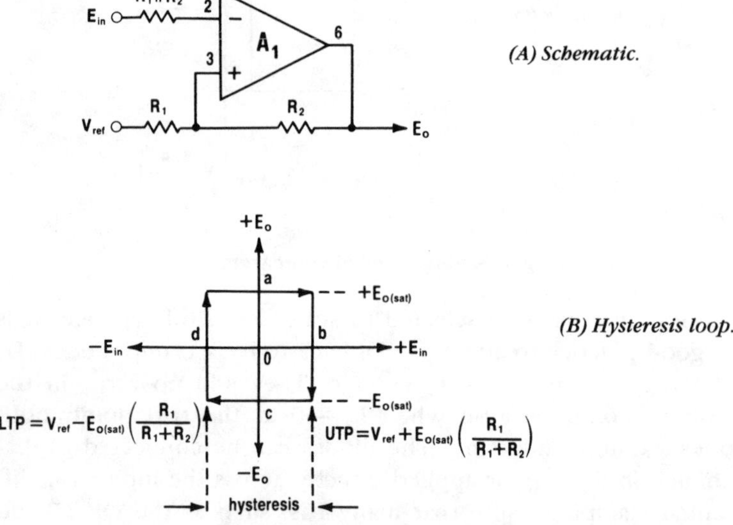


**Fig. 8-4. Differential comparator with hysteresis.**

By introducing $R_2$, however, positive feedback is developed across $R_1$. If the output is high, $R_2$ will feed back a signal, which will be added to the reference voltage. This voltage increment will be

$$\Delta V_{ref} = (E_{o(sat)} - V_{ref}) \frac{R_1}{R_1 + R_2}$$

making the new reference voltage

$$V_{ref+} = V_{ref} + \Delta V_{ref}$$

$$= V_{ref} + (E_{o(sat)} - V_{ref}) \frac{R_1}{R_1 + R_2}$$

Therefore, a new trip level is introduced, high by the amount of positive voltage fed back from $E_o$. As $E_{in}$ crosses this voltage, the output will start to fall. Immediately, the falling condition is sensed across $R_1$, and the noninverting input is also driven negative. This feedback is regenerative, and the output quickly "snaps" to the opposite state. This effect is prevalent regardless of the rate of change in $E_{in}$, giving the circuit a constant output transition time for any speed of input variation. Since $E_o$ is now at negative saturation, the voltage fed back to $R_1$ is

$$\Delta V_{ref} = (-E_{o(sat)} - V_{ref}) \frac{R_1}{R_1 + R_2}$$

making a new reference voltage,

$$V_{ref-} = V_{ref} + \Delta V_{ref}$$

$$= V_{ref} + (-E_{o(sat)} - V_{ref}) \frac{R_1}{R_1 + R_2}$$

Positive feedback has introduced new terms, and there are now two threshold points rather than one; these are called the *upper threshold point* (UTP) and the *lower threshold point* (LTP). The difference between these two thresholds is the ***hysteresis***.

Graphically, the effect of hysteresis is depicted in Fig. 8-4B. For simplicity, let $V_{ref} = 0$. Assume the output is already in a high state and the input is rising, approaching the UTP. On the figure, this is line *a*. As $E_{in}$ reaches the voltage

$$E_{in} = +E_{o(sat)} \left( \frac{R_1}{R_1 + R_2} \right)$$

the output voltage snaps from $+E_{o(sat)}$ to $-E_{o(sat)}$ (line *b*). As $E_{in}$ decreases toward the LTP (line *c*), $E_o$ will remain at $-E_{o(sat)}$ until

$$E_{in} = -E_{o(sat)} \left( \frac{R_1}{R_1 + R_2} \right)$$

At this point, the output voltage snaps back to $+E_{o(sat)}$ (line *d*), thus returning to the original state. The input threshold region is a voltage span determined by the amount of feedback:

$$\begin{aligned} E_{hys} &= V_{ref+} - V_{ref-} \\ &= \left[ V_{ref} + (+E_{o(sat)} - V_{ref}) \frac{R_1}{R_1 + R_2} \right] \\ &\quad - \left[ V_{ref} + (-E_{o(sat)} - V_{ref}) \frac{R_1}{R_1 + R_2} \right] \end{aligned}$$

If $V_{ref} = 0$, this simplifies to

$$E_{hys} = \frac{R_1}{R_1 + R_2} [+E_{o(sat)} - (-E_{o(sat)})]$$

This threshold region is the hysteresis region shown on the graph as the horizontal span from the LTP to the UTP. Hysteresis is always some span about $V_{ref}$, although the span is not always symmetrical. Note that if $V_{ref}$ is above or below ground, the hysteresis region will not be symmetrical as was the case for $V_{ref} = 0$. If the respective saturation voltages of the op amp are not equal, this will also make the hysteresis region asymmetrical.

Hysteresis is a useful feature in comparators for reasons other than reducing the response time. If the input signal has a low-level noise superimposed on it, an open-loop comparator will "chatter" (switch rapidly back and forth) due to noise fluctuations as the input voltage passes through the threshold region. Also, some open-loop circuits may oscillate during a transition due to stray capacitive feedback. One cure for either of these difficulties is a small amount of hysteresis, which provides a "dead zone" over which the comparator will not be responsive. Often, as little as 10 mV of hysteresis, which can be negligible if the voltage being sensed is large, can eliminate these effects.

## 8.4 SINGLE-ENDED COMPARATOR WITH HYSTERESIS

Single-ended comparators can also be used with hysteresis; an example is shown in Fig. 8-5A. In this circuit, amplifier $A_1$ senses the difference between ground potential (at the inverting input) and the junction potential of $R_1$-$R_2$. In this regard, it is similar to Fig. 8-3; in fact, the threshold voltage can be determined the same way by substituting the appropriate voltage representing $E_o$ for $V_{ref}$. In this case,

$$E_{in(th)} = E_o\left(\frac{R_1}{R_2}\right)$$

The difference is that $E_o$ will not be constant here, whereas previously, $V_{ref}$ was constant by definition. The circuit operates as follows: Assuming the output is in a state of negative saturation, a positive voltage applied at $E_{in}$ will raise the potential at $E_a$ until it reaches a potential equal to zero. This is line *a* in Fig. 8-5B. At this point,

$$E_{in(th+)} = -(-E_{o(sat)})\frac{R_1}{R_2}$$

the equation for the UTP. The output then snaps positive due to the feedback through $R_2$ (line *b*). The voltage at $E_a$ is now positive due to the feedback from $+E_{o(sat)}$. When $E_{in}$ becomes negative, it will reduce the voltage at $E_a$ until $E_a$ again approaches zero. This is represented by line *c*. When $E_{in}$ has reached a negative value such that

$$E_{in(th-)} = -(+E_{o(sat)})\frac{R_1}{R_2}$$

the circuit has reached its LTP, and the circuit snaps negative again, tracing section ***d*** of the loop and returning it to its original state.

This circuit does not need a separate reference voltage, although a potential may be applied to the inverting input of $A_1$ as a reference if desired. Using it as in Fig. 8-5A, the output saturation voltages define the trip points in conjunction with $R_1$ and $R_2$. This configuration is generally very useful since it does not require a separate reference voltage and can use an unprotected op amp for the same reasons as in Fig. 8-3. The only drawback is the fact that the positive and negative saturation voltages are not stable potentials—they can vary from device to device and also with loading and temperature. Therefore, if the output voltage is used directly as feedback, as shown in Fig. 8-5A, the threshold voltage will vary directly with the $\pm E_{o(sat)}$ levels. With devices that have output stages that can swing "rail-to-rail," such as the 3130 or the LM10, the *supply voltage(s)* will be the saturation level(s).

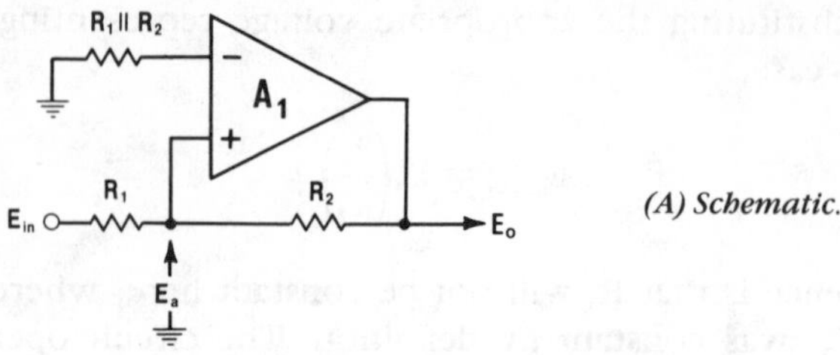

(A) Schematic.

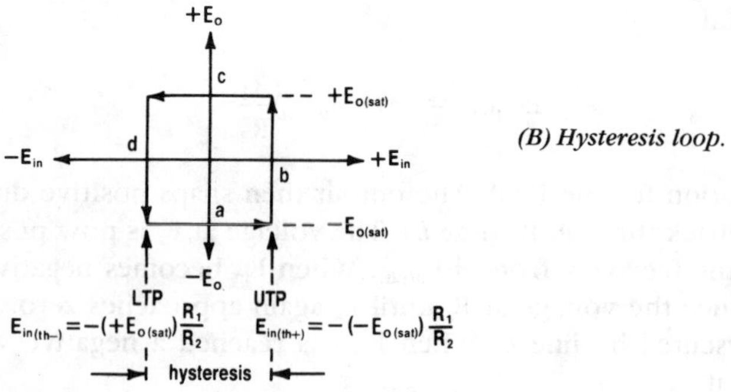

(B) Hysteresis loop.

**Fig. 8-5. Single-ended comparator with hysteresis.**

## 8.5 COMPARATORS WITH CLAMPED FEEDBACK

A solution to the variation in threshold potential is the introduction of a zener clamping network in the positive feedback loop as in Fig. 8-6. In this circuit, back-to-back zeners $D_1$ and $D_2$ clamp the output of $A_1$ to $\pm V_Z$, and this voltage is used as the feedback voltage. The combination of a single zener junction and two forward-biased diodes in series results in a net temperature coefficient close to zero, providing temperature-stabilized thresholds. Another low-TC alternative is to use a stabilized reference diode, such as the LM329, AD589, or LM336, within a bridge. This yields a relatively stable breakdown voltage of $V_Z + 2V_f$ where $V_f$ is the forward voltage of the bridge diodes.

With the equal-value resistors shown for $R_1$ and $R_2$, the input trip level will be equal to the $\pm V_Z$ reference source. Lower or higher input potentials may be realized by scaling $R_1$ appropriately. Two outputs are available—a direct output from pin 6,

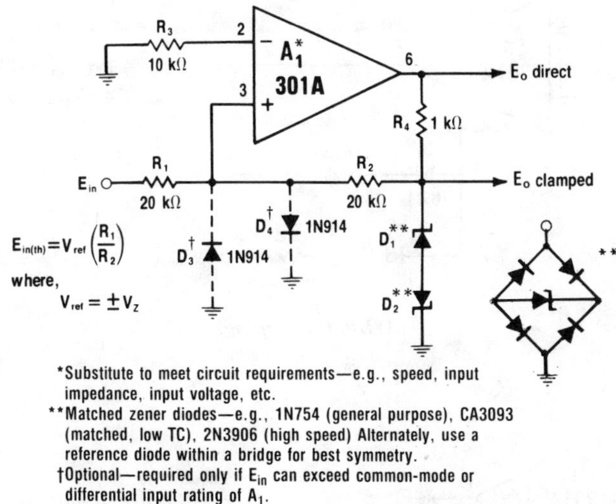


**Fig. 8.6. Single-ended comparator with hysteresis and clamped feedback.**

which will vary from $-E_{o(sat)}$ to $+E_{o(sat)}$, and the clamped reference $\pm V_Z$ across $D_1$-$D_2$. If input voltages beyond the common-mode range of $A_1$ are used, a pair of input clamp diodes, such as $D_3$ and $D_4$, should also be used. Where justified, these diodes should be low-leakage types.

## 8.6 CLAMPED COMPARATORS WITH TYPE-301A AMPLIFIERS

A very useful scheme for defining the output voltage swing of a comparator is possible using type-301A amplifiers. The basic technique is shown in Fig. 8-7.

From the LM101 schematic of Chapter 2, the internal connection on pin 8 of a 101 or 301A op amp is a relatively high-impedance source for either positive or negative swings. The p-p voltage swing at this point is roughly the same value as that at the output, the main difference being the dc offset of the emitter followers between pin 8 and the output. This point, then, is a convenient one to define the range of swing at the output, since it can be clamped either with zener diodes or with biased signal diodes.

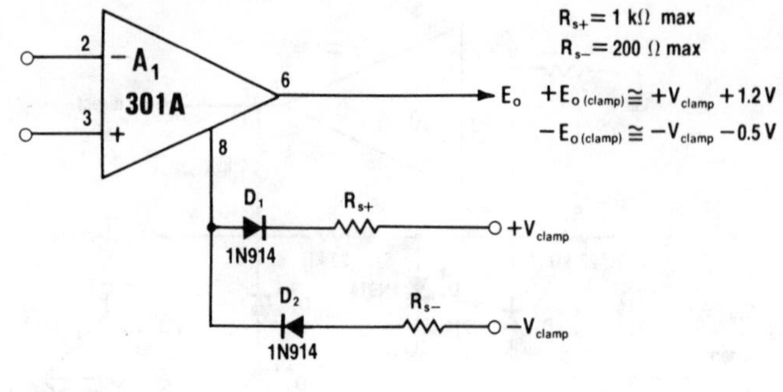


*(A) Diode clamping.*

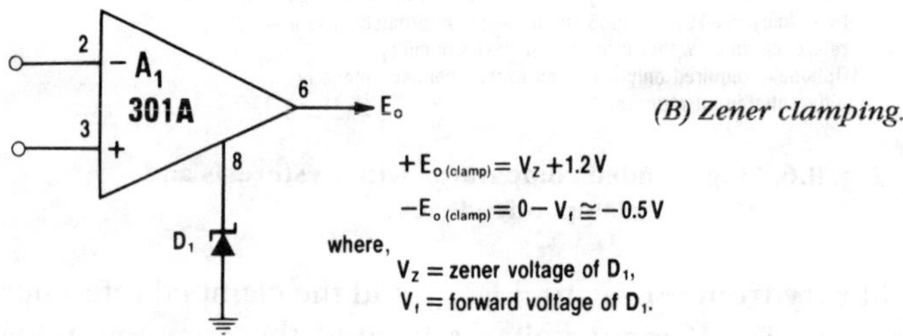


*(B) Zener clamping.*

**Fig. 8-7. Clamped comparators.**

Fig. 8-7A illustrates the requirements for clamping positive or negative voltages. Diode $D_1$ will clamp positive swings to voltage $+E_{o(clamp)} \simeq +V_{clamp} + 1.2$ V, where $+V_{clamp}$ is the bias voltage applied to the cathode of $D_1$. The source resistance of $+V_{clamp}$ should be no greater than about 1 kΩ for this relationship to apply.

Similarly, diode $D_2$ may be used to clamp the negative swing to a negative potential, $-V_{clamp}$. The voltage at the output terminal will be $-E_{o(clamp)} \simeq -V_{clamp} - 0.5$ V. The source resistance of $-V_{clamp}$ should be no more than about 200 Ω to maintain this relation.

An alternate method of accomplishing the same result uses a zener diode connected to pin 8 (Fig. 8-7B). In this case, the zener voltage and ground define the $\pm V_{clamp}$ potentials, and the output swing is the same as in Fig. 8-7A. The zener diode may, of course, be reversed for a predominately negative output swing if desired.

The advantages of voltage clamping are twofold. First, it

makes available the full current output of the op amp by internally clamping at a low-current point. Second, it allows greater operating speed by avoiding saturation of the op-amp voltage-amplification stages. To realize such speed, however, minimal capacitance should be added to the pin-8 node.

A circuit for a general-purpose comparator with a clamped 5-V logic output is shown in Fig. 8-8. $R_1$ and $R_2$ provide a +3.8-V bias for $D_1$, clamping the positive output to +5 V; $D_2$ limits the negative output excursion to −0.5 V. It should be noted that this is still an open-loop circuit, so the output voltage will change in proportion to the load current (the open loop $r_o$ of $A_1$ is 75 Ω). Externally compensated FET-input devices, such as the TL070, TL080, and CA080, will also generally work well in this circuit.

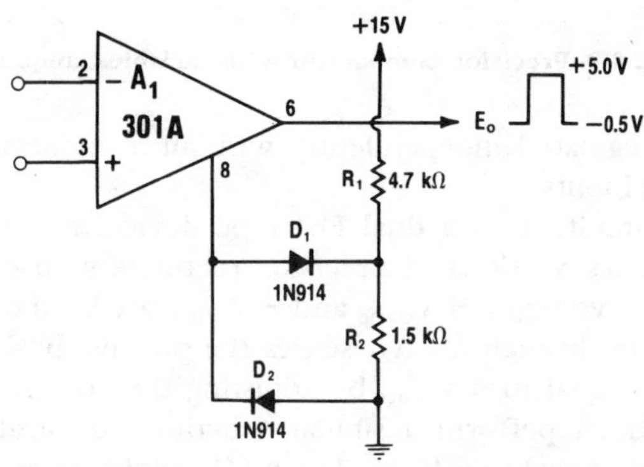


**Fig. 8-8. General-purpose comparator with 5-V logic output.**

## 8.7 PRECISION VARIABLE CLAMPING

The circuit of Fig. 8-8 is very useful for driving DTL or TTL logic devices, which do not require great accuracy in voltage levels. It is entirely adequate for this application, although the output voltage is neither temperature-compensated nor regulated in amplitude.

More demanding applications may require a tight control of positive and negative amplitude limits, and regulation of these amplitudes with load and temperature. A circuit that accomplishes this is shown in Fig. 8-9—a comparator that has both output-volt-

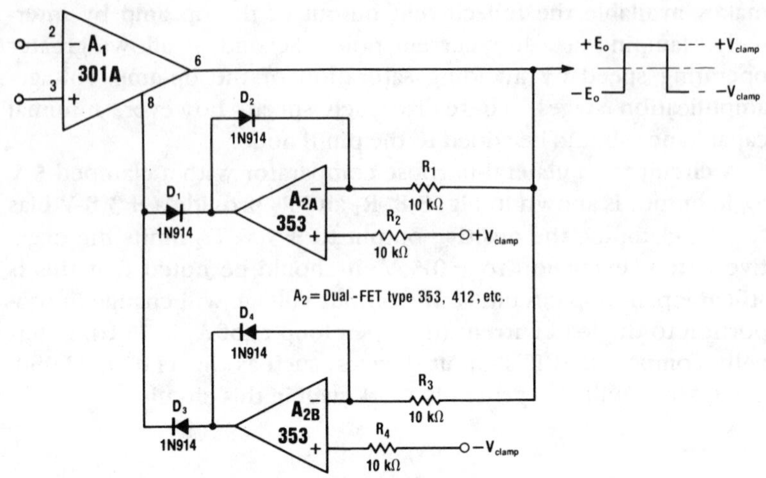


**Fig. 8-9. Precision comparator with variable clamping.**

age limits regulated independently without a connection to the comparison inputs.

In this circuit, $A_2$ is a dual FET-type device. $A_{2A}$ and $A_{2B}$ are complementary versions of precision rectifiers whose independent reference voltages, $+V_{clamp}$ and $-V_{clamp}$, are both operating in a closed loop through $A_1$. $A_{2A}$ senses the positive peak of $E_o$ and maintains it equal to $+V_{clamp}$ by adjusting the voltage applied to $D_1$. $A_{2B}$ and $D_3$ perform a similar function on negative peaks, adjusting $-E_o$ equal to $-V_{clamp}$. $D_2$ and $D_4$ minimize response time by preventing saturation of $A_{2A}$ and $A_{2B}$.

The feedback network around the output stage of $A_1$ regulates the output voltage independent of the inputs to $A_1$. These inputs can be used in any of the previous comparator configurations, with the obvious virtues of a precisely maintained output voltage and stable $\pm V_{clamp}$ potentials. For example, the circuit can be used in a single-ended comparator, such as Fig. 8-6, with improved results. Also, the threshold voltages can be made independently variable, if desired, by separate adjustments for $+V_{clamp}$ and $-V_{clamp}$; operating speed is determined primarily by $A_{2A}$ and $A_{2B}$. Note that bias-compensation resistances $R_2$ and $R_4$ are not necessary for FET amplifiers as shown, but they are appropriate for bipolar types. Note that the dc precision of clamping in this circuit will be generally as good as the amplifier used for $A_2$.

## 8.8 ZERO-CROSSING DETECTOR

One form of comparator finding wide use is the zero-crossing detector. Usually, it takes the form of a high-gain amplifier that changes state each time the input signal changes direction. For an ac signal, this occurs when the signal value is zero, hence the name *zero-crossing detector*. The circuit amplifies and "squares" the input signal into a series of rectangular output pulses with rising and falling slopes corresponding to the input zero crossing.

A circuit that derives the zero-crossing information of an ac signal is shown in Fig. 8-10. This is an ac logarithmic amplifier, with feedback current $I_f$ creating a logarithmic output voltage due to diodes $D_1$ and $D_2$. Amplifier $A_1$ is a high-SR FET type, to optimize speed and minimize phase error at high frequencies. The output voltage is nominally $\pm V_f$, where $V_f$ is the forward voltage drop of either diode.

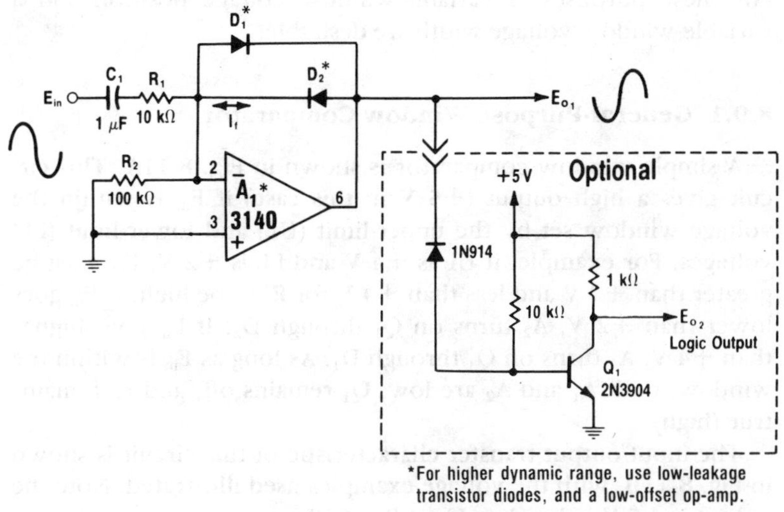


**Fig. 8-10. Zero-crossing detector.**

This basic circuit has a dynamic range of over 70 dB, which is realized by using low-leakage transistor diodes for $D_1$ and $D_2$. Should the nominal voltage drop be insufficient as an output, or if a constant output is desired, the optional connection shown in the dashed-line inset can be used. This is a saturated switch, delivering a 0- to +5-V output at $E_{o_2}$. (Note: an alternate variation on this circuit is discussed in Chapter 5, as a "bipolar limiter.")

## 8.9 WINDOW COMPARATORS

A window comparator, as its name implies, is a specialized form of comparator designed to detect the presence of a voltage between two prescribed voltage limits—that is, within a voltage "window." This is accomplished by logically combining the outputs of two single-ended comparators, one indicating "greater than" a lower limit, the other indicating "less than" an upper limit. If both comparators indicate a "true" condition, the output is true. If either is "not true," the output is not true.

Window comparators (also referred to as *double-ended* comparators) are useful in the grading and selection of components or in the production testing of circuits to a specified output tolerance. For these purposes, a variable-window voltage position and a variable-window voltage width are desirable.

### 8.9.1 General-Purpose Window Comparator

A simple window comparator is shown in Fig. 8-11A. This circuit gives a high output (+5 V in this case) if $E_{in}$ is within the voltage window set by the upper-limit (UL) and lower-limit (LL) voltages. For example, if UL is +4 V and LL is +2 V, $E_{in}$ must be greater than +2 V and less than +4 V for $E_o$ to be high. If $E_{in}$ goes lower than +2 V, $A_2$ turns on $Q_1$ through $D_2$. If $E_{in}$ goes higher than +4 V, $A_1$ turns on $Q_1$ through $D_1$. As long as $E_{in}$ is within the window, both $A_1$ and $A_2$ are low, $Q_1$ remains off, and $E_o$ remains true (high).

The input/output transfer characteristic of this circuit is shown in Fig. 8-11B, with the voltage examples used illustrated. Note the output is +5 V only when $LL \le E_{in} \le UL$.

A type-1458 (dual 741) op amp is convenient to use for $A_1$ and $A_2$ in this circuit if its differential-input rating is observed. The resistor-diode input clamping network shown by dotted lines is recommended here for other op amps which have lower differential ratings. For applications requiring greater speed, a FET dual, such as the 3240, 353, TL072, or 772, may be substituted and the resistor-diode clamping omitted. Very high dc precision would be

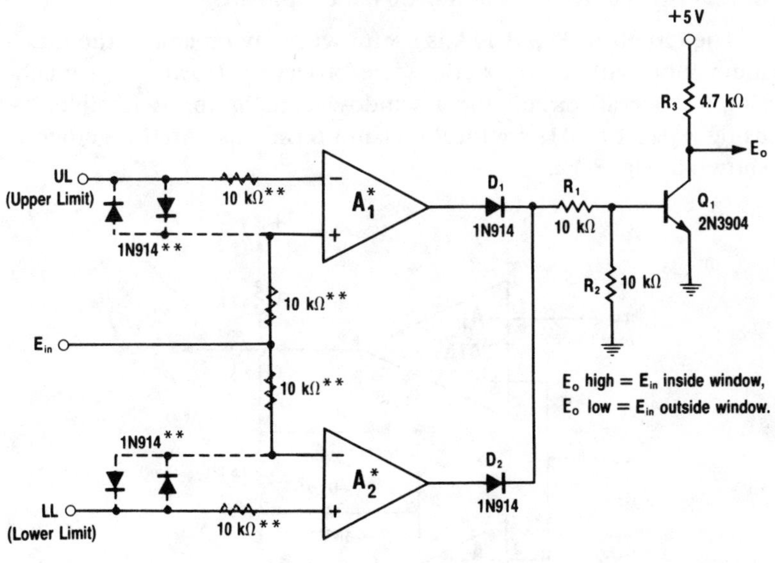


*(A) Schematic.*

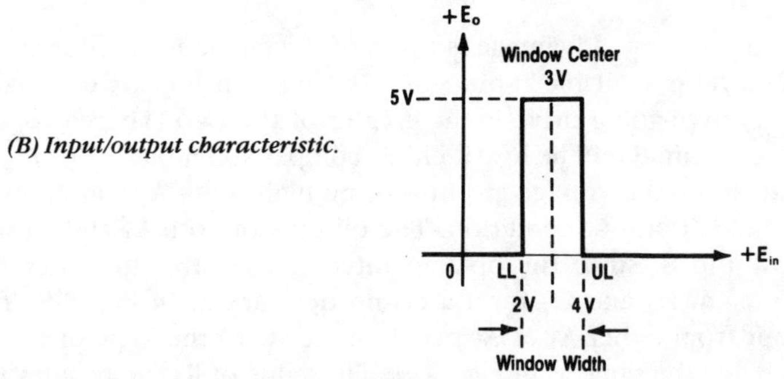


*(B) Input/output characteristic.*

**Fig. 8-11. General-purpose window comparator.**

served by such FET types as the OPA2111, the AD647, or the AD712. $R_3$ is shown as a fixed load resistor, but may be a relay or a lamp if desired. These types of loads will be covered in a following section.

## 8.9.2 Internally Gated Window Comparator

The circuit of Fig. 8-11A is useful with any op amp if the maximum input-voltage restrictions are observed. However, a much-simpler overall circuit for a window comparator is feasible, by using a pair of 301As with the clamp terminals wired together as shown in Fig. 8-12.

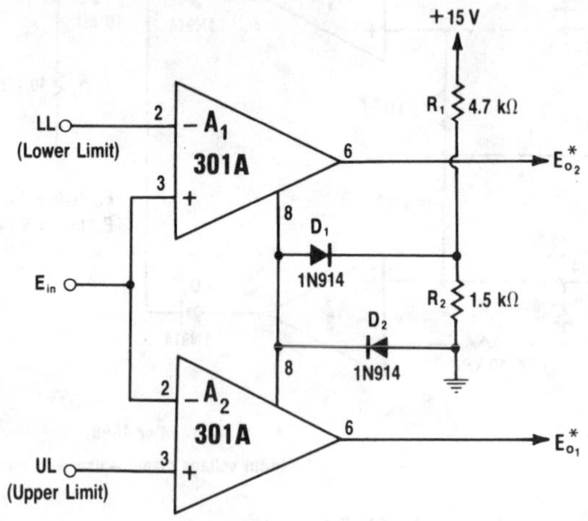


**Fig. 8-12. Internally gated window comparator.**

In this circuit, advantage is taken of the fact that the source and sink currents available at pin 8 of a 301-type op amp are unequal—the negative-going drive is the greater of the two. Therefore, the voltage at pin 8 will be low if either comparison input ($A_1$ or $A_2$) so dictates; for the voltage at pin 8 to be high, *both* $A_1$ and $A_2$ must have a high output condition. The outputs of both $A_1$ and $A_2$ will follow pin 8, since the op-amp internal circuitry thereafter has unity gain. $D_1$ and $D_2$ form a clamp network as in Fig. 8-8. The output from either $A_1$ or $A_2$ may be used, with the logic of operation being the same as in Fig. 8-11. The value of $R_2$ can be adjusted to accommodate other output logic levels. For example, if $R_2$ is 10 kΩ, the output is 10-V CMOS compatible.

## 8.9.3 Comparator With Variable Window

As mentioned previously, a variable-width window is very useful, allowing a single comparator to be programmed for a wide variety of tasks. Fig. 8-13 is such a circuit, generating variable UL and LL voltages for use with either of the previous comparator circuits.

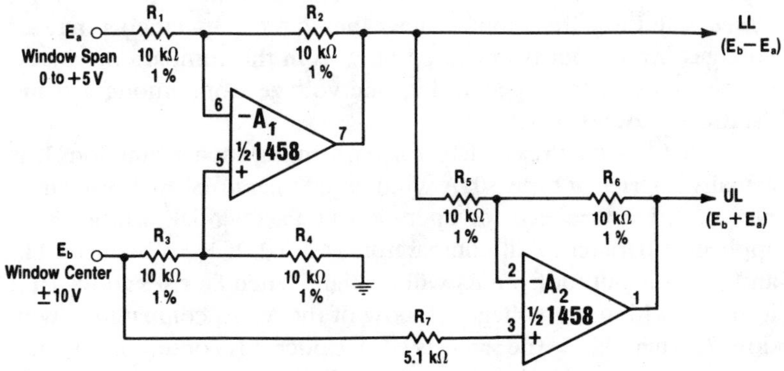


**Fig. 8-13. Comparator with variable window.**

This circuit combines two variable voltages, $E_b$ (window center) and $E_a$ (window span), such that $E_b$ positions the window and voltage $E_a$ is added to $E_b$ to set the width of the window. $A_1$ is a subtractor, generating a voltage of $E_b - E_a$, which is used as the LL voltage. This voltage is also added to $2E_b$ at $A_2$ to derive $E_b + E_a$, the UL voltage. $E_b$ may be any voltage within the comparator common-mode range of ±10 V. $E_a$ is a positive voltage that is added to $E_b$ at the input; so, the combination of $E_b + E_a$ or $E_b - E_a$ must not exceed ±10 V for any combination of $E_a$ and $E_b$. Normally, this will not be a restriction, since $E_a$ is commonly a small percentage of $E_b$—for example, +5 V ±5%.

For applications where $E_{in}$ is much greater than the common-mode range of ±10 V, it may be scaled with a voltage divider to bring it into the comparison range, allowing window-voltage comparison of much higher voltages. Although a general-purpose dual device is shown (the 1458), high precision will require a well-matched, low-offset unit, such as the OP-207 or OP-227, or a FET dual, such as the OPA2111 or the AD712. Note the matched resistors are preferably contained in a common array.

### 8.9.4 Staircase Window Comparator

The most basic comparator tells only one thing about an unknown voltage—whether it is above or below a reference voltage. The window comparator tells a little more about the unknown voltage—that it is between two specified voltage limits. But, in certain situations, even this may not be enough. A third type of comparator, the staircase detector, tells where (within *which* window) the input voltage lies over a specified range of voltages. An output is indicated only from the comparator closest to the input voltage, generating one voltage from among a rising "staircase" of levels.

Fig. 8-14 is a staircase detector performing such a function. It is actually a series of type-301A window comparators, each similar to Fig. 8-12, but arranged for operation in a sequential fashion. $E_{in}$ is applied in parallel to all comparators ($A_1$–$A_6$). If $E_{in}$ is between $LL_1$ and $UL_1$, the output from $A_2$ will be high. When $E_{in}$ rises above $UL_1$, it moves within the voltage window of the $A_3$–$A_4$ comparator (window 2, where $LL_2$ is the same as $UL_1$). Under this condition, the output of $A_2$ drops and $A_4$ switches high. Should $E_{in}$ rise further (above $UL_2$, for instance), it moves into window 3, actuating $A_6$ to a high state with $A_4$ going low. This "staircase" of windows can be continued indefinitely using as many comparators as necessary to provide the required resolution. An output indicator (lamp, relay, etc.) added to each of the $E_{o_1}$, $E_{o_2}$, and $E_{o_3}$ lines will indicate the presence of $E_{in}$ within the corresponding window.

## 8.10 STANDARD IC COMPARATOR DEVICES

The discussion on comparators thus far has been either generalized or oriented toward how IC op amps can be used as comparators. However, as alluded to initially, it may often be undesirable to attempt to use a standard op amp as a comparator for many reasons. Standard comparator ICs are readily available for voltage-comparison uses, in a variety of styles and performance features, and in single, dual, and quad versions.

Pinouts for two representative families of these comparators are illustrated in Fig. 8-15. Fig. 8-15A shows the type-LM311 comparator, which operates from dual power supplies and has an open-collector output that is compatible with all logic families. The 311 is a workhorse device and is, perhaps, more suitable for the broadest range of applications than any other device. It can sink or source up to 50 mA and operates from power supplies with outputs from as low as 5 volts up to 36 volts. It can handle

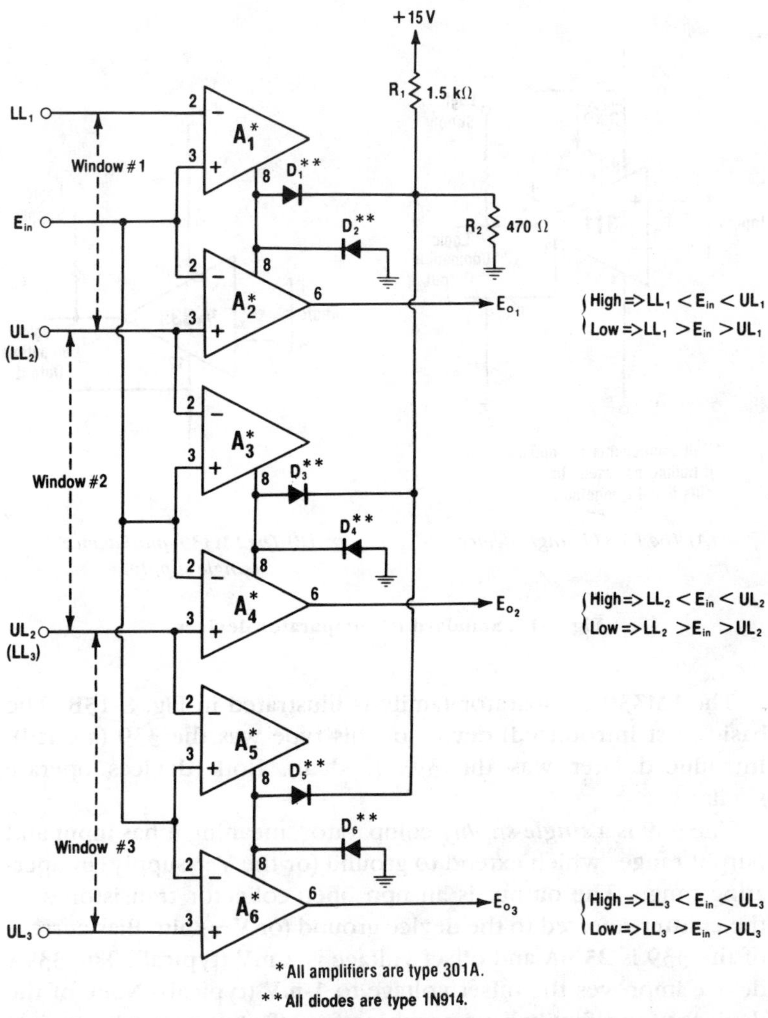


**Fig. 8-14. Staircase window comparator.**

differential inputs of up to ±30 V and has a bias current of 100 nA and an offset voltage of 2 mV (typical figures), and can respond in 250 nanoseconds. It is nulled (optionally) as shown in the illustration of Fig. 8-15. The 311 is available in a dual version as the LH2311 and in a (single) FET-input version as the LF311. A low-power version is also available, the LP311.

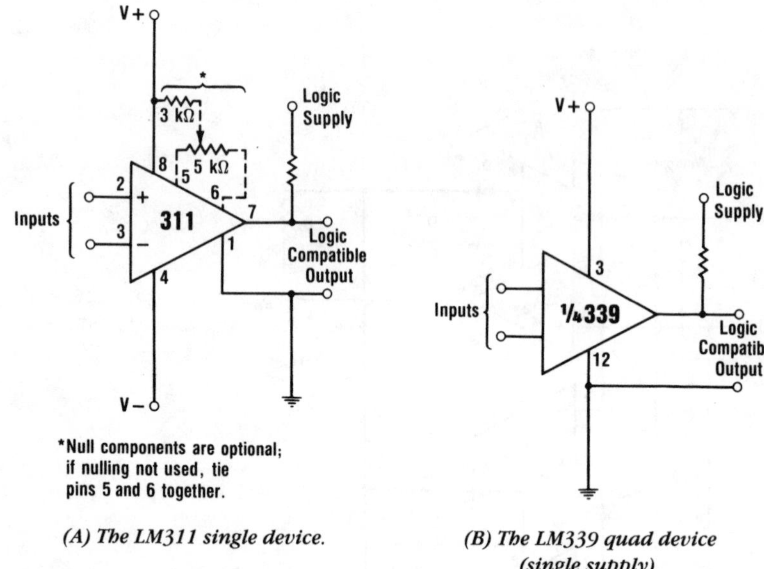


*(A) The LM311 single device.*

*(B) The LM339 quad device (single supply).*

**Fig. 8-15. Standard IC comparator devices.**

The LM339 comparator family is illustrated in Fig. 8-15B. The basic (first introduced) device of this type was the 339 (a quad). Introduced later was the 393 (a dual). Both devices operate similarly.

The 339 is a *single-supply* comparator, meaning it has input and output ranges which extend to ground (or the V− supply) in operating range. The output is an npn open-collector transistor with the emitter referred to the device ground (or V−) pin. Bias current of the 339 is 25 nA and offset voltage is 2 mV (typical). The 339A device improves the offset voltage to 1 mV (typical). None of the devices of this family have provision for offset voltage adjustment.

Although IC comparators are not intended for feedback uses, they must also sometimes be stabilized. Stray coupling between output and input should be minimized, impedances kept low and leads short, and supplies well bypassed for ac stability.

In addition to the performance features mentioned, comparators designed specifically for the voltage-comparison function have features of flexibility which can be optimized unlike a standard op amp. For example, using the open-collector outputs of multiple (dual or quad) devices allows easy implementation of such functions as window comparators or other such amplitude classifiers. Two examples of this are shown in Fig. 8-16, using single-supply comparators.

Fig. 8-16A shows a basic window comparator which compares a voltage, $V_{in}$, to high- and low-limit reference voltages. When the voltage is in the window, both of the npn outputs of the 339 will be off and the 2N2222 switching transistor will be held on. When either window limit is passed, one output will then be on and the transistor and the indicator lamp will be turned off.

The MOSFET-input comparator of Fig. 8-16B uses the CA3290 dual comparator, which is a single-supply device with MOSFET input transistors. It has very low currents, on the order of only a few picoamperes, making it very useful for very-high-impedance sources. The application illustrated here is similar to the 339 in function, and turns on the LED when the input voltage is within the comparison window.

Detailed application coverage of the numerous comparator devices is simply beyond the scope of this chapter, and we only hope to make the reader aware that there are many of them to choose from. Some devices which warrant investigation (in addition to those already mentioned) are the Fairchild 710, the National 306, 319, 360, and 361, and the Precision Monolithics CMP-01, CMP-02, and CMP-05, and the Linear Technology 1016.

## 8.11 PRECISION OP-AMP COMPARATOR

One application area where IC op amps *are* more useful as voltage comparators than standard comparator devices is when very high dc precision is required. Such a case may be when ultralow bias currents are required, and very low offset voltages and/or very high common-mode rejection. Standard comparator devices in particular fall short with regard to very low offset voltage (and drift) and CMRR. In fact, as comparator speed is extended, dc accuracy usually suffers, with higher offset voltages, drifts, etc.

If an externally compensated (high-speed) instrumentation-grade op amp is used as the voltage comparator, it is possible to achieve voltage offsets of 500 $\mu$V or less, drifts of 1 $\mu$V/°C, and a CMRR of well over 100 dB. Removal of the compensation capacitance (when possible) will allow maximum speed to be realized from the device.

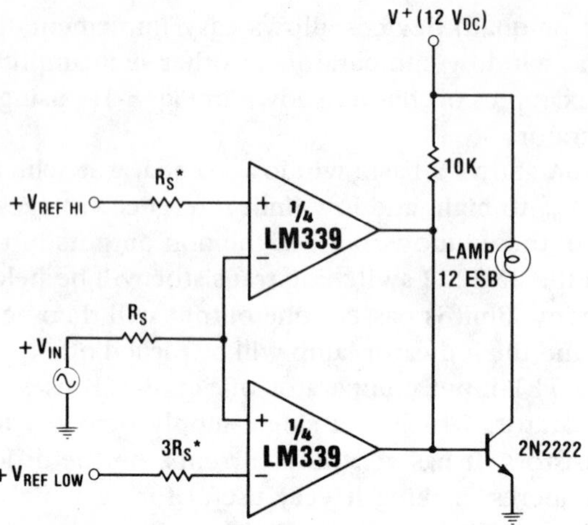


*(A) Using the LM339 bipolar-input device.*

*(Courtesy National Semiconductor Corp.)*

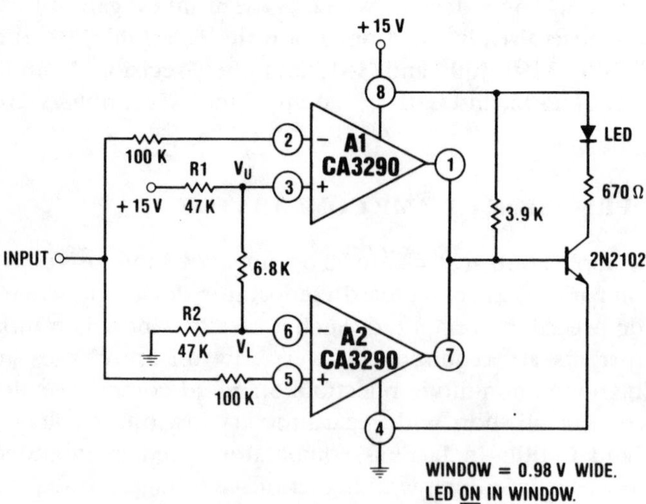


*(B) Using the CA3290 MOSFET-input device.*

*(Courtesy RCA Solid State Division)*

**Fig. 8-16. Window comparators using multiple devices (single supply).**

With some units, for example the SSS725* and others (such as the OP-37 and similar devices), response times of several microseconds can be achieved with such dc precision.

A general circuit optimized for maximum dc precision is shown in Fig. 8-17 and, as can be noted, can generally employ any of the devices listed in the table in the drawing. In actual practice, the specific device that will be declared most suitable will be governed by how much speed and dc accuracy are desired.

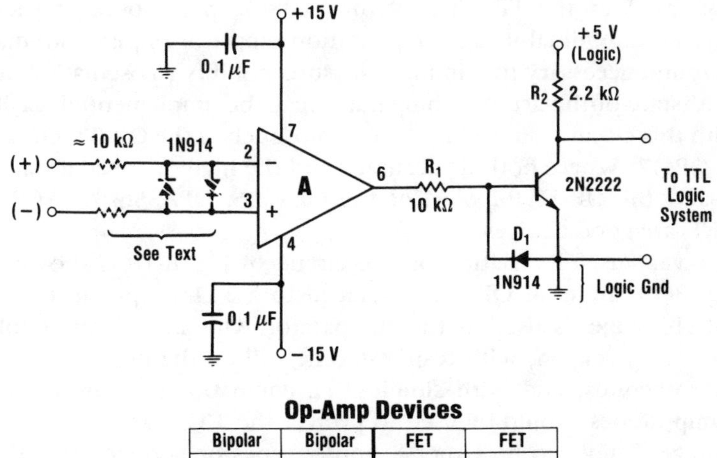


**Op-Amp Devices**

| Bipolar | Bipolar | FET | FET |
|---|---|---|---|
| SSS725* | OP-37* | OP-17* | OPA111* |
| AD504* | 5534 | AD547* | TL070 |
| 308A* | 2625 | AD711 | 1022 |
| LM11* | 3100 | OPA356 | 1056 |
| OP-07* | 709 | OPA606 | |
| OP-27* | 301A,748 | | |

**Fig. 8-17. Precision op-amp comparator.**

In using this circuit, input-current-limiting resistances should be used in series with the amplifier inputs to limit the current when protection diodes are used. Note that these diodes must be used to protect bipolar input stages, either within the IC or added externally (with the 709, for example). Then, the resistors chosen can be no larger than necessary, and the 10-kΩ value indicated should

* See reference 3 at the end of the chapter.

be the maximum. The diodes should be low-leakage units (see Chapter 3).

The output buffer serves two purposes—to minimize loading on the op amp and thus preserve its gain and drift characteristics, and also provide a high-current TTL logic output. It will add minimal delay (compared to the op amp).

The op amps in the table in Fig. 8-17 that are marked with asterisks (*) are those capable of the most dc precision and are not necessarily the fastest of those shown. It is worth noting that even general-purpose units such as the 301A perform well in this circuit, as does the FET TL070 (or TL080). With some of these devices, a minimal-value compensation capacitor (5 pF or so) may be found necessary to minimize the susceptibility to oscillation.

A state-of-the-art dc comparator can be implemented easily with this circuit using a bipolar op amp, such as the OP-07, OP-27, or OP-37. Where both dc accuracy and the highest speed are necessary, the OP-37 (bipolar) or the OPA606, OPA356, or AD711 (FET) are good choices.

A very useful variation on the circuit of Fig. 8-18 is shown in Fig. 8-19. Here an OP-37, interfaced to a TTL output using the 74LS14 stage, is used as the comparator. This allows microvolt-level dc precision, with response time still on the order of a few microseconds, and with simple TTL compatibility. Low-voltage clamp diodes should be used to protect the TTL stage from overvoltage. They can be either Schottky types or germanium. $D_1$ is necessary to prevent overdriving $A_2$, but $D_2$ can be deleted if the op amp positive supply is +5 V. A CMOS stage such as the 74C14 will also work well in this circuit.

For comparison of dc levels with an error of a few microvolts, a chopper-stabilized op-amp comparator, such as shown in Fig. 8-19, can be used. Here a 1052 chopper amp is used, operating from ±5-V supplies. It can compare voltages from −5 V up to +2.7 V (as shown). The bipolar output swings ±5 V, and will drive one of the buffers of Fig. 8-20 directly. A unipolar (0 to +5 V) CMOS output can easily be derived, using a 74C14 inverter. Response time for overdrive values of 10 μV is on the order of 80 μs (or less).

## 8.12 GENERAL-PURPOSE INTERFACING CIRCUITS

Fig. 8-20 illustrates a few ways in which any op-amp comparator with a bipolar output swing can be used to drive high-powered

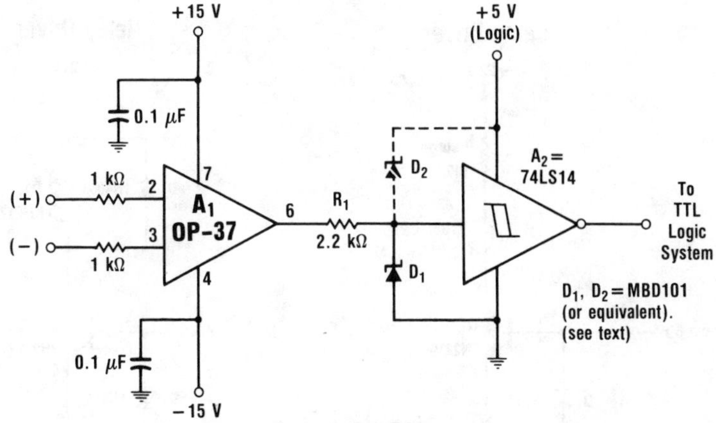


**Fig. 8-18. High-speed precision op-amp comparator (TTL interface).**

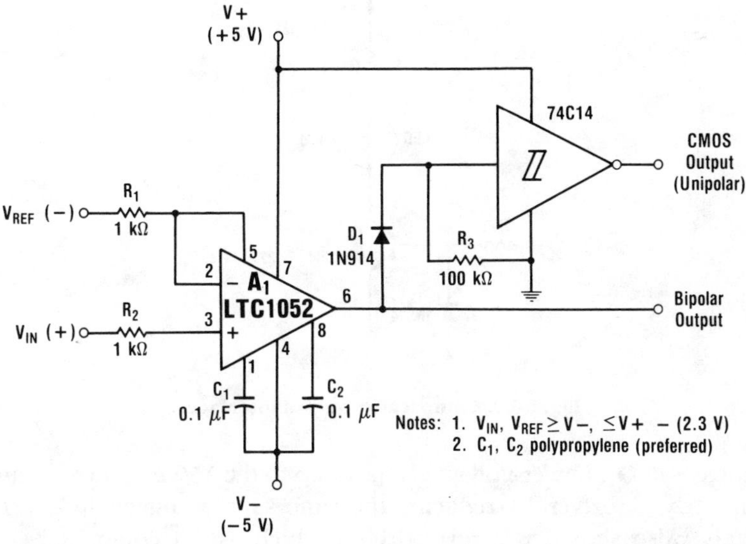


**Fig. 8-19. High-dc-precision op-amp comparator (CMOS interface).**

loads, other than the logic output stages shown in Figs. 8-17, 8-18, and 8-19. All three circuits interface with an op-amp output (pin 6) through an input resistor ($R_1$) of 5 kΩ to 10 kΩ, and thus can be driven from any of the op amps mentioned (including the chopper amp).

In the lamp-driver circuit, $R_{surge}$ is added to limit the cold in-rush

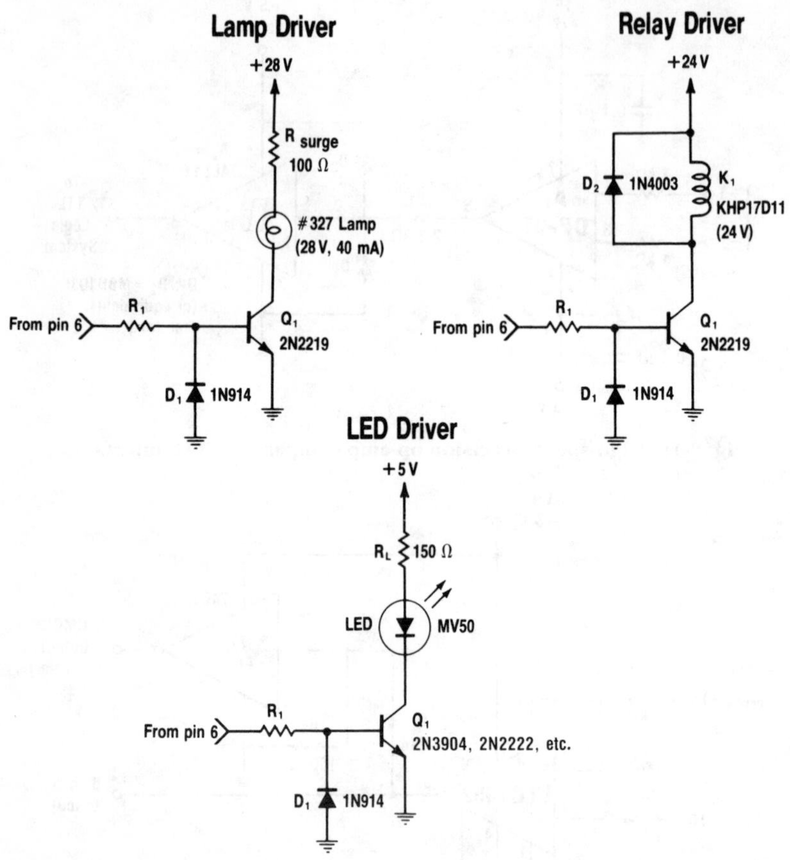


**Fig. 8-20. Comparator output interfaces.**

current in $Q_1$. Lower-voltage lamps of up to the 150-mA current rating may be driven by reducing the lamp supply voltage appropriately. Also shown is a relay driver, which uses a common 24-V relay. Diode $D_2$ clamps the back emf of $K_1$ after turn-off. Again, lower-voltage relays may be driven, as long as $I_{max}$ is 150 mA or less.

Finally, this circuit is easily adapted as an LED indicator driver, using a 2N3904 switch rather than the high-powered 2N2219. Maximum diode current is set by the supply voltage (minus approximately 1.5 V, the LED forward drop) and by $R_L$; in this case, $I_{max}$ is about 20 mA.

## 8.13 REFERENCES


1. *Applications Manual for Computing Amplifiers,* 2nd ed. Philbrick/Nexus Research, Dedham, MA, 1966, 1984.
2. Dobkin, R. C. "Comparators Can Do More Than Just Compare." *EDN,* November 1, 1972.
3. Givens, S. D. "The 725 Op Amp As a Low Level Comparator." *Precision Monolithics Application Note AN-25,* February 1978. Precision Monolithics, Inc., Santa Clara, CA.
4. Korn, G. A., and Korn, T. M. *Electronic Analog and Hybrid Computers,* 2nd ed. McGraw-Hill Book Co., New York, 1972.
5. Smathers, R. T., Frederiksen, T. M., and Howard, W. M. "LM139/LM239/LM339—A Quad of Independently Functioning Comparators." *National Semiconductor Application Note AN-74,* January 1973. National Semiconductor Corp., Santa Clara, CA.
6. Widlar, R. J. "An IC Voltage Comparator for High Impedance Circuitry." *National Semiconductor Linear Brief LB-12,* January 1970. National Semiconductor Corp., Santa Clara, CA.
7. _________. "Designing With Super-Beta Transistor Op-Amp ICs." *EEE,* December 1969, February and March 1970.
8. _________. "IC Op Amp Beats FETs on Input Current." *National Semiconductor Application Note AN-29,* December 1969. National Semiconductor Corp., Santa Clara, CA.
9. _________. "Monolithic Operational Amplifiers—The Universal Linear Component." *National Semiconductor Application Note AN-4,* April 1968. National Semiconductor Corp., Santa Clara, CA.
10. _________. "Precision IC Comparator Runs From 5V Logic Supply." *National Semiconductor Application Note AN-41,* October 1970. National Semiconductor Corp., Santa Clara, CA.
11. Manufacturers' Data Sheets:

| | | |
|---|---|---|
| Intersil | ICL7652 | Op Amp |
| Linear Technology Corp. | LTC1052 | Op Amp |
| National Semiconductor | LM139/339 | Comparator |
| | LM193/393 | Comparator |
| | LM311 | Comparator |
| RCA | CA3290 | Comparator |

NINE

# *The Integrator and the Differentiator*

Previously, we have seen how modifications of the basic inverting amplifier give rise to the integrator and the differentiator. Both of these circuits are based primarily on the relationship between voltage and current in a capacitor. This chapter explores the practical points for optimum use of these two basic op-amp circuits.

## 9.1 THE INTEGRATOR

Fig. 9-1 illustrates the operation of an ideal integrator. If a constant current, I (amperes), is applied to capacitor C (farads), the voltage across the capacitor will rise linearly at a rate of

$$\frac{\Delta E}{\Delta t} = \frac{I}{C} \text{ volts per second}$$

Thus, a current of 1 $\mu$A applied to a capacitor of 1 $\mu$F produces a rising voltage across the capacitor, and this change occurs at the rate of 1 V/s.

In Fig. 9-1B, the 1-$\mu$A current is applied at time t = 0 and continues for a period of 4 seconds. At t = 4, the current is interrupted and the capacitor voltage (Fig. 9-1C) remains at the value accumulated up to that time (4 V). This mechanism is the basis for the op-amp integrator shown in Fig. 9-2. Capacitor C has one ter-

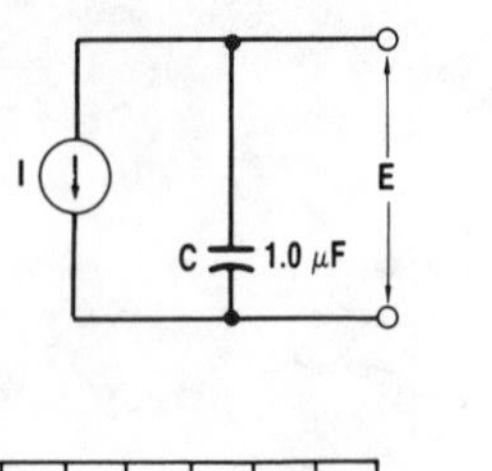


*(A) Test circuit.*

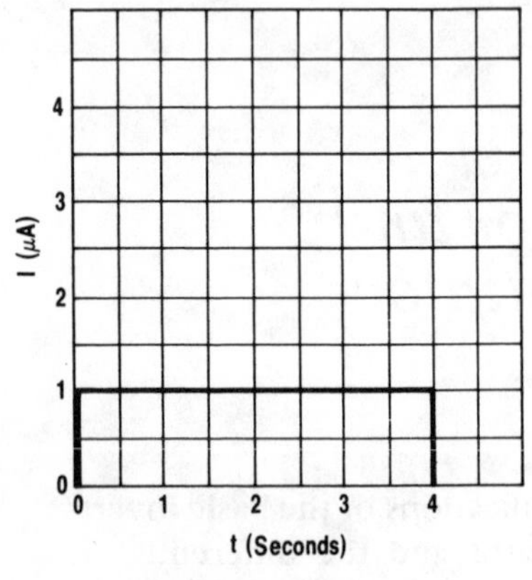


*(B) Applied current versus time.*

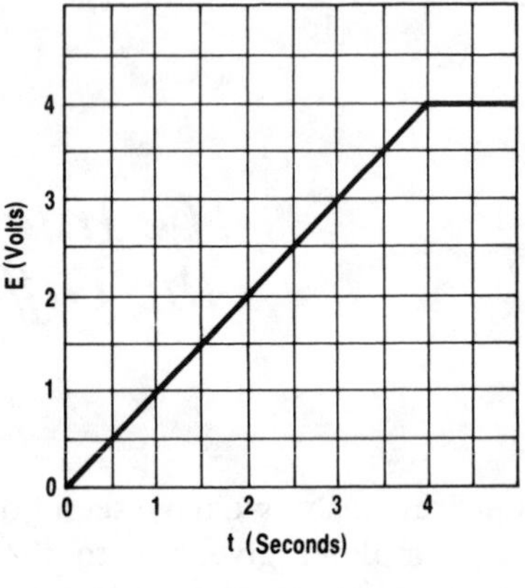


*(C) Resultant capacitor voltage.*

**Fig. 9-1. Dc current/voltage relationships for a capacitor.**

minal at the summing point, the other at the output; thus, the capacitor voltage is also the output voltage. In effect, the op amp simulates an ideal current source, with one important difference—the output impedance of the op amp is low. So, while it provides a constant current drive to C, it simultaneously makes the voltage across the capacitor available at a low-impedance level.

The integrator output cannot be described by a simple algebraic

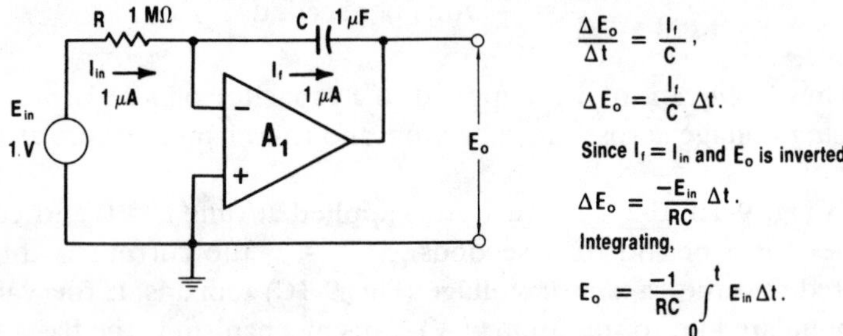


**Fig. 9-2. Basic op-amp integrator.**

relation; as we have seen, for a fixed (nonzero) input, the output voltage changes at a rate determined by $E_{in}$, R, and C. Thus, the output voltage cannot be determined without also specifying the duration of time an input has been applied. If the output rate of change is expressed in terms of the feedback current, $I_f$, then

$$\frac{\Delta E_o}{\Delta t} = \frac{I_f}{C}$$

or

$$\Delta E_o = \frac{I_f}{C} \Delta t$$

Since $I_f = I_{in}$ and, due to inversion,

$$\Delta E_o = \frac{-E_{in}}{RC} \Delta t$$

after integration from 0 to t seconds,

$$E_o = \frac{-1}{RC} \int_o^t E_{in} \Delta t$$

which is the output of the inverting integrator shown in Fig. 9-2. The 1/RC term of this expression is often called the integrator "gain," as it determines the slope of the output change. In the circuit shown,

$$\frac{1}{RC} = \frac{1}{10^6 \times 10^{-6}} = 1$$

so this time constant of 1 second produces a 1-V/s rate of change per volt of input. From this, it can be seen that integrators make convenient circuits for the measurement of elapsed time—apply a known input to a calibrated integrator, and the voltage output can be read directly as a measure of elapsed time.

### 9.1.1 Realization of Practical Integrators

As may be noticed from the idealized integrator of Fig. 9-2, this configuration has an inherent drawback if $E_{in}$ is left connected indefinitely: $E_o$ will continue to rise until it reaches the saturation limits of $A_1$. This can again be appreciated by noting that the integrator is operating as an open-loop amplifier for dc inputs.

Integrators that allow operation in view of this limitation can take on various forms; for applications requiring precisely timed periods of integration, the circuits of Figs. 9-3 and 9-4 can be used. Both are sequentially operated integrators and have defined modes of operation. Since integration begins with whatever voltage is on the capacitor initially, the capacitor voltage must be reset to the desired value (usually zero) before each integration. Fig. 9-3B illustrates how this is done.

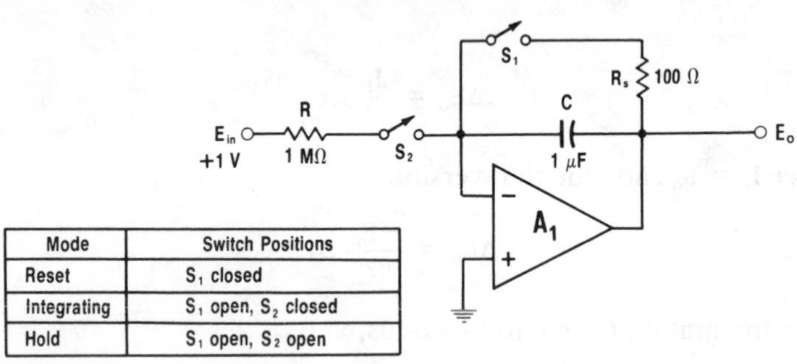


| Mode | Switch Positions |
|---|---|
| Reset | $S_1$ closed |
| Integrating | $S_1$ open, $S_2$ closed |
| Hold | $S_1$ open, $S_2$ open |

*(A) Test circuit.*

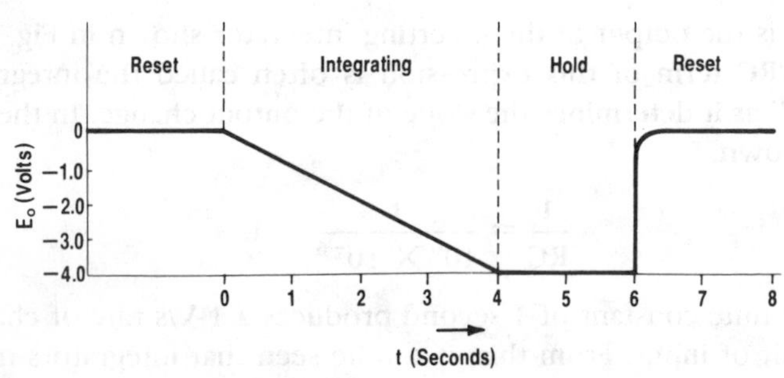


*(B) Operating sequence.*

**Fig. 9-3. Integrator operation.**

Switch $S_1$ is closed during the reset mode, discharging C through $R_S$. This leaves a residual voltage of zero. To begin the integration, $S_1$ is opened and $S_2$ is closed. During this period, the circuit operates as described previously and will charge C to a potential of −4 V in 4 seconds with the values shown. At the end of this period, $S_2$ is opened and the circuit "holds" the voltage on the capacitor for readout and subsequent processing.

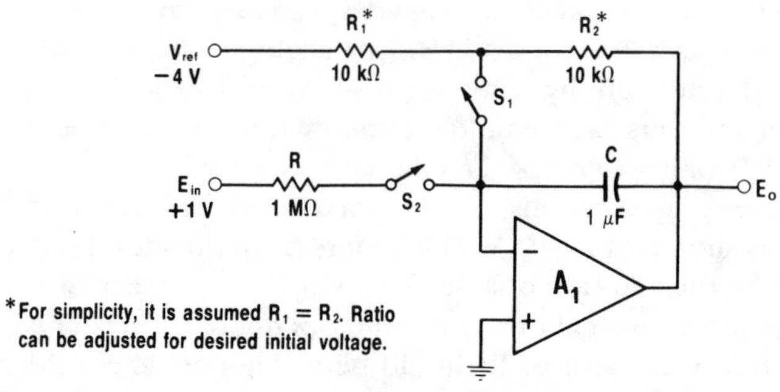


*(A) Test circuit.*

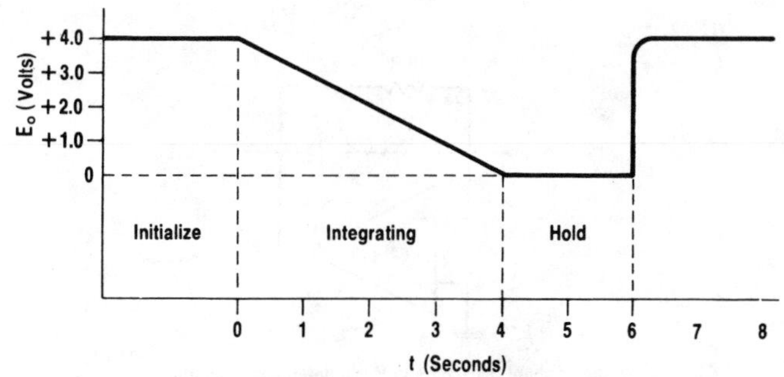


| Mode | Switch Positions |
|---|---|
| Initialize | $S_1$ closed, $S_2$ open |
| Integrating | $S_1$ open, $S_2$ closed |
| Hold | $S_1$ open, $S_2$ open |

*(B) Operating sequence.*

**Fig. 9-4. Integrator with initial voltage applied.**

An alternate form of the basic circuit is shown in Fig. 9-4. This circuit converts the reset cycle to an "initialize" cycle by the addition of $R_1$ and $R_2$. With $S_1$ closed (and $S_2$ open), $R_1$, $R_2$, and $A_1$ form an inverter, which charges C to an inverted multiple of the reference voltage. In this example, the multiple is unity, thus C is initialized to +4 V. The integrating and hold modes are the same as before; the only difference is that C starts the integration with a +4 V potential. For example, a +1 V input will diminish the potential on C to zero in 4 seconds.

In either circuit, switches $S_1$ and $S_2$ can be relays or (more often) solid-state switches such as FETs or analog switches. Switching is directed externally by some form of control logic. Very-low-integration currents will employ ultralow-leakage switches, such as MOSFETs or glass-enclosed reed relays.

In many applications, a simplified form of integrator can be used, as shown in Fig. 9-5. This differs from the ideal form of integrator by the addition of a dc "gain stop" resistor across C, reducing the integrator gain from the full open-loop value. The ratio of this shunt resistance to R should be as high as dc considerations allow so that ideal integrator behavior is not compromised. This form of integrator can be used successfully in such applications as square-wave to triangular-wave conversion.

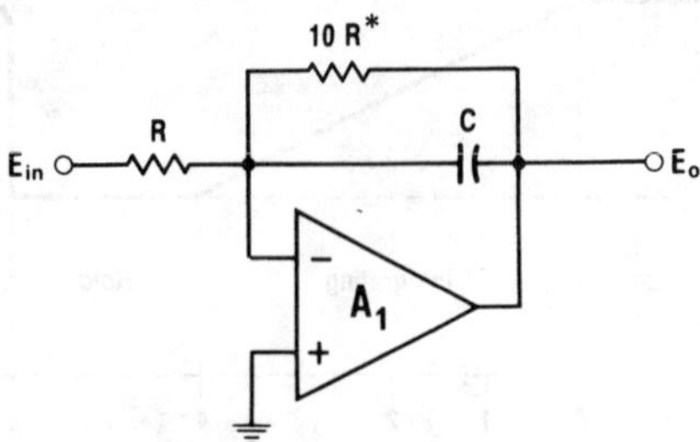


*Ratio of shunt resistance to R will determine integrator accuracy. If possible (with dc stability), make shunt resistance 100R or even 1000 R.

**Fig. 9-5. Continuous integrator.**

The use of the integrator in its unmodified form (Fig. 9-2) is generally avoided because some means of introducing dc feedback must be used to stabilize the operating point and prevent saturation. However, this "classic" form of the integrator may sometimes appear as part of an overall feedback loop.

### 9.1.2 Error Sources in the Integrator

Fig. 9-6 illustrates the sources of error that must be considered and minimized for optimum integrator performance. Capacitor $C_t$ must be a high-quality type with a leakage current lower than the bias current of the amplifier. Typical examples are Teflon, polystyrene, polypropylene, polycarbonate, or polyester types. The capacitor used should have very low dielectric absorption as well as very low leakage. Teflon, polystyrene, and polypropylene types are superior in both regards, having leakages down to the picoampere level.

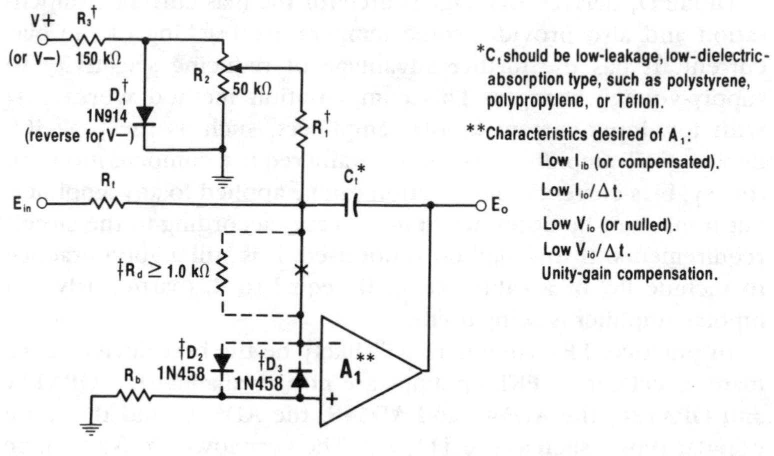


**Fig. 9-6. Error sources and practical problems in the integrator.**

The amplifier must be compensated for unity gain and have good-to-excellent dc characteristics, depending on the maximum integration time. For long integration times, $R_t$ and $C_t$ will become large, which will reduce the integration current and make the error due to input bias current important. Any offset will be seen as an input current by the circuit and will be integrated along with the signal. Offset at the summing point can originate from either $V_{io}$ or from $I_{ib} \times R_t$. $V_{io}$ should be minimized by amplifier selection or nulling; $I_{ib}$ can be minimized by choosing an op amp with high input impedance, such as an FET or a low-bias-current bipolar type. In some applications, $I_{ib}$ can be compensated satisfactorily by using the method shown in Fig. 9-6.

When necessary, compensation of bias current is supplied by $R_1$, which provides a current to the summing point equal to the nominal bias current. This compensation is optimized by adjusting $R_2$ for minimal output drift with zero input voltage applied to $R_t$.

The adjustment should be done after the input offset voltage has been nulled. See Chapter 7 for details of alternate bias-compensation techniques.

Diode $D_1$ derives a voltage source for the bias-current compensation and also provides some temperature tracking of the bias current. It has the further advantage of reducing sensitivity to supply-voltage changes. This compensation method works best with low-input-current bipolar amplifiers, such as the type-308 device. (FET amplifiers do not generally require compensation.) In theory, bias-current compensation can be applied to any amplifier, but it may not be necessary in many cases according to the circuit requirements. If this method is not used, it is still a good practice to include $R_b$, of a value nominally equal to $R_t$ (particularly if a bipolar amplifier is being used).

In practice, FET amplifiers will likely be the best device, since many excellent IC FET op amps are now available—the OPA111 and OPA128, the AD547 and AD548, the AD549, and the more popular types, such as the 411, etc. The very lowest offset will be achieved with chopper-stabilized types, such as the 7650, 7652, and the 1052.

Possible failure of the amplifier input stage is a problem in integrators, particularly if $C_t$ is greater than about 0.1 $\mu$F. In such cases, it is advisable to include a series resistance ($R_d$) to limit the discharge rate, and clamping diodes ($D_2$ and $D_3$) are recommended if the amplifier is unprotected differentially. They should be low-leakage diodes, if used (see Chapter 3).

Integrators may also be required to operate at high speeds—for instance, in the generation of fast ramps and triangular waveforms. Such applications require a high slew rate and a high $f_t$ for the op amp but normally with less stringent dc requirements. These applications are satisfied by such high-speed devices as the various 356 types, or types such as the AD711, the 34081 series, etc.

### 9.1.3 Expanding Integrator Flexibility and Capability

Beyond the basic points just described, there are a number of ways in which integrator performance can be improved.

One of the most basic limitations of the integrator is the practical problem of attaining long time constants. The problems stem mainly from the very real difficulties in obtaining high-quality resistors with values above about 10 megohms, and capacitors with values above a few microfarads. If high accuracy is desired, these components should be stable types.

While film capacitors of good quality are available with values up to about 50 μF, they are generally undesirable from the standpoints of size, cost, and high-temperature leakage. There will also be procurement problems since they will have to be obtained through special order. On the other hand, small precision resistors are generally available up to 10 or 20 megohms (although they may not be shelf items either).

The circuit of Fig. 9-7 multiplies the time constant of an integrator circuit by replacing $R_t$ with a T network, which raises the effective input resistance as shown; in this case, to 100 megohms. A 100-second time constant can then be realized with only a 1-μF integrator capacitor, which is readily available in small sizes. For this technique to be most effective, the amplifier used should have low offset voltage (and drift), such as the AD548 shown in Fig. 9-7, or a similar type.

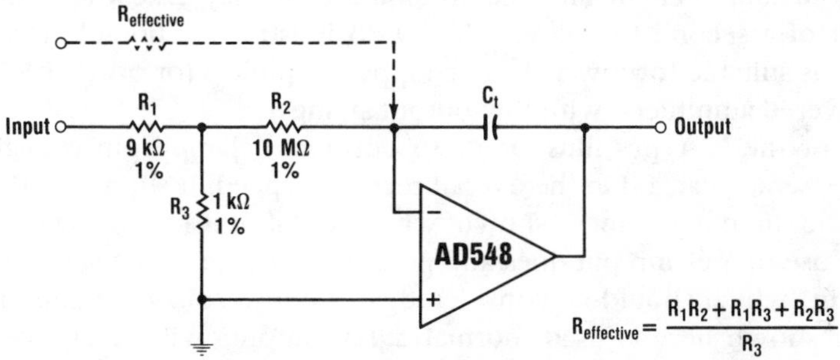


**Fig. 9-7. Multiplying integrator time constant with a T input.**

Note that virtually any combination of $R_1$, $R_2$, and $R_3$ values can be used to "multiply" $R_t$, but $R_1$ and $R_3$ should preferably be relatively low-value film units as shown. Since all three resistors enter into the time-constant expression, they all should be stable units. $R_3$ (or $R_1$) can be trimmed for calibration, if desired.

### *Integrator Bound Circuits*

Fig. 9-8 illustrates two circuits that are useful with virtually any integrator—a reliable limit, or "bound," circuit. These circuits are a virtual necessity for many integrator uses, and prevent the (very) long saturation delays associated with an integrator that is driven into saturation.

Fig. 9-8A is a more or less general-purpose bound circuit, useful with the standard integrator connection, and using ±15-V supplies and ±10-V output swings.

Here, $R_t$, $C_t$, and $A_1$ might be any such integrator, and the diodes are added across $C_t$ to clamp $A_1$ *before* it reaches saturation. Low-leakage diodes for the bridge and for coupling to the summing point will generally be desirable, but the degree of their specification limits will depend on the individual application. The shunt bleeder resistance ($R_1$) will minimize diode leakage problems. The zener should, of course, be selected for a voltage ***higher*** than the greatest linear output from $A_1$, minus two diode voltage drops. Note that for many practical applications, this circuit can be scaled down considerably (for example, a simple back-to-back diode pair across $C_t$, for low voltages).

Some particular amplifier types require even more special considerations for use in integrators. One example is the chopper-stabilized amplifier—if allowed to saturate, it may take the better part of a second to recover. Fig. 9-8B illustrates a bound circuit that is suitable for low-voltage chopper amplifiers (or other ±5-V-powered amplifiers, with low-output swings).

For the IC types illustrated, an optional "clamp" pin is available, which can aid in the overall recovery speed. It should be tied to the summing point as shown, when, and if, it is used. However, the use of a clamp pin does add potential leakage to the summing point, which should be considered. If the low-voltage bound circuit shown here is used, normal ±1-V outputs will be allowed, but limiting will occur at under ±3 V.

LEDs and low-leakage JFETs are used for this low-voltage limiter. Note that for best performance, the green LEDs must be light-tight to prevent photocurrent-related leakage. The JFET diodes can be a variety of types, dependent upon availability, inasmuch as the 30-pA bias current of the chopper amp obviates sub-picoampere leakages for this pair. Selected transistor collector-base junctions (such as 2N5088s or 5089s) can also work well (see Chapter 3).

### *Noninverting Integrators*

One of the applications "hang-ups" of the traditional integrator is the loading of the input resistor, as well as the signal inversion

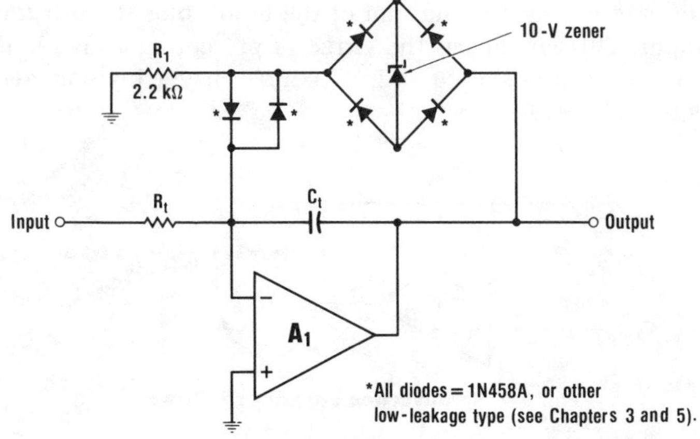


*(A) General-purpose circuit.*

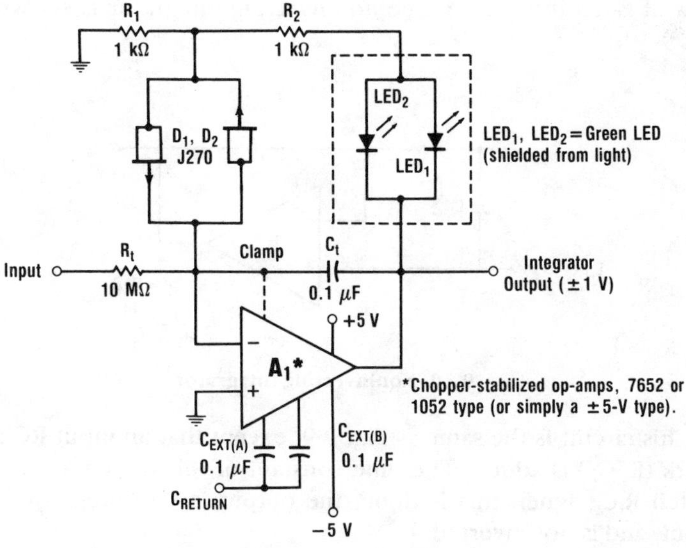


*(B) Low-voltage (±5 V) circuit (for chopper-stabilized integrator).*

**Fig. 9-8. Bound integrator circuits.**

of the stage. Noninverting forms of integration require a few more components to implement, but get around these problems.

A simple variant on the basic inverting integrator is the circuit of Fig. 9-9, an "integrator/follower." In this circuit, the amplifier has an output which is the sum of the input, plus the integral of the input. This circuit has the virtue of presenting a high impedance to the signal source and, of course, has a low-impedance output.

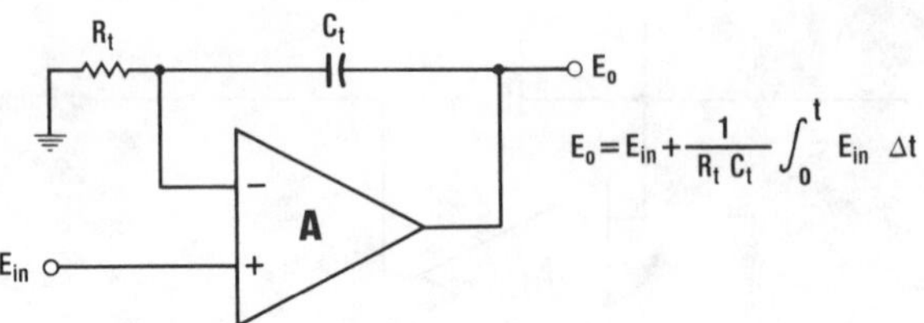


**Fig. 9-9. Noninverting integrator/follower.**

The drawback of the integrator/follower lies in the fact that it is not a "true" integrator, inasmuch as the output contains the input as well as its integral. A true noninverting integrator is shown in Fig. 9-10.

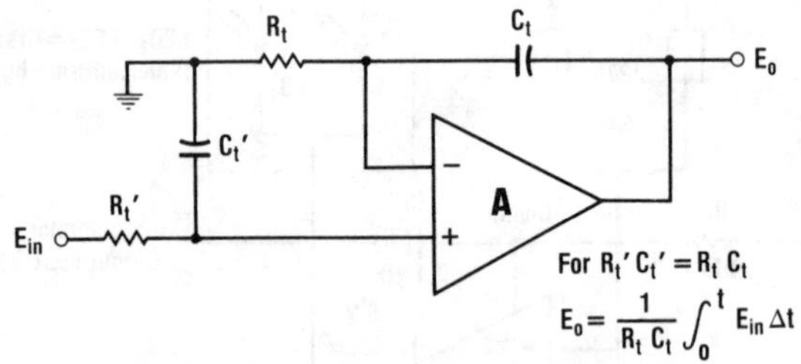


**Fig. 9-10. Noninverting integrator.**

This circuit is the same as Fig. 9-9, except that an input RC network ($R_t'C_t'$) is added. The time constant of this network is set to match $R_tC_t$; when this is done, the output is the integral of the input, and is not inverted.

Disadvantages of this circuit are that the time constants need to be matched, which requires either trimming or close tolerances. This is a serious problem, particularly for capacitance. In addition, it loads the signal source with the series impedance of $R_t'$ and $C_t'$. Also, like any noninverting amplifier, it will be sensitive to CM voltage, as limited by the amplifier used. Resetting of the capacitor(s) to a known voltage is more difficult—there are two of them in Fig. 9-10, with one floating.

### *Differential Integrators*

Sometimes it is necessary to integrate the *difference* between two voltages (or a floating voltage). The circuits of Fig. 9-11 can accomplish this.

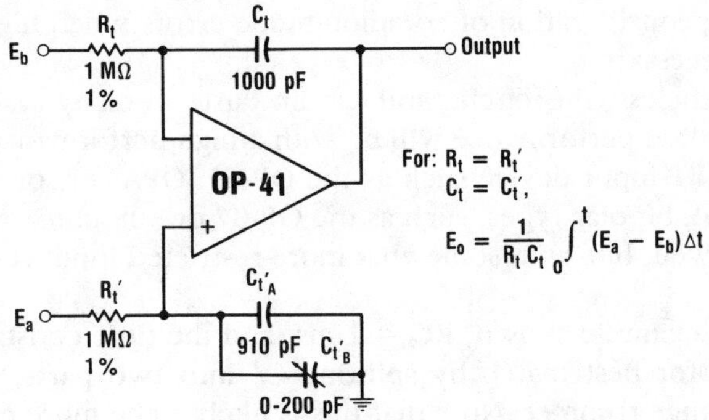


*(A) Using differential amplifier.*

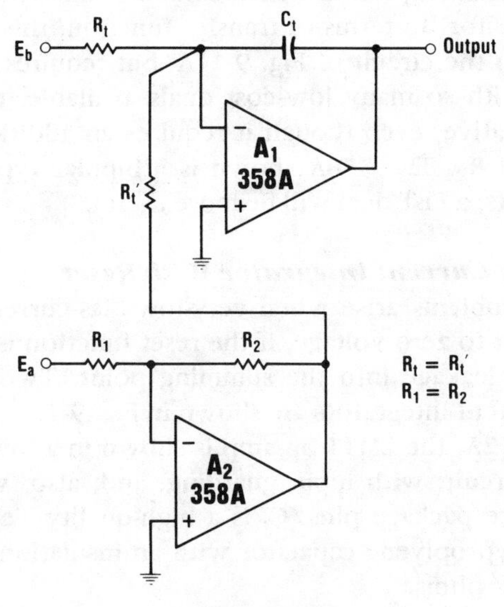


*(B) Using inverters only.*

**Fig. 9-11. Differential integrators.**

The circuit of Fig. 9-11A is the classic approach and requires a pair of matched resistors and capacitors. This circuit can be recognized as a variant of Fig. 9-10, with some similar problems. Again, while matched resistors are readily available, capacitors usually are not, and the circuit tends to be impractical from this standpoint. Also, it applies common-mode voltage to the amplifier, requiring consideration of common-mode errors when high accuracy is necessary.

Nevertheless, this circuit still can find uses in many cases. Best all-around dc performance will be with a high-performance, high-CMRR, FET-input device such as the OP-41, OPA111, or AD548. In general, bipolar types, such as the OP-07 or 308 units, will also do very well, but with somewhat more-restricted input-resistance ranges.

In the example shown, RC = 1 ms, and the time constants are trimmed for best match by splitting $C_t'$ into two parts, using a ±10% range trimmer. Note that this is likely to be more practical than matching 1% capacitors.

The alternate circuit of Fig. 9-11B uses an inverter to invert one of the differential inputs, and then combines this voltage in a combining integrator. In terms of transfer function, this yields results equivalent to the circuit of Fig. 9-11A, but requires an additional amplifier. With so many low-cost duals available, this may be a viable alternative, even though it requires an additional matched pair—$R_1$ and $R_2$. The 358A shown is a bipolar type; for lowest input currents, a FET dual will be more useful.

### *Low-Bias-Current Integrator With Reset*

Special problems arise when very-low-bias-current integrators must be reset to zero voltage, if the reset function is not to introduce excess leakage into the summing point. Two examples of low-bias-current integrators are shown in Fig. 9-12.

In Fig. 9-12A, the LM11 op amp is shown in a low-bias-current integrator circuit, with input guarding, and, also, with guarding on the device package pins. $C_t$ is a high-quality Teflon, polystyrene, or polyproplyene capacitor with an insulation resistance in excess of $10^{12}$ ohms.

$Q_1$ and $Q_2$ are low-leakage p-channel MOSFETs, which are used for the reset function of $C_t$. They are used in a T configuration, with bleeder resistance $R_3$ absorbing the leakage of $Q_2$. Reset of the integrator to zero volts out is accomplished with a negative voltage of −5 V to the RESET terminal.

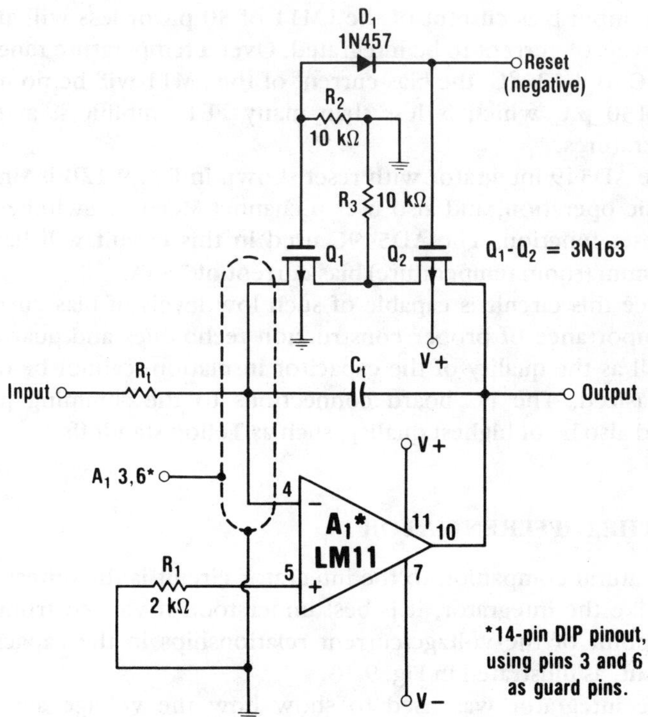


*(A) LM11 circuit.* (Courtesy National Semiconductor Corp.)

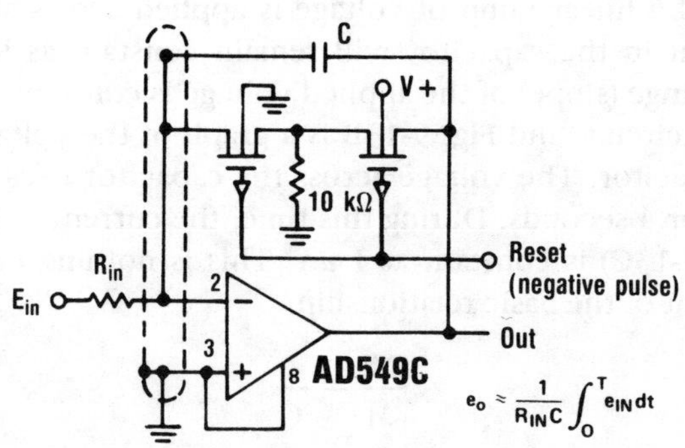


*(B) AD549 circuit.* (Courtesy Analog Devices, Inc.)

**Fig. 9-12. Low-bias-current integrators with reset.**

The input bias current of the LM11 of 50 pA or less will allow low levels of current to be integrated. Over a temperature range of −55°C to +125°C, the bias current of the LM11 will be no more than 150 pA, which is less than many FET amplifiers, at high temperatures.

The AD549 integrator with reset shown in Fig. 9-12B is similar in basic operation, and also uses p-channel MOSFET switches for the reset function. The AD549C used in this circuit will have a maximum (room temperature) bias current of 75 fA.

Since this circuit is capable of such low levels of bias current, the importance of proper construction techniques and guarding, as well as the quality of the capacitor insulation, cannot be overemphasized. The PC board connections to the summing point should also be of highest quality, such as Teflon standoffs.

## 9.2 THE DIFFERENTIATOR

A natural companion to the integrator circuit is the differentiator. Like the integrator, it is best understood if viewed from the standpoint of the voltage/current relationships in the capacitive element, as illustrated in Fig. 9-13.

The integrator was used to show how the voltage across a capacitor rises linearly if the current in the capacitor is maintained constant. The reverse of this concept forms the basis of the differentiator. If a linear ramp of voltage is applied across a capacitor, the current in the capacitor will remain constant as long as the rate of change (slope) of the applied voltage is constant. Fig. 9-13A is the test circuit, and Fig. 9-13B is a graph of the voltage applied to the capacitor. The voltage across the capacitor rises at the rate of 1 V/s for 4 seconds. During this time, the current in the capacitor (Fig. 9-13C) is constant at 1 μA. This is nothing more than a restatement of the basic relationship:

$$\frac{\Delta E}{\Delta t} = \frac{I}{C}$$

which may be stated as

$$I = C\frac{\Delta E}{\Delta t}$$

Thus, in the example shown, a 1-V/s rate of change across the

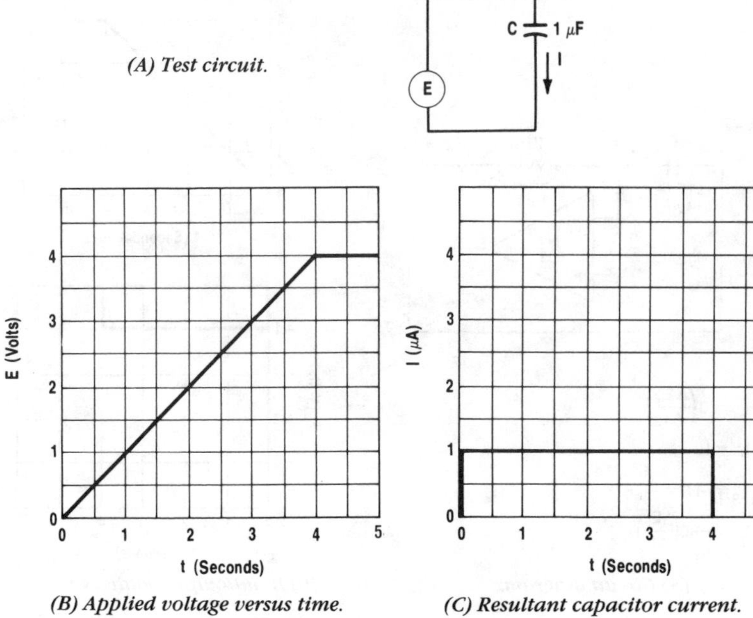


*(A) Test circuit.*

*(B) Applied voltage versus time.*

*(C) Resultant capacitor current.*

**Fig. 9-13. Dc voltage/current relationships for a capacitor.**

1-$\mu$F capacitor results in a 1-$\mu$A current in the capacitor, as shown in Fig. 9-13C.

When the rate of voltage change across the capacitor falls to zero, the current in the capacitor also falls to zero. This is shown on the graph as the time beyond 4 seconds where $\Delta E = 0$ and $I = 0$.

In the op-amp differentiator of Fig. 9-14, capacitor C is the input element of the configuration and resistor R is the feedback element. As in the integrator, one terminal of the capacitor is held at ground potential by the summing point.

Input current $I_{in}$ can be expressed in terms of C and $E_{in}$ as

$$I_{in} = C\left(\frac{\Delta E_{in}}{\Delta t}\right)$$

and since $I_f = I_{in}$, we may substitute and add the sign inversion of $A_1$ to write an expression for $E_o$:

$$E_o = -I_f R = -I_{in} R$$

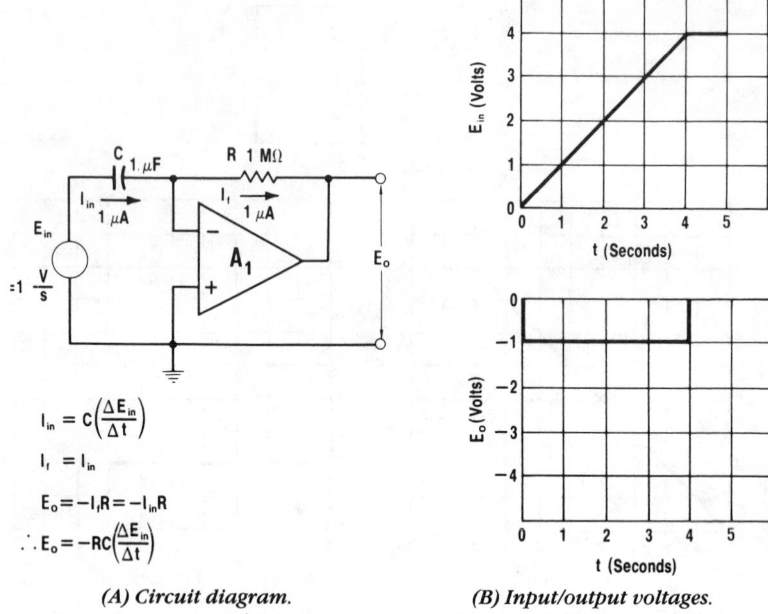


*(A) Circuit diagram.*

*(B) Input/output voltages.*

**Fig. 9-14. Basic op-amp differentiator.**

which, when substituted for $I_{in}$, becomes

$$E_o = -RC\left(\frac{\Delta E_{in}}{\Delta t}\right)$$

Graphically, the differentiator is illustrated by the sketches, which show the output to be an inverted product of the capacitor current ($I_{in}$) and R.

Since the differentiator responds only to changes in the input voltage (a capacitor will not carry current unless the applied voltage is changing), it will yield zero output voltage for a fixed (static) input voltage. The faster the input voltage changes, the greater will be the output voltage. In terms of frequency, the differentiator gives a rising output voltage with increasing frequency, as would be expected from the decreasing input impedance due to the falling reactance of the capacitor.

In practical terms, the differentiator also has problems when applied in basic form (as in Fig. 9-14). The most apparent drawback is illustrated in Fig. 9-15. An earlier discussion of stability noted that, for stability to exist, the ideal $1/\beta$ curve must cross the open-loop gain curve with a net slope of less than $-12$ dB/octave. In Fig. 9-15A, it will be noted that the ideal $1/\beta$ line intersects the gain curve at $f_o$, with a net slope between the curves of $-12$ dB/octave; thus, the characteristic is unstable. To make the $1/\beta$ line intersect the gain curve with a net slope of $-6$ dB/octave (required for stability), it must be modified to form the curve shown as $1/\beta'$. If $f_o$ is the frequency at which the ideal $1/\beta$ curve intersected, a breakout must be introduced in the $1/\beta$ curve at $f_1$, where $f_1 = f_o/3.16$. This will cause the new curve ($1/\beta'$) to intersect the Bode plot at a frequency $f_2 = 3.16\ f_o$. The $1/\beta'$ curve, therefore, has a full decade ($f_2 = 10f_1$) to assume a zero phase characteristic, which ensures stability.

To accomplish the frequency breakout, a resistive "stop" ($R_2$) is connected in series with $C_1$, as shown in Fig. 9-15B. The value of $R_2$ is chosen so that it is 3.16 times the reactance of $C_1$ at frequency $f_o$, or

$$R_2 = \frac{3.16}{2\pi f_o C_1}$$

This gives the minimum value of $R_2$ required for stability, but, in practice, the next higher standard value is satisfactory. From the Bode plot, it can be seen that such a modification results in little sacrifice of the useful dynamic range of the differentiator.

Another problem inherent in the differentiator is the 6-dB/octave rise in gain with increasing frequency. In practice, this effect greatly increases high-frequency noise and, in some cases, the noise may mask the desired signal if the signal has a small amplitude and is low in frequency. The noise problem can be dealt with by modifying the differentiator characteristic in another fashion.

If a modified curve ($1/\beta''$) is introduced at frequency $f_3$, the high-frequency gain is reduced drastically from this frequency upward, while the differentiator characteristic remains unaltered below $f_3$. In actuality, a $1/\beta''$ characteristic produces a combination of differentiator and integrator action, with the crossover occurring at $f_3$.

To realize the $1/\beta''$ curve, $C_2$ is added across $R_1$, with $C_2$ chosen to be equal in reactance to $R_1$ at the crossover frequency, $f_3$, or

$$C_2 = \frac{1}{2\pi f_3 R_1}$$

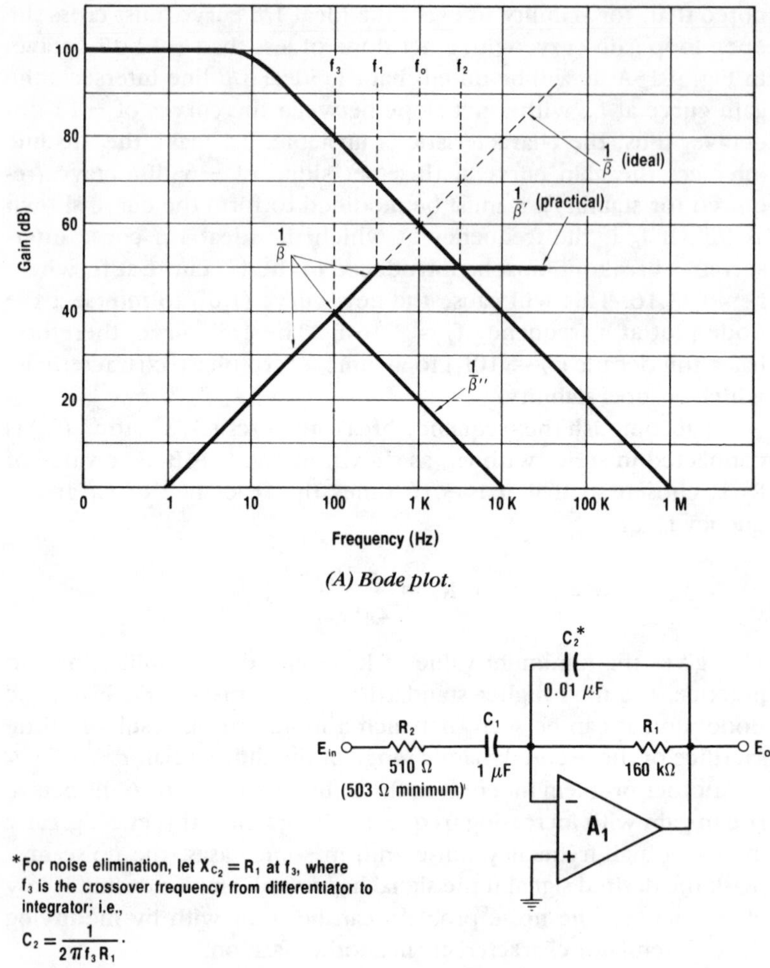


*(A) Bode plot.*

*(B) Practical circuit.*

**Fig. 9-15. Stability conditions in the differentiator.**

With this $1/\beta''$ curve, unity-gain stability is now required of $A_1$, whereas this was not necessarily so in the case of $1/\beta'$.

With regard to the types of devices that are best suited for differentiator use, the high-frequency amplitude-rise characteristic tends to limit the choice to devices with high slew rate and GBP. High-slew-rate FET units, such as the 356-family types, will be found useful, as will such units as the 318, and slew-rate enhanced devices like the 535.

## 9.3 REFERENCES


1. *Applications Manual for Computing Amplifiers,* 2nd ed. Philbrick/ Nexus Research, Dedham, MA, 1966, 1984.
2. Dobkin, R. C. "Op Amp Circuit Collection." *National Semiconductor Application Note AN-31,* February 1970. National Semiconductor Corp., Santa Clara, CA.
3. Giles, J. N. *Fairchild Semiconductor LIC Handbook.* Fairchild Semiconductor, Mountain View, CA, 1967.
4. Korn, G. A., and Korn, T. M. *Electronic Analog and Hybrid Computers,* 2nd ed. McGraw-Hill Book Co., New York, 1972.
5. "FET Circuit Applications." *National Semiconductor Application Note AN-32,* February 1970. National Semiconductor Corp., Santa Clara, CA.
6. Roberge, J. K. *Operational Amplifiers—Theory and Practice.* John Wiley & Sons, Inc., New York, 1975.
7. Rudin, M. "Applying Microvolt/Picoamp Instrumentation Operational Amplifiers." *Precision Monolithics Application Note.* Precision Monolithics, Inc., Santa Clara, CA.
8. Sheingold, D. H. *Analog-Digital Conversion Handbook,* 3rd Edition. Analog Devices, Inc., Norwood, MA, 1986.
9. Tobey, G. E., Graeme, J. G., and Huelsman, L. P. (eds.). *Operational Amplifiers—Design and Applications.* McGraw-Hill Book Co., New York, 1971.
10. Widlar, R. J. "Designing With Super-Beta Transistor Op-Amp ICs." *EEE,* December 1969, February and March 1970.
11. _________. "Drift Compensation Techniques for Integrated DC Amplifiers." *National Semiconductor Application Note AN-3,* November 1967. National Semiconductor Corp., Santa Clara, CA.
12. _________. "IC Op Amp Beats FETs on Input Current." *National Semiconductor Application Note AN-29,* December 1969. National Semiconductor Corp., Santa Clara, CA.
13. _________. "Monolithic Operational Amplifiers—The Universal Linear Component." *National Semiconductor Application Note AN-4,* April 1968. National Semiconductor Corp., Santa Clara, CA.

14. Manufacturers' Data Sheets:

| | | |
|---|---|---|
| Analog Devices | AD547 | Operational Amplifier |
| | AD548 | Operational Amplifier |
| | AD549 | Operational Amplifier |
| Burr-Brown | OPA111 | Operational Amplifier |
| | OPA128 | Operational Amplifier |
| Intersil | ICL7650 | Operational Amplifier |
| | ICL7652 | Operational Amplifier |
| Linear Technology Corp. | LTC1052 | Operational Amplifier |
| National Semiconductor | LM11 | Operational Amplifier |
| Precision Monolithics Inc. | OP-41 | Operational Amplifier |

TEN

# *Signal-Generation Circuits*

This chapter deals with the generation, shaping, and control of various kinds of signal waveforms. Discussed are signal-generation circuits such as sine-wave oscillators; multivibrators; triangle, ramp, and sawtooth generators; pulse generators; voltage-controlled oscillators; and voltage-controlled timers. These circuits will involve both new material and much of the circuit knowledge gained in previous chapters.

## 10.1 SINUSOIDAL OSCILLATORS

The most familiar waveform in electronics is the sine wave, and sine-wave signals of different frequencies are as fundamental to circuit work as dc supply potentials are. Generating a sine wave is not hard to do in concept, as we have already seen by the ready tendency of op amps to oscillate when amplification is really the prime intent. This section concerns itself with the controlled form of sine-wave oscillator, one which has a defined and predictable frequency, amplitude, and, in some cases, phase.

### 10.1.1 Wien-Bridge Oscillators

One of the most familiar types of sine-wave generators is the Wien-bridge circuit shown in Fig. 10-1. This circuit is a classic one in electronics and has been used since the days when it was realized in vacuum-tube versions. The Wien network, composed of $R_1$-$C_1$ and $R_2$-$C_2$, provides a positive feedback path around $A_1$, while $R_3$ and $L_1$ provide negative feedback. At some fundamental frequency, $f_o$, the overall phase shift is zero and transmission at this frequency is maximum in relation to all others. Positive feedback causes the circuit to oscillate at $f_o$, where

$$f_o = \frac{1}{2\pi R_1 C_1}$$

and $R_1 = R_2$, $C_1 = C_2$.

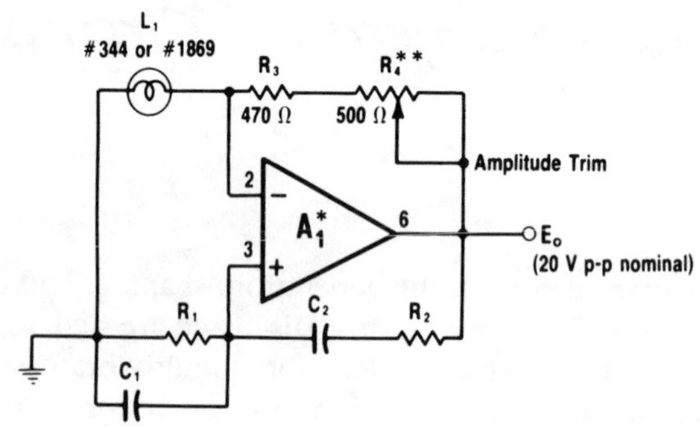


Set $R_1 = R_2$, $C_1 = C_2$; then $f_o = \frac{1}{2\pi R_1 C_1}$

*Select $A_1$ according to operating frequency and value of $R_1$; see text.

**Use $R_4$ to correct for lamp tolerance and set output level, or fix $R_3 + R_4$ at 750 Ω.

**Fig. 10-1. Wien-bridge oscillator.**

The attenuation of the Wien network at the frequency of oscillation is equal to 3. If the attenuation in the negative feedback path ($R_3$-$L_1$) is also 3, the bridge will be in balance and the differential input to $A_1$ will be small. It is this condition that must be maintained because, if the positive feedback is greater, oscillation will quickly build up until amplifier saturation is reached. On the other hand, if the negative feedback is greater, the oscillations will die out. In a practical Wien-bridge oscillator, it is therefore necessary to provide some means of automatically balancing the feedback so that the amplitude of the oscillations can be regulated. In this state, the purity of the output waveform will be at a maximum, as will the frequency stability. Another requisite is that the amplifier gain be large at the frequency of oscillation—the input error will then be small, which guarantees that the Wien network operates close to its natural frequency and is stable.

Any means of stabilizing the amplitude of a Wien-bridge oscillator must involve adjustment of the negative feedback divider to maintain the attenuation at 3. Typically, this can take the form of nonlinear elements: thermal types such as lamps or thermistors, nonlinear resistors, diode clippers, zener diodes, and so on. In Fig. 10-1, the nonlinear resistance of a lamp ($L_1$) is used to regulate the amount of negative feedback. $L_1$ responds to the average (not instantaneous) output level of $A_1$, adjusting its terminal resistance in inverse proportion. If the output level of $A_1$ rises, $L_1$ increases resistance, counteracting the rise. Similarly, a reduction in output results in a decrease of lamp resistance, stabilizing the oscillation level.

Lamp and thermistor stabilization schemes are popular ones, due mainly to their simplicity. They have several drawbacks, however, such as an inherent thermal time constant that limits usefulness at the lower frequencies.

Resistors $R_3$-$R_4$ trim the negative feedback to accommodate lamp tolerances. If precise amplitude control or optimum stability for individual lamps is not necessary, let $R_3 + R_4 = 750\ \Omega$. For frequency-tunable generators, either $R_1$-$R_2$ or $C_1$-$C_2$ can be ganged as a pair (with good tracking characteristics). If $C_1$-$C_2$ are selected as the tunable elements, large values for $R_1$-$R_2$ are required to cover the lower frequencies, which raises the network impedance. In such cases, $C_1$-$C_2$ should be in a shielded enclosure and stray capacitance to ground should be minimized for best results.

The circuit of Fig. 10-1 is quite useful as a general-purpose Wien oscillator; with the appropriate amplifier, it can be used over a wide range of frequencies. Amplifiers most suitable (over the widest frequency range) are high-slew-rate FET-input units such as the TL071, 411, 3140, 771, etc., and the very-high-SR types, such as the 356-improved units like the LT1056 and the OPA606, the 34081 series, and the AD711. These devices can be useful up to 100 kHz or more. Lowest distortion, along with lowest output noise, will be realized with high-slew-rate, high-GBP, low-noise types, such as the 5534, the OP-37, and similar units.

An example of this general type of sine-wave oscillation, optimized for lab bench use, is shown in Fig. 10-2. This circuit uses a 1037 (or OP-37) op amp as the $A_1$ oscillation stage, optionally buffered by $A_2$, an LT1010 unity-gain follower. The use of the buffer stage is not absolutely necessary, but it does allow over $\pm 100$ mA of peak output, with low distortion. If the buffer is used, it should be bypassed locally with low-impedance capacitors in both supplies.

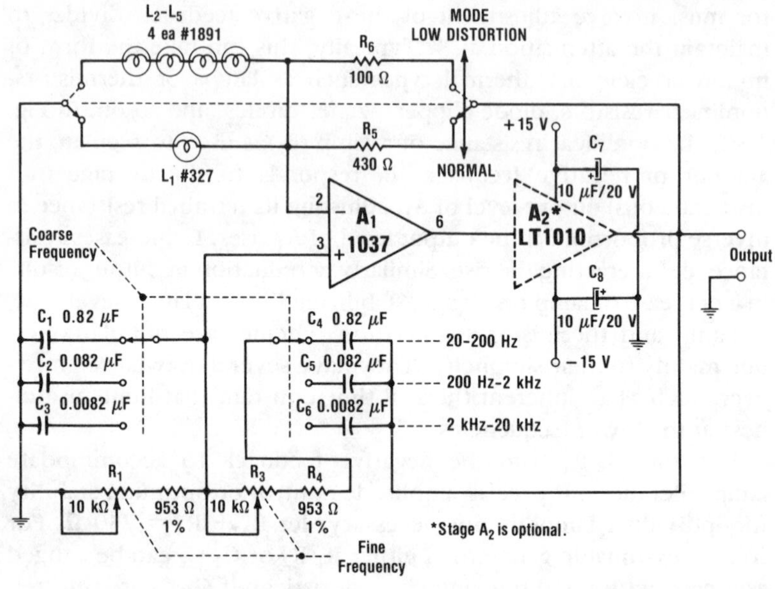


**Fig. 10-2. High-performance lab oscillator.**

In this circuit, matched capacitor sets are selected for the coarse range, and a dual-tracking potentiometer is used for fine-frequency control. This circuit is capable of performance in proportion to the care put into it, so polypropylene film capacitors are recommended, with outside foils connected as shown. Circuit layout should be compact, without excessive lead lengths.

Two methods of amplitude control are possible; "NORMAL" which uses lamp $L_1$, and "LOW DISTORTION" which uses lamp $L_2$–$L_5$. The former is suitable for general use down to 200 Hz at distortion levels of 0.01%, while the latter allows distortion of similarly low levels down to 20 Hz (at the expense of reduced amplitude and longer settling time).

### *Easily Tuned Wien Oscillator*

One of the great difficulties of the Wien oscillator lies in tuning. Brokaw* has shown a technique for tuning the circuit that uses

* See reference 2 at the end of the chapter.

the addition of a second amplifier and a gain-compensation network. This form of the circuit is shown in Fig. 10-3.

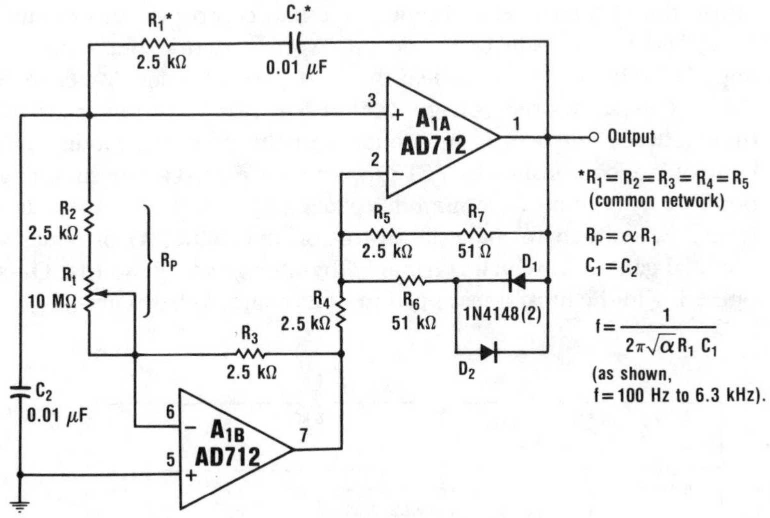


**Fig. 10-3. Easily tuned Wien oscillator.**

In this circuit, the network resistance, $R_p$, is returned to the summing point of a second amplifier, rather than to ground. With resistance values set as noted, this allows resistor $R_p$ to be variable for tuning. Most importantly, the gain compensation allows tuning without changing the gain, so the output amplitude is stable.

In this example, the $D_1$-$D_2$ diode-pair provides amplitude limiting, to stabilize the output. The use of a high-performance dual-FET op amp, such as the AD712, allows the tuning resistance $R_p$ to be as high as 10 megohms. For simplicity and stability, the matched resistances should be contained in a single network. The circuit as shown here is tunable from 100 Hz to 6 kHz.

### 10.1.2 FET-Stabilized Wien Oscillators

Another popular method of achieving amplitude control is by using the variable channel resistance of a field-effect transistor. By sampling and rectifying the output voltage, a dc signal proportional to an output amplitude can be made to control the resistance of an FET. This technique works well within limits, but the channel resistance of the FET itself is nonlinear at high signal levels, and appreciable distortion can occur.

Fig. 10-4 is a Wien-bridge oscillator rearranged to effectively utilize the FET channel resistance as a gain control. In this circuit, $R_1$-$R_2$ and $C_1$-$C_2$ again comprise the Wien network, but with the impedance of the legs scaled so that $R_1$-$C_1$ see a voltage lower than $R_2$-$C_2$. This allows the voltage level at $R_1$-$C_1$ to be of lower amplitude. The circuit uses two amplifiers in the inverting mode; gain control is accomplished by FET $Q_1$, which is placed at the summing point of $A_2$ (a point of minimum voltage). $Q_1$ can then operate at a level low enough to make its distortion negligible. $A_2$ provides a nominal gain of 2, which is trimmed by the agc and $R_5$ so that $Q_1$ is operating in the most linear portion of its characteristic curve.

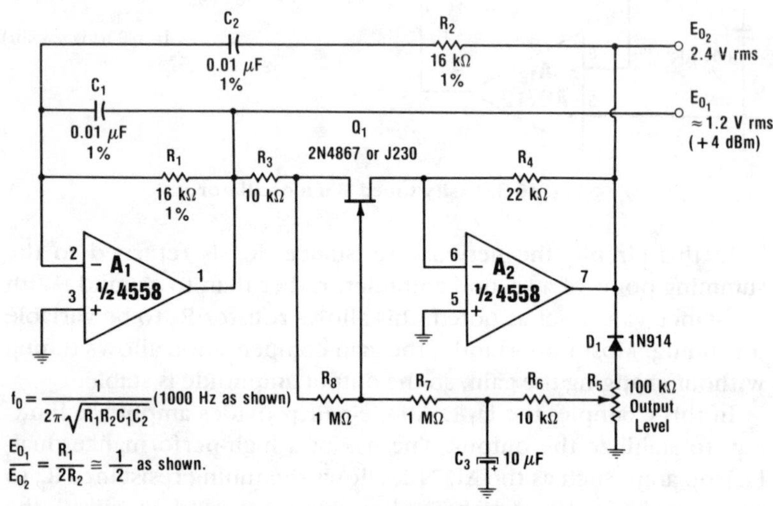


**Fig. 10-4. Wien-bridge oscillator with FET stabilization.**

The circuit has dual outputs available with two amplifiers; level differences are determined by the Wien network ($R_1$-$R_2$). $E_{o_1}$ will have less residual distortion than $E_{o_2}$ because it operates at a lower level and is filtered by the Wien network prior to being controlled by $Q_1$. The high output level of $A_2$ is sampled by $R_5$, which derives the bias to control $Q_1$. This resistor can be used as an output-level control to adjust output level linearly. Resistor $R_4$ can be trimmed for low distortion, which is necessary due to differences among individual FETs, if distortion is to be at the lowest level.

The amplifiers used should follow the guidelines set down previously. For general-purpose use, dual devices are obviously attractive. It should be noted that the equation for operating frequency is the general form, due to the difference in $R_1$-$R_2$ and $C_1$-$C_2$. The ratio used in the bridge may be altered from that shown, if desired, but it should be appreciated that as $R_2$ increases above $R_1$, stage $A_2$ must supply additional gain, thus making its distortion contribution more significant, particularly at high frequencies. The ratio shown provides a good compromise between low-level operation of $Q_1$ and minimal gain for the $A_2$ stage. When properly trimmed, the oscillator has less than 0.05% distortion at both outputs with $A_2$ levels up to ±10 V p-p. When the oscillator is operated over a wide range of frequencies, some adjustment of $C_3$ may be necessary at low frequencies. Dual FET op amps, such as the TL072, 411, or 3240, can be quite effective in this circuit. The general-purpose 4558 device shown is useful up to about 20 kHz (the values shown yield 1 kHz).

The FET-stabilized oscillator of Fig. 10-5 also uses a dual op amp for its implementation, but has only a single output available. In this case, the second amplifier is used in a more sophisticated agc system, employing an active integrator in a comparison loop.

In this circuit, $A_1$ is a Wien-bridge oscillator somewhat similar to that of Fig. 10-1 except for the agc method used. Here, $Q_1$ is the agc element, an n-channel JFET with less than 100 ohms of "on" resistance and a relatively low cutoff voltage. $Q_1$ forms a variable resistance in series with $R_3$, with a value such that the value of $R_4 + R_5$ is twice that of $R_3$ + the resistance of $Q_1$. This provides the necessary gain of 3, which is automatically stabilized by the agc loop.

The agc loop consists of $D_1$ to rectify the output, forming a dc voltage on $R_8$, which is compared (through $R_9$) to the −15-V supply as a reference. The output level is set by the ratio of $R_9$ to $R_8$; in this case, it results in a ±10-V (7-V rms) output. Other levels may be realized by adjusting $R_9$.

$A_2$ forms an integrator to smooth the ripple on the rectified dc and develop an appropriate dc control level for the gate of $Q_1$. Resistors $R_6$ and $R_7$ provide local ac feedback for $Q_1$, for lowest distortion.

This circuit is capable of very low distortion, on the order of 0.01% when optimized. $R_5$ can be trimmed for lowest distortion, if this is desired. This will likely be necessary to accommodate different FETs, due to their production spread. This adjustment is an

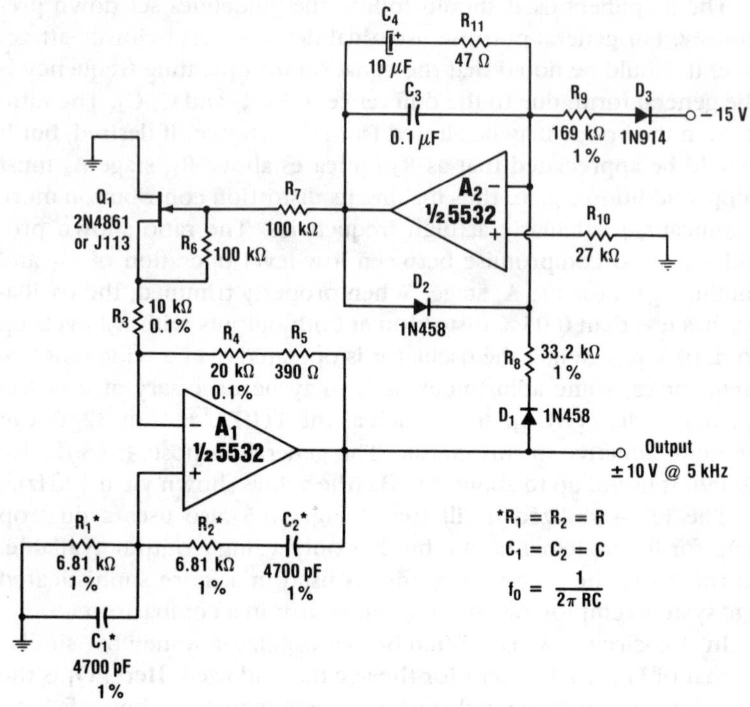


**Fig. 10-5. Low-distortion, FET-stabilized, Wien-bridge oscillator.**

optional "fine tweak," not necessary for the basic function, but to optimize performance in terms of distortion.

As shown, the frequency of the circuit is 5 kHz, as set by $R_1$-$C_1$ and $R_2$-$C_2$, which should, of course, be close-tolerance units. The 5532 op amp will yield very low distortion at frequencies up to about 50 kHz. Other suitable op amps are the low-noise duals, such as the 833 and the OP-227. If the operating frequency is to be widely separated from that shown, the values of $C_3$ and $C_4$ may require proportional adjustment.

### 10.1.3 Quadrature (Sine/Cosine) Oscillator

In electronic systems, a need often arises for two sine waves in quadrature, or with 90° of phase difference between them. This function is fulfilled by the oscillator of Fig. 10-6, which generates sine and cosine waveforms. The circuit is essentially two integrators in cascade, with positive feedback. Because the phase shift of an integrator is 90°, outputs $E_{o_1}$ and $E_{o_2}$ differ in phase by 90°. Thus, with the loop oscillating, these outputs deliver sine and cosine waves.

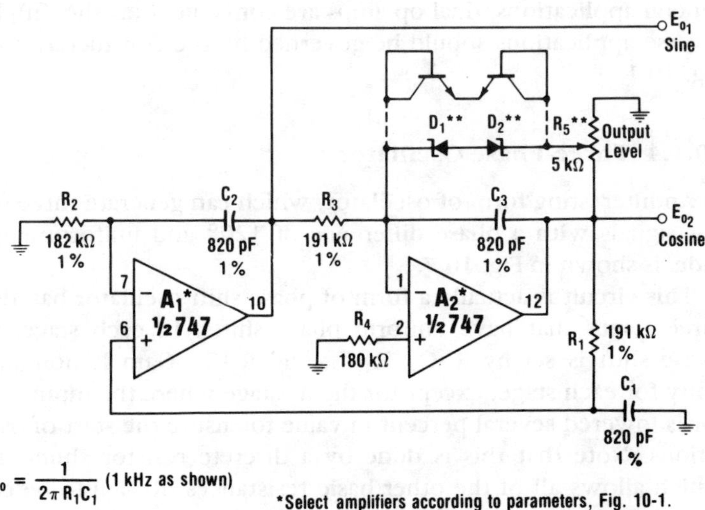


**Fig. 10-6. Quadrature (sine/cosine) oscillator.**

In theory, all resistances and capacitances are equal in value, but, in practice, one resistance (typically $R_2$) is slightly lowered in value to ensure starting. The remaining timing components assume close tolerance values. With $R_2$ slightly lowered in value, the oscillations tend to grow in amplitude; if unchecked, they will stabilize at the saturation limits of $A_2$. To prevent this and to exercise predictable control on output amplitude, a limiting network, consisting of $D_1$, $D_2$, and $R_5$, is used around $A_2$. With $D_1$-$D_2$ directly across $A_2$, the outputs will stabilize at $\pm[V_z + V_f]$. $R_5$ is used to set the output at any level above the zener limits of $D_1$-$D_2$. Monolithic zeners are recommended in this circuit for good stability of clipping and best symmetry.

The distortion of the output waveform is reasonably small, in the vicinity of 1%. The sine output terminal is lower in distortion than the cosine output because additional filtering of the sine signal occurs after limiting by $D_1$-$D_2$.

This circuit is useful where fixed-frequency sine waves in quadrature are needed. Because of the number of precision components in the loop, it does not adapt well to tuned applications. For general applications, dual op amps are convenient (as shown), but diverse applications should be governed by the considerations of Fig. 10-1.

### 10.1.4 Three-Phase Oscillator

An interesting form of oscillator, which can generate three output signals with a phase difference of 120° and uniform amplitude, is shown in Fig. 10-7.

This circuit is actually a form of phase-shift oscillator based on three stages that have uniform phase shifts. In each stage, the phase shift is set by $R_1$-$C_1$, $R_2$-$C_2$, and $R_3$-$C_3$. Gain is nominally unity for each stage, except for the $A_3$ stage where the input resistor is lowered several percent in value to ensure the start of oscillations. Note that this is done by a discrete resistor shunt ($R_7$), which allows all of the other basic resistances ($R_1$–$R_6$) to be contained in a single array, if desired.

Oscillations are stabilized by the symmetrical zener clamp $Q_1$-$Q_2$, with the amplitude set by $R_8$ and $R_9$. The output(s) are stabilized at $\pm 10$ V, and the frequency is set at 400 Hz. For such a relatively low frequency, a 348 or other similar op amp performs well.

A point that should be noted about this oscillator is that the limiting which takes place on output 1 unavoidably involves distortion. However, outputs 2 and 3 constitute low-pass filters for this distortion and are, therefore, cleaner. For example, 0.5% distortion at output 1 will be reduced to below 0.1% at output 2. Even more impressive results in terms of distortion are possible if an active agc scheme (such as that of Fig. 10-5) is employed, using the fourth op amp of the quad IC.

### 10.1.5 Oscillators Based on the State Variable Filter (SVF)

Some interesting, useful, and very-high-performance oscillators can be realized by using a *state variable filter* as the frequency-determining element of a sine-wave oscillator. Two examples of this technique are shown in Figs. 10-8 and 10-9.

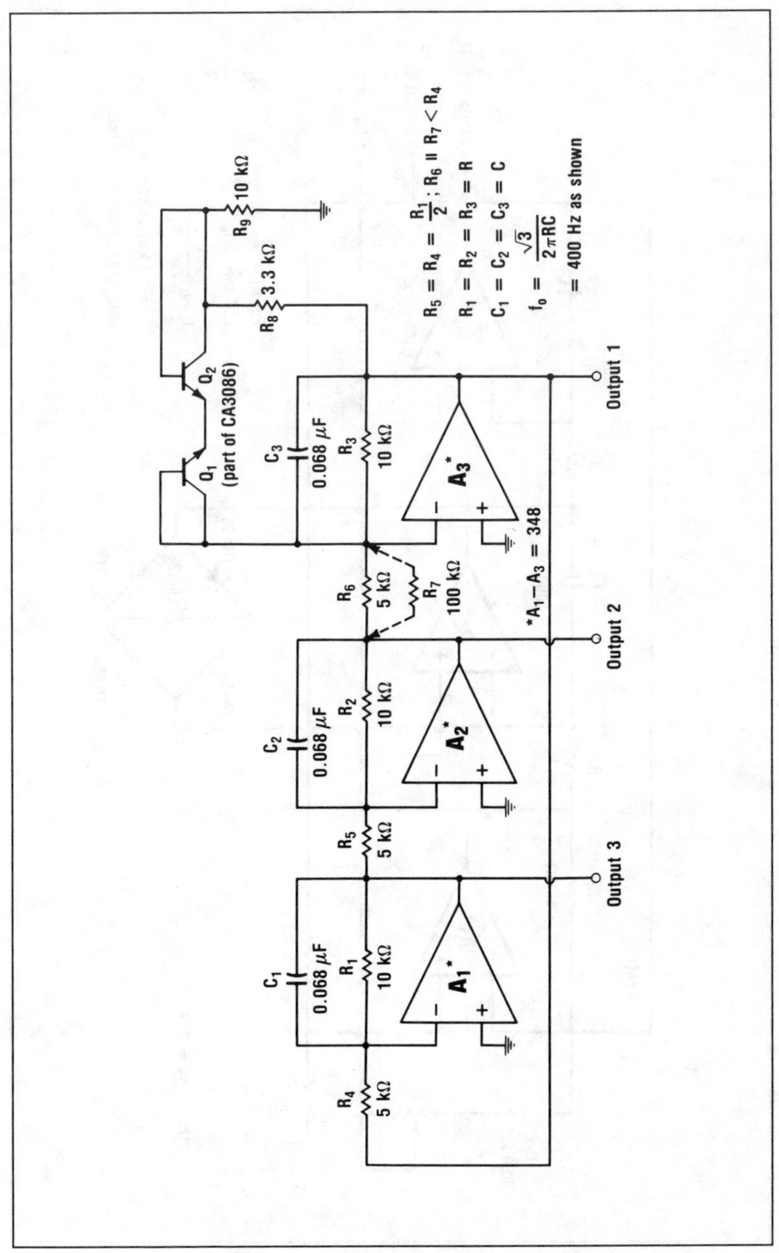


**Fig. 10-7. Three-phase oscillator.**

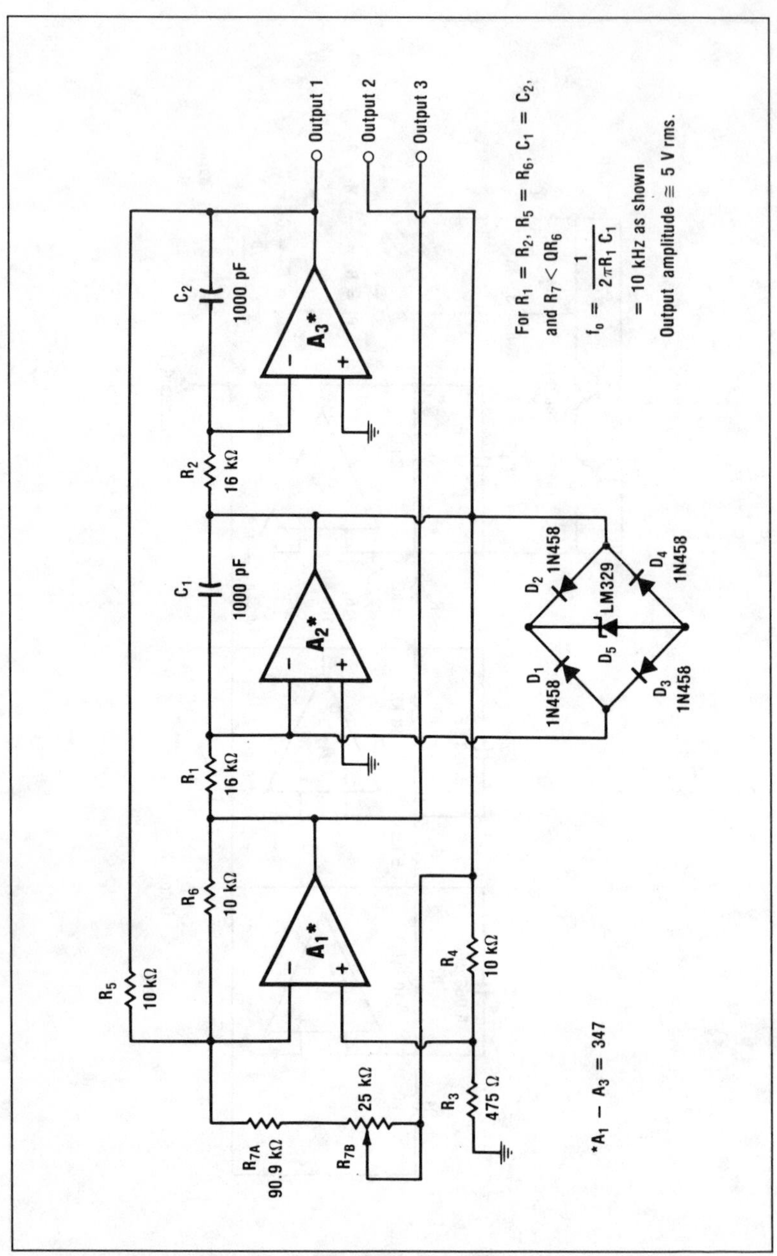


Fig. 10-8. Zener-stabilized oscillator using a state variable filter.

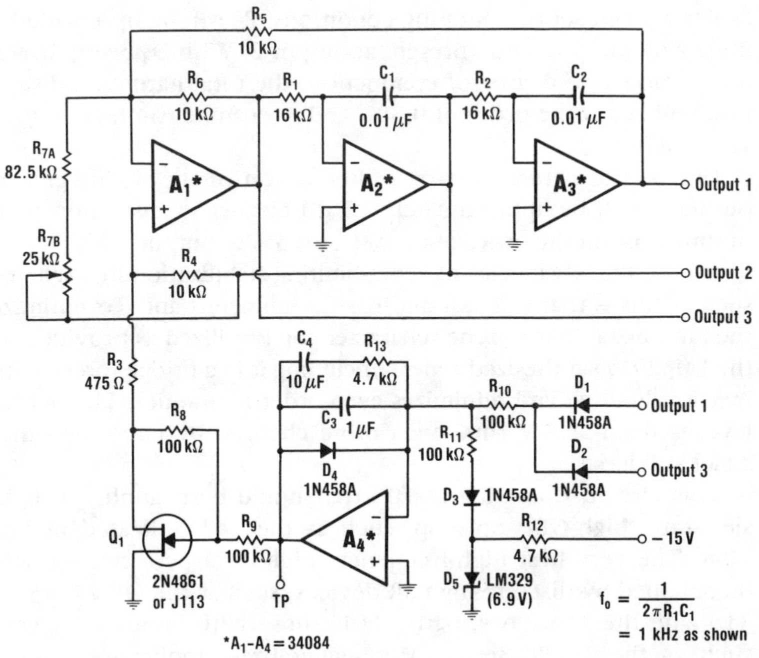


**Fig. 10-9. FET-stabilized oscillator using a state variable filter.**

In Fig. 10-8, a state variable oscillator is shown which is amplitude stabilized by the zener diode clamping one of the integrators, $A_2$. This has the effect of rigidly defining the peak amplitude swing of this stage and, in turn, defines the output amplitude of all outputs. At resonance, all three outputs are of equal amplitude, but differ in phase.

Aside from the zener-amplitude limiter, the other salient difference of this state variable filter is resistor $R_7$. This resistor provides the necessary positive feedback to initiate and sustain oscillation. The selection of $R_7$ is somewhat critical if maximum performance is to be realized. If it is too large, insufficient positive feedback is attained and oscillation will not start. If it is too small, the excessive positive feedback drives $D_5$ into hard conduction and creates excessive distortion.

Trimmer $R_{7B}$ is selected to provide a gain of just in excess of the Q, as defined by $(R_3 + R_4)/2R_3$. Here, the Q is 11, which indicates that $R_7$ should be 10 kΩ × 11 = 110 kΩ. To guarantee self-starting, $R_{7A}$ is purposely made several percent lower, or 90.9 kΩ.

For a given set of operating conditions, $R_7$ can be optimized by observing the distortion present at output 2. With $D_5$ being driven into a moderate degree of conduction, the total harmonic distortion will be on the order of 0.3%, and $R_{7B}$ can be trimmed to provide this.

Due to the natural low-pass-filter action of the $A_3$ integrator, output 1 will be approximately 10 dB cleaner in distortion than output 2, or on the order of 0.1% (for 0.3% at output 2).

One of the advantages of zener-limiting stabilization in a scheme such as this is that it is fast and has no time constant. To optimize the limiting, a "hard" zener characteristic is utilized as provided by the LM329 (a synthesized zener). Including it in a bridge forces symmetrical limiting and minimizes even-order harmonics. The output level is fixed at 5-V rms, but can be changed by choosing other zener voltages.

This circuit, operating at 10 kHz, should best employ a high-slew-rate, high-GBP op amp, such as the 347 indicated in Fig. 10-8. The very best high-frequency, high-level performance will be obtained by discrete high-SR devices, such as the OP-27, 5534, 318, and the 356 types, particularly those with improved speed, such as the 34080 series. More-generalized applications could employ general-purpose quad devices, such as the 348, 4136, 4156, 4741, etc. All resistors should be stable film types having 1% tolerances.

This same basic form of state variable oscillator can be modified by applying active agc to achieve much lower levels of distortion. This is shown in Fig. 10-9.

In this circuit, gain-determining resistor $R_3$ is adjusted by FET $Q_1$ which, in turn, is driven by an integrator/rectifier consisting of op amp $A_4$ and diodes $D_1$, $D_2$. The nonlinearity of the FET is minimized by local feedback, and the "infinite" dc gain of the integrator renders FET $V_{GS}$ variations negligible. The rectifier senses outputs 1 and 3, which are 180° out of phase, providing full-wave rectification and easier filtering. Resistor $R_{11}$ and reference diode $D_5$ provide a 70-$\mu$A reference current (set by $R_{11}$) which determines the stabilized output level. Resistor $R_{11}$ can, of course, be adjusted for other output levels, if desired.

Distortion in this oscillator is extremely low, due to the linear method of stabilization. Output 2 distortion will be 0.005% or less when operating properly, and output 1 distortion will be 0.002% or less. Settling time for frequency changes, or at switch-on, is 5 seconds or less, and is determined by the feedback network of the integrator, $C_3$ and $C_4$-$R_{13}$. These components may require adjustment for frequencies largely separated from 1 kHz (which is shown).

Distortion is, of course, lowest at output 1, but the extremely low levels quoted should be preserved by the use of a buffer amplifier, such as one of those described in Chapter 7. The circuit will operate with either FET as specified, but distortion can be optimized to match the exact characteristics of the FET by trimming $R_7$. As $R_7$ is increased above 82 kΩ, distortion lowers toward 0.002% (output 2). The dc bias at the test point should be −1 V to −2 V, an indication of linear operation.

All resistors in this circuit (with the exception of $R_8$, $R_9$, $R_{12}$, and $R_{13}$) should be stable, close-tolerance, metalized-film units. The 34084 op amp specified is useful for frequencies up to 20 kHz or more. Also, discrete high-performance units could be used for $A_1$–$A_3$.

## 10.2 MULTIVIBRATORS

### 10.2.1 Astable Multivibrator

Another op-amp oscillator circuit that must be termed as a classic is the astable oscillator circuit of Fig. 10-10A. This circuit, with minor modification, can be used with virtually any op amp and is useful over a very wide range of frequencies, limited only by the devices used. The output square wave can either be taken as the full saturated swing of the device ($E_{o_1}$) or it can be clamped with diodes for defined limits ($E_{o_2}$). The operating frequency is determined by $R_t$-$C_t$ and the $R_1$-$R_2$ positive feedback network, and is relatively independent of supply voltages and output swing, particularly if output clamping diodes are used.

Operation is best understood by examining the time relationships in Fig. 10-10B. The circuit uses both positive and negative feedback, with regeneration supplied by the $R_1$-$R_2$ network and integration by the $R_t$-$C_t$ network. The op amp compares the voltage output from these two points and changes states as their relative states change. (Operation as described here is applicable to the use of the clamping diodes, but is also generally true if they are not used.)

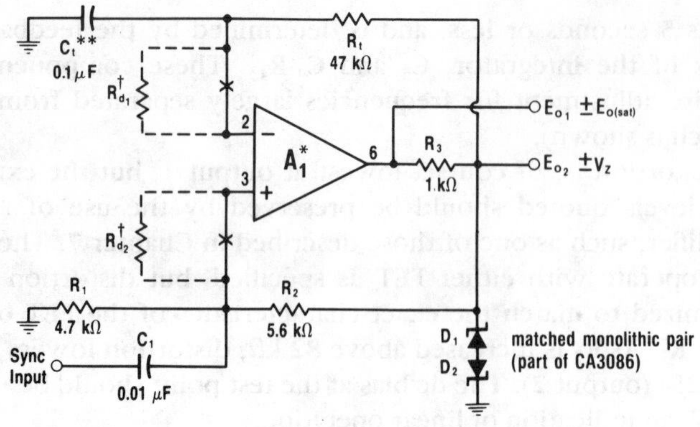


T = oscillation period

$$T = 2R_tC_t \log_e\left(\frac{1+\beta}{1-\beta}\right),$$

where $\beta = \dfrac{R_1}{R_1+R_2}$.

If $\beta = 0.462$, $T = 2R_tC_t$,

$$f_o = \frac{1}{T} = \frac{1}{2R_tC_t} \text{ (100 Hz as shown)}$$

*Any op amp with required differential input rating; ***uncompensated*** unit preferred.

Suggestions:
- general-purpose—301A, TL070
- vlf to medium hf—357
- hf—709, 725, 3100
- single supply—TLC271 series, 3130, etc.

**$C_1$ should be returned to V+ for polarized capacitors.

† Use $R_{d_1}$–$R_{d_2}$ for differential input protection if necessary, or use modified values for $R_1$–$R_2$ (see text).

*(A) Circuit diagram.*

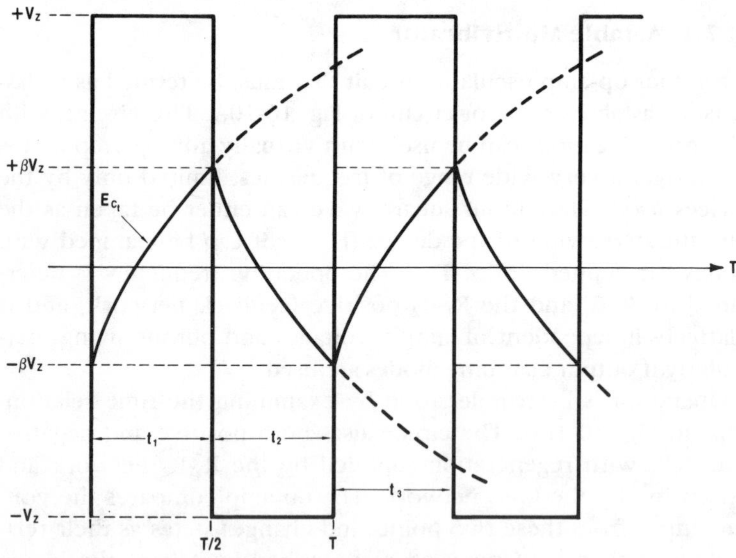


*(B) Voltage/time relationships.*

**Fig. 10-10. Astable multivibrator.**

Assume the $R_1$-$R_2$ junction is initially at $+\beta V_Z$, which means the output ($E_{o_2}$) is at $+V_Z$. The voltage across $C_t$ ($E_{C_t}$) was previously at $-\beta V_Z$, but it is now charging toward $+V_Z$ through $R_t$ ($t_1$). As $E_{C_t}$ rises and reaches the potential of $+\beta V_Z$, the threshold potential of the amplifier is reached. Since $E_{C_t}$ is applied to the (−) input, this positive rise causes the output to go negative. The negative-going output is immediately transmitted through $R_2$-$R_1$, and this positive feedback path causes the output to snap rapidly from positive to negative saturation. With $E_{o_2}$ at $-V_Z$, $E_{C_t}$ now begins to change toward this potential ($t_2$). The $R_1$-$R_2$ junction is now at $-\beta V_Z$, and as $E_{C_t}$ decreases and crosses this potential, the circuit flips positive by regeneration, beginning a second cycle of oscillation ($t_3$); $t_1$ and $t_2$ comprise one period of the oscillation cycle, which is of duration T. Since $\beta$ is a fixed fraction of $V_Z$ (or $E_{o(sat)}$ if clamping is not used) and $E_{C_t}$ is timed against $V_Z$ and compared to $\beta V_Z$, $V_Z$ is eliminated from the timing expression for symmetrical values of $V_Z$ (or $E_{o(sat)}$). If $t_1$ and $t_2$ are the half-periods, the full period T is $t_1 + t_2$, which can be shown to be

$$T = 2R_tC_t\log_e\left(\frac{1+\beta}{1-\beta}\right)$$

This is the *general* expression for the period of the oscillator, from which it is apparent that the period (or frequency) can be controlled by either $R_t$, $C_t$, or $\beta$($R_1$ and $R_2$).

It can also be seen that if the natural log factor on the right of this expression is made equal to unity, the log term will disappear, which considerably simplifies time calculations. If $\beta$ is made equal to 0.462, this term becomes very close to unity, which allows the use of the simplified expression. In practice, $R_1$ and $R_2$ are chosen for a $\beta$ of 0.462, then $R_t$ and $C_t$ are chosen from the simplified equation. It should be remembered that for low offset, $R_t$ and $R_1 \parallel R_2$ should be chosen for minimum input offset error. It is, in fact, the low input offset of the op amp that lends the circuit its innate timing precision, for as long as $V_{io}$ is much less than $\beta V_Z$, the timing error due to amplifier offset will be negligible. With FET-input amplifiers, there is (generally) no requirement for matched source resistances, due to their much lower bias currents.

There are a few practical problems associated with this circuit that may or may not be significant, depending on the exact amplifier chosen. Since the function of the amplifier is essentially that of a comparator, performance is governed by many of the same considerations (see Chapter 8). The amplifier should *not* be frequency compensated, as compensation is not necessary for stability and it reduces high-frequency capability. The inputs to the amplifier can be subjected to large differential voltages; therefore, protection should be considered for unprotected types, and internally clamped types, such as the 308 and 318, will need a series resistance ($R_d$) to prevent excessive differential current flow—the resistance used should be much greater than either $R_t$ or $R_1 \parallel R_2$ for minimum degradation of timing characteristics.

At low frequencies, or when $C_t$ becomes large indicating use of a polarized type, two problems can occur. First, the capacitor must be polarized *properly*—this is satisfied by returning it to a supply voltage rather than ground, V+ usually being preferred. Also, $C_t$ represents a potentially destructive element on switchoff because it looks directly into the amplifier input. For capacitance values greater than 0.1 μF, some series resistance to the (−) input is desired, usually 1 kΩ or greater, to provide safety for the amplifier (if not internal to the amplifier).

Most of the preceding problems can be dealt with by the selection of the proper device or by the judicious use of protection. For instance, either the 301A or 748 would be a good choice for general-purpose use at frequencies below 10 kHz. The 357 can be operated over an extremely wide range because of its low input current (which allows low frequencies) and wide bandwidth (which allows high frequencies). Generally, the op-amp types that are optimum for comparator use will work best in this circuit.

At the higher frequencies, stray capacitance should be minimized and resistances should be kept low. Also, transistor junction zeners work best for $D_1$-$D_2$ and are recommended for the cleanest waveforms. Op amps that can be clamped at an internal high-impedance node (such as the 301A, 2525, or 2625) allow clamping for good accuracy without excessive power dissipation.

Note that if zener clamping is used to limit $E_{o_2}$ to approximately ±7 V, and if $\beta$ is set at less than 0.5, the differential input to $A_1$ will be less than the differential breakdown potential. This allows use of unprotected types without clamping, such as the 709, etc., or the npn-input comparator ICs.

If desired, this astable multivibrator can be easily synchronized by injecting pulses at the (+) input through $C_1$. The circuit can be synchronized with either positive or negative pulses.

The circuits of Fig. 10-11 illustrate this same basic circuit, as set up to use the general-purpose 311-type comparator. The version of Fig. 10-11A uses ±15-V supplies, while the one of Fig. 10-11B uses a single ±5-V supply. They both have the same frequency expression and, as shown, they operate at 2 and 100 kHz, respectively,

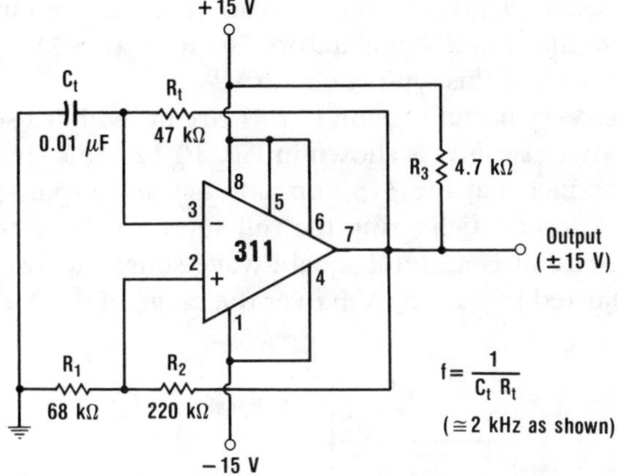


*(A) 311-type device, dual supply.*

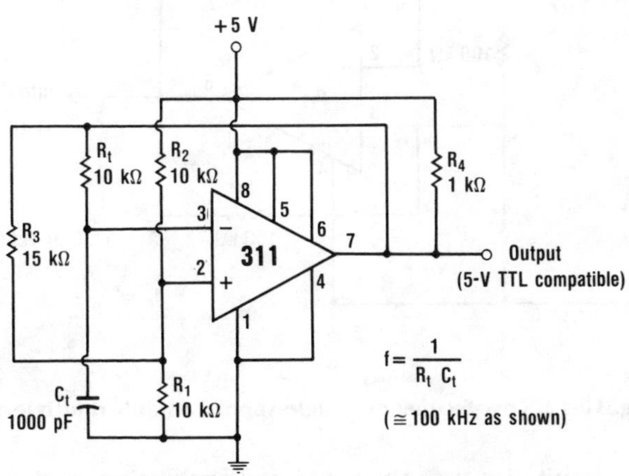


*(B) 311-type device, single (+5 V) supply.*

**Fig. 10-11. Comparator-based astable multivibrator.**

It should be noted that the frequency expression for these circuits is further simplified, as shown. This comes about from changing the $R_2/R_1$ relative resistances to a ratio of 3/1, which results in a $\beta$ of 0.25. This not only results in the simplified expression, but, importantly, it also reduces the maximum input differential to the amplifier to less than the zener breakdown level, even if $R_2$ and $R_t$ are returned directly to the output (no clamping used). Thus, any op amp can be used. Even unprotected high-speed npn-input comparators, such as the 710, 306, 360, etc., work well in this type of circuit.

Another very useful version of this circuit, which also uses only a single power supply, is shown in Fig. 10-12. This circuit capitalizes on the fact that the 3130 op amp has an output stage which swings rail-to-rail, delivering the full supply voltage to the load. Thus, the circuit is a useful square-wave source whose amplitude can be adjusted by varying V+ over the range of +5 V to +15 V.

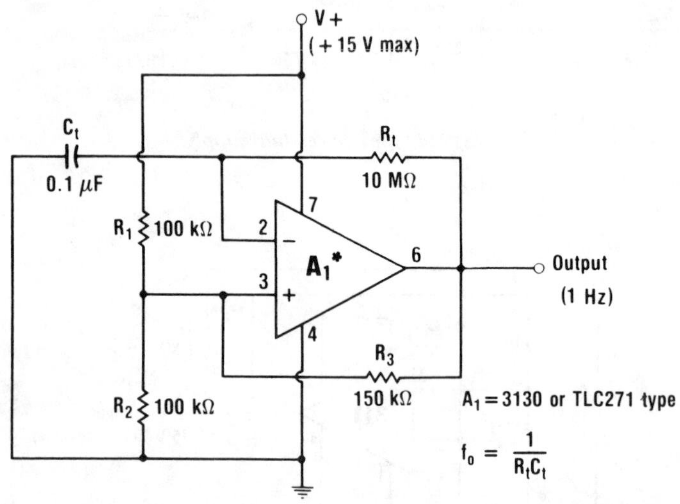


**Fig. 10-12. Low-frequency, single-supply, astable multivibrator.**

This circuit also uses the simplified expression, and is arranged for oscillation at a 1-Hz rate. Since the 3130 has very low input currents, $R_t$ can range up to hundreds of megohms, if desired. $C_t$ should be a low-leakage unit, such as polypropylene.

Other single-supply op amps can also be used in this circuit, but all will not swing rail-to-rail and will operate over a more narrow range of $R_t$. Examples are such units as the 358, 324, 3403, etc.

For medium-to-high supply voltages, almost any op amp can be used in this circuit; they just will not have as high an output swing. It can also be optimized towards micropower use, using such devices as the TLC271 or TLC251 in their low-power mode.

### 10.2.2 Monostable (One-Shot) Multivibrator

A modification of Fig. 10-10A yields a circuit that is stable in only one output state (monostable); this circuit is shown in Fig. 10-13.

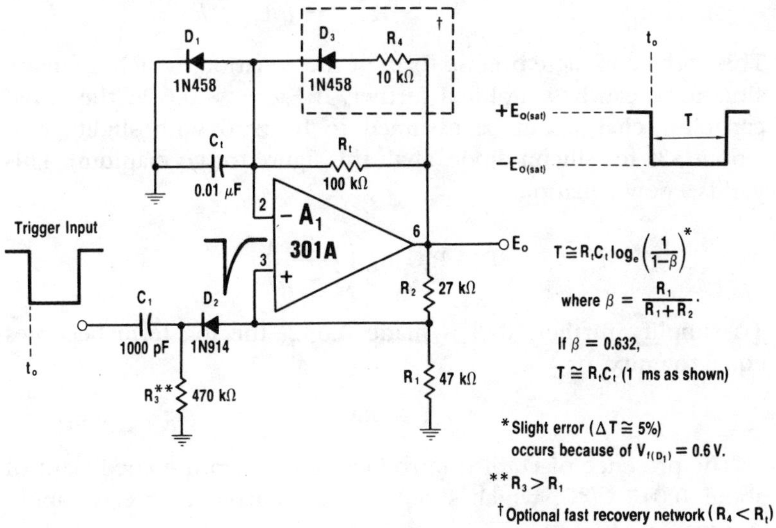


**Fig. 10-13. Monostable multivibrator.**

The addition of $D_1$ across $C_t$ clamps the (−) input of the op amp at +0.6 V, which forces the amplifier to assume a defined output state. Since the positive feedback from $R_1$-$R_2$ can rise above this level, the output latches to the "high" state, or $+E_{o(sat)}$. Triggering is accomplished through $D_2$, which passes the differentiated negative input pulse at $C_1$-$R_3$ to the (+) input. A negative input pulse that pulls the (+) input below +0.6 V causes the output to flip to the negative output state, $-E_{o(sat)}$. This action is regenerative, and the output stays in the negative state because the (−) input cannot immediately follow due to $R_t$-$C_t$. $C_t$ then starts charging toward $-E_{o(sat)}$ through $R_t$.

The voltage at the (+) input at this time is $-\beta E_{o(sat)}$. As the voltage across $C_t$ reaches this voltage, the amplifier flips positive once again, due to regeneration. The circuit does not completely recover, however, until $C_t$ charges back up to +0.6 V, where $D_1$ clamps it in a static state. Because of this relatively long recharge cycle, the recovery time is poor, but it can be improved if desired by adding $R_4$-$D_3$, which provides a shorter time constant for recovery. The time of the output pulse width is determined by the initial capacitor voltage (+0.6 V) and by $-E_{o(sat)}$:

$$T = R_t C_t \log_e\left(\frac{-E_{o(sat)} - (V_{f(D_1)})}{-E_{o(sat)} - (\beta E_{o(sat)})}\right)$$

This is the complete precise form of the equation, which, in many situations, can be simplified further. If $E_{o(sat)} \gg 0.6$ V, the initial capacitor charge can be assumed to be zero with slight error (about 5% for silicon diodes, half this figure for germanium). This yields a new equation,

$$T \cong R_t C_t \log_e\left(\frac{1}{1 - \beta}\right)$$

To simplify further: if $\beta$ is made 0.632, the log term becomes equal to unity, or

$$T \cong R_t C_t$$

The presence of $D_1$ also introduces a temperature coefficient of about 0.017%/°C, which should be considered in precise applications.

The triggering network of $C_1$-$R_3$ may need some alteration to suit different triggering requirements. $R_3$ should be greater than $R_1$ so that timing is not upset. Note that $\beta$ determines the trigger sensitivity since it sets the voltage at the (+) input.

In the example shown, $R_t$ and $C_t$ are chosen for a 1-ms pulse width, and $\beta$ is set at 0.632 by $R_1$-$R_2$. As with the astable multivibrator on which this circuit is based, the timing period is independent of supplies, and zener clamping may be used at the output. Speed is not a prime feature of this circuit, so an uncompensated op amp or comparator device should be used to optimize this factor. The 301A and 748 are good for pulse widths down to approximately 100 μs; below this, one of the faster types mentioned previously, or a comparator such as the 311, should be used. Input protection is a consideration in this circuit as in the astable circuit. The circuit can also be rearranged to deliver positive-going output pulses, if desired, by reversing $D_1$ and $D_2$ (and $D_3$, if used). Note that $D_1$ and $D_3$ (if used) should be low-leakage types if high $R_t$ values are used. If an FET-input device such as the 357 or the high-speed 356 variant is used, very long pulse times are possible. This circuit can also be arranged for single-supply use.

Whether this configuration will be practical for a given application may be debatable. This is because of the presence of so many high-performance one-shot circuits using dedicated timer ICs, such as the 555 (see the author's *IC Timer Cookbook,* 2nd ed.).

### 10.2.3 Bistable Multivibrator

The third general form of multivibrator, the bistable, is shown in Fig. 10-14. This circuit is essentially a comparator with ac input coupling through $C_1$-$R_3$ (see Chapter 8). Input sensitivity can be adjusted by altering $\beta$, since $\pm\beta E_{o(sat)}$ is the voltage required for triggering at the (−) input.

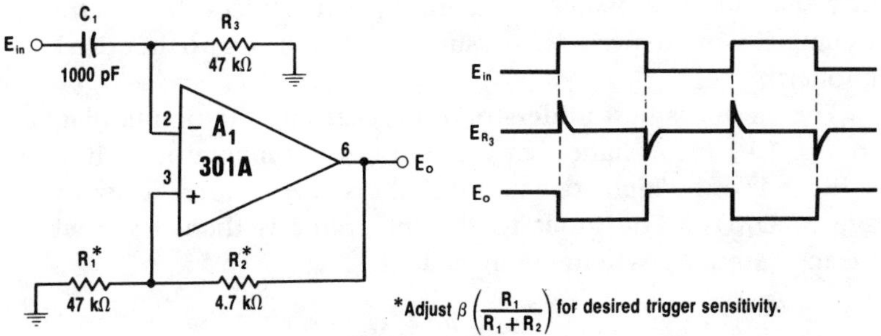


**Fig. 10-14. Bistable multivibrator.**

Although a dual-supply circuit using a 301A is shown in Fig. 10-14, a single-supply version can be arranged simply by applying a V+/2 voltage to the bottom of $R_3$ and making $R_1$ a pair of 100-kΩ resistors–one connected to ground, the other connected to V+. $A_1$ should then be a single-supply device, such as the 358, 324, 3140, etc.

## 10.3 FUNCTION GENERATORS

As will soon be noted, many of the following circuits use voltage comparators in combination with integrators to generate time-dependent functions. To a large degree, many of these circuits will be limited primarily by the device used as a comparator. The response time and slew rate define the upper limit of usable speed; therefore, nearly all the devices recommended will be uncompensated types to take advantage of their greater inherent speed. An optimum selection for a given application must be based on the performance of the op amp as a comparator. It is recommended that the performance of various devices as comparators (Chapter 8) be reviewed to keep the pertinent considerations in mind when making the choice between an op amp and a true comparator device (such as a 311, or other unit).

### 10.3.1 Basic Circuit

One of the most useful op-amp oscillator circuits is the triangle/square-wave function generator of Fig. 10-15A. This circuit, comprised of an "ideal" integrator ($A_2$) and a comparator with hysteresis ($A_1$), simultaneously generates ultralinear triangle waveforms and symmetrical square waves. Operation over an extremely wide range of frequencies is possible using appropriate devices, from ultralow frequencies with periods measured in hours, up to hundreds of kilohertz.

The circuit is best understood by regarding the timing diagram in Fig. 10-15C. Assume that the output of comparator $A_1$ has just switched to the high state and that $E_{o_2}$ is at $+V_Z$ (breakdown voltage of $D_1$-$D_2$). The input to the integrator is then $+V_Z$, which means output $E_{o_1}$ will integrate at a rate

$$\frac{E_{in}}{RC} = \frac{V_Z}{R_tC_t} \text{ V/s}$$

$A_1$ compares the sum of the voltages $E_{o_1}$ and $E_{o_2}$ against the ground reference on its (−) input. As $E_{o_1}$ ramps negative, the junction of $R_1$-$R_2$ decreases toward zero ($t_1$). When $E_{o_1}$ crosses $-V_Z$ (or when $E_{o_1}$ is equal and opposite to $E_{o_2}$), the voltage at the (+) input of $A_1$ crosses zero and $A_1$ changes state rapidly, aided by positive feedback. $E_{o_2}$ is now at $-V_Z$, and $A_2$ begins to integrate positive toward $+V_Z$ ($t_2$). When $E_{o_1}$ crosses $+V_Z$, $A_1$ changes back to the high state and the cycle repeats.

The output voltage of $A_1$ is clamped at $\pm V_Z$, which determines the output amplitudes. The positive and negative ramp slopes are determined by $\pm V_Z$, $R_t$, and $C_t$, and are equal if the zener voltages are symmetrical (usually the case).

The unclamped direct output from $A_1$ ($E_{o_3}$) is available for external use. This output can be used to feed $A_2$ (in lieu of the clamping network), but it will introduce side effects. First, the frequency will vary as the saturation voltages or supply voltages of $A_1$ change. Second, if $A_1$ is used to feed $A_2$ directly, $R_1$ must always be greater than $R_2$ to set $E_{o_1} < E_{o_3}$. $A_2$ cannot be allowed to saturate, which would happen if $R_2 = R_1$ (or $E_{o_1} = E_{o_3}$). This modification may be useful for noncritical applications, however, as it does eliminate $R_3$ and $D_1$-$D_2$.

Since the circuit is actually a linear ramp generator made to oscillate between two voltage limits, an expression for the operating period (or frequency) must include both slopes of the ramp and the voltage limits. The pertinent timing data is shown in Fig. 10-15C, where $t_1$ and $t_2$ are the half-periods and T is the period of a single cycle.

The time for the first half-cycle is

$$t_1 = \frac{E_o t}{\Delta E_o}$$

Since

$$\frac{\Delta E_o}{t} = \frac{E_{in}}{RC}$$

then

$$t_1 = \frac{E_o RC}{E_{in}}$$

or

$$t_1 = \frac{E_{o_1} R_t C_t}{V_Z}$$

where

$E_{o_1}$ is the desired peak-to-peak amplitude from $A_2$.

Similarly, for the second half-cycle,

$$t_2 = \frac{E_{o_1} R_t C_t}{V_Z}$$

Since the zener voltages (hence the times) are symmetrical, the period T is then

$$T = \frac{2E_{o_1}R_tC_t}{V_Z}$$

This equation can be used in designing the circuit for specific times or specific output voltages. The output voltage ($E_{o_1}$) is scaled to $V_Z$ by a factor of 2:1; therefore, the relationship used to select $R_1$ and $R_2$ is

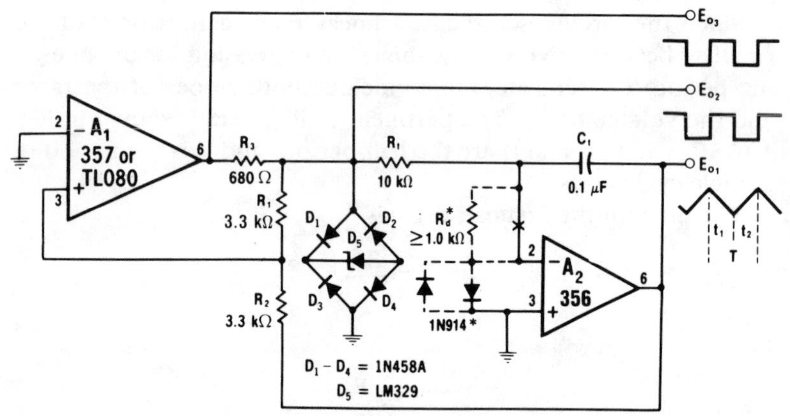


$E_{o_1}$ = p-p output from $A_2$

$= \pm V_Z\left(\frac{R_2}{R_1}\right)$

$E_{o_2} = \pm V_Z$

$E_{o_3} = \pm E_{o(sat)}$

$t_1 = t_2$ and $T = t_1 + t_2 = \frac{2E_{o_1}R_tC_t}{V_Z}$ (4 ms as shown)

* Input overvoltage protection, for $A_2$ if required.

*(A) Circuit diagram.*

**Fig. 10-15.**

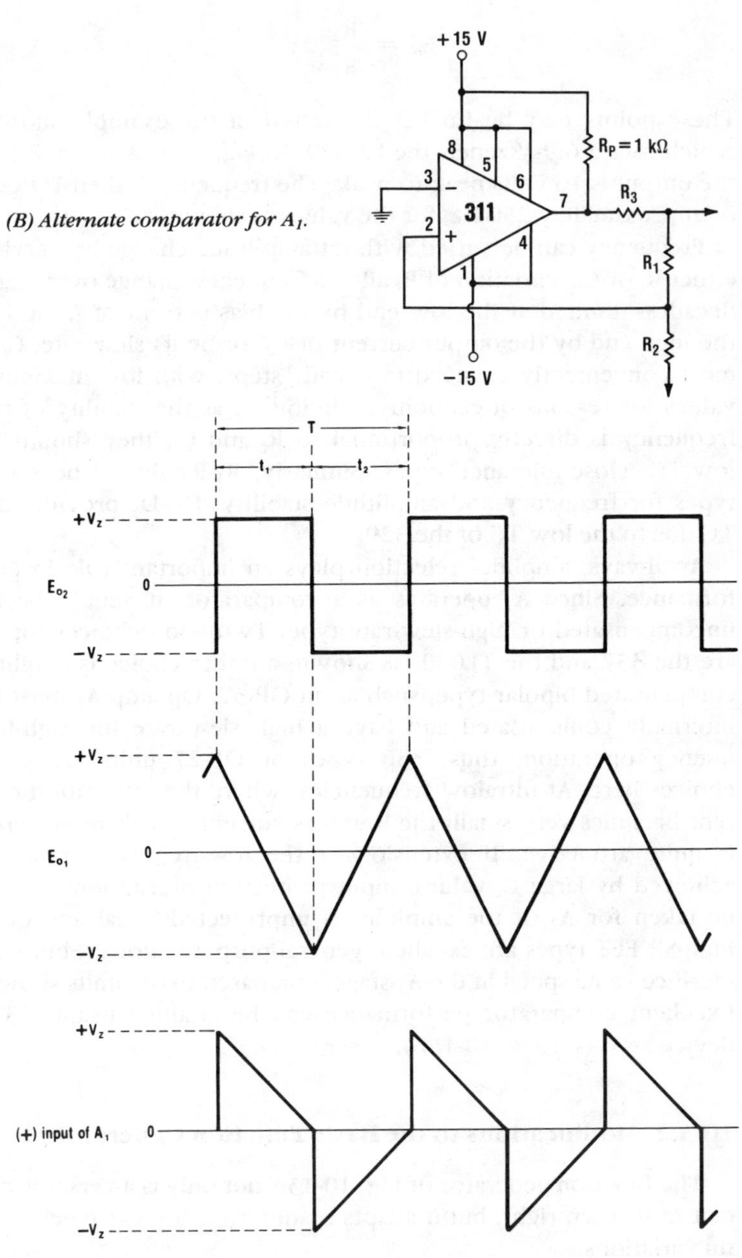


*(B) Alternate comparator for $A_1$.*

*(C) Output characteristics.*

**Function generator.**

$$E_{o_1} = \frac{R_2}{R_1} 2V_Z$$

These points may be further illustrated in the example shown, which uses a 6.9-V zener, the LM329. In Fig. 10-15A, $R_1 = R_2$, so the output is 16 volts peak-to-peak. The frequency is then $1/4R_tC_t$, or approximately 250 Hz, for the values as shown.

Frequency can be varied without amplitude change by varying either $R_t$ or $C_t$. Variation of $R_t$ allows frequency change over many decades—limited at the low end by the bias current of $A_2$ and at the high end by the output current of $A_2$, or by its slew rate. $C_t$ is most conveniently changed in decade steps, with low maximum values for reasons of economy. Obviously, as the stability of the frequency is directly proportional to $R_t$ and $C_t$, they should be low-TC, close-tolerance types. Similarly, $R_1$-$R_2$ should be stable types for frequency and amplitude stability. $D_1$–$D_5$ provide low TC due to the low TC of the 329.

As always, amplifier selection plays an important role in performance. Since $A_1$ operates as a comparator, it should be an uncompensated or high-slew-rate type. Two good choices for $A_1$ are the 357 and the TL080, as shown; another choice is a lightly compensated bipolar type, such as an OP-37. Op amp $A_2$ must be internally compensated and have a high slew rate for high-frequency operation; thus, 356 types or OP-27 units are good choices here. At ultralow frequencies, where the integration current becomes very small, the low bias current of FET-input types is quite attractive. If extension of the low-frequency range is achieved by large $C_t$ values, input protection precautions should be taken for $A_2$, if the amplifier is unprotected. Dual and quad high-SR FET types are excellent general-purpose choices, but may sacrifice some speed in the $A_1$ stage, compared to the units shown. Excellent comparator performance can be attained using a 311 device as $A_1$ (see Fig. 10-15B).

### 10.3.2 Modifications to the Basic Function Generator

The function generator of Fig. 10-15A not only is a versatile circuit in its own right, but it adapts readily to many extremely useful variations.

As discussed previously, the frequency of the function generator is controllable by varying either $R_t$ or $C_t$. Variation of $R_t$ is the simplest means of frequency change, but there is an alternate method that is attractive for other reasons. This circuit (Fig. 10-16) introduces a divider network ($\beta$) composed of $R_5$ and $R_6$, and placed between the clamping diodes and $R_t$. $R_5$ and $R_6$ vary the fraction of $\pm V_Z$ that is fed to $R_t$; thus, the input to the integrator in this circuit is $\pm \beta V_Z$. If the term $\beta V_Z$ is substituted in the expression for $t_1$ and $t_2$, the new timing equation for T is

$$t_1 = t_2 = \frac{E_{o_1} R_t C_t}{\beta V_Z}$$

and

$$T = t_1 + t_2 = \frac{2E_{o_1} R_t C_t}{\beta V_Z}$$

The use of $R_5$-$R_6$ represents a reduction in input current to $A_2$, the fraction being equal to $\beta$. Thus, similar to large values of $R_t$, it places a limitation on $A_2$ (in terms of input current) for very low frequencies. Dynamic range of frequency is optimized by selecting $R_t$ such that, when $\beta$ is minimum, the input current is still much greater than the bias current of $A_2$. Frequency variations of 1000:1 are relatively easy to achieve, and ratios of 10,000:1 are possible.

The reason behind the $\beta$ voltage divider may already be obvious—the control of gain from unity to some fraction of the input can next be provided by an analog multiplier, as discussed in Chapter 6. Thus, if $\beta$ ($R_5$ plus $R_6$) is replaced by a two-quadrant multiplier, the frequency may be made electronically variable, with a dynamic range proportional to the range of the multiplier.

In the example shown, the dynamic range is 3 decades or more; $R_t$ and $C_t$ are chosen for an upper frequency limit of 20 kHz, and the oscillator can be adjusted from 20 Hz to 20 kHz using $R_5$. FET-input devices, such as the OPA606 or the AD711 (or other high-SR, low-offset devices), are recommended for best high-frequency operation and low input current. This will yield the best overall frequency capability and widest dynamic range.

A point that should be appreciated is that the offset voltage of $A_2$ *must* be low for a wide dynamic range of operation. This can be accomplished by specifying a low offset device, or by careful offset trimming. As an alternative, a composite ultralow offset-voltage connection can be employed for the $A_2$ function (see Chapter 7). High performance in the comparator is achieved with a 311 device.

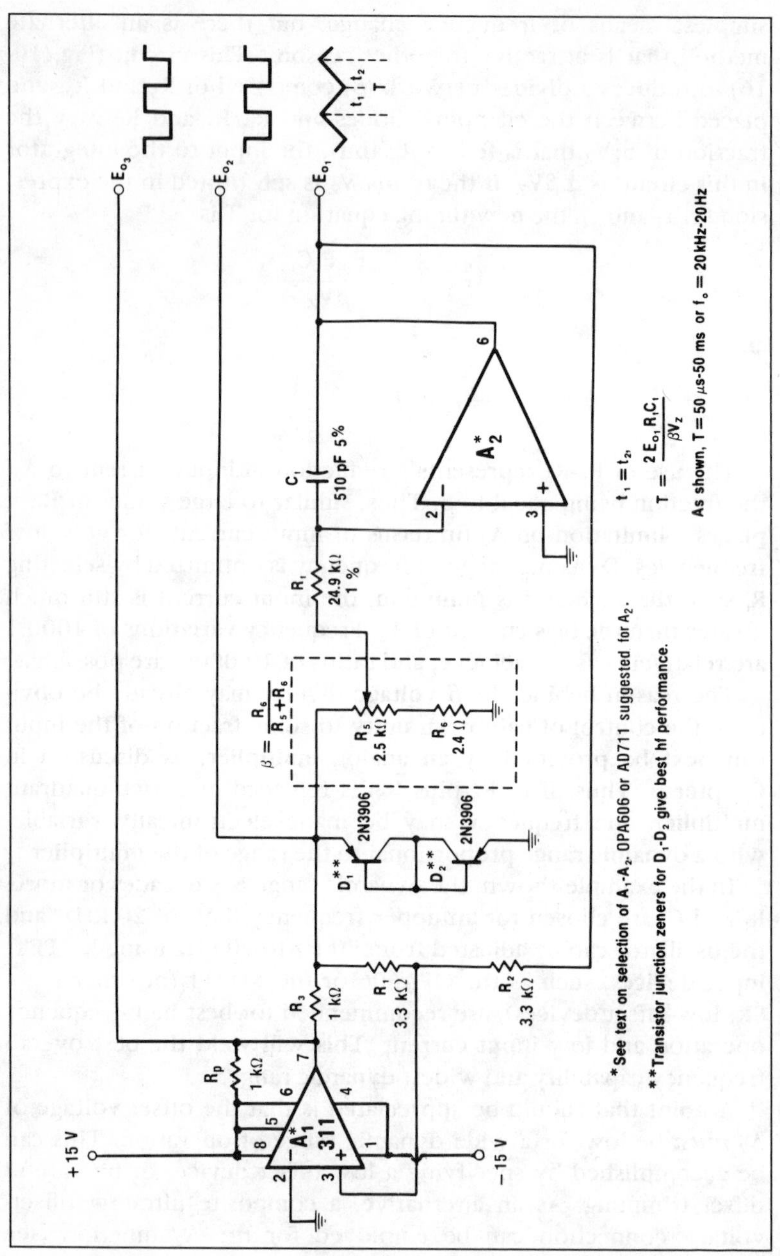


**Fig. 10-16. Variable-frequency function generator (20 Hz to 20 kHz).**

### 10.3.3 Function Generator With Independently Variable Slopes

Another modification to the basic function generator allows independent control of positive and negative slopes, both of which have a dynamic range of control limited only by the integrating op amp. With individual ramp control, the basic triangular output wave can be modified into a sawtooth waveform, either positive-going or negative-going.

The circuit shown in Fig. 10-17 uses a diode bridge at the summing point of integrator $A_2$, which acts as a current-mode gate under control of $A_1$. If the output of $A_1$ is high, $D_1$ and $D_4$ are switched on, causing $R_{t_1}$ to inject current into the summing point from V+ through $D_1$. When the output of $A_1$ is low, $D_2$ and $D_3$ switch on and $R_{t_2}$ conducts current from V− to the summing point of $A_2$, thus controlling positive ramp slopes. If these currents are termed $I_{i_1}$ and $I_{i_2}$, respectively, new expressions for the output slopes of $A_2$ can be written. If the negative slope has duration $t_1$ (Fig. 10-17) and the positive slope has duration $t_2$, the negative slope, then, is

$$\frac{\Delta E_o}{\Delta t} = \frac{I_{i_1}}{C_t}$$

or

$$t_1 = \frac{E_{o_1} C_t}{I_{i_1}}$$

where

$E_{o_1}$ is the desired peak-to-peak output swing from $A_2$.

For the positive slope,

$$\frac{\Delta E_o}{\Delta t} = \frac{I_{i_2}}{C_t}$$

or

$$t_2 = \frac{E_{o_1} C_t}{I_{i_2}}$$

and

$$T = t_1 + t_2$$

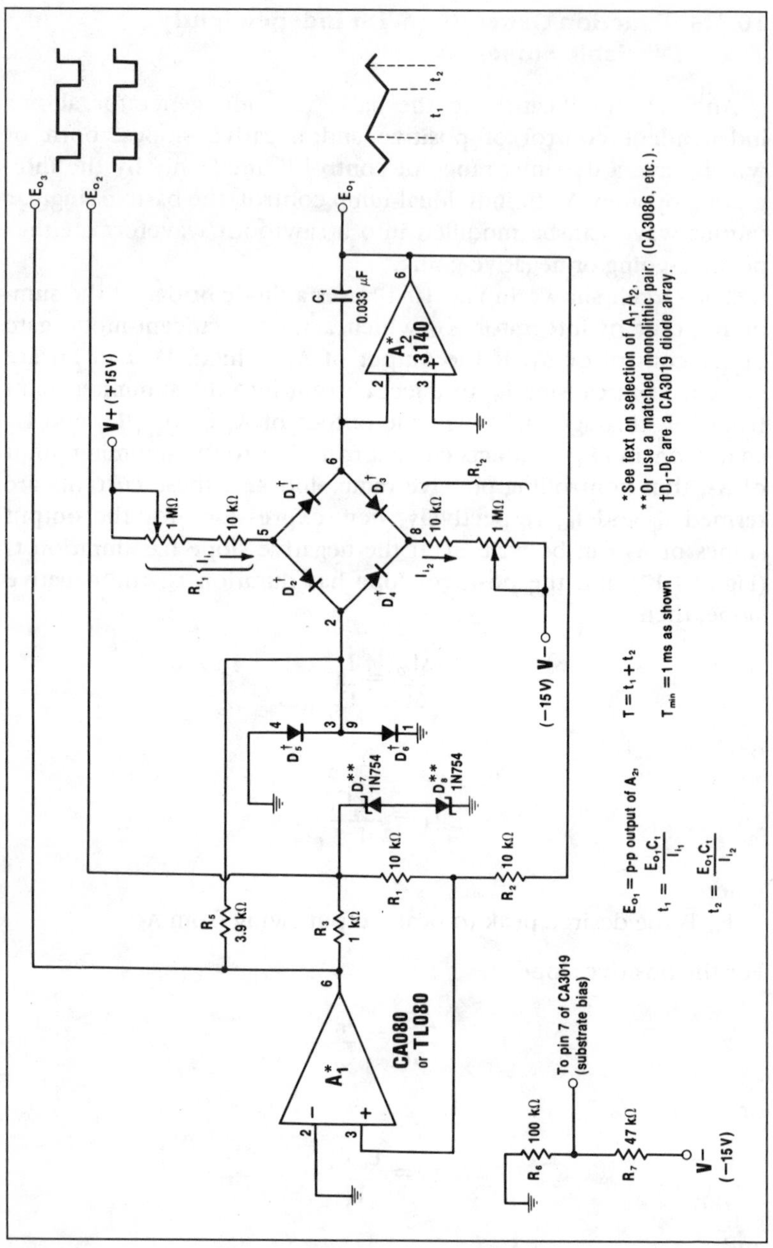


**Fig. 10-17. Function generator with independently variable slopes.**

In practice, the control of $I_{i_1}$ and $I_{i_2}$ to vary ramp slopes can involve any of several methods: potentiometers for $R_{t_1}$ and $R_{t_2}$, fixed resistors returned to variable voltages, or current sources. If $R_{t_1}$ and $R_{t_2}$ are potentiometers, the most stable method of control is a variable resistance to V+ or V− because diodes $D_1$ and $D_3$ introduce some temperature dependence into $I_{i_1}$ and $I_{i_2}$, and this variance is minimized when the voltage is high. Ideally, $I_{i_1}$ and $I_{i_2}$ are generated by constant-current sources such as those described in Chapter 4. If the current source is made voltage-controlled, it allows voltage control of either the wave period or, in the case of dual (complementary) current sources, the frequency at a constant duty cycle.

This circuit can also be used as a variable-period pulse generator with precise control of pulse width and frequency. A wide dynamic range of operation is available (many decades). $C_t$ can be switch-selected for decade ranges, and $R_{t_1}$ and $R_{t_2}$ can be made calibrated potentiometers.

The considerations in this circuit differ little from the previous one. $D_1$–$D_4$ are, ideally, a matched diode bridge as shown. This has the bonus of allowing the use of "extra" diodes ($D_5$-$D_6$) as clamps for the input from $A_1$. $R_5$ is chosen so that it can supply adequate current to absorb $I_{i_1}$ or $I_{i_2}$ at their highest levels. Any excess current provided by $R_5$ is absorbed in the clamps, $D_5$-$D_6$. Amplifier selection is governed by the previous restrictions—the devices shown are for medium-speed use. $R_{t_1}$, $R_{t_2}$, and $C_t$, in this circuit example, are chosen to yield a minimum total period of 1 ms. $R_{t_1}$ and $R_{t_2}$ allow the period to be lengthened to a total of 100 ms. This circuit can serve as a useful source of adjustable ramp, triangular, pulse, and square waveforms.

### 10.3.4 Triangle-Wave to Sine-Wave Converters

The formation of sine waves in a function generator is normally a conversion process, one that synthesizes approximate sine curves from a triangular input wave. Many sine-wave generators use biased diodes to "round" and shape the peaks of triangle waves. The circuits of Fig. 10-18 are much simpler in concept, yet produce excellent results. Either can be used with any triangle wave that has constant amplitude and a 50% duty cycle, such as those just described.

The circuit of Fig. 10-18A uses a differential amplifier ($A_1$) and a matched transistor pair ($Q_1$-$Q_2$) with a controlled amount of emitter coupling. As has been previously described in the section on multipliers and instrumentation amplifiers (Chapters 6 and 7), an undegenerated, emitter-coupled amplifier is linear only for small signal levels. However, if operation is held to a region of controlled nonlinearity, an emitter-coupled pair can convert a linear input voltage, such as a triangle wave, into a good approximation of a sine-wave output. This is what $Q_1$-$Q_2$ of Fig. 10-18 accomplish, and $A_1$ converts the sinusoidal output current of $Q_1$-$Q_2$ to a buffered sine-wave output voltage.

Since this scheme involves precise control over the conduction of $Q_1$ and $Q_2$, it is critically dependent upon both the input ampli-

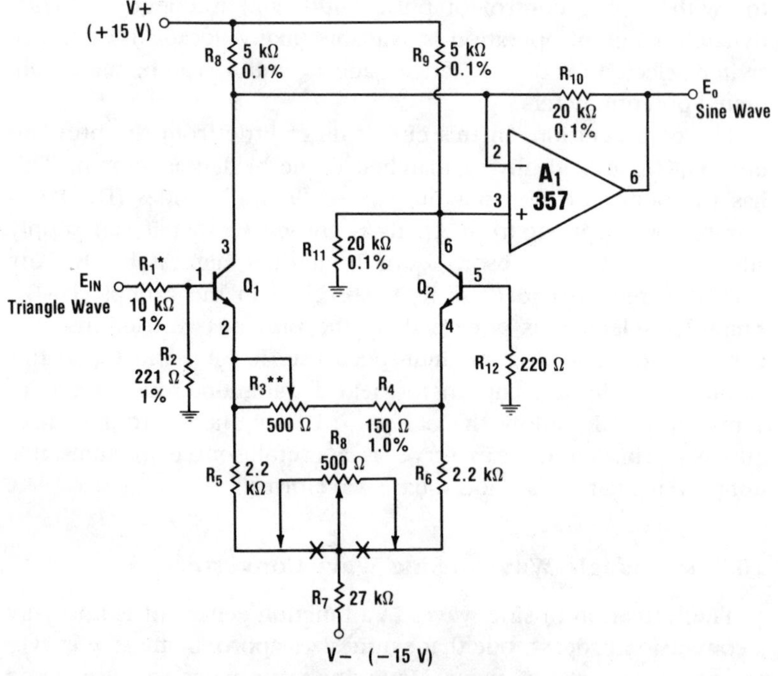


*As shown, the values for divider $R_1$-$R_2$ are selected for a triangle wave of $E_{in} \cong \pm 8$ V p-p. For other input amplitudes, select $R_1$-$R_2$ for 300-400 mV p-p at $Q_1$. $Q_1$-$Q_5$ are part of CA3096 array.

**Trim for minimum THD.

*(A) Differential pair.*

**Fig. 10-18. Triangle-wave**

tude and the degree of emitter coupling. $R_1$ and $R_2$ are chosen for approximately 300- to 400-mV p-p drive at $Q_1$. With levels in this range, $R_3$ can be set for a point of minimum output distortion. This is best done with a distortion analyzer, but an oscilloscope can also be used. Once $R_3$ is set, the input level must not change, or the distortion in the output will climb rapidly. For this reason,

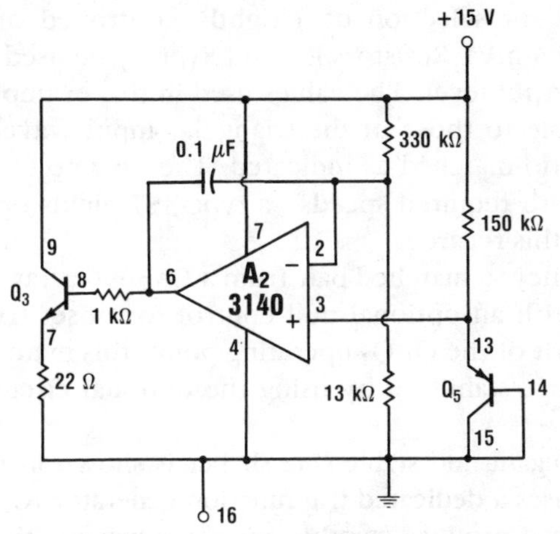


*(B) Temperature stabilization option.*

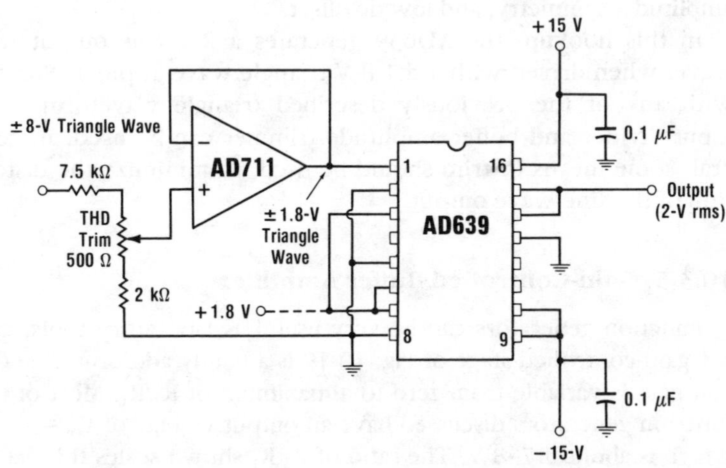


*(C) Precision sine shaper using a trig function generator.*

to sine-wave converters.

high-stability components are preferred for $R_1$–$R_4$; $R_3$ should preferably be a high-resolution trimmer, such as a 10-turn potentiometer.

Distortion is also affected to a smaller degree by the match of $Q_1$-$Q_2$. A matched transistor pair can be used without nulling and distortion will still be reasonably low ($\leq 1\%$). However, optimum performance (total harmonic distortion $\simeq$ 0.3%) is obtained by offset nulling or selection of a tightly controlled offset-voltage pair ($V_{io} < 0.5$ mV). Resistors $R_{10}$ and $R_{11}$ may be used to scale the sine-wave output level. The values used in this example yield levels comparable to those of the triangular input waveforms; they should be ratio-matched as indicated. The op amp ($A_1$) should be consistent with required speeds—a type-357 yields optimum performance in this regard.

For efficiency, a matched pair from a CA3096 array can be used for $Q_1$-$Q_2$, with an optional null control for offset trim ($R_8$). For minimum drift of the $Q_1$-$Q_2$ operating point, this matched pair can be temperature stabilized by using the optional circuit shown in Fig. 10-18B.

A more elegant and stable sine shaper is shown in Fig. 10-18C. This circuit uses a dedicated trig function-generator IC, the AD639. This device can produce (synthesize) a sine wave with a distortion level of less than 0.1%, given an input triangle wave of correct amplitude, symmetry, and low dc offset.

In this hookup, the AD639 generates a 2-V rms output sine wave, when driven with a ±1.8-V triangle wave at pin 1. For use with any of the previously described triangle waveforms, the input divider and buffer amplitude trimmer can be used. In general, some means of trim should be used, to minimize the distortion of the sine-wave output.

### 10.3.5 Gain-Controlled Buffer Amplifier

Function generators can be very useful as laboratory tools, and the gain-controlled stage of Fig. 10-19 is a handy addition. The circuit gain is variable from zero to a maximum of $R_2/R_1$. Most of the function generators discussed have an output voltage of $E_o = \pm V_Z$, which is about ±7–8 V. The ratio of $R_2/R_1$ shown scales this value continuously from zero to ±10 V or so, and allows linear calibration of the output. Minimum rise and fall times are governed by the amplifier—so, choose the highest-performance, unity-gain, stable type. For example, a type-318 or 2520, or any other high $f_t$ and high-SR device can be used. The output impedance remains low for any setting of $R_2$. For greater output current (up to ±50 mA), use an inside-the-loop unity-gain buffer (such as the BUF-03, LT1010, LH0033, or LH0002) between $A_7$ and $E_o$ (see Chapter 7).

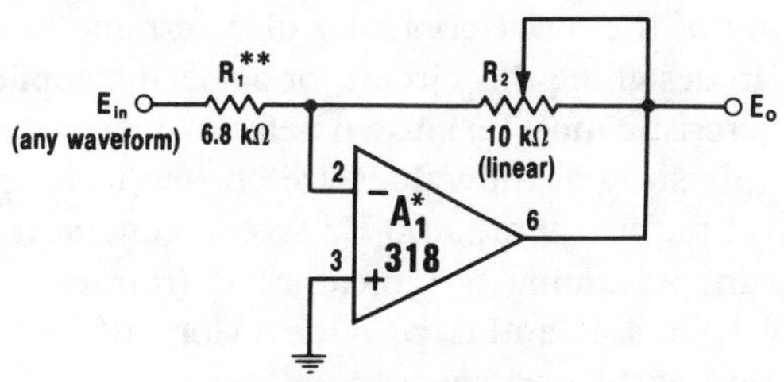


$E_o = \pm 10$ V maximum, reducible to zero by adjusting $R_2$.

*Select for required bandwidth and slew rate. $A_1$ must be stable at unity gain.

**Trim if necessary for exact full-scale output desired.

**Fig. 10-19. Buffer with feedback gain control.**

## 10.4 SAWTOOTH GENERATORS

One common circuit requirement is the sawtooth generator, which is frequently used to generate ramp voltages for timing purposes. The previous function generator with independent slope adjustments can be used to generate sawtooth waveforms, but circuit requirements often dictate a more specialized type of generator. The waveform itself can have either a linear or an exponential slope, depending on the application. Other variations are in the form of control: driven, triggered, free-running, or synchronized.

### 10.4.1 Basic Low-Frequency Sweep Generator

The sawtooth generator of Fig. 10-20 may be recognized as an integrator with clamping transistor $Q_1$ added—to reset the capacitor to zero at the end of each timing cycle. The circuit requires an external control voltage ($E_c$) to command the state of the ramp generator. If $E_c$ is high, the circuit is in the *hold* mode. In this state, $D_1$ is reverse biased and $Q_1$ is in saturation in parallel with $C_t$. This makes the output voltage essentially zero, which establishes the output dc base line. The circuit will remain in this state indefinitely as long as $E_c$ is high. When $E_c$ is taken low, $D_1$ pulls the gate of $Q_1$ low; this turns $Q_1$ off.

With $Q_1$ off, the circuit begins to integrate and the output voltage rises at the rate of $I_t/C_t$ volts per second. This rise continues until $E_c$ again goes high, turning on $Q_1$ and discharging $C_t$. From this, it may be seen that although $R_t$ and $C_t$ set the slope of the output waveform, the timing of the input control will determine the final endpoint voltage. In designing this circuit for a specific application, the $E_c$ timing characteristic must be known beforehand.

In the example shown, the values are chosen to be compatible with one typical requirement, a 60-Hz sweep generator for tv-raster vertical scan. Assuming a typical hold (retrace) interval of approximately 1.5 ms, $R_t$ and $C_t$ provide a slope of 10 V in 15 ms, the active portion of the vertical scan.

This sweep generator is best confined to low-frequency use as shown, generally below 1 kHz. For these applications, the 741 and OP-07 types are adequate. The circuit can be used at higher frequencies with a high-speed FET-input type, such as the 356. A subsequent circuit shows an alternate form of sweep that allows the use of a low-frequency device for timing. Modification of this circuit for negative-going output waveforms is accomplished by substituting a p-channel FET, such as a 2N5116, for $Q_1$ and returning $R_t$ to V+. Also, $D_1$ is reversed, as is the sense of the logic control. For the basic ramp circuit of Fig. 10-20, the controlling voltage, $E_c$, can easily be generated by the one-shot multivibrator shown in the inset. This circuit senses voltage $E_o$ from $A_1$ and produces a 1.5-ms positive pulse when $E_o$ reaches +10 V (adjustable by $R_3$). $A_1$ and $A_2$ can be dual 741s for slow speeds; dual FET types are preferable for higher speeds.

### 10.4.2 Fast Precision Sweep Generator

Due to slew-rate and bandwidth limitations, the sweep generator of Fig. 10-20 must use a high-speed op amp above about 1 kHz. The circuit of Fig. 10-21, however, produces fast sweeps of over 10 V/$\mu$s using a low-frequency op amp. $A_1$ and $Q_1$ form a precision current source, $I_t$ (Chapter 4), and $Q_2$ is a saturated switch that holds $C_t$ near zero volts when $E_c$ is high. When $E_c$ falls low, $Q_2$ switches off and $C_t$ begins to charge toward +15 V, forming a linear ramp with a slope of $I_t/C_t$ (V/s). This voltage rise across $C_t$ is buffered by $A_2$, a low-input-current, high-speed follower. The

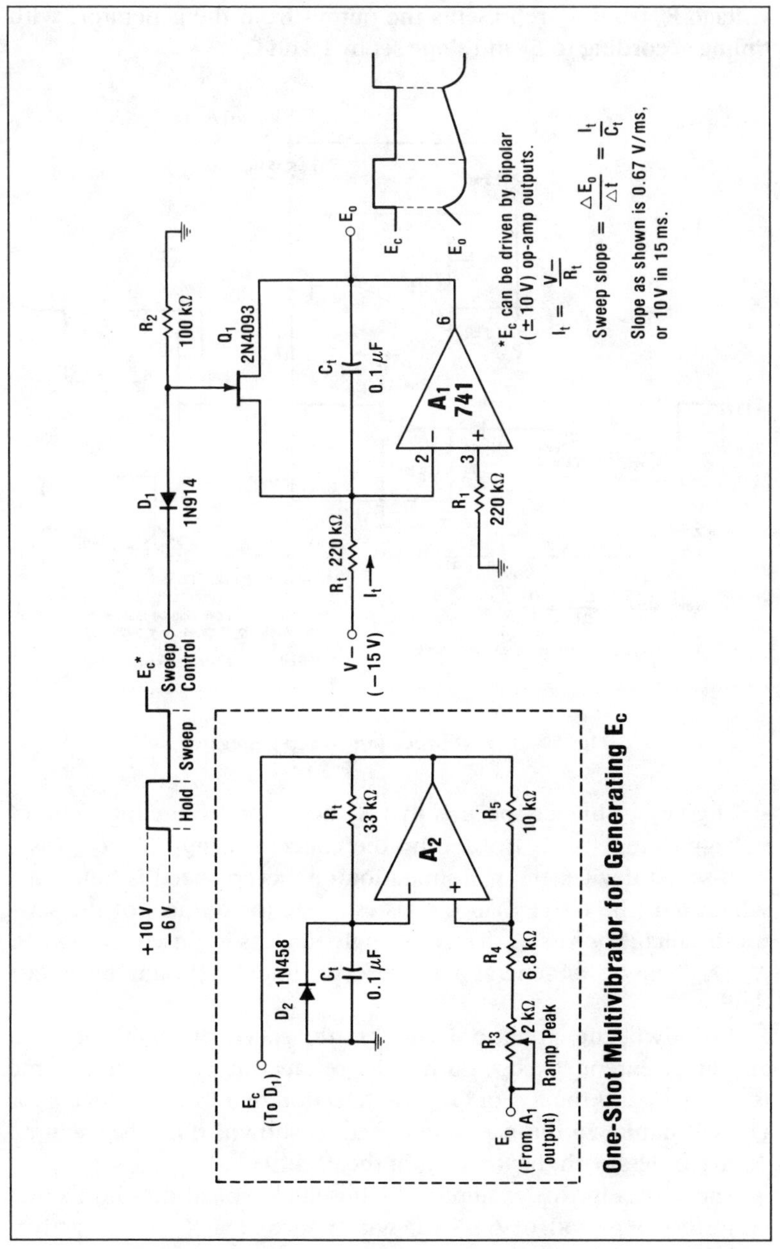


**Fig. 10-20. Low-frequency sweep generator.**

voltage $E_o$ from $A_2$ represents the output from the generator, with timing according to $E_c$ and slope set by $I_t$ and $C_t$.

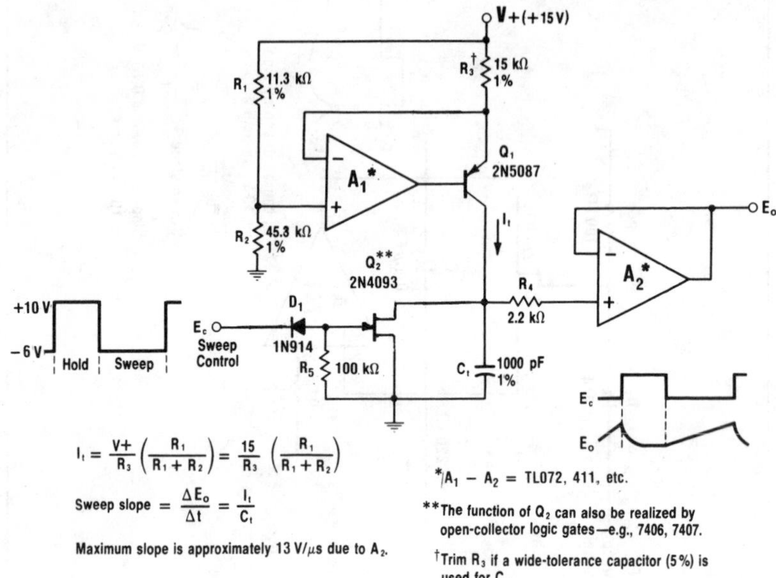


**Fig. 10-21. Fast precision sweep generator.**

The key to higher speed is that $A_1$ does not see the high rate of voltage change—it is isolated by the collector of $Q_1$. Since $Q_2$ is a high-speed device, the real limitation on sweep speed is buffer $A_2$, which must be faster than the fastest trace (or retrace) of the sawtooth voltage across $C_t$. Thus, if single devices are used for $A_1$ and $A_2$, $A_1$ can be a general-purpose type but $A_2$ should be a fast device.

As shown, the circuit does not tax the speed capability of $A_2$ to any great extent, except during the retrace interval. Retrace time is set by $C_t$ and the $R_{on}$ of $Q_2$. The selection of a low-$R_{on}$ device for $Q_2$ will minimize the base-line offset; as shown, the offset will be 16 mV or less with the 80-Ω $R_{on}$ of the 2N4093.

The values in this example are chosen as typical of a horizontal tv raster—a period of 63.5 μs, with approximately 50 μs active. Thus, $E_o$ has a 0.2-V/μs rate of rise, or 10 V in 50 μs. The indicated values are precise, but $R_3$ may be trimmed to set the output slope.

If independence of slope from the +15-V line is desired, $R_1$ may be changed to a 2.5-V zener reference, in which case $R_3$ should be 12.5 kΩ.

### 10.4.3 Long-Period Timers

The two previous circuits are variations of the same type—i.e., driven sawtooth generators. The next circuit type differs in that it responds to input triggers; after triggering, it generates a single sawtooth of prescribed duration, then returns to the standby state. It is similar to the monostable multivibrator, but, in this case, the timing ramp is buffered for external use. The circuit is also capable of extremely short recovery times, thus allowing immediate retriggering.

A dual-supply version of this timer is shown in Fig. 10-22. It consists basically of a transistor switch ($Q_1$), a timing network ($R_t$-$C_t$), a buffer ($A_1$), and a comparator ($A_2$). The state of switch $Q_1$ is controlled by the comparator, which receives trigger pulses from an external source.

In the standby state, output $E_{o_2}$ is high, clamped at $+V_Z$ by $D_1$-$D_2$. This voltage holds $Q_1$ in the "on" state via $D_5$ and $R_6$, and output $E_{o_1}$ remains near zero volts. In this condition, the feedback path through $R_1$-$R_2$ scales the $+V_Z$ voltage at $E_{o_2}$ to $+V_ZR_2(R_1 + R_2)$ at the (+) input of $A_2$; for this example, the input voltage of $A_2$ is $+V_Z/2$. Thus, a trigger pulse of $+(V_Z/2 + V_f)$ volts at $D_3$ (or $D_4$) is required to overcome the bias on $A_2$, where $V_f$ is the diode voltage drop. This positive pulse switches $A_2$ to the low state, which disables clamp $Q_1$. $C_t$ then begins timing toward +15 V with a time constant of $R_tC_t$. The voltage at $E_{o_1}$ required to reset $A_2$ will be $-(-V_ZR_2/R_1)$; in this case, simply $+V_Z$. When $E_{o_1}$ reaches $+V_Z$, $A_2$ switches to the high state, shorting $C_t$ and resetting the circuit to its standby state. The action is extremely fast due to the small time constant formed by $R_5C_t$ and the fact that there is some regeneration for the reset mode (whereas none exists for the trigger mode). The circuit will thus respond almost instantly to a second trigger pulse after reset.

The timing equation for this circuit is dependent upon V+, zener voltage $V_Z$, and $R_tC_t$ according to

$$t = R_tC_t\log_e\left(\frac{V+}{(V+) - (R_2/R_1)V_Z}\right)$$

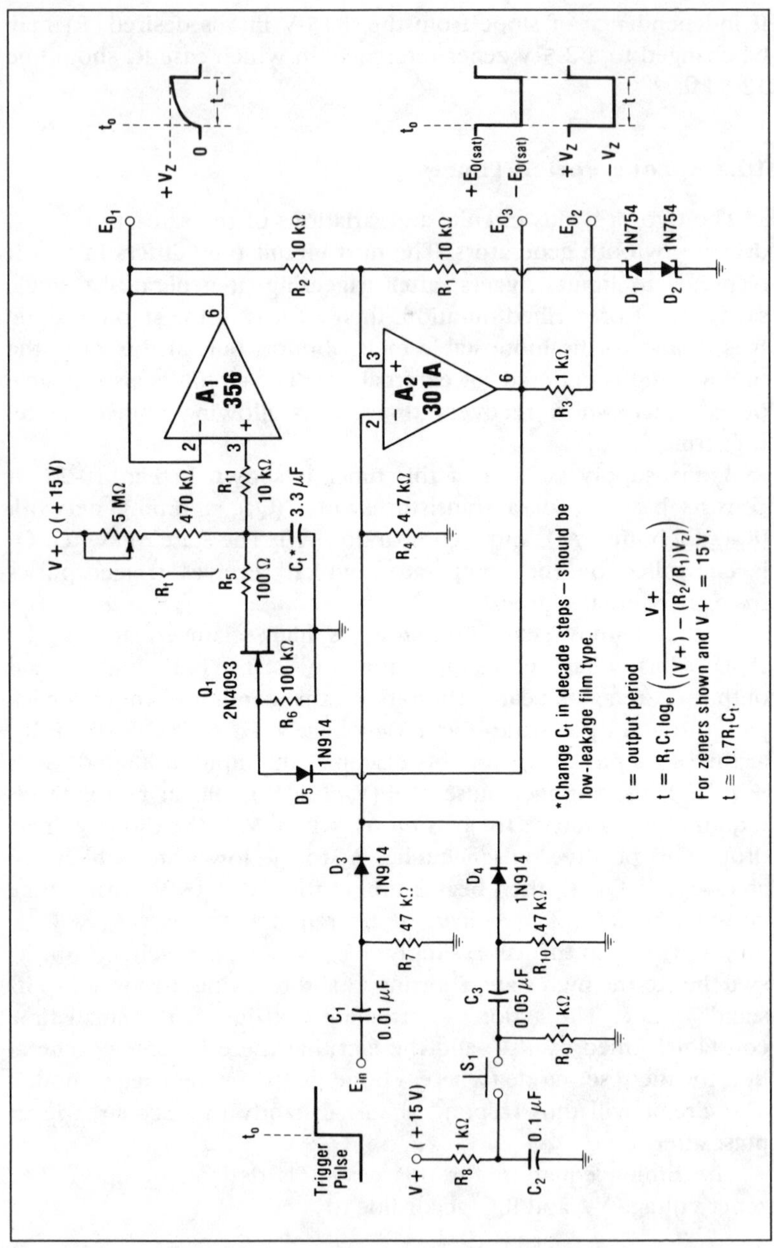


**Fig. 10-22. Dual-supply timer circuit.**

In the present example, $V+ = 15$ V, $V_Z = 7.5$ V, and $R_1 = R_2$, so the timing expression becomes

$$t = 0.7\ R_tC_t$$

As shown, the circuit is suitable for timing periods of up to 10 seconds with $R_t$ maximum, but larger values of $C_t$ can be used for longer periods. $R_t$ can also be increased, but there is an upper limit of about 5 MΩ in standard potentiometer values. $C_t$ can be switch-selected for decade timing changes—good-quality capacitors should be used, such as plastic-film types. The 356 is a good choice for $A_1$ in general-purpose applications, but $A_2$ should be uncompensated to optimize its triggering capability. The push-button trigger circuit shown is designed to minimize the effects of switch bounce; $S_1$ triggers only on actuation, not on release.

The three circuit outputs consist of two timing gates from $A_2$, which are negative during the timing interval, and the sawtooth output, which rises exponentially toward $+V_Z$. Generally, for a timer, the gate is the working output, but the sawtooth output may also be useful.

For single-supply uses, the timer shown in Fig. 10-23 will be found useful. This circuit uses a 555 timer as the monostable,* and a 3140 single-supply FET-input op amp to buffer the timing ramp across $C_t$. A low-leakage CMOS switch ($A_3$, a 4016) is used across $C_t$.

In this circuit, the timing range is extended without using large resistances or capacitances. As can be noted, the V+ voltage is divided by $R_1$ and $R_2$ prior to being applied to $R_t$. In this case, $R_1$ and $R_2$ yield a division ratio of 90:1. This, when multiplied by the basic 555 timing period of $1.1R_tC_t$ achieves a new timing period of $T = 100R_tC_t$.

This circuit is most useful with single supplies in the range of +10 V to +15 V. Timing can be trimmed with $R_2$, or with optional control $R_6$. Although a timing interval of 20 minutes is shown, it can range much longer if desired. The circuit is flexible since it has two outputs, pin 7 of the 555 having been made available as an uncommitted second output. It also has an enable input, if one is desired. If the long timing feature is not necessary, $R_1$ and $R_2$ can be eliminated and $R_t$ returned directly to V+.

---

* See reference 9 at the end of the chapter.

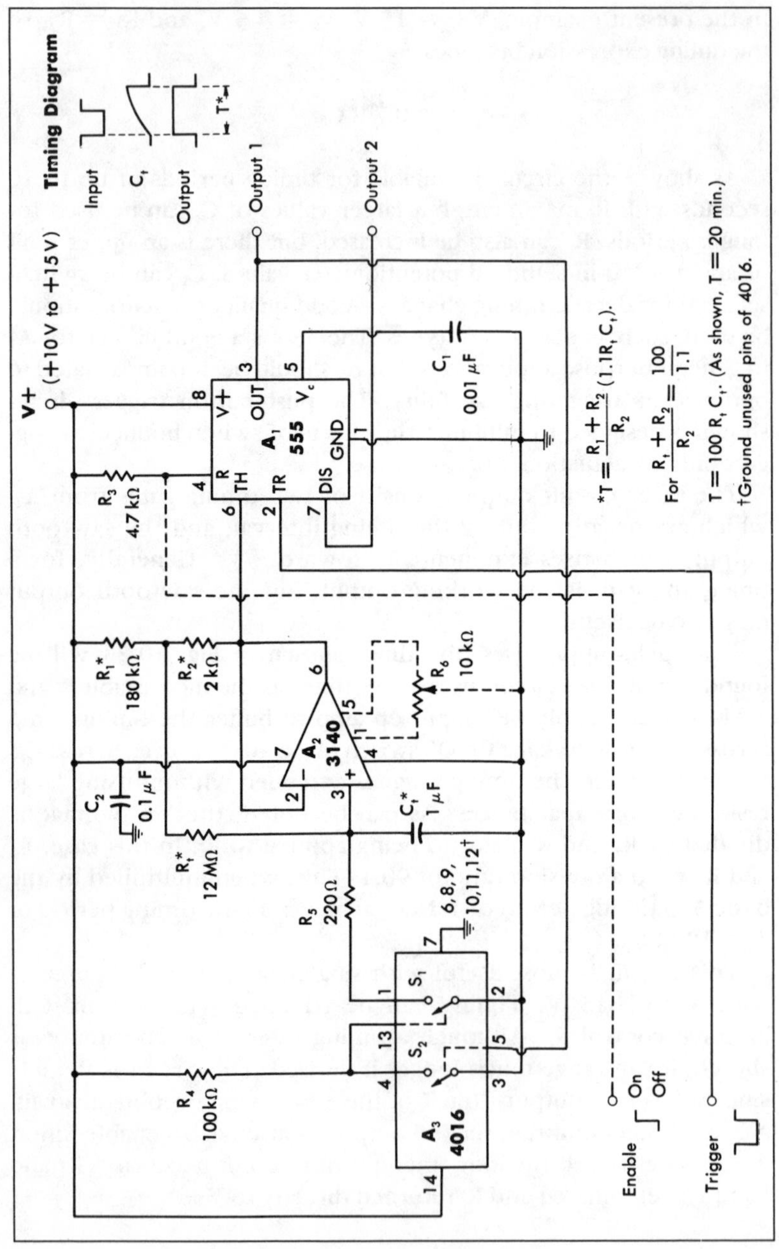


**Fig. 10-23. Single-supply timer circuit.**

### 10.4.4 Synchronized Sawtooth Oscillator

The sawtooth generators described thus far have been either driven or triggered types, both of which require an external oscillator for excitation. Drive sources for these circuits can be made from any of the astable oscillators described previously, but it is sometimes advantageous to make the sweep-generating functions part of the same circuit. An example of this is the sawtooth oscillator of Fig. 10-24, which is a source of both sawtooth and pulse waveforms, capable of being externally synchronized.

This circuit may be recognized as similar to the function generator of Fig. 10-15A, with suitable modifications for the sawtooth asymmetry, tunability, and synchronization. The use of $D_3$ and $D_4$ in series with $R_{t_1}$ and $R_{t_2}$ allows the negative and positive slopes to be set simply and independently. Tuning of the circuit is set by the variable-$\beta$ network, $R_5$-$R_6$. If the negative output slope has duration $t_1$ and the positive slope has duration $t_2$, the design equations for these periods are

$$t_1 = \frac{E_{o_1} R_{t_1} C_t}{\beta V_Z - V_{f_3}}$$

and

$$t_2 = \frac{E_{o_1} R_{t_2} C_t}{\beta V_Z - V_{f_4}}$$

where $V_{f_3}$ and $V_{f_4}$ are the forward voltage drops of $D_3$ and $D_4$, respectively.

The circuit components here are chosen to be compatible with a 60-Hz sweep generator. $R_{t_1}$ is set at $R_{t_2}/15$, yielding a trace/retrace ratio of 15 ($t_2 = 15$ ms and $t_1 = 1$ ms). $R_5$ provides a variation of 0.5:1 to allow tuning over a 2:1 range. In practice, frequency control $R_5$ is set for a natural frequency slightly lower than the incoming sync pulses. $R_{t_1}$ and $R_{t_2}$ may be altered to yield virtually any desired degree of asymmetry for various other applications.

The sync input, $E_s$, may be applied to either the (−) or (+) input of $A_1$, depending on the type of pulse available. Synchronization for either input occurs at the beginning of the $t_1$ interval.

Since this chosen design example is one of low frequency, an internally compensated amplifier, such as the 4558, may be used for $A_1$. As previously, three outputs are available: the sawtooth, $E_{o_1}$, and the two rectangular outputs, $E_{o_2}$ and $E_{o_3}$. This circuit is a

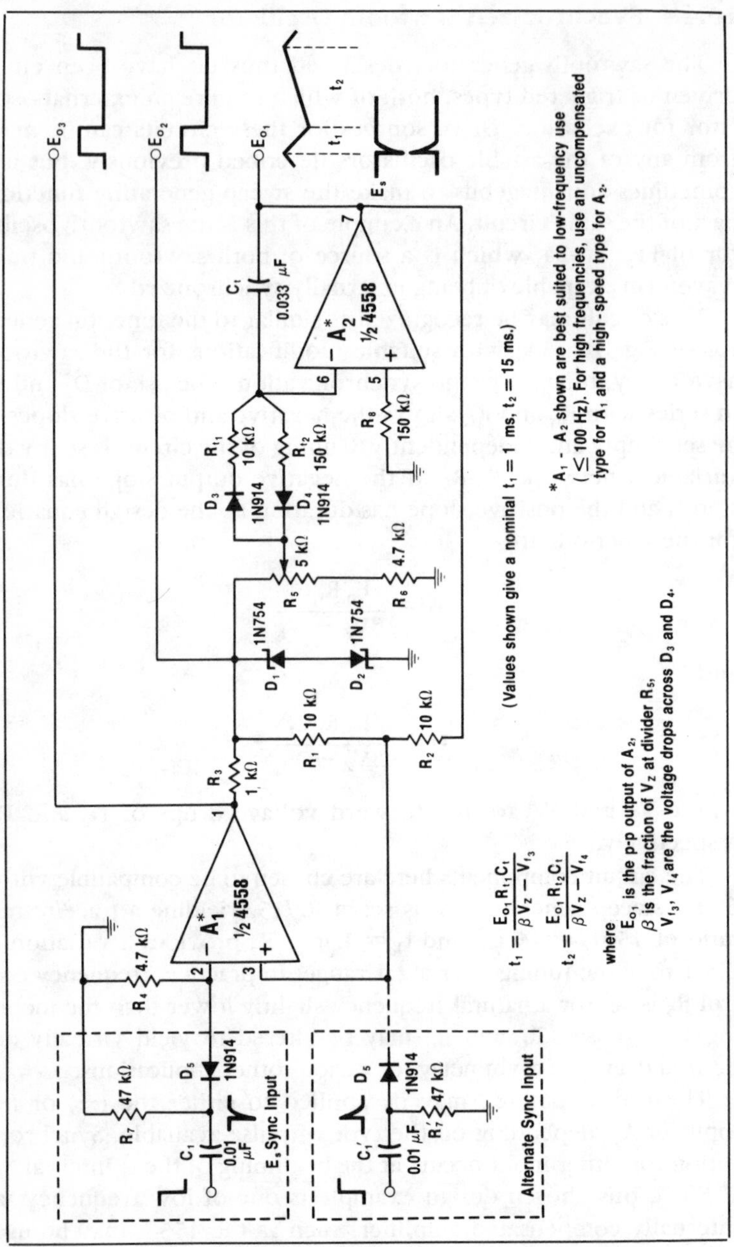


**Fig. 10-24. Synchronized sawtooth oscillator.**

very useful one, and the substitution of high-speed op amps will allow its application over a wide range of frequencies.

## 10.5 VOLTAGE-CONTROLLED SIGNAL GENERATION AND MODULATION

An extension of basic op-amp circuit design is the adaptation to voltage control. Voltage-controlled functions and modulation techniques are extremely useful in instrumentation and control systems. By making some function of a periodic waveform vary according to another voltage, it is possible to transmit dc or low-frequency information as a modulated carrier, often with greater accuracy and greater convenience. This section examines basic op-amp techniques used to produce amplitude modulation, pulse-width modulation, and frequency modulation.

### 10.5.1 Pulse-Amplitude Modulator

A simple form of encoding a low-frequency signal on a carrier is by pulse-amplitude modulation, as in Fig. 10-25. This circuit processes a bipolar pulse carrier, $E_c$, by clamping its positive and negative excursions with diodes $D_1$ and $D_2$. $D_2$ clamps the negative peak of $E_o$ to a voltage $-0.6$ V lower than the bias voltage of $D_3$ or zero volts, since the forward drops of $D_2$ and $D_3$ cancel. $D_1$ and $D_4$ perform a similar function for the positive peaks, clamping them to the voltage at the anode of $D_4$, which is $E_m$, provided by voltage follower $A_1$. The connection of $D_1$ and $D_4$ is arranged to cancel their forward drops and temperature coefficients, and to clamp $E_o$ to $E_m$. Thus, as $E_m$ varies from 0 to $+10$ V, the positive peaks of $E_o$ vary in accordance, accomplishing amplitude modulation of the $E_o$ pulses.

This circuit by nature has a high output impedance, essentially the value of $R_1$. A buffer may be required for low-impedance loads. The input voltage, $E_c$, should always be greater than the most positive peak of $E_m$ for effective clamping. If $E_c$ is small, some amplification may be necessary. This can be provided by a comparator such as $A_2$, which regenerates the input pulses to full amplitude with uniform rise and fall times.

As shown, the circuit responds to positive modulating voltages and delivers positive output pulses. For negative-going outputs, reverse all diodes, change the bias to $-15$ V, and let $E_m$ vary from 0 V to $-10$ V. If minimum temperature drift of the clamping volt-

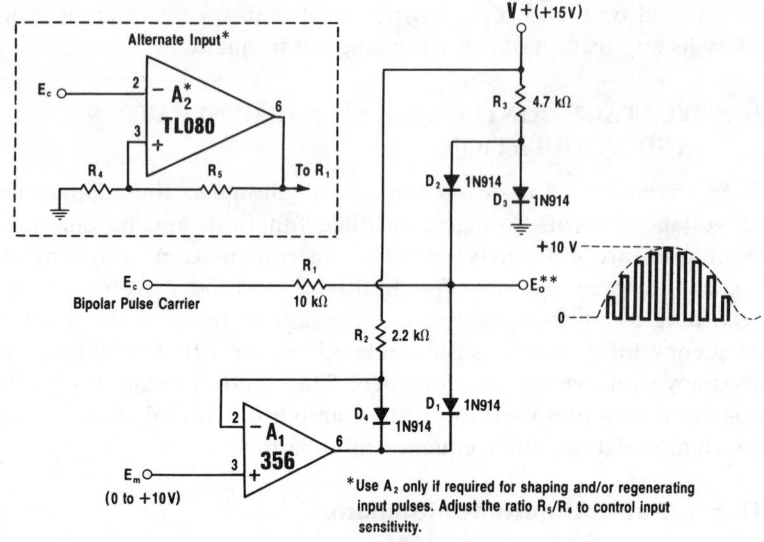


**Fig. 10-25. Pulse-amplitude modulator.**

ages is not required, the circuit can be simplified by wiring $A_1$ as a standard follower, with $D_1$ connected to its output and $D_2$ connected directly to ground.

## 10.5.2 Voltage-Controlled Monostable Circuit (Voltage to Pulse-Width Converter)

A monostable multivibrator that delivers an output pulse width proportional to an input voltage can be used as a pulse-width modulator (or voltage to pulse-width converter); Fig. 10-26 is such a circuit. The highly linear and precisely controlled ramp used to time the output pulse width is made available as an output. Hence, this circuit can also be used as a triggered voltage-controlled sweep generator. The circuit consists of two basic parts: a voltage-controlled integrator ($A_1$) and a threshold comparator ($A_2$). $A_1$ generates a voltage ramp with slope controlled by $E_{in}$, $R_t$, and $C_t$. $A_2$ controls the ramp-limit voltage. Pulse width $t_c$ is determined by

$$t_c = \frac{E_{o_1} R_t C_t}{E_{in}}$$

In the quiescent state, $A_2$ is high, holding $Q_1$ on and clamping the

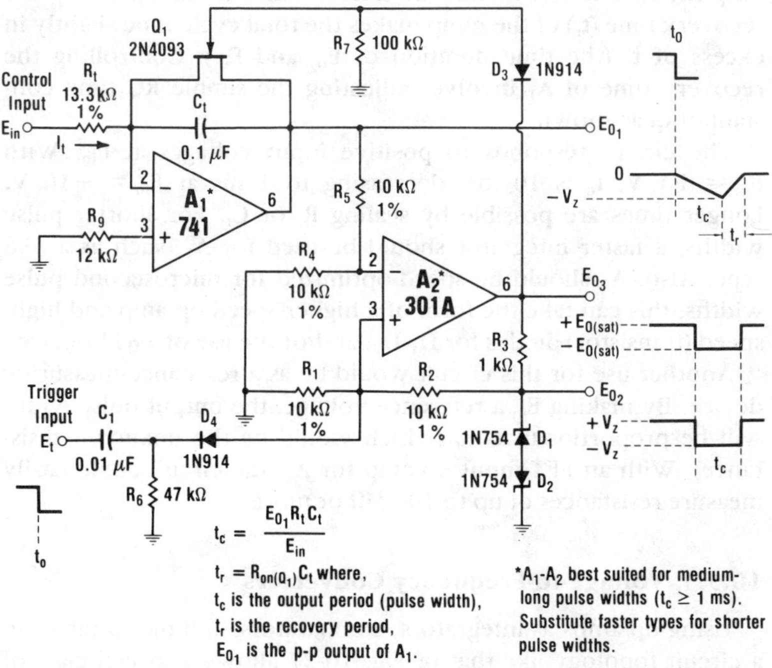


**Fig. 10-26. Voltage-controlled monostable circuit (voltage to pulse-width converter).**

output of $A_1$ at a voltage near ground. The circuit is stable in this state, since positive feedback from $E_{o_2}$ holds the (+) input of $A_2$ at approximately $+V_Z R_1/(R_1 + R_2)$, a voltage greater than the ground level on its (−) input.

A negative trigger pulse at $E_t$, greater than this bias, will flip the comparator to the negative output state, which turns off $Q_1$ and starts a negative ramp of voltage from $A_1$. The output from $A_1$ decreases linearly with a slope of $R_t C_t / E_{in}$ volts per second. This voltage appears at the (−) input of $A_2$, attenuated by a factor of $R_4/(R_4 + R_5)$. Furthermore, the negative threshold of $A_2$ is at $-V_Z R_1/(R_1 + R_2)$. Thus, if the ratios $R_2/R_1$ and $R_5/R_4$ are made equal, the comparator threshold voltage is simply $-V_Z$. When $E_{o_1}$ crosses $-V_Z$, $A_2$ flips back to the positive (stable) state, which turns $Q_1$ on, discharging $C_t$. This returns the circuit to its original state, ready for a second trigger pulse. The circuit has three outputs: the ramp output from $A_1$ (or $E_{o_1}$), and the two comparator outputs, $E_{o_2}$ and $E_{o_3}$. As may be noted from the timing of $E_{o_1}$, the recovery time ($t_r$) of the ramp makes the total cycle time slightly in excess of $t_c$ (the time duration of $E_{o_2}$ and $E_{o_3}$). Controlling the recovery time of $A_1$ involves adjusting the simple RC time constant of $t_r$, as shown.

The circuit responds to positive input voltages at $E_{in}$. With $E_{in} = +1$ V, $t_c$ is 10 ms, decreasing to 1 ms at $E_{in} = +10$ V. Longer times are possible by scaling $R_t$ or $C_t$. For shorter pulse widths, a faster integrator should be used for $A_1$, such as a 356 type. Also, $A_2$ should be speed-optimized for microsecond pulse widths; this can take the form of a higher-speed op amp and high-speed (transistor) diodes for $D_1$-$D_2$; and/or the use of a 311 device.

Another use for this circuit would be as a resistance-measuring device. By making $E_{in}$ a reference voltage, the output pulse width will be proportional to $R_t$ (which would be the unknown resistance). With an FET-input op amp for $A_1$, the circuit could easily measure resistances of up to 100 MΩ or more.

### 10.5.3 Voltage-to-Frequency Converters

Using op amps as integrators, comparators, and monostables in a circuit topology like that of Fig. 10-27 allows a special class of a/d converter to be implemented—the voltage-to-frequency converter. The voltage-to-frequency, or V/F, converter is capable of extremely high accuracies over a very wide dynamic range—on the order of 0.01% and 10,000/1 (80 dB), respectively. These figures would equate to 12 bits of conversion (or more). Full-scale frequencies can range as high as 1 MHz or more.

All of the functional blocks depicted in Fig. 10-27 can be implemented by using op amps as they have been described in this book. However, it is most likely not efficient or practical to do so, since there already exist IC V/F devices which achieve this function economically and efficiently with very high accuracy. Accordingly, some representative V/F converter circuits using IC V/F devices will be presented here. The section following also discusses other V/F circuit types.

A good example of the high performance that can be realized with V/F techniques is the precision V/F converter circuit of Fig. 10-28. Here, a 411-type op amp is used as an integrator, driving either an RC4151 (Raytheon) or an LM331 (National Semiconductor) V/F IC, $A_1$. The V/F IC provides the block functions

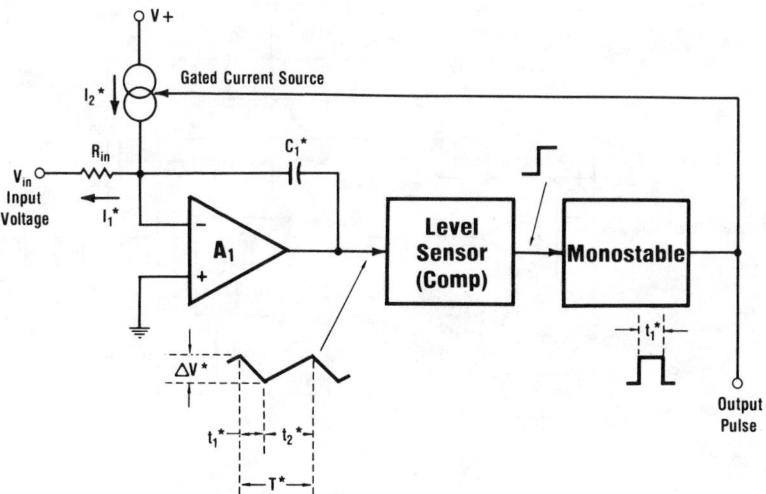


***Basic V/F Relationships:**

$$t_1 = \frac{\Delta V C_1}{I_2 - I_1}, \quad t_2 = \frac{\Delta V C_1}{I_1}$$

Thus,

$$\Delta V C_1 = t_1 (I_2 - I_1)$$

and

$$\Delta V C_1 = t_2 I_1$$

so

$$t_1 (I_2 - I_1) = t_2 I_1$$

Then in terms of $t_1$,

$$t_2 = \frac{t_1 (I_2 - I_1)}{I_1}$$

Solving for total period, T :

$$T = t_1 + t_2 = t_1 + \left( \frac{t_1 (I_2 - I_1)}{I_1} \right)$$

Simplifying,

$$T = t_1 \left( \frac{I_2}{I_1} \right)$$

In terms of frequency,

$$f = \frac{1}{T} = \frac{I_1}{t_1 I_2}$$

In terms of $V_{in}$,

$$f = \frac{V_{in}}{R_{in} t_1 I_2}$$

**Fig. 10-27. Block diagram of a charge-balance type of voltage-to-frequency converter.**

(shown in Fig. 10-27) of the comparator, the monostable, and the gated current source. To set up the V/F for a given input-voltage range, which is to be made to correspond to an output-frequency range, the relationships shown in the illustration are used.

$I_1$ is the full-scale input, which should be 100 μA; thus the desired full-scale voltage will define $R_{in}$. Here, $R_{in}$ is 100 kΩ with $V_{in(max)}$ equal to −10 V. $I_2$ is the V/F reference current, which is set to 135 μA by virtue of the 1.9-V V/F reference voltage and $R_s$. Time period $t_1$ is the monostable pulse width, set by $R_o$ and $C_o$ as

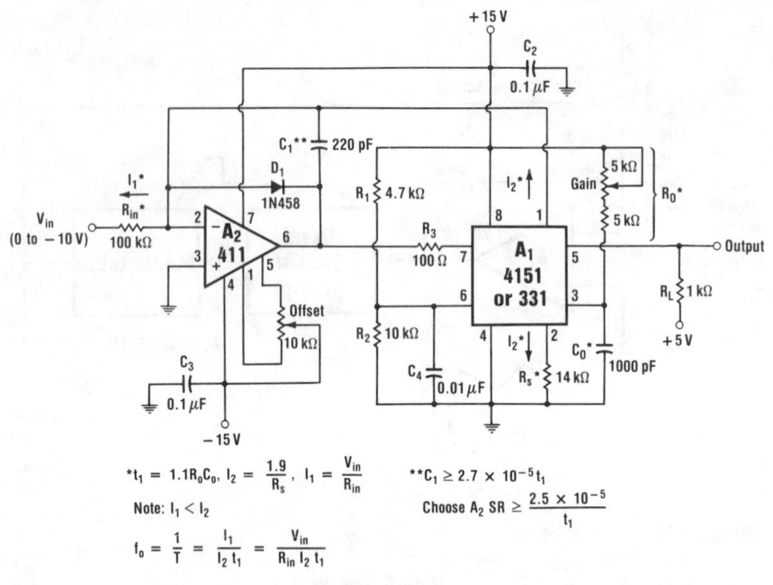


**Fig. 10-28. Wide-range precision V/F converter (0 to −10 V; 0 to 100 kHz).**

shown. By making $R_o$ variable, overall calibration can be accomplished.

Output from this circuit is a train of TTL-compatible pulses, and calibration is achieved by adjusting $R_o$ for 100 kHz with a −10-V (exact) input. The low-frequency end of the range is calibrated by trimming the op-amp offset for an output frequency of 100 Hz with a −10-mV input voltage. Alternately, if a 411A-type (or a similar low-offset device) is used, the offset trim can be eliminated.

Due to the very high dynamic range and high slew rate of the FET-input op amp, this circuit is capable of a linearity error of less than 0.1% of full scale, with errors appreciably smaller at lower full-scale portions of the range. Useful dynamic range of the circuit will be 10,000/1, or more, if offset trim is used for $A_2$ and a 331A device for the V/F converter.

A single-supply V/F circuit useful for 0- to +10-V inputs and a 0- to 10-kHz output is shown in Fig. 10-29. This circuit is set up very similar to that of Fig. 10-28, but op amp $A_2$ is used as a noninverting integrator in this case. Time period $t_1$ of the V/F converter is lowered for the 10-kHz range.

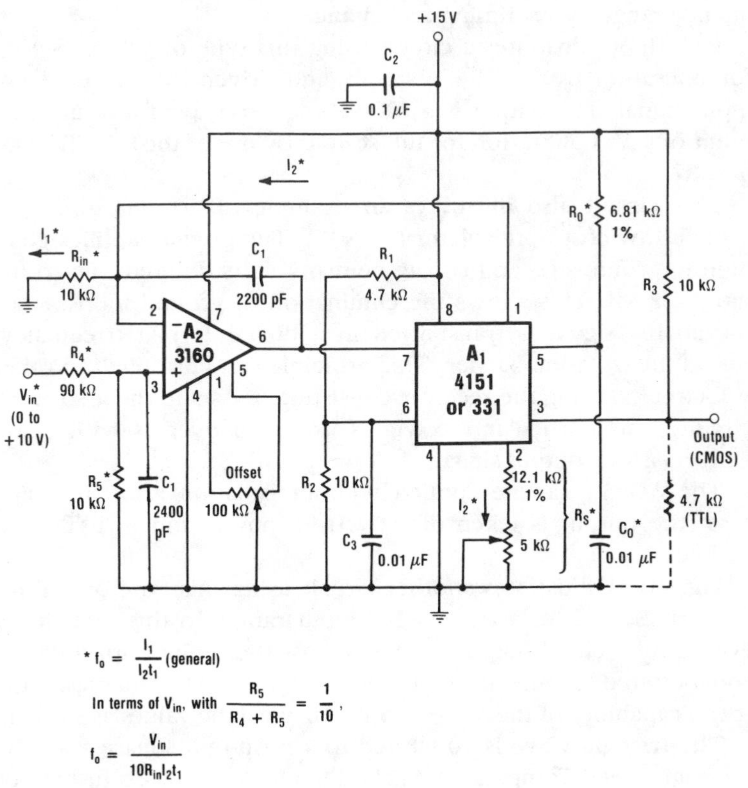


**Fig. 10-29. Single-supply V/F converter (0 to +10 V; 0 to 10 kHz).**

In this circuit, $R_4$ and $R_5$ reduce the 10-V maximum input to 1 V, as applied to $A_2$, an FET-input, single-supply op amp. Calibration is for a 10-kHz output with a +10-V input by means of $R_S$. The offset of $A_2$ is trimmed also, for a 10-Hz output with a 10-mV input. Linearity error of this circuit is 0.05%, or less, over the 10-kHz range. Note that other single-supply op amps are also usable for $A_2$, such as the 358 or 324 types. However, their limitation at low scale will be their offset error, if no external means of trim is provided.

Another type of V/F IC operates internally on the principle of a precision multivibrator. In this type of V/F, represented by the AD537 and AD654 types, the output frequency is directly proportional to an input voltage (or current). Input range scaling is accomplished by the choice of the timing resistor, and the frequency range by the timing capacitance.

Fig. 10-30 illustrates a circuit using this type of V/F IC, set up for operation from a +5-V supply and driven by a 0- to +1-V input signal. The output is a 100-kHz square wave for a full-scale input of 1 V. Calibration for full scale is by use of the SCALE TRIM pot, $R_3$.

This circuit also illustrates an example of *ground isolation,* using a low-cost optoisolator (a 4N37). The power supplies common to grounds $G_1$ and $G_2$ can be many volts different (up to the rating of $A_2$). This allows the elimination of ground loops, since the analog signal is transmitted in a digital format (frequency) across the isolation barrier. This principle can be applied to other V/F circuits, using the receiver consisting of $A_2$ and the associated circuit. This particular receiver is useful to over 100 kHz, and delivers a TTL output signal.

The AD654 can be applied without isolation also, in which case, the output is taken directly from pin 1, and is TTL compatible.

Fig. 10-31 illustrates another circuit using this type of V/F IC, similarly scaled for a 0- to +1-V input range. In this case, however, a differential output is taken across the timing capacitor, $C_t$, and buffered by the push-pull FETs, $Q_1$-$Q_2$. This increases the speed capability of the AD654 to 1 MHz with the values as shown.

The triangle wave is converted to a push-pull square wave by the high-speed comparator $A_2$. Following $A_2$ is an exclusive—OR stage, $A_3$. This is designed to produce a narrow output pulse for each input transition. This effectively doubles the frequency, to 2 MHz for a +1-V full-scale input to the V/F converter. With this higher full-scale frequency, data can be processed more quickly.

Fig. 10-32 illustrates a V/F circuit which operates on the charge-pump principle.* This type of design features relative simplicity, with excellent linearity for frequencies below 100 kHz.

In this circuit, an LTC1043 switching device is used, which charges the 0.01-$\mu$F capacitor from the −2.5-V reference diode. This charge is periodically dumped into the summing point, at a rate which is in proportion to the input voltage. Another section of the same switch commutates between +5 V and ground, furnishing a logic output signal.

---

* See reference 18 at the end of the chapter.

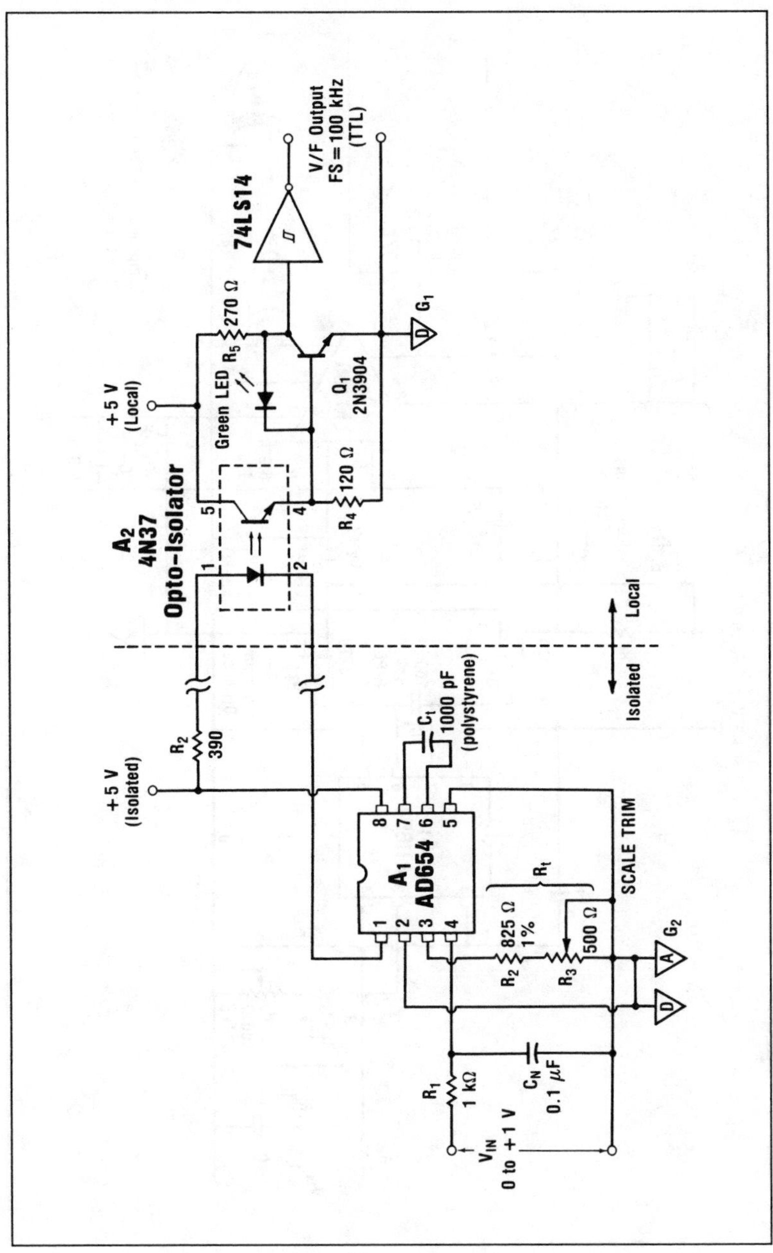


**Fig. 10-30. A 100-kHz isolated V/F converter.**

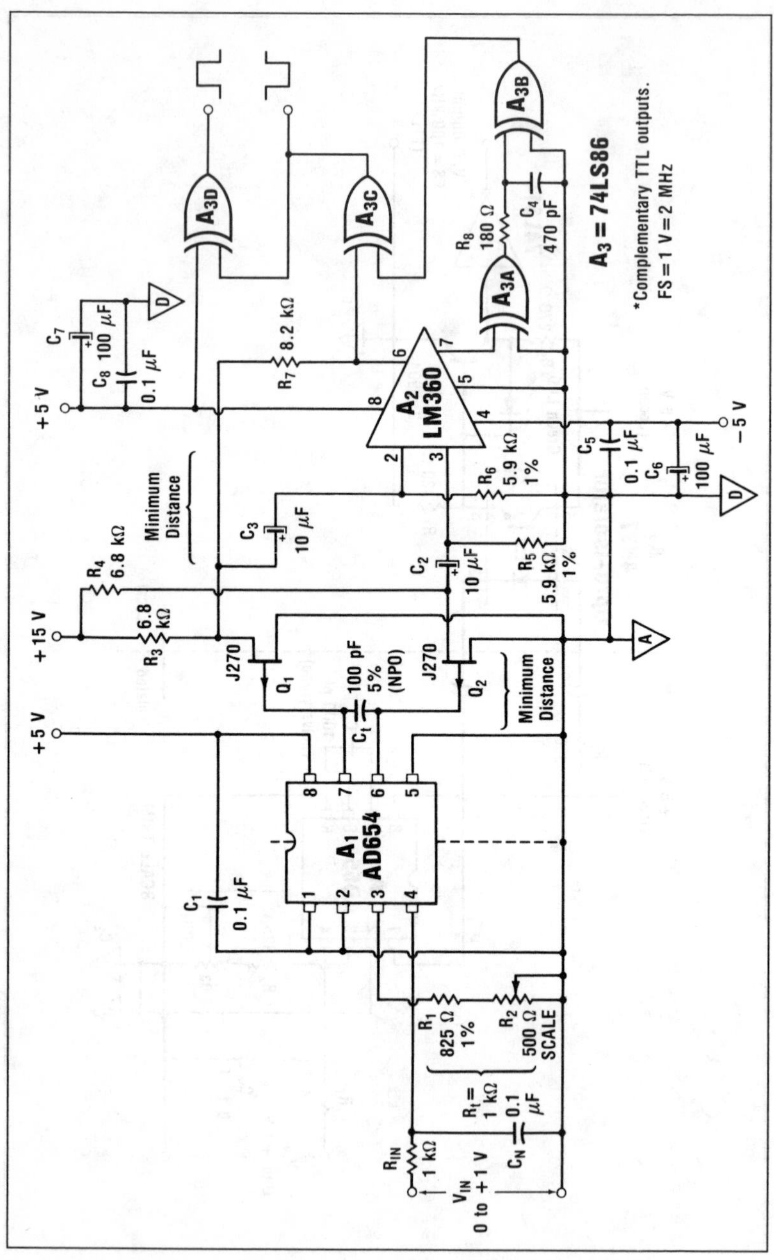


Fig. 10-31. A 2-MHz frequency-doubling V/F converter.

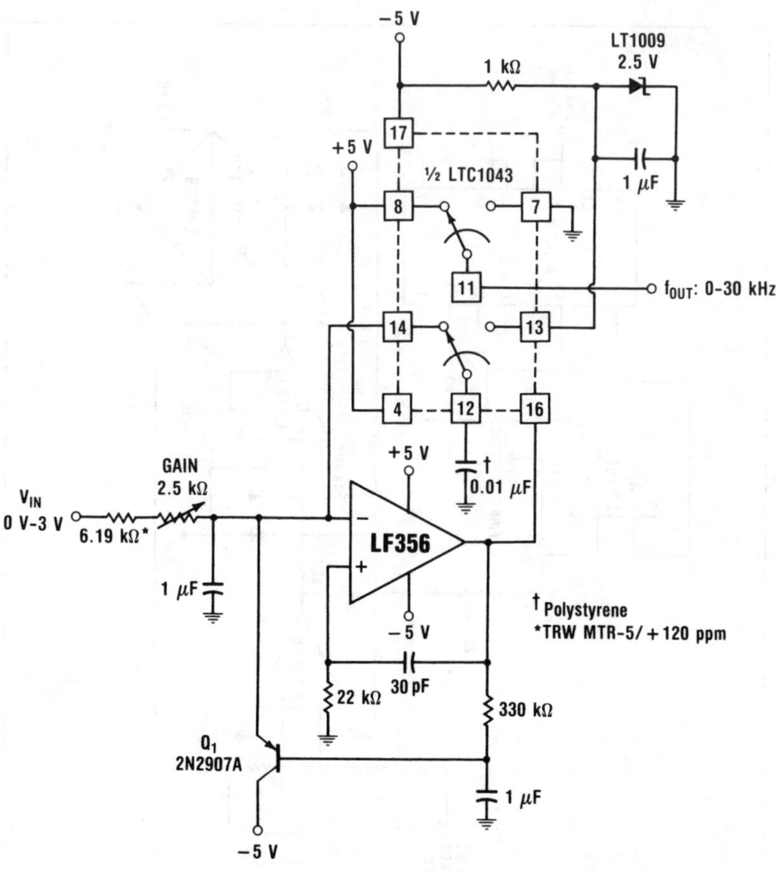


**Fig. 10-32. A 30-kHz charge-pump V/F circuit.**
*(Courtesy Linear Technology Corp.)*

For low overall TC, the polystyrene capacitor's TC should be balanced against an opposite TC in the input resistor, using the type(s) shown. Trim for full-scale calibration is by adjustment of this resistor, for 30 kHz out with a 3-V input. ($Q_1$ and the associated components form an antilatch network, and are not part of the basic V/F proper.) Offset error of this circuit will be basically that of the op amp used, and can be improved with devices such as the 1056 types.

The V/F circuit of Fig. 10-33 is a variation of the basic triangle waveform generator, insofar as its fundamental operation. Refinements have been added to enhance the speed and linearity, and an

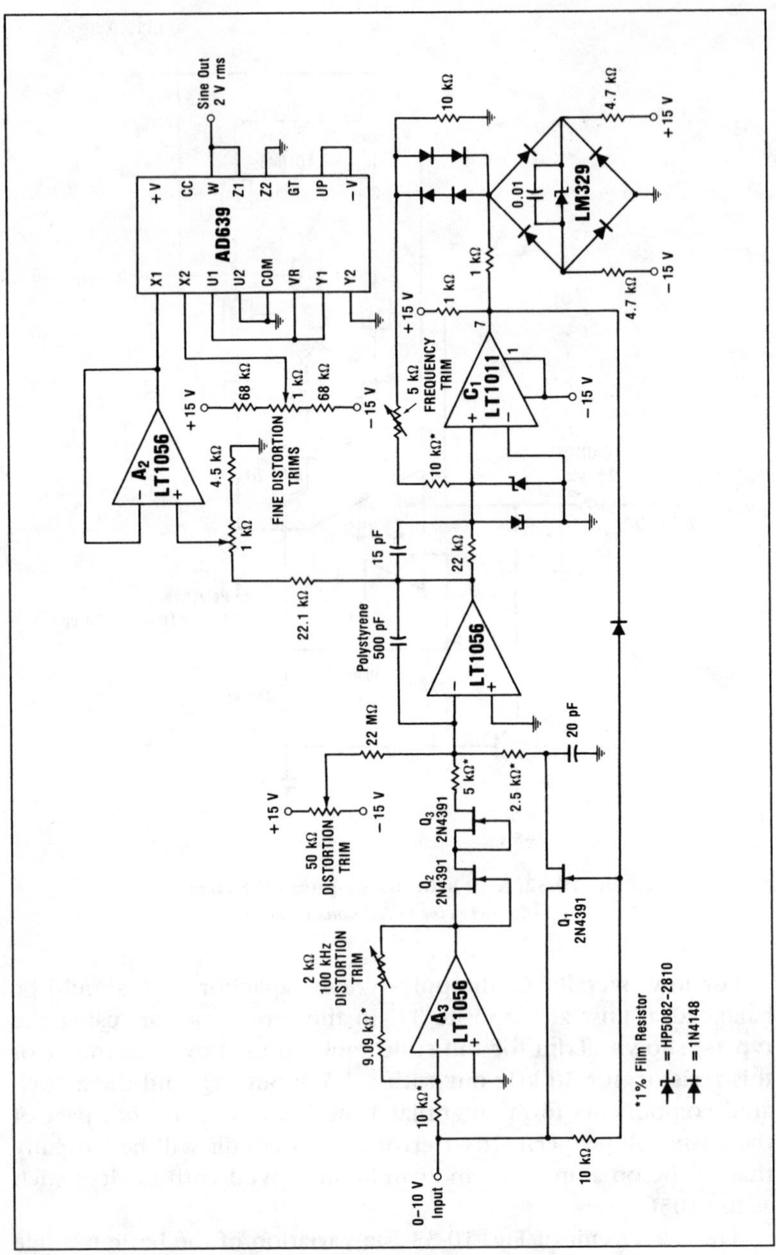


**Fig. 10-33. A 100-kHz sine-wave output V/F circuit.**

*(Courtesy Linear Technology Corp.)*

output sine-wave converter is also used. This circuit produces 2-V rms sine waves at frequencies up to 100 kHz, controlled by a 0- to +10-V full-scale input.

A number of trims are used, to optimize overall performance. The 5-kΩ potentiometer sets the full-scale frequency, while the 2-kΩ pot trims symmetry (and distortion) at this frequency. "Fine" distortion trims are accomplished by the two 1-kΩ pots associated with $A_2$ and the AD639, for the lowest possible THD in the sine-wave conversion.

When trimmed properly, distortion is no more than 0.2% over the entire range, and less than 0.1% below 10 kHz.

## 10.6 REFERENCES


1. *Applications Manual for Computing Amplifiers.* Philbrick/Nexus Research, Dedham, MA, 1966, 1984.
2. Brokaw, P. "FET Op Amp Adds New Twist to an Old Circuit." *EDN,* June 5, 1974.
3. Burwen, R. "Ultra Low Distortion Oscillator." *EDN/EEE,* June 1, 1971.
4. Dobkin, R. C. "Op Amp Circuit Collection." *National Semiconductor Application Note AN-31,* February 1970. National Semiconductor Corp., Santa Clara, CA.
5. Gilbert, B. A. "Monolithic Analog Trig Function Generator." *Analog Dialogue,* Vol. 18, No. 3, 1984.
6. Giles, J. N. *Fairchild Semiconductor LIC Handbook.* Fairchild Semiconductor, Mountain View, CA, 1967.
7. Grebene, A. B. "Monolithic Waveform Generation." *IEEE Spectrum,* April 1972.
8. Jung, W. G. *Audio IC Op-Amp Applications,* 3rd ed. Howard W. Sams & Co., Inc., Indianapolis, IN, 1986.
9. ________. *IC Timer Cookbook,* 2nd ed. Howard W. Sams & Co., Inc., Indianapolis, IN, 1983.
10. ________. "LEDs Do Dual Duty in Sine Wave Oscillator." *EDN,* August 20, 1976.
11. ________. "Low Distortion Oscillator Uses State Variable Filter." *Electronics,* February 5, 1976.
12. ________. "The Signal Path: Function Generators." *db, The Sound Engineering Magazine,* December 1975.

13. _________. "The Signal Path: Sine Wave Oscillators." *db, The Sound Engineering Magazine,* July 1976.

14. _________. "Operation and Applications of the AD654 IC V-to-F Converter." *Analog Devices Application Note,* July 1985, Analog Devices, Inc., Norwood, MA.

15. Korn, G. A., and Korn, T. M. *Electronic Analog and Hybrid Computers,* 2nd ed. McGraw-Hill Book Co., New York, 1972.

16. Miler, G. G. "A Simple Square-Triangle Waveform Generator." *Harris Semiconductor Application Note 510/A,* October 1970. Harris Semiconductor, Melbourne, FL.

17. Olson, H. "Use Diodes for Amplitude Control in 0.001 Hz Wien Bridge Oscillator." *Electronic Design,* April 1, 1972.

18. Pease, R.; Teledyne Inc., *Amplitude to Frequency Converter.* US Patent #3,746,968, filed September 1972.

19. Routh, S. W. "An Applications Guide for Operational Amplifiers." *National Semiconductor Application Note AN-20,* February 1969. National Semiconductor Corp., Santa Clara, CA.

20. Smith, J. I. *Modern Operational Circuit Design.* John Wiley & Sons, Inc., New York, 1971.

21. Tobey, G. E., Graeme, J. G., and Huelsman, L. P. (eds.). *Operational Amplifiers—Design and Applications.* McGraw-Hill Book Co., New York, 1971.

22. Widlar, R. J. "Drift Compensation Techniques for Integrated DC Amplifiers." *National Semiconductor Application Note AN-3,* November 1967, National Semiconductor Corp., Santa Clara, CA.

23. _________. "IC Op Amp Beats FETs on Input Current." *National Semiconductor Application Note AN-29,* December 1969, National Semiconductor Corp., Santa Clara, CA.

24. _________. "Monolithic Operational Amplifiers—The Universal Linear Component." *National Semiconductor Application Note AN-4,* April 1968, National Semiconductor Corp., Santa Clara, CA.

25. Williams, J. "Thermal Techniques in Measurement and Control Circuitry." *Linear Technology Application Note AN5,* December 1984, Linear Technology Corp.

26. _________. "Applications for a Switched-Capacitor Instrumentation Building Block." *Linear Technology Application Note AN3,* December 1984, Linear Technology Corp.

27. _________. "Designs for High Performance Voltage-to-Frequency Converters." *Linear Technology Application Note AN14,* July 1985, Linear Technology Corp.

# III

# *Appendix*

APPENDIX

# *Manufacturers' Data Sheets for General-Purpose Op Amps*

This appendix includes catalog specification sheets for selected general-purpose op amps that will be useful in a large percentage of applications. These devices are well accepted as industry standards, and the data sheets reproduced here are generally those of the original manufacturers.

Space does not permit the inclusion of data sheets for the more-specialized devices; thus, it is suggested that the reader assemble catalog information for these types by writing the various manufacturers listed. The manufacturers listed in Table A-1 are the original sources of the respective devices listed. Table A-2 lists some additional IC op-amp sources.

### Table A-1. Original Source Manufacturers and Their Respective Devices

| Manufacturer | Prefix | Part No. Example(s) | Devices (Generic Form) |
|---|---|---|---|
| Analog Devices, Inc.<br>Route One Industrial Park<br>P.O. Box 280<br>Norwood, MA 02062<br>(617) 329-4700 | "AD" | AD542J<br>AD542K<br>AD542L | 504, 509, 510, 515, 517, 518, 540, 542, 544, 545, 548, 549, 642, 644, 647, 648, 711, 712 |
| Burr-Brown<br>International Airport Industrial Park<br>P.O. Box 11400<br>Tucson, AZ 85734<br>(602) 746-1111 | None<br><br><br>"OPA" | 3500A<br>3500B<br>3500C<br>OPA111AM<br>OPA111BM<br>OPA111SM | 3500, 3500MP, 3521, 3528, 3580, 3581<br>OPA111, OPA121, OPA2111, OPA27, OPA37, OPA606 |
| Fairchild Semiconductor<br>464 Ellis Street<br>Mountain View, CA 94042<br>(415) 962-5011 | "μA" | μA741<br>μA741C | 702, 709, 714, 715, 725, 740, 741, 747, 748, 759, 771, 772, 774, 776, 791, 798 |
| Harris Semiconductor<br>P.O. Box 883<br>Melbourne, FL 32901<br>(305) 727-5430 | "HA" | HA2600<br>HA2602<br>HA2605 | 2500, 2510, 2520, 2539, 2540, 2541, 2542, 2600, 2620, 2640, 2900, 4622, 4741, 5100, 5130, 5160, 5170, 5180, 5190 |
| Intersil, Inc.<br>10710 North Tantau Avenue<br>Cupertino, CA 95014<br>(408) 996-5000 | "ICL" | ICL7650<br>ICL7652 | 7611, 7612, 7613, 7614, 7621, 7622, 7631, 7641, 7650, 7652, 8007, 8008, 8043, 8500 |
| Linear Technology Corp.<br>1630 McCarthy Blvd.<br>Milpitas, CA 95035<br>(408) 942-0810 | "LT" | LT1001M<br>LT1001C<br>LT1001AM<br>LT1001AC | 1001, 1002, 1007, 1008, 1012, 1013, 1014, 1022, 1037, 1052, 1056 |
| Motorola Semiconductor Products, Inc.<br>P.O. Box 20912<br>Phoenix, AZ 85036<br>(602) 244-6900 | "MC" | MC3303<br>MC3403<br>MC3503 | 1436, 1437, 1456, 1458, 1741S, 3403, 3458, 3471, 4741, 34001, 34002, 34080, 34081, 34082, 34084 |

**Table A-1 (cont). Original Source Manufacturers and Their Respective Devices**

| Manufacturer | Prefix | Part No. Example(s) | Devices (Generic Form) |
|---|---|---|---|
| National Semiconductor Corp.<br>2900 Semiconductor Drive<br>Santa Clara, CA 95051<br>(408) 737-5000 | "LF"<br><br><br>"LH"<br><br>"LM" | LF356<br>LF256<br>LF156<br>LH0002<br>LH0002C<br>LM301A<br>LM201A<br>LM101A | 301A, 307, 308, 308A, 310, 312, 316, 318, 321, 324, 343, 344, 347, 348, 349, 351, 353, 354, 355, 356, 357, 358, 411, 412, 441, 442, 444, 2902, 2904, 0002, 0044, LM10, LM11 |
| Precision Monolithics, Inc.<br>1500 Space Park Drive<br>Santa Clara, CA 95050<br>(408) 727-9222 | "OP" | OP-07<br>OP-07A<br>OP-07C<br>OP-07D<br>OP-07E | OP-01, OP-02, OP-04, OP-05, OP-07, OP-08, OP-09, OP-10, OP-11, OP-12, OP-14, OP-15, OP-16, OP-17, OP-20, OP-21, OP-27, OP-37, OP-41, OP-50 |
| Raytheon Semiconductor<br>350 Ellis Street<br>Mountain View, CA 94042<br>(415) 968-9211 | "RC"<br>"RM" | RC4558<br>RM4558 | 4131, 4132, 4136, 4156, 4558, 4559 |
| RCA Solid State Division<br>Route 202<br>Somerville, NJ 08876<br>(201) 685-6000 | "CA" | CA3140<br>CA3140A<br>CA3140B | 3100, 3130, 3140, 3160, 3193, 3240, 3260, 3420, 3440 |
| Signetics Corp.<br>811 East Arques Avenue<br>Sunnyvale, CA 94086<br>(408) 739-7700 | "NE"<br>"SE" | NE5534<br>SE5534 | 530, 531, 532, 538, 5530, 5532, 5533, 5534, 5535 |
| Texas Instruments Incorporated<br>P.O. Box 225012<br>Dallas, TX 75265<br>(214) 238-2011 | "TL"<br><br><br>"TLC" | TL071<br>TL071A<br>TL071B<br>TLC251 (series)<br>TLC271 (series) | TL060, TL061, TL062, TL064, TL070, TL071, TL072, TL074, TL075, TL080, TL081, TL082, TL084, TL087, TL088, TL287, TL288 |

**Table A-2. Additional IC Op-Amp Manufacturers**

| | |
|---|---|
| Advanced Micro Devices<br>901 Thomson Place<br>Sunnyvale, CA 94086<br>(408) 732-2400 | Micro Power Systems<br>3100 Alfred Street<br>Santa Clara, CA 95050<br>(408) 727-5350 |
| Analog Systems<br>Box 35879<br>Tucson, AZ 85740<br>(602) 299-9831 | Silicon General Inc.<br>11651 Monarch Street<br>Garden Grove, CA 92641<br>(714) 892-5531 |
| Apex Microtechnology Corp.<br>1130 E. Pennsylvania Street<br>Tucson, AZ 85714<br>(602) 746-0849 | Siliconix<br>2201 Laurelwood Road<br>Santa Clara, CA 95059<br>(408) 988-8000 |
| Exar Integrated Systems, Inc.<br>750 Palomar Avenue<br>Sunnyvale, CA 94088<br>(408) 732-7970 | Teledyne Semiconductor<br>1300 Terra Bella Avenue<br>P.O. Box 7267<br>Mountain View, CA 94039<br>(415) 968-9241 |
| Maxim Integrated Products<br>510 N. Pastoriza Avenue<br>Sunnyvale, CA 94086<br>(408) 737-7600 | |

Inasmuch as a number of device types are featured in this section, the data sheets are presented in abbreviated form to conserve space. Complete performance information is included, but (in some cases) specific application examples are omitted. The reader is encouraged to assemble complete (and most current) catalog information by writing directly to the manufacturer in question. Please mention this book.

In some cases, the data sheet presented is an example of the forerunner of a generic type of op amp. Related and similarly performing amplifiers (such as duals or quads) are listed in Table A-3.

For simplicity, the device numbers used in most of the circuits in this book are presented in *generic form,* as listed in the right-hand column of Table A-1. Thus, for example, a "301A" op amp implies a National Semiconductor LM301A device. Since each generic type listed may actually fan-out to as many as five different specific devices (excluding package options), it is easy to appreciate why this must be done (see OP-07 for an example).

## Table A-3. Basic Device Types and Their Generic Relations*

| Basic Type | Generic Relation* | Manufacturer |
|---|---|---|
| 741 | $\mu$A748 (single)<br>$\mu$A747 (dual)<br>MC1458 (dual)<br>MC4741 (quad)<br>LM348 (quad) | Fairchild<br>Fairchild<br>Motorola<br>Motorola<br>National |
| 301A<br>(uncompensated 741) | LM307 (single)<br>LH2301 (dual) | National<br>National |
| 4558 | RC4559 (dual)<br>RC4136 (quad) | Raytheon<br>Raytheon |
| 324 | LM358 (dual)<br>MC3403 (quad) | National<br>Motorola |
| 3140 | CA3130 (single)<br>CA3160 (single)<br>CA3240 (dual) | RCA<br>RCA<br>RCA |
| 411 | LF411 (single)<br>LF412 (dual)<br>AD611 (single) | National<br>National<br>Analog Devices |
| 355/356/357 | LF356 (series)<br>OP-15/OP-16/OP-17 (series)<br>OPA356A (series)<br>OPA606 (series)<br>LT1056 (series)<br>(all are singles) | National<br>Precision Monolithics<br>Burr-Brown<br>Burr-Brown<br>Linear Technology |
| OP-07 | OP-07 (single)<br>OP-207 (dual)<br>ADOP-07 (single)<br>LT1001 (single)<br>LT1002 (dual) | Precision Monolithics<br>Precision Monolithics<br>Analog Devices<br>Linear Technology<br>Linear Technology |
| OP-27 | OP-27 (single)<br>OP-227 (dual)<br>OPA27 (single)<br>ADOP-27 (single)<br>LT1007 (single) | Precision Monolithics<br>Precision Monolithics<br>Burr-Brown<br>Analog Devices<br>Linear Technology |

*This table lists devices that are generically related to the standard types. Complete information on these devices may be obtained by writing the manufacturer in question. Please mention this book.

In some cases, it is not appropriate (and may actually cause confusion) to delete the prefix used by a manufacturer. For example, both Signetics and Analog Devices use a "500" series generic device number system, with many identically appearing generic numbers for *different* devices. Thus, an Analog Devices AD542 op amp should be understood to be entirely different from a Signetics NE542, but *both* are actual devices that are available. Further confusion can arise when a prefix preceding a part number that begins with a zero is dropped; an OP-07 would then be an 07 (which is ambiguous). Even worse, an OP-10 and an LM10, without prefixes, would appear identical (which they are not, of course). In such cases, we have endeavored to use complete part numbers, such as TL071, OP-02, LM10, etc.

As noted at the beginning of this appendix, most of the manufacturers of the devices listed in Table A-1 are the original sources of those devices. In addition, most manufacturers supply, or "second-source," standard devices that are not of their own origin. A second-source cross-reference listing (available from most manufacturers) will be helpful in identifying alternate sources of the various devices. However, some caution should be extended to the reader as to what constitutes a *true* alternate-source device. Generally, if a device is listed as a "direct" or "pin-for-pin" replacement, it is an electrical and mechanical duplicate of the original. If, however, a device is listed as a "functional" equivalent, it is not necessarily a direct substitute mechanically or electrically in all details. In these cases, consult the manufacturer's data sheets in order to be assured of suitability. In all cases, alternate-source types should be qualified by in-circuit testing as a final verification of suitability.

In one sense of the word, the above discussion is meant as a precaution, to warn the user that "look-a-likes" may not perform exactly like an original. However, this is not necessarily bad; in fact, many "look-a-like" op amps far outperform the original, both in better behavior in general, but also with tighter specs. The informed user can take advantage of this, and often obtain better overall system performance at little cost penalty.

# LM741/LM741A/LM741C/LM741E Operational Amplifier

## General Description

The LM741 series are general purpose operational amplifiers which feature improved performance over industry standards like the LM709. They are direct, plug-in replacements for the 709C, LM201, MC1439 and 748 in most applications.

The amplifiers offer many features which make their application nearly foolproof: overload protection on the input and output, no latch-up when the common mode range is exceeded, as well as freedom from oscillations.

The LM741C/LM741E are identical to the LM741/LM741A except that the LM741C/LM741E have their performance guaranteed over a 0°C to +70°C temperature range, instead of −55°C to +125°C.

## Schematic and Connection Diagrams (Top Views)

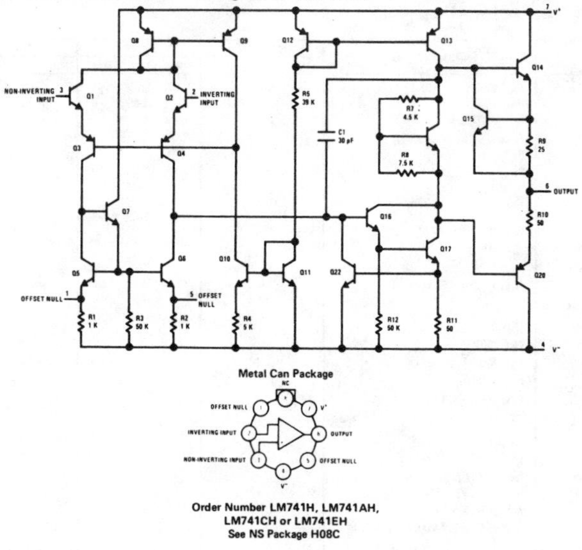

**Metal Can Package**

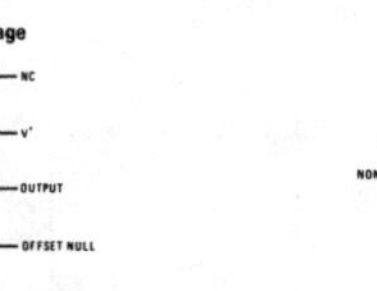

**Order Number LM741H, LM741AH, LM741CH or LM741EH**
**See NS Package H08C**

**Dual-In-Line Package**

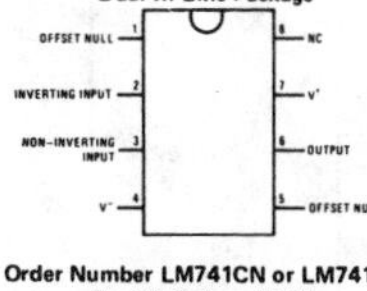

**Order Number LM741CN or LM741EN**
**See NS Package N08B**
**Order Number LM741CJ**
**See NS Package J08A**

**Dual-In-Line Package**

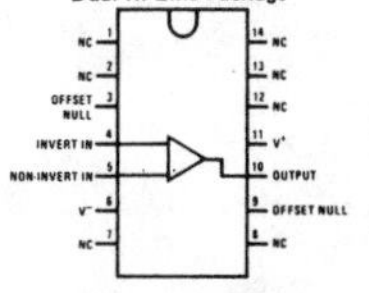

**Order Number LM741CN-14**
**See NS Package N14A**
**Order Number LM741J-14, LM741AJ-14 or LM741CJ-14**
**See NS Package J14A**

*Courtesy National Semiconductor Corp.*

## Absolute Maximum Ratings

| | LM741A | LM741E | LM741 | LM741C |
|---|---|---|---|---|
| Supply Voltage | ±22V | ±22V | ±22V | ±18V |
| Power Dissipation (Note 1) | 500 mW | 500 mW | 500 mW | 500 mW |
| Differential Input Voltage | ±30V | ±30V | ±30V | ±30V |
| Input Voltage (Note 2) | ±15V | ±15V | ±15V | ±15V |
| Output Short Circuit Duration | Indefinite | Indefinite | Indefinite | Indefinite |
| Operating Temperature Range | −55°C to +125°C | 0°C to +70°C | −55°C to +125°C | 0°C to +70°C |
| Storage Temperature Range | −65°C to +150°C | −65°C to +150°C | −65°C to +150°C | −65°C to +150°C |
| Lead Temperature (Soldering, 10 seconds) | 300°C | 300°C | 300°C | 300°C |

## Electrical Characteristics (Note 3)

| PARAMETER | CONDITIONS | LM741A/LM741E | | | LM741 | | | LM741C | | | UNITS |
|---|---|---|---|---|---|---|---|---|---|---|---|
| | | MIN | TYP | MAX | MIN | TYP | MAX | MIN | TYP | MAX | |
| Input Offset Voltage | $T_A = 25°C$ | | | | | | | | | | |
| | $R_S \leq 10\ k\Omega$ | | | | | 1.0 | 5.0 | | 2.0 | 6.0 | mV |
| | $R_S \leq 50\Omega$ | | 0.8 | 3.0 | | | | | | | mV |
| | $T_{AMIN} \leq T_A \leq T_{AMAX}$ | | | | | | | | | | |
| | $R_S \leq 50\Omega$ | | | 4.0 | | | | | | | mV |
| | $R_S \leq 10\ k\Omega$ | | | | | | 6.0 | | | 7.5 | mV |
| Average Input Offset Voltage Drift | | | | 15 | | | | | | | μV/°C |
| Input Offset Voltage Adjustment Range | $T_A = 25°C$, $V_S = \pm 20V$ | ±10 | | | | ±15 | | | ±15 | | mV |
| Input Offset Current | $T_A = 25°C$ | | 3.0 | 30 | | 20 | 200 | | 20 | 200 | nA |
| | $T_{AMIN} \leq T_A \leq T_{AMAX}$ | | | 70 | | 85 | 500 | | | 300 | nA |
| Average Input Offset Current Drift | | | | 0.5 | | | | | | | nA/°C |
| Input Bias Current | $T_A = 25°C$ | | 30 | 80 | | 80 | 500 | | 80 | 500 | nA |
| | $T_{AMIN} \leq T_A \leq T_{AMAX}$ | | | 0.210 | | | 1.5 | | | 0.8 | μA |
| Input Resistance | $T_A = 25°C$, $V_S = \pm 20V$ | 1.0 | 6.0 | | 0.3 | 2.0 | | 0.3 | 2.0 | | MΩ |
| | $T_{AMIN} \leq T_A \leq T_{AMAX}$, $V_S = \pm 20V$ | 0.5 | | | | | | | | | MΩ |
| nput Voltage Range | $T_A = 25°C$ | | | | | | | ±12 | ±13 | | V |
| | $T_{AMIN} \leq T_A \leq T_{AMAX}$ | | | | ±12 | ±13 | | | | | V |
| Large Signal Voltage Gain | $T_A = 25°C$, $R_L \geq 2\ k\Omega$ | | | | | | | | | | |
| | $V_S = \pm 20V$, $V_O = \pm 15V$ | 50 | | | | | | | | | V/mV |
| | $V_S = \pm 15V$, $V_O = \pm 10V$ | | | | 50 | 200 | | 20 | 200 | | V/mV |
| | $T_{AMIN} \leq T_A \leq T_{AMAX}$, $R_L \geq 2\ k\Omega$, | | | | | | | | | | |
| | $V_S = \pm 20V$, $V_O = \pm 15V$ | 32 | | | | | | | | | V/mV |
| | $V_S = \pm 15V$, $V_O = \pm 10V$ | | | | 25 | | | 15 | | | V/mV |
| | $V_S = \pm 5V$, $V_O = \pm 2V$ | 10 | | | | | | | | | V/mV |
| Output Voltage Swing | $V_S = \pm 20V$ | | | | | | | | | | |
| | $R_L \geq 10\ k\Omega$ | ±16 | | | | | | | | | V |
| | $R_L \geq 2\ k\Omega$ | ±15 | | | | | | | | | V |
| | $V_S = \pm 15V$ | | | | | | | | | | |
| | $R_L \geq 10\ k\Omega$ | | | | ±12 | ±14 | | ±12 | ±14 | | V |
| | $R_L \geq 2\ k\Omega$ | | | | ±10 | ±13 | | ±10 | ±13 | | V |
| Output Short Circuit Current | $T_A = 25°C$ | 10 | 25 | 35 | | 25 | | | 25 | | mA |
| | $T_{AMIN} \leq T_A \leq T_{AMAX}$ | 10 | | 40 | | | | | | | mA |
| Common-Mode Rejection Ratio | $T_{AMIN} \leq T_A \leq T_{AMAX}$ | | | | | | | | | | |
| | $R_S \leq 10\ k\Omega$, $V_{CM} = \pm 12V$ | | | | 70 | 90 | | 70 | 90 | | dB |
| | $R_S \leq 50\ k\Omega$, $V_{CM} = \pm 12V$ | 80 | 95 | | | | | | | | dB |

*Courtesy National Semiconductor Corp.*

## Electrical Characteristics (Continued)

| PARAMETER | CONDITIONS | LM741A/LM741E | | | LM741 | | | LM741C | | | UNITS |
|---|---|---|---|---|---|---|---|---|---|---|---|
| | | MIN | TYP | MAX | MIN | TYP | MAX | MIN | TYP | MAX | |
| Supply Voltage Rejection Ratio | $T_{AMIN} \le T_A \le T_{AMAX}$, $V_S = \pm 20V$ to $V_S = \pm 5V$ | | | | | | | | | | |
| | $R_S \le 50\Omega$ | 86 | 96 | | | | | | | | dB |
| | $R_S \le 10\ k\Omega$ | | | | 77 | 96 | | 77 | 96 | | dB |
| Transient Response | $T_A = 25°C$, Unity Gain | | | | | | | | | | |
| Rise Time | | | 0.25 | 0.8 | | 0.3 | | | 0.3 | | µs |
| Overshoot | | | 6.0 | 20 | | 5 | | | 5 | | % |
| Bandwidth (Note 4) | $T_A = 25°C$ | 0.437 | 1.5 | | | | | | | | MHz |
| Slew Rate | $T_A = 25°C$, Unity Gain | 0.3 | 0.7 | | | 0.5 | | | 0.5 | | V/µs |
| Supply Current | $T_A = 25°C$ | | | | | 1.7 | 2.8 | | 1.7 | 2.8 | mA |
| Power Consumption | $T_A = 25°C$ | | | | | | | | | | |
| | $V_S = \pm 20V$ | | 80 | 150 | | | | | | | mW |
| | $V_S = \pm 15V$ | | | | | 50 | 85 | | 50 | 85 | mW |
| LM741A | $V_S = \pm 20V$ | | | | | | | | | | |
| | $T_A = T_{AMIN}$ | | | 165 | | | | | | | mW |
| | $T_A = T_{AMAX}$ | | | 135 | | | | | | | mW |
| LM741E | $V_S = \pm 20V$ | | | 150 | | | | | | | mW |
| | $T_A = T_{AMIN}$ | | | 150 | | | | | | | mW |
| | $T_A = T_{AMAX}$ | | | 150 | | | | | | | mW |
| LM741 | $V_S = \pm 15V$ | | | | | | | | | | |
| | $T_A = T_{AMIN}$ | | | | | 60 | 100 | | | | mW |
| | $T_A = T_{AMAX}$ | | | | | 45 | 75 | | | | mW |

**Note 1:** The maximum junction temperature of the LM741/LM741A is 150°C, while that of the LM741C/LM741E is 100°C. For operation at elevated temperatures, devices in the TO-5 package must be derated based on a thermal resistance of 150°C/W junction to ambient, or 45°C/W junction to case. The thermal resistance of the dual-in-line package is 100°C/W junction to ambient.

**Note 2:** For supply voltages less than ±15V, the absolute maximum input voltage is equal to the supply voltage.

**Note 3:** Unless otherwise specified, these specifications apply for $V_S = \pm 15V$, $-55°C \le T_A \le +125°C$ (LM741/LM741A). For the LM741C/LM741E, these specifications are limited to $0°C \le T_A \le +70°C$.

**Note 4:** Calculated value from: BW (MHz) = 0.35/Rise Time(µs).

*Courtesy National Semiconductor Corp.*

LM124/LM224/LM324, LM124A/LM224A/LM324A, LM2902

# Operational Amplifiers/Buffers

## LM124/LM224/LM324, LM124A/LM224A/LM324A, LM2902 low power quad operational amplifiers

### general description

The LM124 series consists of four independent, high gain, internally frequency compensated operational amplifiers which were designed specifically to operate from a single power supply over a wide range of voltages. Operation from split power supplies is also possible and the low power supply current drain is independent of the magnitude of the power supply voltage.

Application areas include transducer amplifiers, dc gain blocks and all the conventional op amp circuits which now can be more easily implemented in single power supply systems. For example, the LM124 series can be directly operated off of the standard +5 $V_{DC}$ power supply voltage which is used in digital systems and will easily provide the required interface electronics without requiring the additional ±15 $V_{DC}$ power supplies.

### unique characteristics

- In the linear mode the input common-mode voltage range includes ground and the output voltage can also swing to ground, even though operated from only a single power supply voltage.
- The unity gain cross frequency is temperature compensated.
- The input bias current is also temperature compensated.

### advantages

- Eliminates need for dual supplies
- Four internally compensated op amps in a single package
- Allows directly sensing near GND and $V_{OUT}$ also goes to GND
- Compatible with all forms of logic
- Power drain suitable for battery operation

### features

- Internally frequency compensated for unity gain
- Large dc voltage gain 100 dB
- Wide bandwidth (unity gain) (temperature compensated) 1 MHz
- Wide power supply range: Single supply 3 $V_{DC}$ to 30 $V_{DC}$ or dual supplies ±1.5 $V_{DC}$ to ±15 $V_{DC}$
- Very low supply current drain (800µA) – essentially independent of supply voltage (1 mW/op amp at +5 $V_{DC}$)
- Low input biasing current (temperature compensated) 45 $nA_{DC}$
- Low input offset voltage 2 $mV_{DC}$ and offset current 5 $nA_{DC}$
- Input common-mode voltage range includes ground
- Differential input voltage range equal to the power supply voltage
- Large output voltage swing 0 $V_{DC}$ to $V^+$ − 1.5 $V_{DC}$

### connection diagram

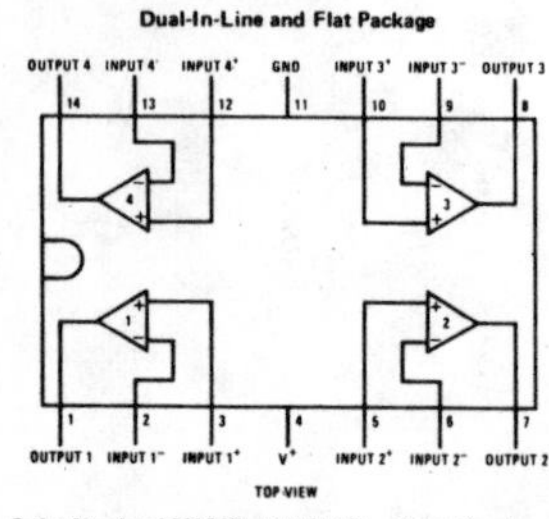


**Order Number LM124D, LM124AD, LM224D or LM224AD**
**See NS Package D14E**

**Order Number LM124F, LM124AF, LM224F or LM224AF**
**See NS Package F14A**

**Order Number LM124J, LM124AJ, LM224J, LM224AJ, LM324J, LM324AJ or LM2902J**
**See NS Package J14A**

**Order Number LM324N, LM324AN or LM2902N**
**See NS Package N14A**

### schematic diagram (Each Amplifier)

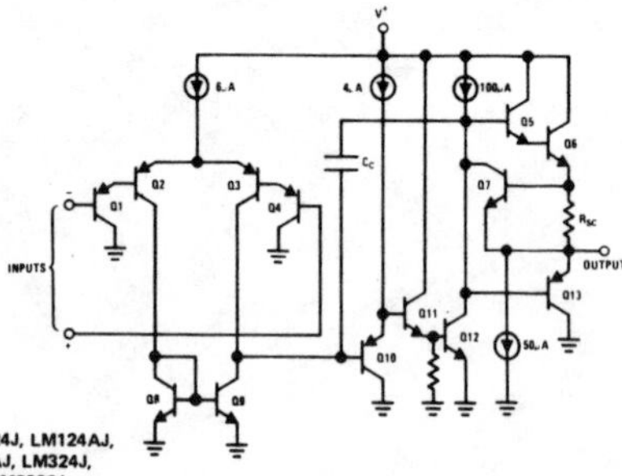

*Courtesy National Semiconductor Corp.*

# LM124/LM224/LM324, LM124A/LM224A/LM324A, LM2902

## absolute maximum ratings

| | LM124/LM224/LM324<br>LM124A/LM224A/LM324A | LM2902 |
|---|---|---|
| Supply Voltage, $V^+$ | 32 $V_{DC}$ or ±16 $V_{DC}$ | 26 $V_{DC}$ or ±13 $V_{DC}$ |
| Differential Input Voltage | 32 $V_{DC}$ | 26 $V_{DC}$ |
| Input Voltage | −0.3 $V_{DC}$ to +26 $V_{DC}$ | −0.3 $V_{DC}$ to +26 $V_{DC}$ |
| Power Dissipation (Note 1) | | |
| Molded DIP | 570 mW | 570 mW |
| Cavity DIP | 900 mW | |
| Flat Pack | 800 mW | |
| Output Short-Circuit to GND (One Amplifier) (Note 2)<br>$V^+ \leq 15\ V_{DC}$ and $T_A = 25°C$ | Continuous | Continuous |

| | LM124/LM224/LM324<br>LM124A/LM224A/LM324A | LM2902 |
|---|---|---|
| Input Current ($V_{IN} < -0.3\ V_{DC}$) (Note 3) | 50 mA | 50 mA |
| Operating Temperature Range | | −40°C to +85°C |
| LM324/LM324A | 0°C to +70°C | |
| LM224/LM224A | −25°C to +85°C | |
| LM124/LM124A | −55°C to +125°C | |
| Storage Temperature Range | −65°C to +150°C | −65°C to +150°C |
| Lead Temperature (Soldering, 10 seconds) | 300°C | 300°C |

## electrical characteristics ($V^+ = +5.0\ V_{DC}$, Note 4)

| PARAMETER | CONDITIONS | LM124A | | | LM224A | | | LM324A | | | LM124/LM224 | | | LM324 | | | LM2902 | | | UNITS |
|---|---|---|---|---|---|---|---|---|---|---|---|---|---|---|---|---|---|---|---|---|
| | | MIN | TYP | MAX | MIN | TYP | MAX | MIN | TYP | MAX | MIN | TYP | MAX | MIN | TYP | MAX | MIN | TYP | MAX | |
| Input Offset Voltage | $T_A = 25°C$, (Note 5) | | 1 | 2 | | 1 | 3 | | 2 | 3 | | ±2 | ±5 | | ±2 | ±7 | | ±2 | ±7 | $mV_{DC}$ |
| Input Bias Current (Note 6) | $I_{IN(+)}$ or $I_{IN(-)}$, $T_A = 25°C$ | | 20 | 50 | | 40 | 80 | | 45 | 100 | | 45 | 150 | | 45 | 250 | | 45 | 250 | $nA_{DC}$ |
| Input Offset Current | $I_{IN(+)} - I_{IN(-)}$, $T_A = 25°C$ | | 2 | 10 | | 2 | 15 | | 5 | 30 | | ±3 | ±30 | | ±5 | ±50 | | ±5 | ±50 | $nA_{DC}$ |
| Input Common-Mode Voltage Range (Note 7) | $V^+ = 30\ V_{DC}$, $T_A = 25°C$ | 0 | | $V^+-1.5$ | 0 | | $V^+-1.5$ | 0 | | $V^+-1.5$ | 0 | | $V^+-1.5$ | 0 | | $V^+-1.5$ | 0 | | $V^+-1.5$ | $V_{DC}$ |
| Supply Current | $R_L = \infty$, $V_{CC} = 30V$, (LM2902 $V_{CC} = 26V$)<br>$R_L = \infty$ On All Op Amps<br>Over Full Temperature Range | | 1.5<br>0.7 | 3<br>1.2 | | 1.5<br>0.7 | 3<br>1.2 | | 1.5<br>0.7 | 3<br>1.2 | | 1.5<br>0.7 | 3<br>1.2 | | 1.5<br>0.7 | 3<br>1.2 | | 1.5<br>0.7 | 3<br>1.2 | $mA_{DC}$<br>$mA_{DC}$ |
| Large Signal Voltage Gain | $V^+ = 15\ V_{DC}$ (For Large $V_O$ Swing)<br>$R_L \geq 2\ k\Omega$, $T_A = 25°C$ | 50 | 100 | | 50 | 100 | | 25 | 100 | | 50 | 100 | | 25 | 100 | | | 100 | | V/mV |
| Output Voltage Swing | $R_L = 2\ k\Omega$, $T_A = 25°C$ (LM2902 $R_L \geq 10\ k\Omega$) | 0 | | $V^+-1.5$ | 0 | | $V^+-1.5$ | 0 | | $V^+-1.5$ | 0 | | $V^+-1.5$ | 0 | | $V^+-1.5$ | 0 | | $V^+-1.5$ | $V_{DC}$ |
| Common-Mode Rejection Ratio | DC, $T_A = 25°C$ | 70 | 85 | | 70 | 85 | | 65 | 85 | | 70 | 85 | | 65 | 70 | | 50 | 70 | | dB |
| Power Supply Rejection Ratio | DC, $T_A = 25°C$ | 65 | 100 | | 65 | 100 | | 65 | 100 | | 65 | 100 | | 65 | 100 | | 50 | 100 | | dB |
| Amplifier-to-Amplifier Coupling (Note 8) | f = 1 kHz to 20 kHz, $T_A = 25°C$ (Input Referred) | | −120 | | | −120 | | | −120 | | | −120 | | | −120 | | | −120 | | dB |
| Output Current | | | | | | | | | | | | | | | | | | | | |
| Source | $V_{IN}^+ = 1\ V_{DC}$, $V_{IN}^- = 0\ V_{DC}$, $V^+ = 15\ V_{DC}$, $T_A = 25°C$ | 20 | 40 | | 20 | 40 | | 20 | 40 | | 20 | 40 | | 20 | 40 | | 20 | 40 | | $mA_{DC}$ |
| Sink | $V_{IN}^- = 1\ V_{DC}$, $V_{IN}^+ = 0\ V_{DC}$, $V^+ = 15\ V_{DC}$, $T_A = 25°C$ | 10 | 20 | | 10 | 20 | | 10 | 20 | | 10 | 20 | | 10 | 20 | | 10 | 20 | | $mA_{DC}$ |
| | $V_{IN}^- = 1\ V_{DC}$, $V_{IN}^+ = 0\ V_{DC}$, $T_A = 25°C$, $V_O = 200\ mV_{DC}$ | 12 | 50 | | 12 | 50 | | 12 | 50 | | 12 | 50 | | 12 | 50 | | | | | $\mu A_{DC}$ |
| Short Circuit to Ground | $T_A = 25°C$, (Note 2) | | 40 | 60 | | 40 | 60 | | 40 | 60 | | 40 | 60 | | 40 | 60 | | 40 | 60 | $mA_{DC}$ |

*Courtesy National Semiconductor Corp.*

# LM124/LM224/LM324, LM124A/LM224A/LM324A, LM2902

## electrical characteristics (con't)

| PARAMETER | CONDITIONS | LM124A | | | LM224A | | | LM324A | | | LM124/LM224 | | | LM324 | | | LM2902 | | | UNITS |
|---|---|---|---|---|---|---|---|---|---|---|---|---|---|---|---|---|---|---|---|---|
| | | MIN | TYP | MAX | MIN | TYP | MAX | MIN | TYP | MAX | MIN | TYP | MAX | MIN | TYP | MAX | MIN | TYP | MAX | |
| Input Offset Voltage | (Note 5) | | | 4 | | | 4 | | | 5 | | | ±7 | | | ±9 | | | ±10 | $mV_{DC}$ |
| Input Offset Voltage Drift | $R_S = 0\Omega$ | | 7 | 20 | | 7 | 20 | | 7 | 30 | | 7 | | | 7 | | | 7 | | $\mu V/^\circ C$ |
| Input Offset Current | $I_{IN(+)} - I_{IN(-)}$ | | | 30 | | | 30 | | | 75 | | | ±100 | | | ±150 | | 45 | ±200 | $nA_{DC}$ |
| Input Offset Current Drift | | | 10 | 200 | | 10 | 200 | | 10 | 300 | | 10 | | | 10 | | | 10 | | $pA_{DC}/^\circ C$ |
| Input Bias Current | $I_{IN(+)}$ or $I_{IN(-)}$ | | 40 | 100 | | 40 | 100 | | 40 | 200 | | 40 | 300 | | 40 | 500 | | 40 | 500 | $nA_{DC}$ |
| Input Common-Mode Voltage Range (Note 7) | $V^+ = 30\ V_{DC}$ | 0 | | $V^+-2$ | 0 | | $V^+-2$ | 0 | | $V^+-2$ | 0 | | $V^+-2$ | 0 | | $V^+-2$ | 0 | | $V^+-2$ | $V_{DC}$ |
| Large Signal Voltage Gain | $V^+ = +15\ V_{DC}$ (For Large $V_O$ Swing) $R_L \geq 2\ k\Omega$ | 25 | | | 25 | | | 15 | | | 25 | | | 15 | | | 15 | | | V/mV |
| Output Voltage Swing $V_{OH}$ | $V^+ = +30\ V_{DC}$, $R_L = 2\ k\Omega$ | 26 | | | 26 | | | 26 | | | 26 | | | 26 | | | 22 | | | $V_{DC}$ |
| | $R_L \geq 10\ k\Omega$ | 27 | 28 | | 27 | 28 | | 27 | 28 | | 27 | 28 | | 27 | 28 | | 23 | 24 | | $V_{DC}$ |
| $V_{OL}$ | $V^+ = 5\ V_{DC}$, $R_L \leq 10\ k\Omega$ | | 5 | 20 | | 5 | 20 | | 5 | 20 | | 5 | 20 | | 5 | 20 | | 5 | 100 | $mV_{DC}$ |
| Output Current Source | $V_{IN}^+ = +1\ V_{DC}$, $V_{IN}^- = 0\ V_{DC}$, $V^+ = 15\ V_{DC}$ | 10 | 20 | | 10 | 20 | | 10 | 20 | | 10 | 20 | | 10 | 20 | | 10 | 20 | | $mA_{DC}$ |
| Sink | $V_{IN}^- = +1\ V_{DC}$, $V_{IN}^+ = 0\ V_{DC}$, $V^+ = 15\ V_{DC}$ | 10 | 15 | | 5 | 8 | | 5 | 8 | | 5 | 8 | | 5 | 8 | | 5 | 8 | | $mA_{DC}$ |
| Differential Input Voltage | (Note 7) | | | $V^+$ | | | $V^+$ | | | $V^+$ | | | $V^+$ | | | $V^+$ | | | $V^+$ | $V_{DC}$ |

**Note 1:** For operating at high temperatures, the LM324/LM324A, LM2902 must be derated based on a +125°C maximum junction temperature and a thermal resistance of 175°C/W which applies for the device soldered in a printed circuit board, operating in a still air ambient. The LM224/LM224A and LM124/LM124A can be derated based on a +150°C maximum junction temperature. The dissipation is the total of all four amplifiers—use external resistors, where possible, to allow the amplifier to saturate or to reduce the power which is dissipated in the integrated circuit.

**Note 2:** Short circuits from the output to $V^+$ can cause excessive heating and eventual destruction. The maximum output current is approximately 40 mA independent of the magnitude of $V^+$. At values of supply voltage in excess of +15 $V_{DC}$, continuous short-circuits can exceed the power dissipation ratings and cause eventual destruction. Destructive dissipation can result from simultaneous shorts on all amplifiers.

**Note 3:** This input current will only exist when the voltage at any of the input leads is driven negative. It is due to the collector-base junction of the input PNP transistors becoming forward biased and thereby acting as input diode clamps. In addition to this diode action, there is also lateral NPN parasitic transistor action on the IC chip. This transistor action can cause the output voltages of the op amps to go to the $V^+$ voltage level (or to ground for a large overdrive) for the time duration that an input is driven negative. This is not destructive and normal output states will re-establish when the input voltage, which was negative, again returns to a value greater than −0.3 $V_{DC}$.

**Note 4:** These specifications apply for $V^+ = +5\ V_{DC}$ and $-55^\circ C \leq T_A \leq +125^\circ C$, unless otherwise stated. With the LM224/LM224A, all temperature specifications are limited to $-25^\circ C \leq T_A \leq +85^\circ C$, the LM324/LM324A temperature specifications are limited to $0^\circ C \leq T_A \leq +70^\circ C$, and the LM2902 specifications are limited to $-40^\circ C \leq T_A \leq +85^\circ C$.

**Note 5:** $V_O \cong 1.4\ V_{DC}$, $R_S = 0\Omega$ with $V^+$ from 5 $V_{DC}$ to 30 $V_{DC}$; and over the full input common-mode range (0 $V_{DC}$ to $V^+ - 1.5\ V_{DC}$).

**Note 6:** The direction of the input current is out of the IC due to the PNP input stage. This current is essentially constant, independent of the state of the output so no loading change exists on the input lines.

**Note 7:** The input common-mode voltage or either input signal voltage should not be allowed to go negative by more than 0.3V. The upper end of the common-mode voltage range is $V^+ -1.5V$, but either or both inputs can go to +32 $V_{DC}$ without damage (+26 $V_{DC}$ for LM2902).

**Note 8:** Due to proximity of external components, insure that coupling is not originating via stray capacitance between these external parts. This typically can be detected as this type of capacitive increases at higher frequencies.

*Courtesy National Semiconductor Corp.*

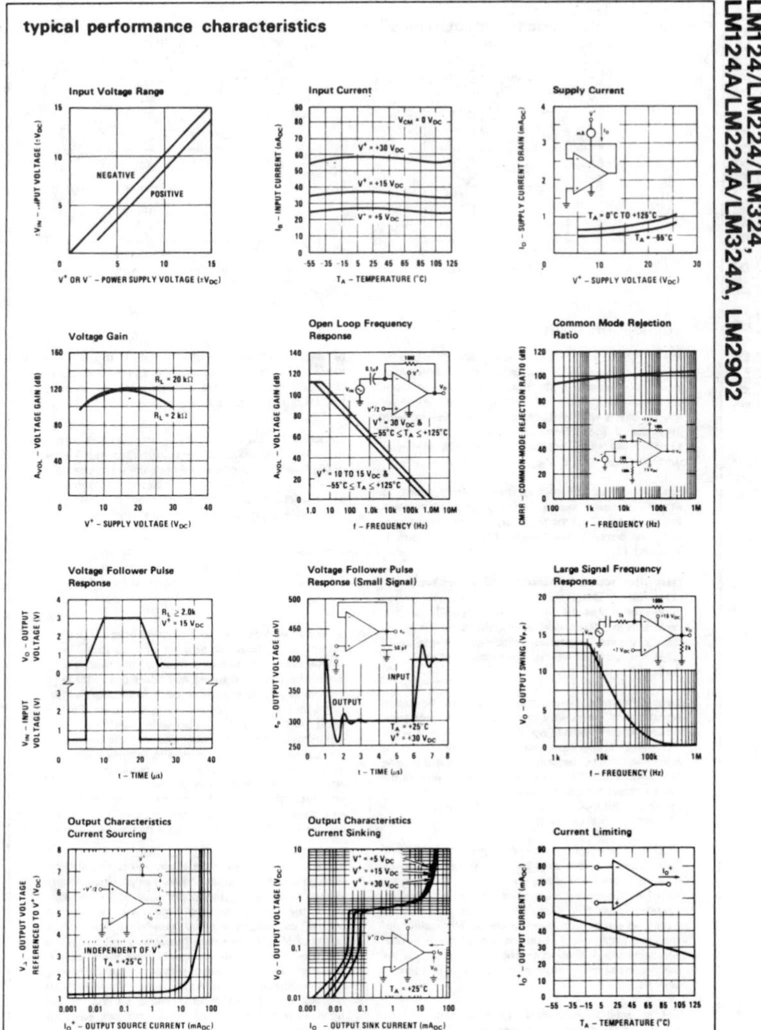


*Courtesy National Semiconductor Corp.*

## typical performance characteristics (LM2902 only)

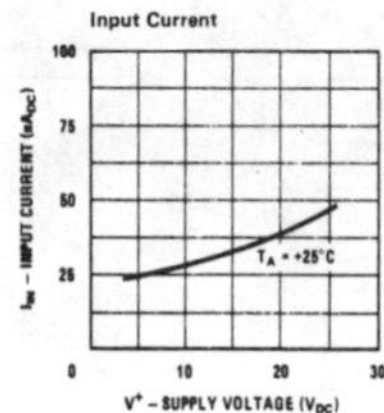


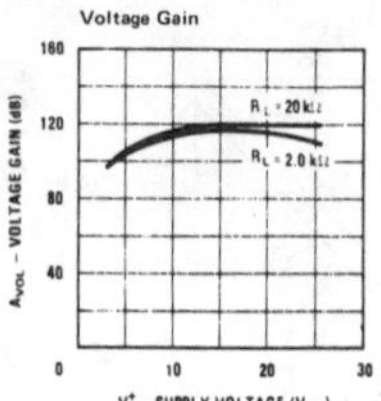


## application hints

The LM124 series are op amps which operate with only a single power supply voltage, have true-differential inputs, and remain in the linear mode with an input common-mode voltage of 0 $V_{DC}$. These amplifiers operate over a wide range of power supply voltage with little change in performance characteristics. At 25°C amplifier operation is possible down to a minimum supply voltage of 2.3 $V_{DC}$.

The pinouts of the package have been designed to simplify PC board layouts. Inverting inputs are adjacent to outputs for all of the amplifiers and the outputs have also been placed at the corners of the package (pins 1, 7, 8, and 14).

Precautions should be taken to insure that the power supply for the integrated circuit never becomes reversed in polarity or that the unit is not inadvertently installed backwards in a test socket as an unlimited current surge through the resulting forward diode within the IC could cause fusing of the internal conductors and result in a destroyed unit.

Large differential input voltages can be easily accommodated and, as input differential voltage protection diodes are not needed, no large input currents result from large differential input voltages. The differential input voltage may be larger than $V^+$ without damaging the device. Protection should be provided to prevent the input voltages from going negative more than −0.3 $V_{DC}$ (at 25°C). An input clamp diode with a resistor to the IC input terminal can be used.

To reduce the power supply current drain, the amplifiers have a class A output stage for small signal levels which converts to class B in a large signal mode. This allows the amplifiers to both source and sink large output currents. Therefore both NPN and PNP external current boost transistors can be used to extend the power capability of the basic amplifiers. The output voltage needs to raise approximately 1 diode drop above ground to bias the on-chip vertical PNP transistor for output current sinking applications.

For ac applications, where the load is capacitively coupled to the output of the amplifier, a resistor should be used, from the output of the amplifier to ground to increase the class A bias current and prevent crossover distortion. Where the load is directly coupled, as in dc applications, there is no crossover distortion.

Capacitive loads which are applied directly to the output of the amplifier reduce the loop stability margin. Values of 50 pF can be accommodated using the worst-case non-inverting unity gain connection. Large closed loop gains or resistive isolation should be used if larger load capacitance must be driven by the amplifier.

The bias network of the LM124 establishes a drain current which is independent of the magnitude of the power supply voltage over the range of from 3 $V_{DC}$ to 30 $V_{DC}$.

Output short circuits either to ground or to the positive power supply should be of short time duration. Units can be destroyed, not as a result of the short circuit current causing metal fusing, but rather due to the large increase in IC chip dissipation which will cause eventual failure due to excessive junction temperatures. Putting direct short-circuits on more than one amplifier at a time will increase the total IC power dissipation to destructive levels, if not properly protected with external dissipation limiting resistors in series with the output leads of the amplifiers. The larger value of output source current which is available at 25°C provides a larger output current capability at elevated temperatures (see typical performance characteristics) than a standard IC op amp.

The circuits presented in the section on typical applications emphasize operation on only a single power supply voltage. If complementary power supplies are available, all of the standard op amp circuits can be used. In general, introducing a pseudo-ground (a bias voltage reference of $V^+/2$) will allow operation above and below this value in single power supply systems. Many application circuits are shown which take advantage of the wide input common-mode voltage range which includes ground. In most cases, input biasing is not required and input voltages which range to ground can easily be accommodated.

*Courtesy National Semiconductor Corp.*

RCA
Solid State
Division

# Linear Integrated Circuits

Monolithic Silicon

# CA3140B, CA3140A, CA3140 Types

File Number 957

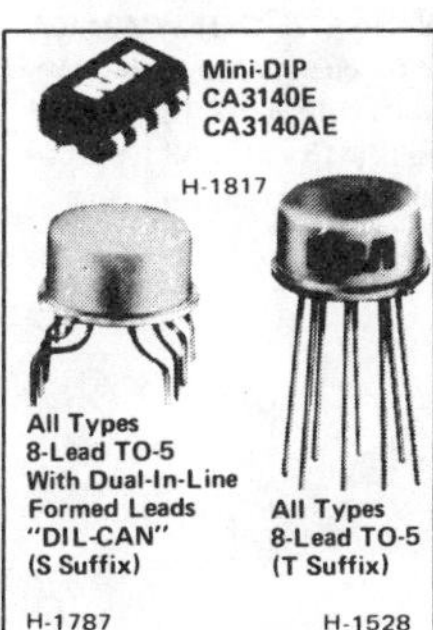


## BiMOS Operational Amplifiers

With MOS/FET Input, Bipolar Output

*Features:*

- **MOS/FET Input Stage**
  - **(a) Very high input impedance ($Z_{IN}$) – 1.5 TΩ typ.**
  - **(b) Very low input current ($I_I$) – 10 pA typ. at ± 15 V**
  - **(c) Low input-offset voltage ($V_{IO}$) – to 2 mV max.**
  - **(d) Wide common-mode input-voltage range ($V_{ICR}$) – can be swung 0.5 volt below negative supply-voltage rail**
  - **(e) Output swing complements input common-mode range**
  - **(f) Rugged input stage – bipolar diode protected**
- **Directly replaces industry type 741 in most applications**
- **Includes numerous industry operational amplifier categories such as general-purpose, FET input, wideband (high slew rate)**
- **Operation from 4-to-44 volts Single or Dual supplies**
- **Internally compensated**
- **Characterized for ± 15-volt operation and for TTL supply systems with operation down to 4 volts**
- **Wide bandwidth – 4.5 MHz unity gain at ± 15 V or 30 V; 3.7 MHz at 5 V**
- **High voltage-follower slew rate – 9 V/μs**
- **Fast settling time – 1.4 μs typ. to 10 mV with a 10-$V_{p-p}$ signal**
- **Output swings to within 0.2 volt of negative supply**
- **Strobable output stage**

*Applications:*

- **Ground-referenced single-supply amplifiers in automobile and portable instrumentation**
- **Sample and hold amplifiers**
- **Long-duration timers/multivibrators (microseconds–minutes–hours)**
- **Photocurrent instrumentation**
- **Peak detectors**
- **Active filters**
- **Comparators**
- **Interface in 5 V TTL systems & other low-supply voltage systems**
- **All standard operational amplifier applications**
- **Function generators**
- **Tone controls**
- **Power supplies**
- **Portable instruments**
- **Intrusion alarm systems**

The CA3140B, CA3140A, and CA3140 are integrated-circuit operational amplifiers that combine the advantages of high-voltage PMOS transistors with high-voltage bipolar transistors on a single monolithic chip. Because of this unique combination of technologies, this device can now provide designers, for the first time, with the special performance features of the CA3130 COS/MOS operational amplifiers and the versatility of the 741 series of industry-standard operational amplifiers.

The CA3140,CA3140A , and CA3140 BiMOS operational amplifiers feature gate-protected MOS/FET (PMOS) transistors in the input circuit to provide very-high-input impedance, very-low-input current, and high-speed performance. The CA3140B operates at supply voltages from 4 to 44 volts; the CA3140A and CA3140 from 4 to 36 volts (either single or dual supply). These operational amplifiers are internally phase-compensated to achieve stable operation in unity-gain follower operation, and, additionally, have access terminals for a supplementary external capacitor if additional frequency roll-off is desired. Terminals are also provided for use in applications requiring input offset-voltage nulling.

The use of PMOS field-effect transistors in the input stage results in common-mode input-voltage capability down to 0.5 volt below the negative-supply terminal, an important attribute for single-supply applications. The output stage uses bipolar transistors and includes built-in protection against damage from load-terminal short-circuiting to either supply-rail or to ground.

The CA3140 Series has the same 8-lead terminal pin-out used for the "741" and other





Printed in USA/3-77

*Supersedes issue dated 5-76*

*Courtesy RCA Solid State Division*

industry-standard operational amplifiers. They are supplied in either the standard 8-lead TO-5 style package (T suffix), or in the 8-lead dual-in-line formed-lead TO-5 style package "DIL-CAN" (S suffix). The CA3140 is available in chip form (H suffix). The CA3140A and CA3140 are also available in an 8-lead dual-in-line plastic package (Mini-DIP-E suffix). The CA3140B is intended for operation at supply voltages ranging from 4 to 44 volts, for applications requiring premium-grade specifications and with electrical limits established for operations over the range from −55°C to +125°C. The CA3140A and CA3140 are for operation at supply voltages up to 36 volts (±18 volts). The CA3140 ages up to 36 volts (±18 volts). All types can be operated safely over the temperature range from −55°C to +125°C.

### TYPICAL ELECTRICAL CHARACTERISTICS

| CHARACTERISTIC | | TEST CONDITIONS $V^+ = +15$ V, $V^- = -15$ V, $T_A = 25^{\circ}$C | CA3140B (T,S) | CA3140A (T,S,E) | CA3140 (T,S,E) | UNITS |
|---|---|---|---|---|---|---|
| Input Offset Voltage Adjustment Resistor | | Typ. Value of Resistor Between Term. 4 and 5 or 4 and 1 to Adjust Max. $V_{IO}$ | 43 | 18 | 4.7 | kΩ |
| Input Resistance | $R_I$ | | 1.5 | 1.5 | 1.5 | TΩ |
| Input Capacitance | $C_I$ | | 4 | 4 | 4 | pF |
| Output Resistance | $R_O$ | | 60 | 60 | 60 | Ω |
| Equivalent Wideband Input Noise Voltage (See Fig. 39) | $e_n$ | BW = 140 kHz, $R_S$ = 1 MΩ | 48 | 48 | 48 | μV |
| Equivalent Input Noise Voltage (See Fig. 10) | $e_n$ | f = 1 kHz, $R_S$ = 100 Ω | 40 | 40 | 40 | nV/$\sqrt{Hz}$ |
| | | f = 10 kHz, $R_S$ = 100 Ω | 12 | 12 | 12 | nV/$\sqrt{Hz}$ |
| Short-Circuit Current to Opposite Supply: Source | $I_{OM}^+$ | | 40 | 40 | 40 | mA |
| Sink | $I_{OM}^-$ | | 18 | 18 | 18 | mA |
| Gain-Bandwidth Product, (See Figs. 5 & 18) | $f_T$ | | 4.5 | 4.5 | 4.5 | MHz |
| Slew Rate, (See Fig. 6) | SR | | 9 | 9 | 9 | V/μs |
| Sink Current From Terminal 8 To Terminal 4 to Swing Output Low | | | 220 | 220 | 220 | μA |
| Transient Response: Rise Time | $t_r$ | $R_L$ = 2 kΩ, $C_L$ = 100 pF | 0.08 | 0.08 | 0.08 | μs |
| Overshoot (See Fig. 37) | | $R_L$ = 2 kΩ, $C_L$ = 100 pF | 10 | 10 | 10 | % |
| Settling Time at 10 $V_{p-p}$, (See Fig. 17): 1 mV | $t_s$ | $R_L$ = 2 kΩ, $C_L$ = 100 pF, Voltage Follower | 4.5 | 4.5 | 4.5 | μs |
| 10 mV | $t_s$ | $R_L$ = 2 kΩ, $C_L$ = 100 pF, Voltage Follower | 1.4 | 1.4 | 1.4 | μs |

*Courtesy RCA Solid State Division*

**ELECTRICAL CHARACTERISTICS FOR EQUIPMENT DESIGN**
**At $V^+$ = 15 V, $V^-$ = 15 V, $T_A$ = 25°C Unless Otherwise Specified**

| CHARACTERISTIC | | LIMITS | | | | | | | | UNITS |
|---|---|---|---|---|---|---|---|---|---|---|
| | CA3140B | | | CA3140A | | | CA3140 | | | |
| | Min. | Typ. | Max. | Min. | Typ. | Max. | Min. | Typ. | Max. | |
| Input Offset Voltage, $\|V_{IO}\|$ | – | 0.8 | 2 | – | 2 | 5 | – | 5 | 15 | mV |
| Input Offset Current, $\|I_{IO}\|$ | – | 0.5 | 10 | – | 0.5 | 20 | – | 0.5 | 30 | pA |
| Input Current, $I_I$ | – | 10 | 30 | – | 10 | 40 | – | 10 | 50 | pA |
| Large-Signal Voltage Gain, $A_{OL}$● (See Figs. 4,18) | 50 k | 100 k | – | 20 k | 100 k | – | 20 k | 100 k | – | V/V |
| | 94 | 100 | – | 86 | 100 | – | 86 | 100 | – | dB |
| Common-Mode Rejection Ratio, CMRR (See Fig.9) | – | 20 | 50 | – | 32 | 320 | – | 32 | 320 | μV/V |
| | 86 | 94 | – | 70 | 90 | – | 70 | 90 | – | dB |
| Common-Mode Input-Voltage Range, $V_{ICR}$ (See Fig.20) | –15 | –15.5 to +12.5 | 12 | –15 | –15.5 to +12.5 | 12 | –15 | –15.5 to +12.5 | 11 | V |
| Power-Supply Rejection Ratio, PSRR (See Fig.11) $\Delta V_{IO}/\Delta V$ | – | 32 | 100 | – | 100 | 150 | – | 100 | 150 | μV/V |
| | 80 | 90 | – | 76 | 80 | – | 76 | 80 | – | dB |
| Max. Output Voltage■ (See Figs.13,20) $V_{OM}^+$ | +12 | 13 | – | +12 | 13 | – | +12 | 13 | – | V |
| $V_{OM}^-$ | –14 | –14.4 | – | –14 | –14.4 | – | –14 | –14.4 | – | V |
| Supply Current, $I^+$ (See Fig.7) | – | 4 | 6 | – | 4 | 6 | – | 4 | 6 | mA |
| Device Dissipation, $P_D$ | – | 120 | 180 | – | 120 | 180 | – | 120 | 180 | mW |
| Input Current, $I_I$▲ (See Fig.19) | – | 10 | 30 | – | 10 | – | – | 10 | – | nA |
| Input Offset Voltage $\|V_{IO}\|$▲ | – | 1.3 | 3 | – | 3 | – | – | 10 | – | mV |
| Input Offset Voltage Temp. Drift, $\Delta V_{IO}/\Delta T$ | – | 5 | – | – | 6 | – | – | 8 | – | μV/°C |
| Large-Signal Voltage Gain, $A_{OL}$▲ (See Figs.4,18) | 20 k | 100 k | – | – | 100 k | – | – | 100 k | – | V/V |
| | 86 | 100 | – | – | 100 | – | – | 100 | – | dB |
| Max. Output Voltage,★ $V_{OM}^+$ | +19 | +19.5 | – | – | – | – | – | – | – | V |
| $V_{OM}^-$ | –21 | –21.4 | – | – | – | – | – | – | – | V |
| Large-Signal Voltage Gain, $A_{OL}$♠★ | 20 k | 50 k | – | – | – | – | – | – | – | V/V |
| | 86 | 94 | – | – | – | – | – | – | – | dB |

● At $V_O$ = 26$V_{p-p}$, +12V, –14V and $R_L$ = 2 kΩ.
■ At $R_L$ = 2 kΩ.
▲ At $T_A$ = –55°C to +125°C, $V^+$ = 15 V, $V^-$ = 15 V, $V_O$ = 26$V_{p-p}$, $R_L$ = 2 kΩ.
♠ At $V_O$ = +19 V, –21 V, and $R_L$ = 2 kΩ.
★ At $V^+$ = 22 V, $V^-$ = 22 V.

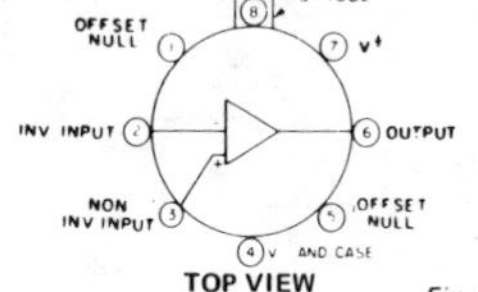

TOP VIEW
S and T Suffixes

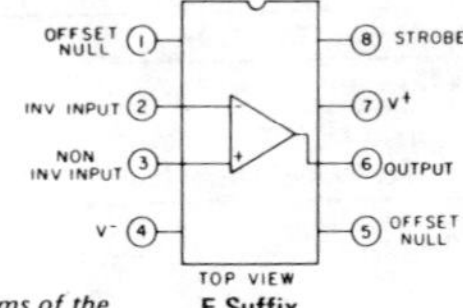

TOP VIEW
E Suffix

*Fig. 1 – Functional diagrams of the CA3140 series.*

*Courtesy RCA Solid State Division*

**MAXIMUM RATINGS,** *Absolute-Maximum Values:*

| | CA3140, CA3140A | CA3140B |
|---|---|---|
| DC SUPPLY VOLTAGE (BETWEEN V+ AND V− TERMINALS) | 36 V | 44 V |
| DIFFERENTIAL-MODE INPUT VOLTAGE | ±8 V | ±8 V |
| COMMON-MODE DC INPUT VOLTAGE | (V+ +8 V) to (V− −0.5 V) | |
| INPUT-TERMINAL CURRENT | 1 mA | |
| DEVICE DISSIPATION: | | |
| WITHOUT HEAT SINK – | | |
| UP TO 55°C | 630 mW | |
| ABOVE 55°C | Derate linearly 6.67 mW/°C | |
| WITH HEAT SINK – | | |
| Up to 55°C | 1 W | |
| Above 55°C | Derate linearly 16.7 mW/°C | |
| TEMPERATURE RANGE: | | |
| OPERATING (ALL TYPES) | −55 to + 125°C | |
| STORAGE (ALL TYPES) | −65 to +150°C | |
| OUTPUT SHORT-CIRCUIT DURATION* | INDEFINITE | |
| LEAD TEMPERATURE (DURING SOLDERING): AT DISTANCE 1/16 ± 1/32 INCH (1.59 ± 0.79 MM) FROM CASE FOR 10 SECONDS MAX. | +265°C | |

* Short circuit may be applied to ground or to either supply.

**TYPICAL ELECTRICAL CHARACTERISTICS FOR DESIGN GUIDANCE**
**At $V^+ = 5$ V, $V^- = 0$ V, $T_A = 25°C$**

| CHARACTERISTIC | | CA3140B (T,S) | CA3140A (T,S,E) | CA3140 (T,S,E) | UNITS |
|---|---|---|---|---|---|
| Input Offset Voltage | $\lvert V_{IO}\rvert$ | 0.8 | 2 | 5 | mV |
| Input Offset Current | $\lvert I_{IO}\rvert$ | 0.1 | 0.1 | 0.1 | pA |
| Input Current | $I_I$ | 2 | 2 | 2 | pA |
| Input Resistance | | 1 | 1 | 1 | TΩ |
| Large-Signal Voltage Gain (See Figs.4,18) | $A_{OL}$ | 100 k | 100 k | 100 k | V/V |
| | | 100 | 100 | 100 | dB |
| Common-Mode Rejection Ratio, | CMRR | 20 | 32 | 32 | μV/V |
| | | 94 | 90 | 90 | dB |
| Common-Mode Input-Voltage Range (See Fig.20) | $V_{ICR}$ | −0.5 | −0.5 | −0.5 | V |
| | | 2.6 | 2.6 | 2.6 | |
| Power-Supply Rejection Ratio | $\Delta V_{IO}/\Delta V^+$ | 32 | 100 | 100 | μV/V |
| | | 90 | 80 | 80 | dB |
| Maximum Output Voltage (See Figs.13,20) | $V_{OM}^+$ | 3 | 3 | 3 | V |
| | $V_{OM}^-$ | 0.13 | 0.13 | 0.13 | |
| Maximum Output Current: | | | | | |
| Source | $I_{OM}^+$ | 10 | 10 | 10 | mA |
| Sink | $I_{OM}^-$ | 1 | 1 | 1 | |
| Slew Rate (See Fig.6) | | 7 | 7 | 7 | V/μs |
| Gain-Bandwidth Product (See Fig.5) | $f_T$ | 3.7 | 3.7 | 3.7 | MHz |
| Supply Current (See Fig.7) | $I^+$ | 1.6 | 1.6 | 1.6 | mA |
| Device Dissipation | $P_D$ | 8 | 8 | 8 | mW |
| Sink Current from Term. 8 to Term. 4 to Swing Output Low | | 200 | 200 | 200 | μA |

*Courtesy RCA Solid State Division*

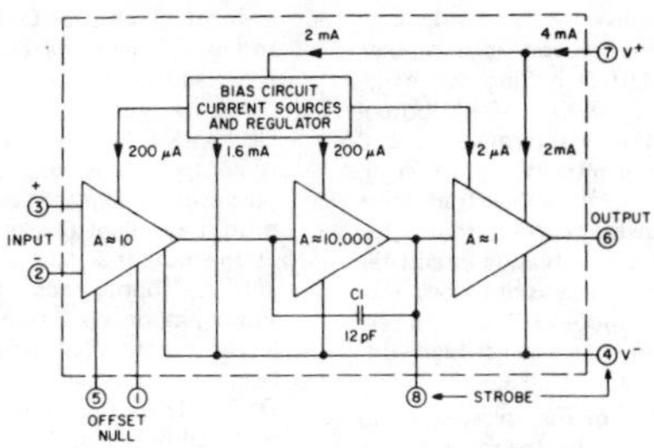


*Fig.2 – Block diagram of CA3140 series.*

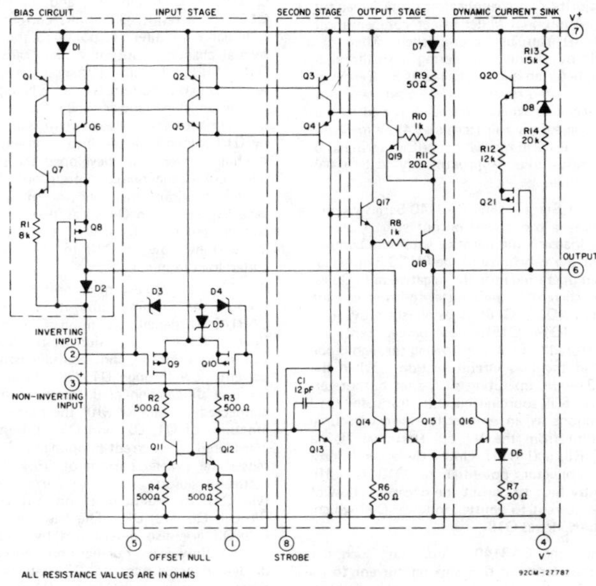


*Fig.3 – Schematic diagram of CA3140 series.*

## CIRCUIT DESCRIPTION

Fig.2 is a block diagram of the CA3140 Series PMOS Operational Amplifiers. The input terminals may be operated down to 0.5 V below the negative supply rail. Two class A amplifier stages provide the voltage gain, and a unique class AB amplifier stage provides the current gain necessary to drive low-impedance loads.

A biasing circuit provides control of cascoded constant-current flow circuits in the first and second stages. The CA3140 includes an on-chip phase-compensating capacitor that is sufficient for the unity gain voltage-follower configuration.

Input Stages – The schematic circuit diagram of the CA3140 is shown in Fig.3. It consists of a differential-input stage using PMOS field-effect transistors (Q9, Q10) working into a mirror pair of bipolar transistors (Q11, Q12) functioning as load resistors together with resistors R2 through R5. The mirror-pair transistors also function as a differen-

*Courtesy RCA Solid State Division*

tial-to-single-ended converter to provide base-current drive to the second-stage bipolar transistor (Q13). Offset nulling, when desired, can be effected with a 10-kΩ potentiometer connected across terminals 1 and 5 and with its slider arm connected to terminal 4. Cascode-connected bipolar transistors Q2, Q5 are the constant-current source for the input stage. The base-biasing circuit for the constant-current source is described subsequently. The small diodes D3, D4, D5 provide gate-oxide protection against high-voltage transients, e.g., static electricity.

Second Stage – Most of the voltage gain in the CA3140 is provided by the second amplifier stage, consisting of bipolar transistor Q13 and its cascode-connected load resistance provided by bipolar transistors Q3, Q4. On-chip phase compensation, sufficient for a majority of the applications is provided by C1. Additional Miller-Effect compensation (roll-off) can be accomplished, when desired, by simply connecting a small capacitor between terminals 1 and 8. Terminal 8 is also used to strobe the output stage into quiescence. When terminal 8 is tied to the negative supply rail (terminal 4) by mechanical or electrical means, the output terminal 6 swings low, i.e., approximately to terminal 4 potential.

Output Stage – The CA3140 Series circuits employ a broadband output stage that can sink loads to the negative supply to complement the capability of the PMOS input stage when operating near the negative rail. Quiescent current in the emitter-follower cascade circuit (Q17, Q18) is established by transistors (Q14, Q15) whose base-currents are "mirrored" to current flowing through diode D2 in the bias circuit section. When the CA3140 is operating such that output terminal 6 is sourcing current, transistor Q18 functions as an emitter-follower to source current from the V+ bus (terminal 7), via D7, R9, and R11. Under these conditions, the collector potential of Q13 is sufficiently high to permit the necessary flow of base current to emitter follower Q17 which, in turn, drives Q18.

When the CA3140 is operating such that output terminal 6 is sinking current to the V– bus, transistor Q16 is the current-sinking element. Transistor Q16 is mirror-connected to D6, R7, with current fed by way of Q21, R12, and Q20. Transistor Q20, in turn, is biased by current-flow through R13, zener D8, and R14. The dynamic current-sink is controlled by voltage-level sensing. For purposes of explanation, it is assumed that output terminal 6 is quiescently established at the potential mid-point between the V+ and V– supply rails. When output-current sinking-mode operation is required, the collector potential of transistor Q13 is driven below its quiescent level, thereby causing Q17, Q18 to decrease the output voltage at terminal 6. Thus, the gate terminal of PMOS transistor Q21 is displaced toward the V– bus, thereby reducing the channel resistance of Q21. As a consequence, there is an incremental increase in current flow through Q20, R12, Q21, D6, R7, and the base of Q16. As a result, Q16 sinks current from terminal 6 in direct response to the incremental change in output voltage caused by Q18. This sink current flows regardless of load; any excess current is internally supplied by the emitter-follower Q18. Short-circuit protection of the output circuit is provided by Q19, which is driven into conduction by the high voltage drop developed across R11 under output short-circuit conditions. Under these conditions, the collector of Q19 diverts current from Q4 so as to reduce the base-current drive from Q17, thereby limiting current flow in Q18 to the short-circuited load terminal.

Bias Circuit – Quiescent current in all stages (except the dynamic current sink) of the CA3140 is dependent upon bias current flow in R1. The function of the bias circuit is to establish and maintain constant-current flow through D1, Q6, Q8 and D2. D1 is a diode-connected transistor mirror-connected in parallel with the base-emitter junctions of Q1, Q2, and Q3. D1 may be considered as a current-sampling diode that senses the emitter current of Q6 and automatically adjusts the base current of Q6 (via Q1) to maintain a constant current through Q6, Q8, D2. The base-currents in Q2, Q3 are also determined by constant-current flow D1. Furthermore, current in diode-connected transistor D2 establishes the currents in transistors Q14 and Q15.

*Courtesy RCA Solid State Division*

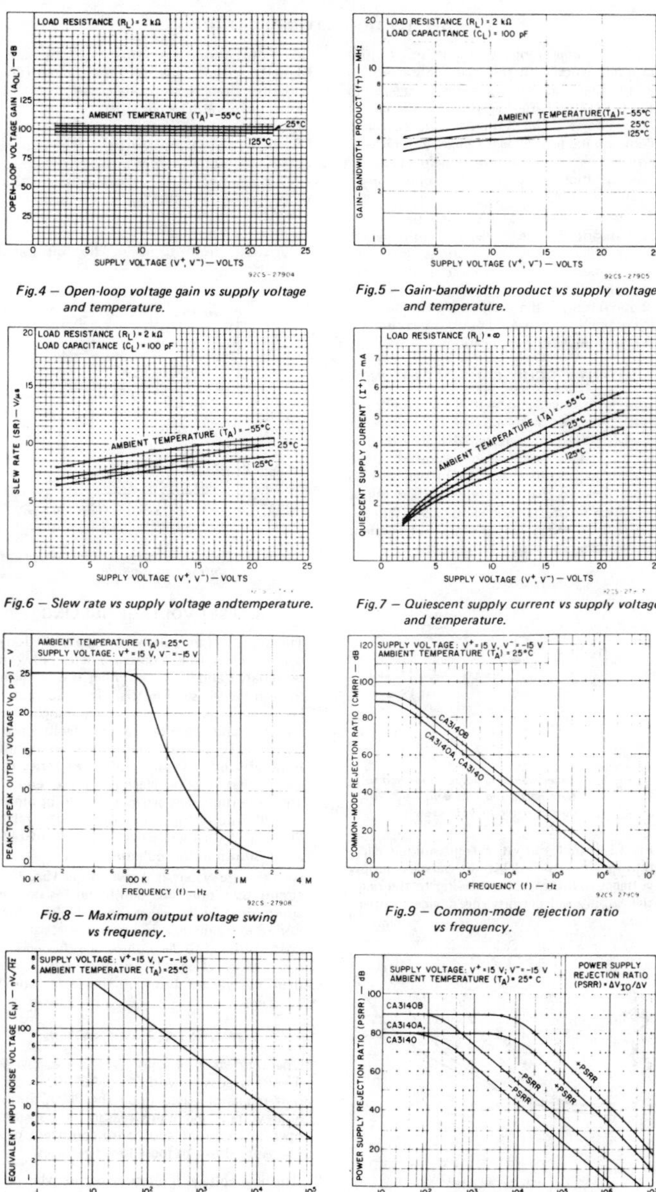


*Fig.4 – Open-loop voltage gain vs supply voltage and temperature.*

*Fig.5 – Gain-bandwidth product vs supply voltage and temperature.*

*Fig.6 – Slew rate vs supply voltage andtemperature.*

*Fig.7 – Quiescent supply current vs supply voltage and temperature.*

*Fig.8 – Maximum output voltage swing vs frequency.*

*Fig.9 – Common-mode rejection ratio vs frequency.*

*Fig.10 – Equivalent input noise voltage vs frequency.*

*Fig.11 – Power supply rejection ratio vs frequency.*

*Courtesy RCA Solid State Division*

## APPLICATIONS CONSIDERATIONS

Wide dynamic range of input and output characteristics with the most desirable high input-impedance characteristic is achieved in the CA3140 by the use of an unique design based upon the PMOS-Bipolar process. Input-common-mode voltage range and output-swing capabilities are complementary, allowing operation with the single supply down to four volts.

The wide dynamic range of these parameters also means that this device is suitable for many single-supply applications, such as, for example, where one input is driven below the potential of terminal 4 and the phase sense of the output signal must be maintained – a most important consideration in comparator applications.

### OUTPUT CIRCUIT CONSIDERATIONS

Excellent interfacing with TTL circuitry is easily achieved with a single 6.2-volt zener diode connected to terminal 8 as shown in Fig.12. This connection assures that the maximum output signal swing will not go more positive than the zener voltage minus two base-to-emitter voltage drops within the CA3140. These voltages are independent of the operating supply voltage.

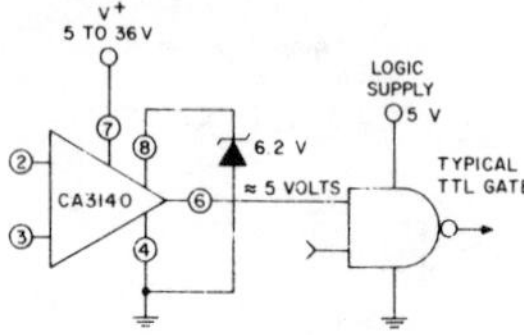


*Fig.12 – Zener clamping diode connected to terminals 8 and 4 to limit CA3140 output swing to TTL levels.*

Fig.13 shows output current-sinking capabilities of the CA3140 at various supply voltages. Output voltage swing to the negative supply rail permits this device to operate both power transistors and thyristors directly without the need for level-shifting circuitry usually associated with the 741 series of operational amplifiers.

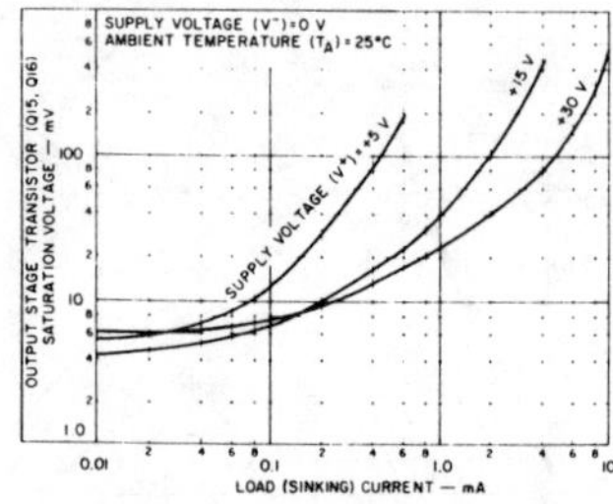


*Fig.13 – Voltage across output transistors Q15 and Q16 vs load current.*

Fig.16 show some typical configurations. Note that a series resistor, $R_L$, is used in both cases to limit the drive available to the driven device. Moreover, it is recommended that a series diode and shunt diode be used at the thyristor input to prevent large negative transient surges that can appear at the gate of thyristors, from damaging the integrated circuit.

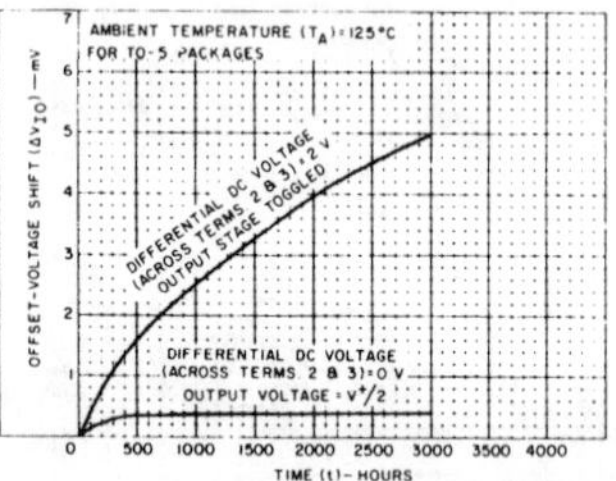


*Fig.14 – Typical incremental offset-voltage shift vs operating life.*

### OFFSET-VOLTAGE NULLING

The input-offset voltage can be nulled by connecting a 10-kΩ potentiometer between terminals 1 and 5 and returning its wiper arm to terminal 4, see Fig.15a. This technique, however, gives more adjustment range than required and therefore, a considerable portion of the potentiometer rotation is not fully utilized. Typical values of series resistors that may be placed at either end of the potentiometer, see Fig.15b, to optimize its utilization range are given in the table "Electrical Characteristics For Design Guidance" shown in this bulletin.

An alternate system is shown in Fig.15c. This circuit uses only one additional resistor of approximately the value shown in the table. For potentiometers, in which the resistance does not drop to zero ohms at either end of rotation, a value of resistance 10% lower than the values shown in the table should be used.

### LOW-VOLTAGE OPERATION

Operation at total supply voltages as low as 4 volts is possible with the CA3140. A current regulator based upon the PMOS threshold voltage maintains reasonable constant operating current and hence consistent performance down to these lower voltages.

The low-voltage limitation occurs when the upper extreme of the input common-mode voltage range extends down to the voltage at terminal 4. This limit is reached at a total

*Courtesy RCA Solid State Division*

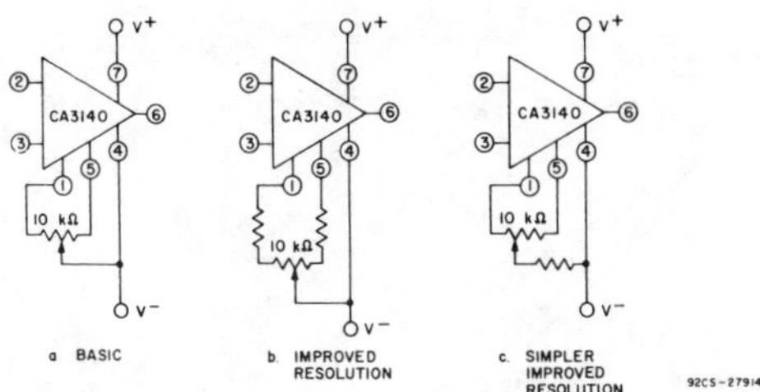


*Fig.15 – Three offset-voltage nulling methods.*

supply voltage just below 4 volts. The output voltage range also begins to extend down to the negative supply rail, but is slightly higher than that of the input. Fig.20 shows these characteristics and shows that with 2-volt dual supplies, the lower extreme of the input common-mode voltage range is below ground potential.

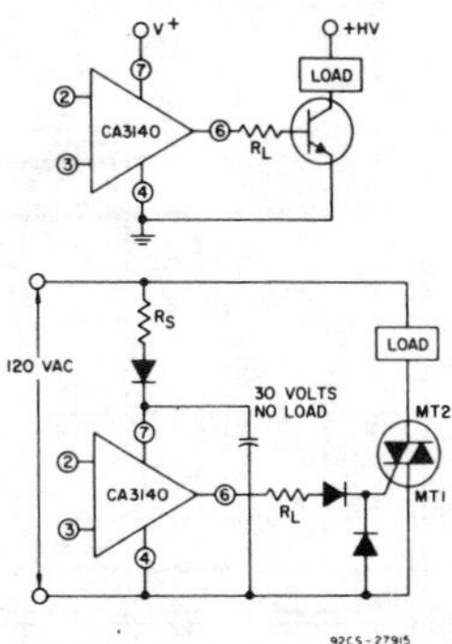


*Fig.16 – Methods of utilizing the $V_{CE}$(sat) sinking-current capability of the CA3140 series.*

## BANDWIDTH AND SLEW RATE

For those cases where bandwidth reduction is desired, for example, broadband noise reduction, an external capacitor connected between terminals 1 and 8 can reduce the open-loop −3 dB bandwidth. The slew rate will, however, also be proportionally reduced by using this additional capacitor. Thus, a 20% reduction in bandwidth by this technique will also reduce the slew rate by about 20%.

Fig.17 shows the typical settling time required to reach 1 mV or 10 mV of the final value for various levels of large signal inputs for the voltage-follower and inverting unity-gain amplifiers. The exceptionally fast settling time characteristics shown in Fig.18 are largely due to the high combination of high gain and wide bandwidth of the CA3140.

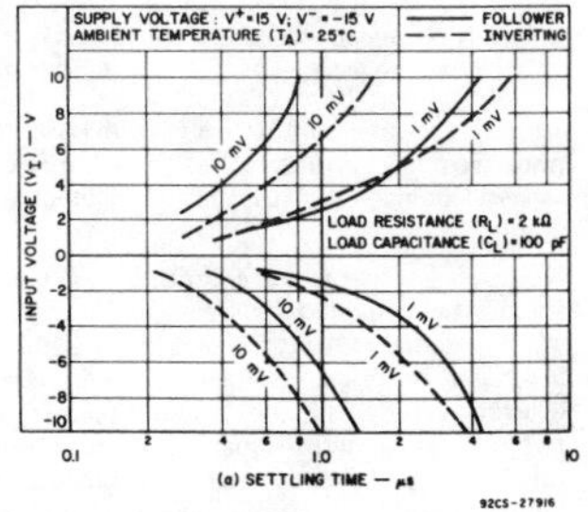


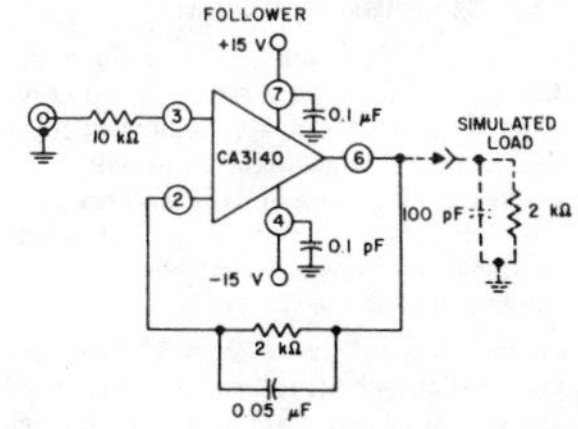


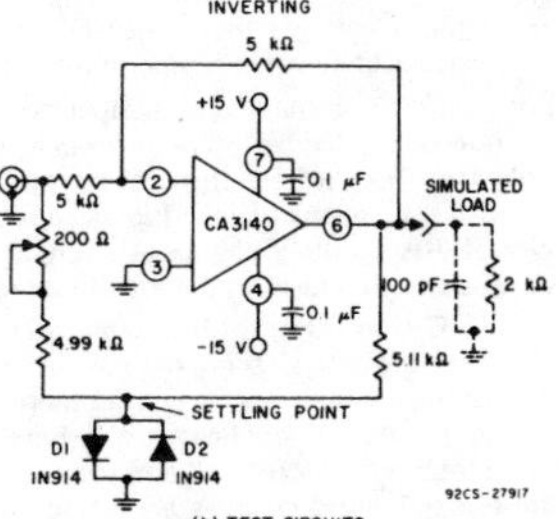


*Fig.17 – Input voltage vs settling time.*

## INPUT CIRCUIT CONSIDERATIONS

As mentioned previously, the amplifier inputs can be driven below the terminal 4 potential, but a series current-limiting re-

*Courtesy RCA Solid State Division*

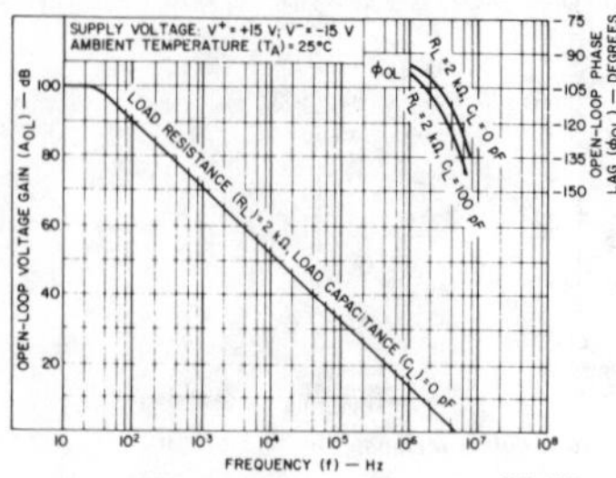


*Fig.18 – Open-loop voltage gain and phase lag vs frequency.*

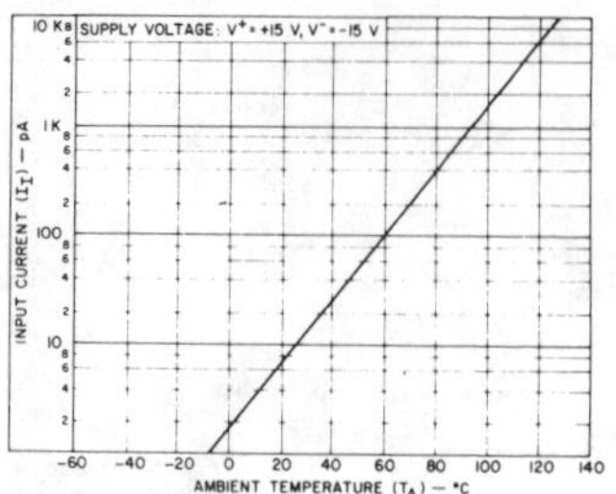


*Fig.19 – Input current vs ambient temperature.*

sistor is recommended to limit the maximum input terminal current to less than 1 mA to prevent damage to the input protection circuitry.

Moreover, some current-limiting resistance should be provided between the inverting input and the output when the CA3140 is used as a unity-gain voltage follower. This resistance prevents the possibility of extremely large input-signal transients from forcing a signal through the input-protection network and directly driving the internal constant-current source which could result in positive feedback via the output terminal. A 3.9-kΩ resistor is sufficient.

The typical input current is in the order of 10 pA when the inputs are centered at nominal device dissipation. As the output supplies load current, device dissipation will increase, raising the chip temperature and resulting in increased input current. Fig.19 shows typical input-terminal current versus ambient temperature for the CA3140.

It is well known that MOS/FET devices can exhibit slight changes in characteristics (for example, small changes in input offset voltage) due to the application of large differential input voltages that are sustained over long periods at elevated temperatures.

Both applied voltage and temperature accelerate these changes. The process is reversible and offset voltage shifts of the opposite polarity reverse the offset. Fig.14 shows the typical offset voltage change as a function of various stress voltages at the maximum rating of 125°C (for TO-5); at lower temperatures (TO-5 and plastic), for example, at 85°C, this change in voltage is considerably less. In typical linear applications, where the differential voltage is small and symmetircal, these incremental changes are of about the same magnitude as those encountered in an operational amplifier employing a bipolar a transistor input stage.

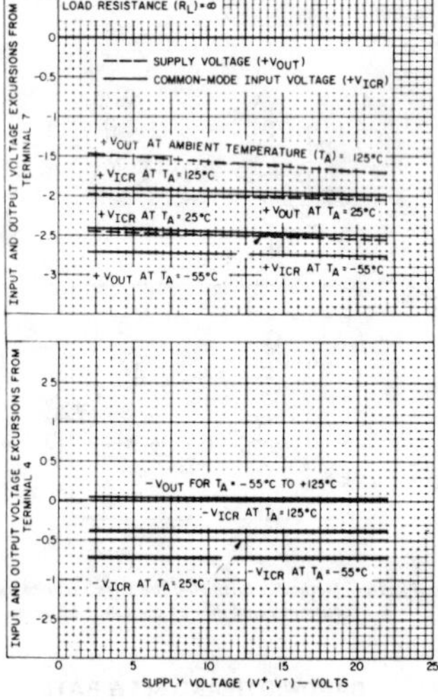


*Fig.20 – Output-voltage-swing capability and common-mode input-voltage range vs supply voltage and temperature.*

*Courtesy RCA Solid State Division*

March 1982

BI-FET II™ Technology

# LF411A/LF411 Low Offset, Low Drift JFET Input Operational Amplifier

## General Description

These devices are low cost, high speed, JFET input operational amplifiers with very low input offset voltage and guaranteed input offset voltage drift. They require low supply current yet maintain a large gain bandwidth product and fast slew rate. In addition, well matched high voltage JFET input devices provide very low input bias and offset currents. The LF411 is pin compatible with the standard LM741 allowing designers to immediately upgrade the overall performance of existing designs.

These amplifiers may be used in applications such as high speed integrators, fast D/A converters, sample and hold circuits and many other circuits requiring low input offset voltage and drift, low input bias current, high input impedance, high slew rate and wide bandwidth.

## Features

- Internally trimmed offset voltage — 0.5 mV (max)
- Input offset voltage drift — 10 μV/°C (max)
- Low input bias current — 50 pA
- Low input noise current — 0.01 pA/$\sqrt{Hz}$
- Wide gain bandwidth — 3 MHz (min)
- High slew rate — 10V/μs (min)
- Low supply current — 1.8 mA
- High input impedance — $10^{12}\Omega$
- Low total harmonic distortion $A_V = 10$, $R_L = 10k$, $V_O = 20$ Vp-p, BW = 20 Hz-20 kHz — <0.02%
- Low 1/f noise corner — 50 Hz
- Fast settling time to 0.01% — 2 μs

## Typical Connection

$R_f$, $R_i$, $V_{CC}$, LF411, 10k, $-V_{EE}$

## Ordering Information

LF411XYZ

X indicates electrical grade
Y indicates temperature range
"M" for military
"C" for commercial
Z indicates package type
"H" or "N"

## Connection Diagrams

**LF411AMH/LF411MH, LF411ACH/LF411CH**
**Metal Can Package**

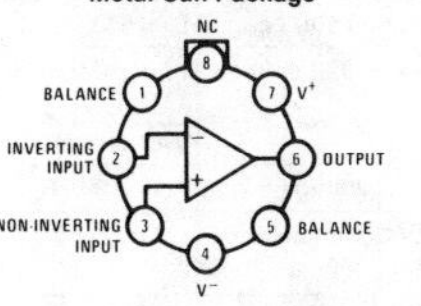


TOP VIEW

Note. Pin 4 connected to case.

**LF411ACN, LF411CN**
**Dual-In-Line Package**

BALANCE 1, INPUT 2, INPUT 3, V⁻ 4, BALANCE 5, OUTPUT 6, V⁺ 7, NC 8

TOP VIEW

## Simplified Schematic

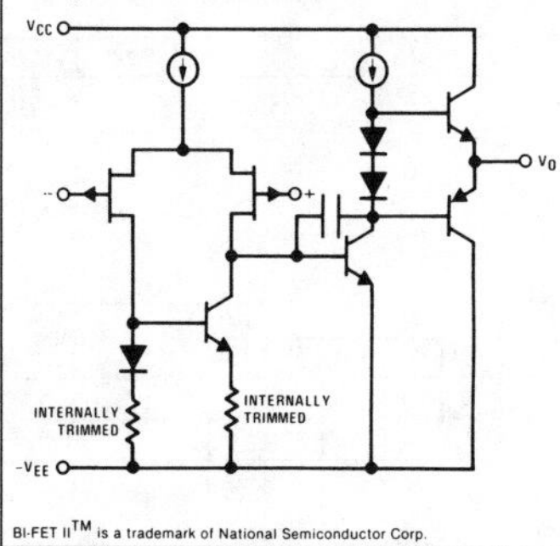


BI-FET II™ is a trademark of National Semiconductor Corp.



IM-B20M32/Printed in U.S.A.

*Courtesy National Semiconductor Corp.*

## Absolute Maximum Ratings

| | LF411A | LF411 |
|---|---|---|
| Supply Voltage | ±22V | ±18V |
| Differential Input Voltage | ±38V | ±30V |
| Input Voltage Range (Note 1) | ±19V | ±15V |
| Output Short Circuit Duration | Continuous | Continuous |

| | H Package | N Package |
|---|---|---|
| Power Dissipation (Note 2) | 670 mW | 500 mW |
| $T_j$ max | 150°C | 115°C |
| $\theta_{jA}$ | 150°C/W | 160°C/W |
| Operating Temperature Range | (Note 3) | (Note 3) |
| Storage Temperature Range | $-65°C \le T_A \le 150°C$ | $-65°C \le T_A \le 150°C$ |
| Lead Temperature (Soldering, 10 seconds) | 300°C | 300°C |

## DC Electrical Characteristics (Note 4)

| Symbol | Parameter | Conditions | | LF411A Min | LF411A Typ | LF411A Max | LF411 Min | LF411 Typ | LF411 Max | Units |
|---|---|---|---|---|---|---|---|---|---|---|
| $V_{OS}$ | Input Offset Voltage | $R_S = 10\ k\Omega$, $T_A = 25°C$ | | | 0.3 | 0.5 | | 0.8 | 2.0 | mV |
| $\Delta V_{OS}/\Delta T$ | Average TC of Input Offset Voltage | $R_S = 10\ k\Omega$ (Note 5) | | | 7 | 10 | | 7 | 20 (Note 5) | μV/°C |
| $I_{OS}$ | Input Offset Current | $V_S = \pm 15V$ Notes 4 and 6 | $T_j = 25°C$ | | 25 | 100 | | 25 | 100 | pA |
| | | | $T_j = 70°C$ | | | 2 | | | 2 | nA |
| | | | $T_j = 125°C$ | | | 25 | | | 25 | nA |
| $I_B$ | Input Bias Current | $V_S = \pm 15V$ Notes 4 and 6 | $T_j = 25°C$ | | 50 | 200 | | 50 | 200 | pA |
| | | | $T_j = 70°C$ | | | 4 | | | 4 | nA |
| | | | $T_j = 125°C$ | | | 50 | | | 50 | nA |
| $R_{IN}$ | Input Resistance | $T_j = 25°C$ | | | $10^{12}$ | | | $10^{12}$ | | Ω |
| $A_{VOL}$ | Large Signal Voltage Gain | $V_S = \pm 15V$, $V_O = \pm 10V$, $R_L = 2k$, $T_A = 25°C$ | | 50 | 200 | | 25 | 200 | | V/mV |
| | | Over Temperature | | 25 | 200 | | 15 | 200 | | V/mV |
| $V_O$ | Output Voltage Swing | $V_S = \pm 15V$, $R_L = 10k$ | | ±12 | ±13.5 | | ±12 | ±13.5 | | V |
| $V_{CM}$ | Input Common-Mode Voltage Range | | | ±16 | +19.5 | | ±11 | +14.5 | | V |
| | | | | | −16.5 | | | −11.5 | | V |
| CMRR | Common-Mode Rejection Ratio | $R_S \le 10k$ | | 80 | 100 | | 70 | 100 | | dB |
| PSRR | Supply Voltage Rejection Ratio | (Note 7) | | 80 | 100 | | 70 | 100 | | dB |
| $I_S$ | Supply Current | | | | 1.8 | 2.8 | | 1.8 | 3.4 | mA |

## AC Electrical Characteristics (Note 4)

| Symbol | Parameter | Conditions | LF411A Min | LF411A Typ | LF411A Max | LF411 Min | LF411 Typ | LF411 Max | Units |
|---|---|---|---|---|---|---|---|---|---|
| SR | Slew Rate | $V_S = \pm 15V$, $T_A = 25°C$ | 10 | 15 | | 8 | 15 | | V/μs |
| GBW | Gain-Bandwidth Product | $V_S = \pm 15V$, $T_A = 25°C$ | 3 | 4 | | 2.7 | 4 | | MHz |
| $e_n$ | Equivalent Input Noise Voltage | $T_A = 25°C$, $R_S = 100\Omega$, f = 1 kHz | | 25 | | | 25 | | nV/√Hz |
| $i_n$ | Equivalent Input Noise Current | $T_A = 25°C$, f = 1 kHz | | 0.01 | | | 0.01 | | pA/√Hz |

*Courtesy National Semiconductor Corp.*

## Notes

**Note 1:** Unless otherwise specified the absolute maximum negative input voltage is equal to the negative power supply voltage.

**Note 2:** For operating at elevated temperature, these devices must be derated based on a thermal resistance of $\theta_{jA}$.

**Note 3:** These devices are available in both the commercial temperature range $0°C \le T_A \le 70°C$ and the military temperature range $-55°C \le T_A \le 125°C$. The temperature range is designated by the position just before the package type in the device number. A "C" indicates the commercial temperature range and an "M" indicates the military temperature range. The military temperature range is available in "H" package only.

**Note 4:** Unless otherwise specified, the specifications apply over the full temperature range and for $V_S = \pm 20V$ for the LF411A and for $V_S = \pm 15V$ for the LF411. $V_{OS}$, $I_B$, and $I_{OS}$ are measured at $V_{CM} = 0$.

**Note 5:** The LF411A is 100% tested to this specification. The LF411 is sample tested to insure at least 90% of the units meet this specification.

**Note 6:** The input bias currents are junction leakage currents which approximately double for every 10°C increase in the junction temperature, $T_j$. Due to limited production test time, the input bias currents measured are correlated to junction temperature. In normal operation the junction temperature rises above the ambient temperature as a result of internal power dissipation, $P_D$. $T_j = T_A + \theta_{jA} P_D$ where $\theta_{jA}$ is the thermal resistance from junction to ambient. Use of a heat sink is recommended if input bias current is to be kept to a minimum.

**Note 7:** Supply voltage rejection ratio is measured for both supply magnitudes increasing or decreasing simultaneously in accordance with common practice.

## Typical Performance Characteristics

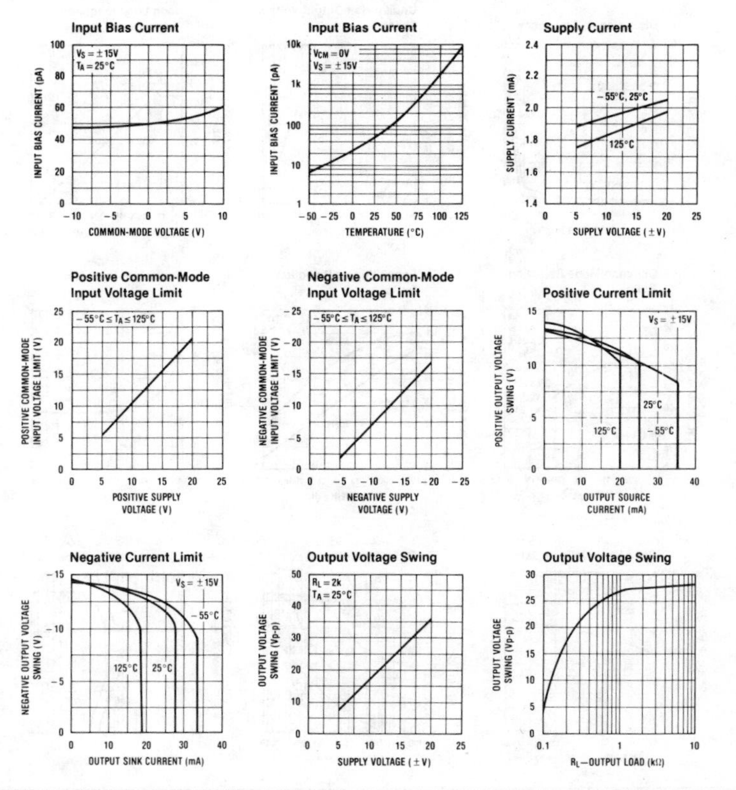


*Courtesy National Semiconductor Corp.*

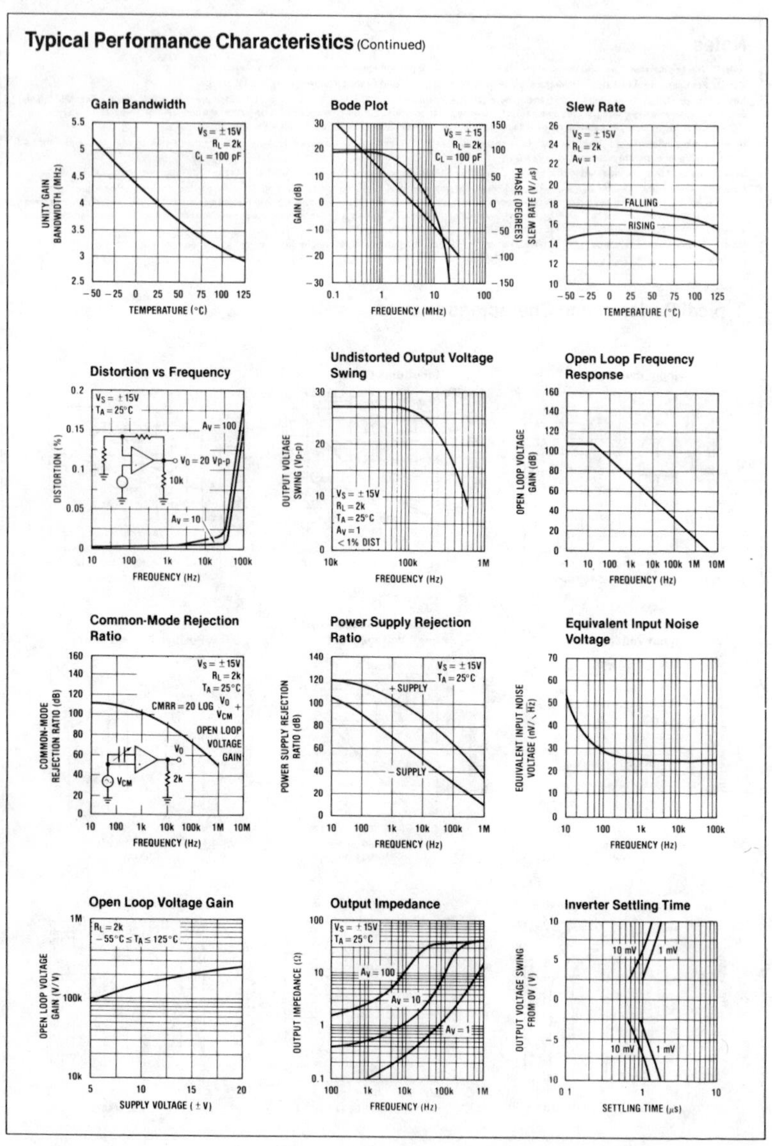


*Courtesy National Semiconductor Corp.*

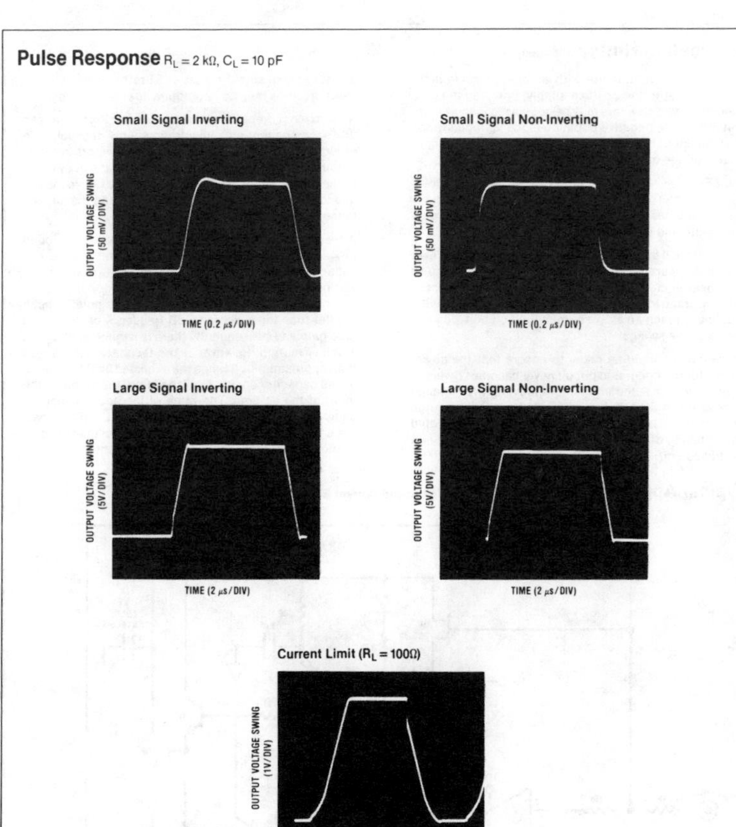


## Application Hints

The LF411 series of internally trimmed JFET input op amps (BI-FET II™) provide very low input offset voltage and guaranteed input offset voltage drift. These JFETs have large reverse breakdown voltages from gate to source and drain eliminating the need for clamps across the inputs. Therefore, large differential input voltages can easily be accommodated without a large increase in input current. The maximum differential input voltage is independent of the supply voltages. However, neither of the input voltages should be allowed to exceed the negative supply as this will cause large currents to flow which can result in a destroyed unit.

Exceeding the negative common-mode limit on either input will cause a reversal of the phase to the output and force the amplifier output to the corresponding high or low state. Exceeding the negative common-mode limit on both inputs will force the amplifier output to a high state. In neither case does a latch occur since raising the input back within the common-mode range again puts the input stage and thus the amplifier in a normal operating mode.

Exceeding the positive common-mode limit on a single input will not change the phase of the output; however, if both inputs exceed the limit, the output of the amplifier may be forced to a high state.

*Courtesy National Semiconductor Corp.*

## Application Hints (Continued)

The amplifier will operate with a common-mode input voltage equal to the positive supply; however, the gain bandwidth and slew rate may be decreased in this condition. When the negative common-mode voltage swings to within 3V of the negative supply, an increase in input offset voltage may occur.

The LF411 is biased by a zener reference which allows normal circuit operation on ±4.5V power supplies. Supply voltages less than these may result in lower gain bandwidth and slew rate.

The LF411 will drive a 2 kΩ load resistance to ±10V over the full temperature range. If the amplifier is forced to drive heavier load currents, however, an increase in input offset voltage may occur on the negative voltage swing and finally reach an active current limit on both positive and negative swings.

Precautions should be taken to ensure that the power supply for the integrated circuit never becomes reversed in polarity or that the unit is not inadvertently installed backwards in a socket as an unlimited current surge through the resulting forward diode within the IC could cause fusing of the internal conductors and result in a destroyed unit.

Because these amplifiers are JFET rather than MOSFET input op amps they do not require special handling.

As with most amplifiers, care should be taken with lead dress, component placement and supply decoupling in order to ensure stability. For example, resistors from the output to an input should be placed with the body close to the input to minimize "pick-up" and maximize the frequency of the feedback pole by minimizing the capacitance from the input to ground.

A feedback pole is created when the feedback around any amplifier is resistive. The parallel resistance and capacitance from the input of the device (usually the inverting input) to AC ground set the frequency of the pole. In many instances the frequency of this pole is much greater than the expected 3 dB frequency of the closed loop gain and consequently there is negligible effect on stability margin. However, if the feedback pole is less than approximately 6 times the expected 3 dB frequency, a lead capacitor should be placed from the output to the input of the op amp. The value of the added capacitor should be such that the RC time constant of this capacitor and the resistance it parallels is greater than or equal to the original feedback pole time constant.

## Typical Applications

**Ultra High Speed Current Booster**

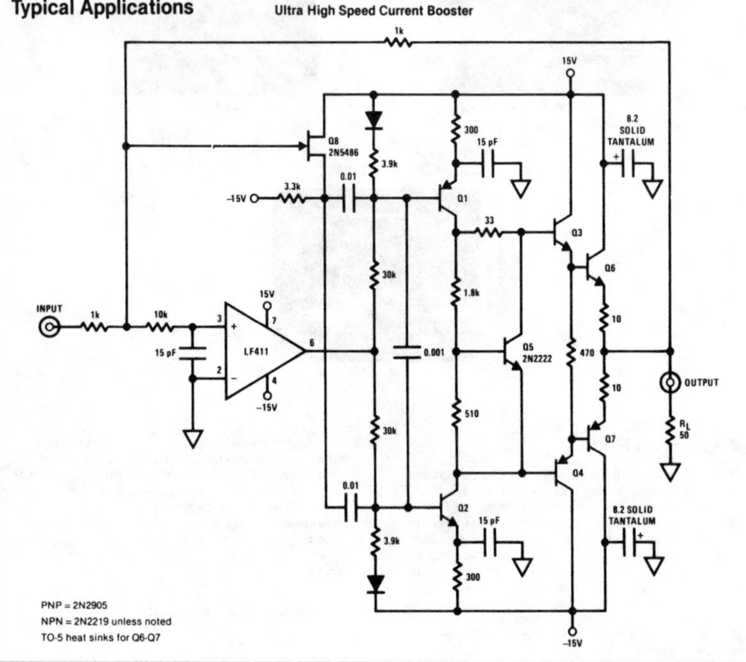


*Courtesy National Semiconductor Corp.*

JUNE 1977

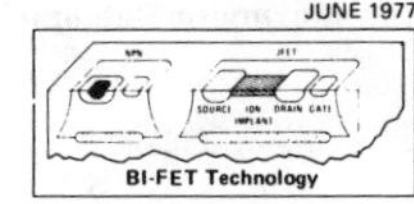


LF155/LF156/LF157 Monolithic JFET Input Operational Amplifiers

# LF155/LF156/LF157 Monolithic JFET Input Operational Amplifiers

**LF155, LF155A, LF255, LF355, LF355A, LF355B low supply current**
**LF156, LF156A, LF256, LF356, LF356A, LF356B wide band**
**LF157, LF157A, LF257, LF357, LF357A, LF357B wide band decompensated ($A_{V_{MIN}} = 5$)**

## General Description

These are the first monolithic JFET input operational amplifiers to incorporate well matched, high voltage JFETs on the same chip with standard bipolar transistors (BI-FET Technology). These amplifiers feature low input bias and offset currents, low offset voltage and offset voltage drift, coupled with offset adjust which does not degrade drift or common-mode rejection. The devices are also designed for high slew rate, wide bandwidth, extremely fast settling time, low voltage and current noise and a low 1/f noise corner.

## Advantages

- Replace expensive hybrid and module FET op amps
- Rugged JFETs allow blow-out free handling compared with MOSFET input devices
- Excellent for low noise applications using either high or low source impedance—very low 1/f corner
- Offset adjust does not degrade drift or common-mode rejection as in most monolithic amplifiers
- New output stage allows use of large capacitive loads (10,000 pF) without stability problems
- Internal compensation and large differential input voltage capability

## Applications

- Precision high speed integrators
- Fast D/A and A/D converters
- High impedance buffers
- Wideband, low noise, low drift amplifiers
- Logarithmic amplifiers
- Photocell amplifiers
- Sample and Hold circuits

## Common Features

(LF155A, LF156A, LF157A)

| | |
|---|---|
| ■ Low input bias current | 30 pA |
| ■ Low Input Offset Current | 3 pA |
| ■ High input impedance | $10^{12}\Omega$ |
| ■ Low input offset voltage | 1 mV |
| ■ Low input offset voltage temperature drift | $3\mu V/°C$ |
| ■ Low input noise current | $0.01\ pA/\sqrt{Hz}$ |
| ■ High common-mode rejection ratio | 100 dB |
| ■ Large dc voltage gain | 106 dB |

## Uncommon Features

| | LF155A | LF156A | LF157A ($A_V = 5$)* | UNITS |
|---|---|---|---|---|
| ■ Extremely fast settling time to 0.01% | 4 | 1.5 | 1.5 | μs |
| ■ Fast slew rate | 5 | 12 | 50 | V/μs |
| ■ Wide gain bandwidth | 2.5 | 5 | 20 | MHz |
| ■ Low input noise voltage | 20 | 12 | 12 | $nV/\sqrt{Hz}$ |

## Simplified Schematic

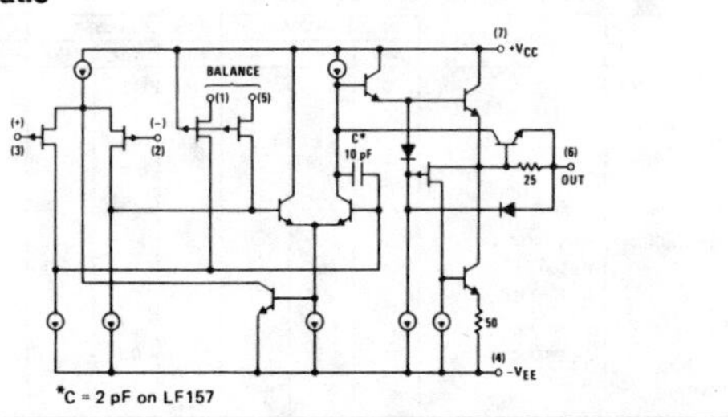


*C = 2 pF on LF157



IM-B20M67/Printed in U.S.A.

*Courtesy National Semiconductor Corp.*

## Absolute Maximum Ratings

| | LF155A/6A/7A | LF155/6/7 | LF355B/6B/7B<br>LF255/6/7<br>LF355B/6B/7B | LF355A/6A/7A<br>LF355/6/7 |
|---|---|---|---|---|
| Supply Voltage | ±22V | ±22V | ±22V | ±18V |
| Power Dissipation ($P_d$ at 25°C) and Thermal Resistance ($\theta_{jA}$) (Note 1) | | | | |
| $T_{jMAX}$ | | | | |
| (H and J Package) | 150°C | 150°C | 115°C | 115°C |
| (N Package) | | | 100°C | 100°C |
| (H Package) $P_d$ | 670 mW | 670 mW | 570 mW | 570 mW |
| $\theta_{jA}$ | 150°C/W | 150°C/W | 150°C/W | 150°C/W |
| (J Package) $P_d$ | 670 mW | 670 mW | 570 mW | 570 mW |
| $\theta_{jA}$ | 140°C/W | 140°C/W | 140°C/W | 140°C/W |
| (N Package) $P_d$ | | | 500 mW | 500 mW |
| $\theta_{jA}$ | | | 155°C/W | 155°C/W |
| Differential Input Voltage | ±40V | ±40V | ±40V | ±30V |
| Input Voltage Range (Note 2) | ±20V | ±20V | ±20V | ±16V |
| Output Short Circuit Duration | Continuous | Continuous | Continuous | Continuous |
| Storage Temperature Range | −65°C to +150°C | −65°C to +150°C | −65°C to +150°C | −65°C to +150°C |
| Lead Temperature (Soldering, 10 seconds) | 300°C | 300°C | 300°C | 300°C |

## DC Electrical Characteristics (Note 3)

| SYMBOL | PARAMETER | CONDITIONS | LF155A/6A/7A | | | LF355A/6A/7A | | | UNITS |
|---|---|---|---|---|---|---|---|---|---|
| | | | MIN | TYP | MAX | MIN | TYP | MAX | |
| $V_{OS}$ | Input Offset Voltage | $R_S = 50\Omega$, $T_A = 25°C$ | | 1 | 2 | | 1 | 2 | mV |
| | | Over Temperature | | | 2.5 | | | 2.3 | mV |
| $\Delta V_{OS}/\Delta T$ | Average TC of Input Offset Voltage | $R_S = 50\Omega$ | | 3 | 5 | | 3 | 5 | μV/°C |
| $\Delta TC/\Delta V_{OS}$ | Change in Average TC with $V_{OS}$ Adjust | $R_S = 50\Omega$, (Note 4) | | 0.5 | | | 0.5 | | μV/°C per mV |
| $I_{OS}$ | Input Offset Current | $T_j = 25°C$, (Notes 3, 5) | | 3 | 10 | | 3 | 10 | pA |
| | | $T_j \leq T_{HIGH}$ | | | 10 | | | 1 | nA |
| $I_B$ | Input Bias Current | $T_J = 25°C$, (Notes 3, 5) | | 30 | 50 | | 30 | 50 | pA |
| | | $T_J \leq T_{HIGH}$ | | | 25 | | | 5 | nA |
| $R_{IN}$ | Input Resistance | $T_J = 25°C$ | | $10^{12}$ | | | $10^{12}$ | | Ω |
| $A_{VOL}$ | Large Signal Voltage Gain | $V_S = \pm 15V$, $T_A = 25°C$ $V_O = \pm 10V$, $R_L = 2k$ | 50 | 200 | | 50 | 200 | | V/mV |
| | | Over Temperature | 25 | | | 25 | | | V/mV |
| $V_O$ | Output Voltage Swing | $V_S = \pm 15V$, $R_L = 10k$ | ±12 | ±13 | | ±12 | ±13 | | V |
| | | $V_S = \pm 15V$, $R_L = 2k$ | ±10 | ±12 | | ±10 | ±12 | | V |
| $V_{CM}$ | Input Common-Mode Voltage Range | $V_S = \pm 15V$ | ±11 | +15.1 | | ±11 | +15.1 | | V |
| | | | | −12 | | | −12 | | V |
| CMRR | Common-Mode Rejection Ratio | | 85 | 100 | | 85 | 100 | | dB |
| PSRR | Supply Voltage Rejection Ratio | (Note 6) | 85 | 100 | | 85 | 100 | | dB |

## AC Electrical Characteristics $T_A = 25°C$, $V_S = \pm 15V$

| SYMBOL | PARAMETER | CONDITIONS | LF155A/355A | | | LF156A/356A | | | LF157A/357A | | | UNITS |
|---|---|---|---|---|---|---|---|---|---|---|---|---|
| | | | MIN | TYP | MAX | MIN | TYP | MAX | MIN | TYP | MAX | |
| SR | Slew Rate | LF155A/6A; $A_V = 1$, | 3 | 5 | | 10 | 12 | | | | | V/μs |
| | | LF157A; $A_V = 5$ | | | | | | | 40 | 50 | | V/μs |
| GBW | Gain Bandwidth Product | | | 2.5 | | 4 | 4.5 | | 15 | 20 | | MHz |
| $t_s$ | Settling Time to 0.01% | (Note 7) | | 4 | | | 1.5 | | | 1.5 | | μs |
| $e_n$ | Equivalent Input Noise Voltage | $R_S = 100\Omega$ | | | | | | | | | | |
| | | f = 100 Hz | | 25 | | | 15 | | | 15 | | $nV/\sqrt{Hz}$ |
| | | f = 1000 Hz | | 25 | | | 12 | | | 12 | | $nV/\sqrt{Hz}$ |
| $i_n$ | Equivalent Input Noise Current | f = 100 Hz | | 0.01 | | | 0.01 | | | 0.01 | | $pA/\sqrt{Hz}$ |
| | | f = 1000 Hz | | 0.01 | | | 0.01 | | | 0.01 | | $pA/\sqrt{Hz}$ |
| $C_{IN}$ | Input Capacitance | | | 3 | | | 3 | | | 3 | | pF |

*Courtesy National Semiconductor Corp.*

## DC Electrical Characteristics (Note 3)

| SYMBOL | PARAMETER | CONDITIONS | LF155/6/7 | | | LF255/6/7 LF355B/6B/7B | | | LF355/6/7 | | | UNITS |
|---|---|---|---|---|---|---|---|---|---|---|---|---|
| | | | MIN | TYP | MAX | MIN | TYP | MAX | MIN | TYP | MAX | |
| $V_{OS}$ | Input Offset Voltage | $R_S = 50\Omega$, $T_A = 25°C$ | | 3 | 5 | | 3 | 5 | | 3 | 10 | mV |
| | | Over Temperature | | | 7 | | | 6.5 | | | 13 | mV |
| $\Delta V_{OS}/\Delta T$ | Average TC of Input Offset Voltage | $R_S = 50\Omega$ | | 5 | | | 5 | | | 5 | | μV/°C |
| $\Delta TC/\Delta V_{OS}$ | Change in Average TC with $V_{OS}$ Adjust | $R_S = 50\Omega$, (Note 4) | | 0.5 | | | 0.5 | | | 0.5 | | μV/°C per mV |
| $I_{OS}$ | Input Offset Current | $T_j = 25°C$, (Notes 3, 5) | | 3 | 20 | | 3 | 20 | | 3 | 50 | pA |
| | | $T_j \leq T_{HIGH}$ | | | 20 | | | 1 | | | 2 | nA |
| $I_B$ | Input Bias Current | $T_J = 25°C$, (Notes 3, 5) | | 30 | 100 | | 30 | 100 | | 30 | 200 | pA |
| | | $T_J \leq T_{HIGH}$ | | | 50 | | | 5 | | | 8 | nA |
| $R_{IN}$ | Input Resistance | $T_J = 25°C$ | | $10^{12}$ | | | $10^{12}$ | | | $10^{12}$ | | Ω |
| $A_{VOL}$ | Large Signal Voltage Gain | $V_S = \pm 15V$, $T_A = 25°C$ $V_O = \pm 10V$, $R_L = 2k$ | 50 | 200 | | 50 | 200 | | 25 | 200 | | V/mV |
| | | Over Temperature | 25 | | | 25 | | | 15 | | | V/mV |
| $V_O$ | Output Voltage Swing | $V_S = \pm 15V$, $R_L = 10k$ | ±12 | ±13 | | ±12 | ±13 | | ±12 | ±13 | | V |
| | | $V_S = \pm 15V$, $R_L = 2k$ | ±10 | ±12 | | ±10 | ±12 | | ±10 | ±12 | | V |
| $V_{CM}$ | Input Common-Mode Voltage Range | $V_S = \pm 15V$ | ±11 | +15.1<br>−12 | | ±11 | +15.1<br>−12 | | ±10 | +15.1<br>−12 | | V<br>V |
| CMRR | Common-Mode Rejection Ratio | | 85 | 100 | | 85 | 100 | | 80 | 100 | | dB |
| PSRR | Supply Voltage Rejection Ratio | (Note 6) | 85 | 100 | | 85 | 100 | | 80 | 100 | | dB |

## DC Electrical Characteristics $T_A = 25°C$, $V_S = \pm 15V$

| PARAMETER | LF155A/155, LF255, LF355A/355B | | LF355 | | LF156A/156, LF256/356B | | LF356A/356 | | LF157A/157 LF257/357B | | LF357A/357 | | UNITS |
|---|---|---|---|---|---|---|---|---|---|---|---|---|---|
| | TYP | MAX | TYP | MAX | TYP | MAX | TYP | MAX | TYP | MAX | TYP | MAX | |
| Supply Current | 2 | 4 | 2 | 4 | 5 | 7 | 5 | 10 | 5 | 7 | 5 | 10 | mA |

## AC Electrical Characteristics $T_A = 25°C$, $V_S = \pm 15V$

| SYMBOL | PARAMETER | CONDITIONS | LF155/255/ 355/355B | LF156/256, LF356B | LF156/256/ 356/356B | LF157/257, LF357B | LF157/257/ 357/357B | UNITS |
|---|---|---|---|---|---|---|---|---|
| | | | TYP | MIN | TYP | MIN | TYP | |
| SR | Slew Rate | LF155/6: $A_V = 1$, | 5 | 7.5 | 12 | | | V/μs |
| | | LF157: $A_V = 5$ | | | | 30 | 50 | V/μs |
| GBW | Gain Bandwidth Product | | 2.5 | | 5 | | 20 | MHz |
| $t_s$ | Settling Time to 0.01% | (Note 7) | 4 | | 1.5 | | 1.5 | μs |
| $e_n$ | Equivalent Input Noise Voltage | $R_S = 100\Omega$ | | | | | | |
| | | f = 100 Hz | 25 | | 15 | | 15 | $nV/\sqrt{Hz}$ |
| | | f = 1000 Hz | 20 | | 12 | | 12 | $nV/\sqrt{Hz}$ |
| $i_n$ | Equivalent Input Current Noise | f = 100 Hz | 0.01 | | 0.01 | | 0.01 | $pA/\sqrt{Hz}$ |
| | | f = 1000 Hz | 0.01 | | 0.01 | | 0.01 | $pA/\sqrt{Hz}$ |
| $C_{IN}$ | Input Capacitance | | 3 | | 3 | | 3 | pF |

*Courtesy National Semiconductor Corp.*

## Notes for Electrical Characteristics

**Note 1:** The maximum power dissipation for these devices must be derated at elevated temperatures and is dictated by $T_{jMAX}$, $\theta_{jA}$, and the ambient temperature, $T_A$. The maximum available power dissipation at any temperature is $P_d = (T_{jMAX} - T_A)/\theta_{jA}$ or the 25°C $P_{dMAX}$, whichever is less.

**Note 2:** Unless otherwise specified the absolute maximum negative input voltage is equal to the negative power supply voltage.

**Note 3:** Unless otherwise stated, these test conditions apply:

| | LF155A/6A/7A LF155/6/7 | LF255/6/7 | LF355A/6A/7A | LF355B/6B/7B | LF355/6/7 |
|---|---|---|---|---|---|
| Supply Voltage, $V_S$ | $\pm15V \le V_S \le \pm20V$ | $\pm15V \le V_S \le \pm20V$ | $\pm15V \le V_S \le \pm18V$ | $\pm15V \le V_S \pm20V$ | $V_S = \pm15V$ |
| $T_A$ | $-55°C \le T_A \le +125°C$ | $-25°C \le T_A \le +85°C$ | $0°C \le T_A \le +70°C$ | $0°C \le T_A \le +70°C$ | $0°C \le T_A \le +70°C$ |
| $T_{HIGH}$ | +125°C | +85°C | +70°C | +70°C | +70°C |

and $V_{OS}$, $I_B$ and $I_{OS}$ are measured at $V_{CM} = 0$.

**Note 4:** The Temperature Coefficient of the adjusted input offset voltage changes only a small amount (0.5μV/°C typically) for each mV of adjustment from its original unadjusted value. Common-mode rejection and open loop voltage gain are also unaffected by offset adjustment.

**Note 5:** The input bias currents are junction leakage currents which approximately double for every 10 C increase in the junction temperature, $T_J$. Due to limited production test time, the input bias currents measured are correlated to junction temperature. In normal operation the junction temperature rises above the ambient temperature as a result of internal power dissipation, Pd. $T_j = T_A + \Theta_{jA}$ Pd where $\Theta_{jA}$ is the thermal resistance from junction to ambient. Use of a heat sink is recommended if input bias current is to be kept to a minimum.

**Note 6:** Supply Voltage Rejection is measured for both supply magnitudes increasing or decreasing simultaneously, in accordance with common practice.

**Note 7:** Settling time is defined here, for a unity gain inverter connection using 2 kΩ resistors for the LF155/6. It is the time required for the error voltage (the voltage at the inverting input pin on the amplifier) to settle to within 0.01% of its final value from the time a 10V step input is applied to the inverter. For the LF157, $A_V = -5$, the feedback resistor from output to input is 2 kΩ and the output step is 10V (See Settling Time Test Circuit, page 9).

## Typical DC Performance Characteristics

Curves are for LF155, LF156 and LF157 unless otherwise specified.

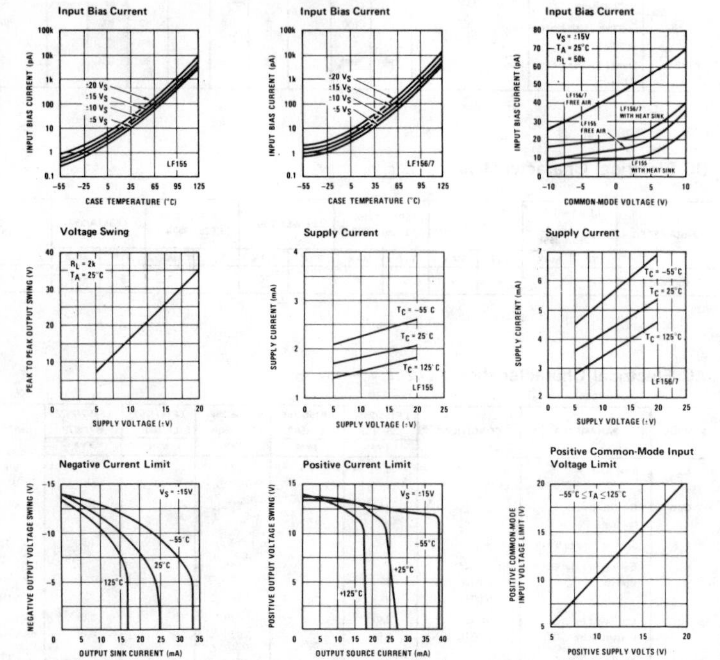


*Courtesy National Semiconductor Corp.*

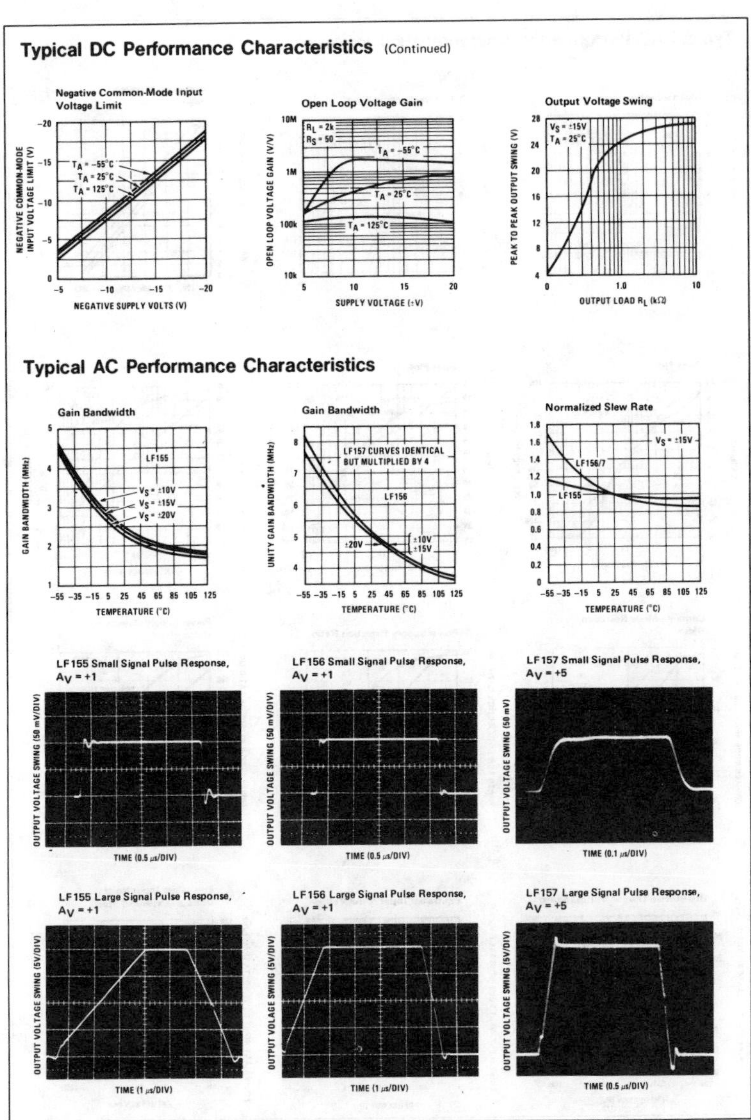


*Courtesy National Semiconductor Corp.*

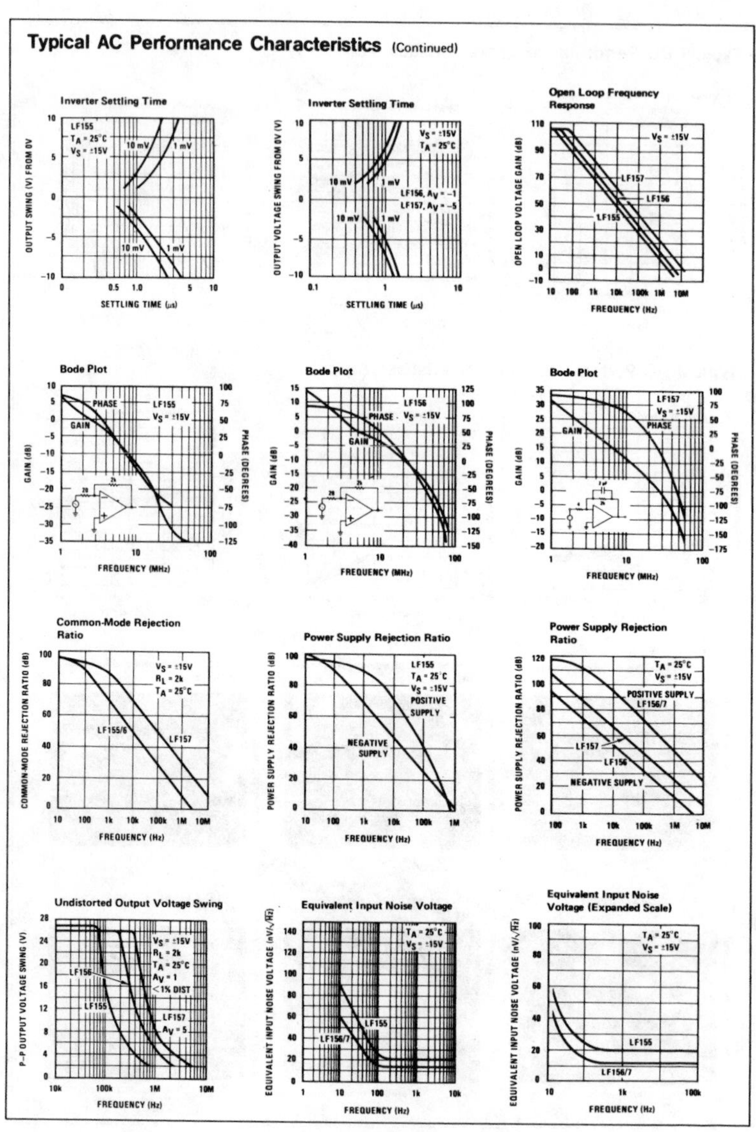


*Courtesy National Semiconductor Corp.*

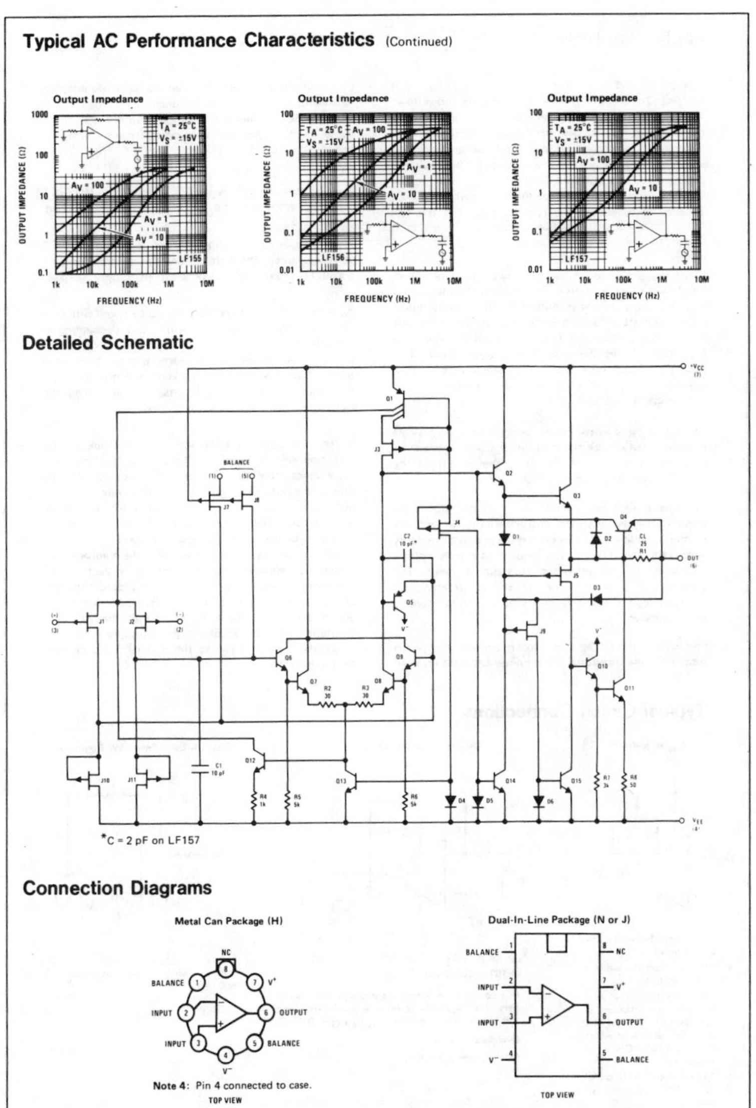


*Courtesy National Semiconductor Corp.*

## Application Hints

The LF155/6/7 series are op amps with JFET input devices. These JFETs have large reverse breakdown voltages from gate to source and drain eliminating the need for clamps across the inputs. Therefore large differential input voltages can easily be accomodated without a large increase in input current. The maximum differential input voltage is independent of the supply voltages. However, neither of the input voltages should be allowed to exceed the negative supply as this will cause large currents to flow which can result in a destroyed unit.

Exceeding the negative common-mode limit on either input will cause a reversal of the phase to the output and force the amplifier output to the corresponding high or low state. Exceeding the negative common-mode limit on both inputs will force the amplifier output to a high state. In neither case does a latch occur since raising the input back within the common-mode range again puts the input stage and thus the amplifier in a normal operating mode.

Exceeding the positive common-mode limit on a single input will not change the phase of the output however, if both inputs exceed the limit, the output of the amplifier will be forced to a high state.

These amplifiers will operate with the common-mode input voltage equal to the positive supply. In fact, the common-mode voltage can exceed the positive supply by approximately 100 mV independent of supply voltage and over the full operating temperature range. The positive supply can therefore be used as a reference on an input as, for example, in a supply current monitor and/or limiter.

Precautions should be taken to ensure that the power supply for the integrated circuit never becomes reversed in polarity or that the unit is not inadvertently installed backwards in a socket as an unlimited current surge through the resulting forward diode within the IC could cause fusing of the internal conductors and result in a destroyed unit.

Because these amplifiers are JFET rather than MOSFET input op amps they do not require special handling.

All of the bias currents in these amplifiers are set by FET current sources. The drain currents for the amplifiers are therefore essentially independent of supply voltage.

As with most amplifiers, care should be taken with lead dress, component placement and supply decoupling in order to ensure stability. For example, resistors from the output to an input should be placed with the body close to the input to minimize "pickup" and maximize the frequency of the feedback pole by minimizing the capacitance from the input to ground.

A feedback pole is created when the feedback around any amplifier is resistive. The parallel resistance and capacitance from the input of the device (usually the inverting input) to ac ground set the frequency of the pole. In many instances the frequency of this pole is much greater than the expected 3 dB frequency of the closed loop gain and consequently there is negligible effect on stability margin. However, if the feedback pole is less than approximately six times the expected 3 dB frequency a lead capacitor should be placed from the output to the input of the op amp. The value of the added capacitor should be such that the RC time constant of this capacitor and the resistance it parallels is greater than or equal to the original feedback pole time constant.

## Typical Circuit Connections

**$V_{OS}$ Adjustment**

- $V_{OS}$ is adjusted with a 25k potentiometer
- The potentiometer wiper is connected to $V^+$
- For potentiometers with temperature coefficient of 100 ppm/°C or less the additional drift with adjust is ≈ 0.5 μV/°C/mV of adjustment
- Typical overall drift: 5 μV/°C ±(0.5 μV/°C/mV of adj.)

**Driving Capacitive Loads**

*LF155/6 R = 5k
LF157 R = 1.25k

Due to a unique output stage design, these amplifiers have the ability to drive large capacitive loads and still maintain stability. $C_{L(MAX)} \cong 0.01\ \mu F$.

Overshoot ≤ 20%

Settling time ($t_s$) ≅ 5 μs

**LF157. A Large Power BW Amplifier**

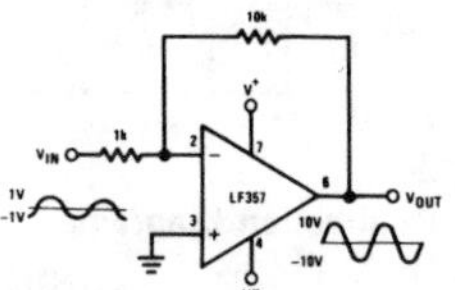


For distortion ≤ 1% and a 20 Vp-p $V_{OUT}$ swing, power bandwidth is: 500 kHz.

*Courtesy National Semiconductor Corp.*

## Typical Applications

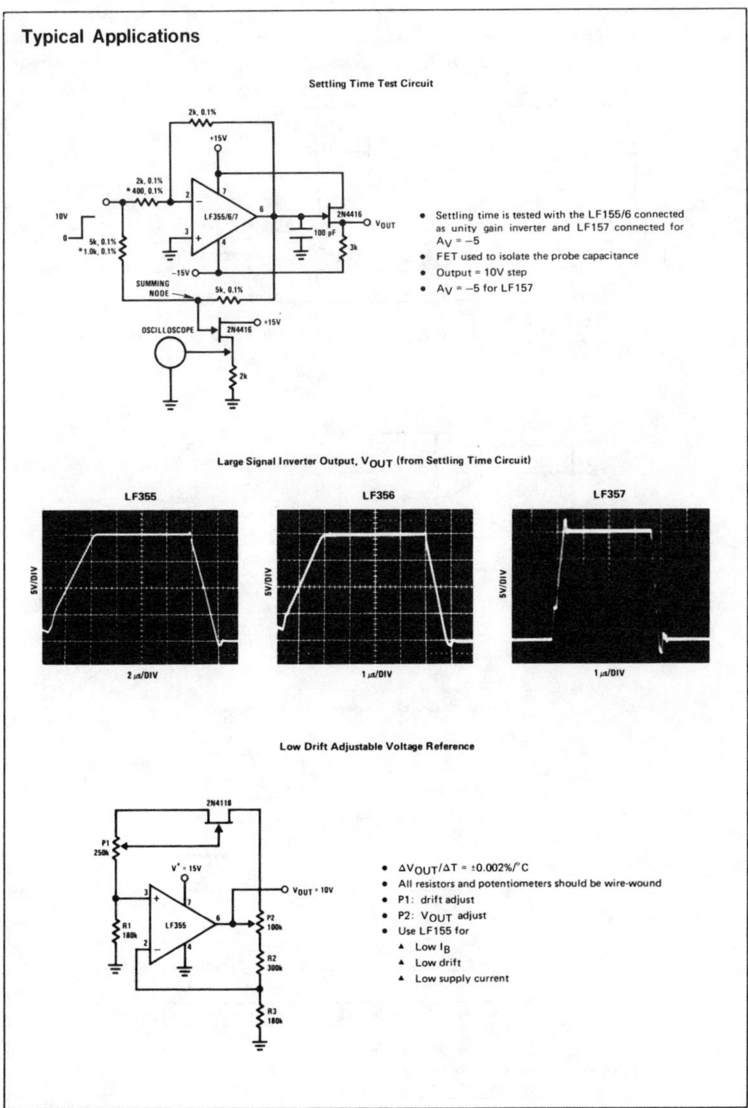


*Courtesy National Semiconductor Corp.*

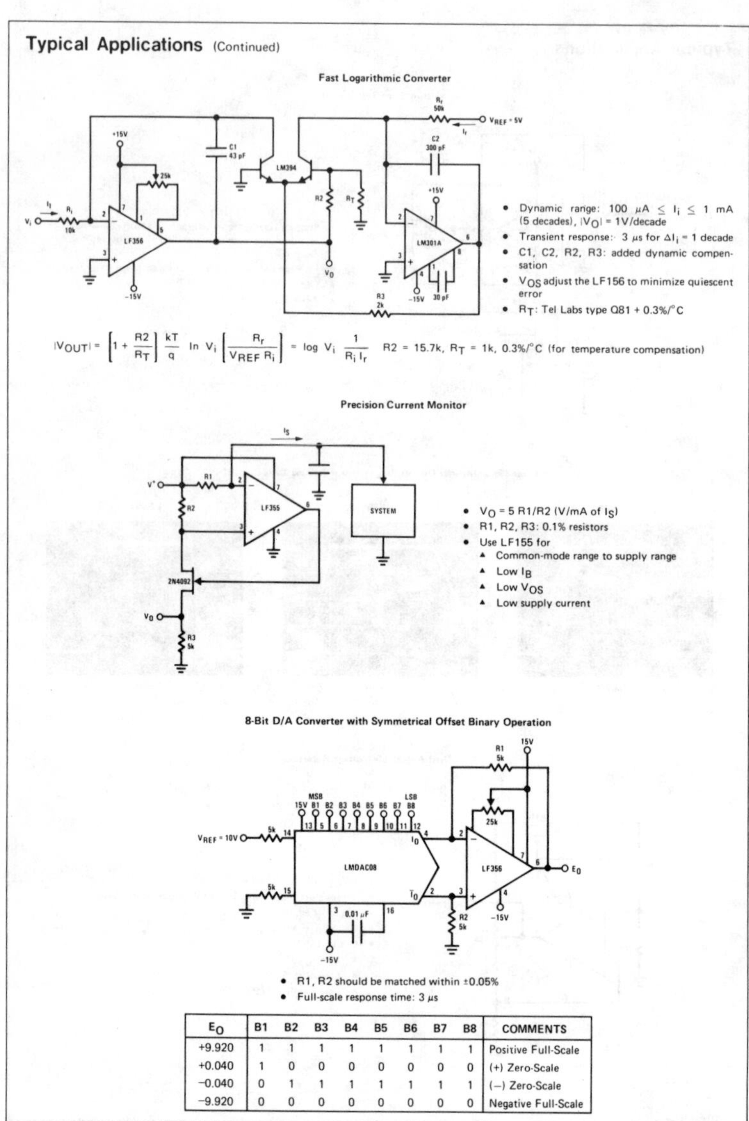


$$|V_{OUT}| = \left[1 + \frac{R2}{R_T}\right] \frac{kT}{q} \ln V_i \left[\frac{R_r}{V_{REF}\,R_i}\right] = \log V_i \frac{1}{R_i\,I_r} \quad R2 = 15.7k,\ R_T = 1k,\ 0.3\%/^{\circ}C \text{ (for temperature compensation)}$$

| $E_O$ | B1 | B2 | B3 | B4 | B5 | B6 | B7 | B8 | COMMENTS |
|---|---|---|---|---|---|---|---|---|---|
| +9.920 | 1 | 1 | 1 | 1 | 1 | 1 | 1 | 1 | Positive Full-Scale |
| +0.040 | 1 | 0 | 0 | 0 | 0 | 0 | 0 | 0 | (+) Zero-Scale |
| −0.040 | 0 | 1 | 1 | 1 | 1 | 1 | 1 | 1 | (−) Zero-Scale |
| −9.920 | 0 | 0 | 0 | 0 | 0 | 0 | 0 | 0 | Negative Full-Scale |

*Courtesy National Semiconductor Corp.*

# ULTRA-LOW OFFSET VOLTAGE OPERATIONAL AMPLIFIER

# OP-07

**PMI**

## FEATURES

- **Low $V_{OS}$** .......... **10μV**
- **Low $V_{OS}$ Drift** .......... **0.2μV/°C**
- **Ultra-Stable vs Time** .......... **0.2μV/Month**
- **Low Noise** .......... **$0.35\mu V_{p\text{-}p}$**
- **Wide Input Voltage Range** .......... **±14V**
- **Wide Supply Voltage Range** .......... **±3V to ±18V**
- **Fits 725, 108A/308A, 741, AD510 Sockets**
- **125°C Temperature-Tested Dice**

## ORDERING INFORMATION†

| $T_A$ = 25°C $V_{OS}$ MAX (μV) | PACKAGE TO-99 8-PIN | PACKAGE HERMETIC DIP 8-PIN | PACKAGE PLASTIC DIP 8-PIN | OPERATING TEMPERATURE RANGE |
|---|---|---|---|---|
| 75 | OP07AJ* | OP07AZ* | | MIL |
| 75 | OP07EJ | OP07EZ | OP07EP | COM |
| 75 | OP07J* | OP07Z* | | MIL |
| 150 | OP07CJ | OP07CZ | OP07CP | COM |
| 150 | OP07DJ | | OP07DP | COM |

*Also available with MIL-STD-883B processing. To order add /883 as a suffix to the part number. Screening Procedure: 1984 Data Book, Section 3.

†All commercial and industrial temperature range parts are available with burn-in per MIL-STD-883. Ordering Information: 1984 Data Book, Section 2.

## GENERAL DESCRIPTION

The OP-07 has very low input offset voltage (25μV max for OP-07A) which is obtained by trimming at the wafer stage. These low offset voltages generally eliminate any need for external nulling. The OP-07 also features low input bias current (±2nA for OP-07A) and high open-loop gain (300V/mV for OP-07A). The low offsets and high open-loop gain make the OP-07 particularly useful for high-gain instrumentation applications.

The wide input voltage range of ±13V minimum combined with high CMRR of 110dB (OP-07A) and high input impedace provides high accuracy in the noninverting circuit configuration. Excellent linearity and gain accuracy can be maintained even at high closed-loop gains.

Stability of offsets and gain with time or variations in temperature is excellent. The accuracy and stability of the OP-07, even at high gain, combined with the freedom from external nulling have made the OP-07 a new industry standard for instrumentation and military applications.

The OP-07 is available in five standard performance grades. The OP-07A and the OP-07 are specified for operation over the full military range of −55°C to +125°C; the OP-07 E, C, and D are specified for operation over the 0°C to +70°C range.

The OP-07 is available in hermetically-sealed TO-99 metal can or ceramic 8-pin Mini-DIP, and in epoxy 8-pin Mini-DIP. It is a direct replacement for 725, 108A, and OP-05 amplifiers; 741-types may be directly replaced by removing the 741's nulling potentiometer. The OP-207, a dual OP-07, is available for applications requiring close matching of two OP-07 amplifiers.

## PIN CONNECTIONS

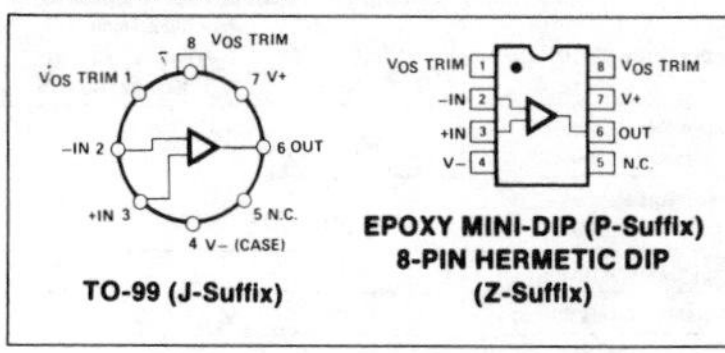


TO-99 (J-Suffix)

EPOXY MINI-DIP (P-Suffix)
8-PIN HERMETIC DIP
(Z-Suffix)

## SIMPLIFIED SCHEMATIC

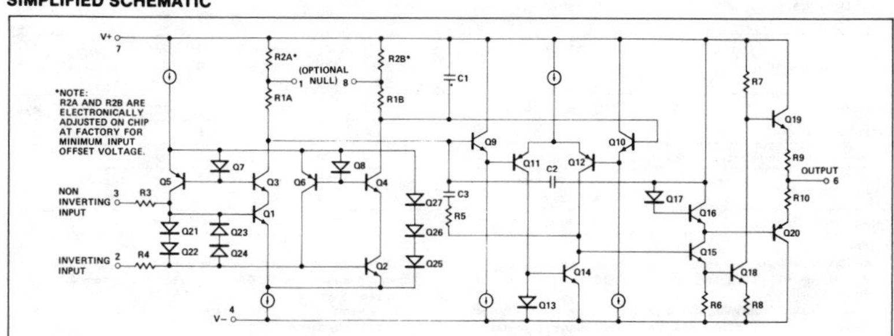


Precision Monolithics Incorporated

*Printed with permission from Precision Monolithics Incorporated.*

## ABSOLUTE MAXIMUM RATINGS (Note 2)

Supply Voltage .................................... ±22V
Internal Power Dissipation (Note 1) .............. 500mW
Differential Input Voltage ........................ ±30V
Input Voltage (Note 3) ............................ ±22V
Output Short-Circuit Duration ............... Indefinite
Storage Temperature Range
J and Z Packages .................... −65° C to +150° C
P Package ........................... −65° C to +125° C
Operating Temperature Range
OP-07A, OP-07 ....................... −55° C to +125° C
OP-07E, OP-07C, OP-07D ............... 0° C to +70° C
Lead Temperature Range (Soldering, 60 sec) ...... 300° C
DICE Junction Temperature ($T_j$) ....... −65° C to +150° C

**NOTES:**

1. See table for maximum ambient temperature rating and derating factor.

| PACKAGE TYPE | MAXIMUM AMBIENT TEMPERATURE FOR RATING | DERATE ABOVE MAXIMUM AMBIENT TEMPERATURE |
|---|---|---|
| TO-99 (J) | 80° C | 7.1mW/° C |
| 8-Pin Hermetic DIP (Z) | 75° C | 6.7mW/° C |
| 8-Pin Plastic DIP (P) | 36° C | 5.6mW/° C |

2. Absolute maximum ratings apply to both packaged parts and DICE, unless otherwise noted.
3. For supply voltages less than ±22V, the absolute maximum input voltage is equal to the supply voltage.

## ELECTRICAL CHARACTERISTICS at $V_S = \pm 15V$, $T_A = 25°C$, unless otherwise noted.

| PARAMETER | SYMBOL | CONDITIONS | OP-07A MIN | OP-07A TYP | OP-07A MAX | OP-07 MIN | OP-07 TYP | OP-07 MAX | UNITS |
|---|---|---|---|---|---|---|---|---|---|
| Input Offset Voltage | $V_{OS}$ | (Note 1) | — | 10 | 25 | — | 30 | 75 | μV |
| Long-Term Input Offset Voltage Stability | $\Delta V_{OS}/Time$ | (Note 2) | — | 0.2 | 1.0 | — | 0.2 | 1.0 | μV/Mo |
| Input Offset Current | $I_{OS}$ | | — | 0.3 | 2.0 | — | 0.4 | 2.8 | nA |
| Input Bias Current | $I_B$ | | — | ±0.7 | ±2.0 | — | ±1.0 | ±3.0 | nA |
| Input Noise Voltage | $e_{np\text{-}p}$ | 0.1Hz to 10Hz (Note 3) | — | 0.35 | 0.6 | — | 0.35 | 0.6 | $\mu V_{p\text{-}p}$ |
| Input Noise Voltage Density | $e_n$ | $f_O$ = 10Hz (Note 3)<br>$f_O$ = 100Hz (Note 3)<br>$f_O$ = 1000Hz (Note 3) | —<br>—<br>— | 10.3<br>10.0<br>9.6 | 18.0<br>13.0<br>11.0 | —<br>—<br>— | 10.3<br>10.0<br>9.6 | 18.0<br>13.0<br>11.0 | $nV/\sqrt{Hz}$ |
| Input Noise Current | $i_{np\text{-}p}$ | 0.1Hz to 10Hz (Note 3) | — | 14 | 30 | — | 14 | 30 | $pA_{p\text{-}p}$ |
| Input Noise Current Density | $i_n$ | $f_O$ = 10Hz (Note 3)<br>$f_O$ = 100Hz (Note 3)<br>$f_O$ = 1000Hz (Note 3) | —<br>—<br>— | 0.32<br>0.14<br>0.12 | 0.80<br>0.23<br>0.17 | —<br>—<br>— | 0.32<br>0.14<br>0.12 | 0.80<br>0.23<br>0.17 | $pA/\sqrt{Hz}$ |
| Input Resistance — Differential-Mode | $R_{IN}$ | (Note 4) | 30 | 80 | — | 20 | 60 | — | MΩ |
| Input Resistance — Common-Mode | $R_{INCM}$ | | — | 200 | — | — | 200 | — | GΩ |
| Input Voltage Range | IVR | | ±13 | ±14 | — | ±13 | ±14 | — | V |
| Common-Mode Rejection Ratio | CMRR | $V_{CM} = \pm 13V$ | 110 | 126 | — | 110 | 126 | — | dB |
| Power Supply Rejection Ratio | PSRR | $V_S = \pm 3V$ to $\pm 18V$ | — | 4 | 10 | — | 4 | 10 | μV/V |
| Large-Signal Voltage Gain | $A_{VO}$ | $R_L \geq 2k\Omega$, $V_O = \pm 10V$<br>$R_L \geq 500\Omega$, $V_O = \pm 0.5V$, $V_S = \pm 3V$ (Note 4) | 300<br>150 | 500<br>400 | —<br>— | 200<br>150 | 500<br>400 | —<br>— | V/mV |
| Output Voltage Swing | $V_O$ | $R_L \geq 10k\Omega$<br>$R_L \geq 2k\Omega$<br>$R_L \geq 1k\Omega$ | ±12.5<br>±12.0<br>±10.5 | ±13.0<br>±12.8<br>±12.0 | —<br>—<br>— | ±12.5<br>±12.0<br>±10.5 | ±13.0<br>±12.8<br>±12.0 | —<br>—<br>— | V |
| Slew Rate | SR | $R_L \geq 2k\Omega$ (Note 3) | 0.1 | 0.3 | — | 0.1 | 0.3 | — | V/μs |
| Closed-Loop Bandwidth | BW | $A_{VCL} = +1$ (Note 3) | 0.4 | 0.6 | — | 0.4 | 0.6 | — | MHz |
| Open-Loop Output Resistance | $R_O$ | $V_O = 0$, $I_O = 0$ | — | 60 | — | — | 60 | — | Ω |
| Power Consumption | $P_d$ | $V_S = \pm 15V$, No Load<br>$V_S = \pm 3V$, No Load | —<br>— | 75<br>4 | 120<br>6 | —<br>— | 75<br>4 | 120<br>6 | mW |
| Offset Adjustment Range | | $R_P = 20k\Omega$ | — | ±4 | — | — | ±4 | — | mV |

**NOTES:**

1. OP-07A grade $V_{OS}$ is measured one minute after application of power. For all other grades $V_{OS}$ is measured approximately 0.5 seconds after application of power.
2. Long-Term Input Offset Voltage Stability refers to the averaged trend line of $V_{OS}$ vs. Time over extended periods after the first 30 days of operation. Excluding the initial hour of operation, changes in $V_{OS}$ during the first 30 operating days are typically 2.5μV — refer to typical performance curves. Parameter is sample tested.
3. Sample tested.
4. Guaranteed by design.

*Printed with permission from Precision Monolithics Incorporated.*

## ELECTRICAL CHARACTERISTICS at $V_S = \pm 15V$, $-55°C \leq T_A \leq +125°C$, unless otherwise noted.

| PARAMETER | SYMBOL | CONDITIONS | OP-07A MIN | OP-07A TYP | OP-07A MAX | OP-07 MIN | OP-07 TYP | OP-07 MAX | UNITS |
|---|---|---|---|---|---|---|---|---|---|
| Input Offset Voltage | $V_{OS}$ | (Note 1) | — | 25 | 60 | — | 60 | 200 | μV |
| Average Input Offset Voltage Drift Without External Trim | $TCV_{OS}$ | (Note 2) | — | 0.2 | 0.6 | — | 0.3 | 1.3 | μV/°C |
| With External Trim | $TCV_{OSn}$ | $R_P = 20k\Omega$ (Note 3) | — | 0.2 | 0.6 | — | 0.3 | 1.3 | μV/°C |
| Input Offset Current | $I_{OS}$ | | — | 0.8 | 4 | — | 1.2 | 5.6 | nA |
| Average Input Offset Current Drift | $TCI_{OS}$ | (Note 2) | — | 5 | 25 | — | 8 | 50 | pA/°C |
| Input Bias Current | $I_B$ | | — | ±1 | ±4 | — | ±2 | ±6 | nA |
| Average Input Bias Current Drift | $TCI_B$ | (Note 2) | — | 8 | 25 | — | 13 | 50 | pA/°C |
| Input Voltage Range | IVR | | ±13 | ±13.5 | — | ±13 | ±13.5 | — | V |
| Common-Mode Rejection Ratio | CMRR | $V_{CM} = \pm 13V$ | 106 | 123 | — | 106 | 123 | — | dB |
| Power Supply Rejection Ratio | PSRR | $V_S = \pm 3V$ to $\pm 18V$ | — | 5 | 20 | — | 5 | 20 | μV/V |
| Large-Signal Voltage Gain | $A_{VO}$ | $R_L \geq 2k\Omega$, $V_O = \pm 10V$ | 200 | 400 | — | 150 | 400 | — | V/mV |
| Output Voltage Swing | $V_O$ | $R_L \geq 2k\Omega$ | ±12 | ±12.6 | — | ±12 | ±12.6 | — | V |

**NOTES:**

1. OP-07A grade $V_{OS}$ is measured one minute after application of power. For all other grades $V_{OS}$ is measured approximately 0.5 seconds after application of power.
2. Sample tested.
3. Guaranteed by design.

### TYPICAL OFFSET VOLTAGE TEST CIRCUIT

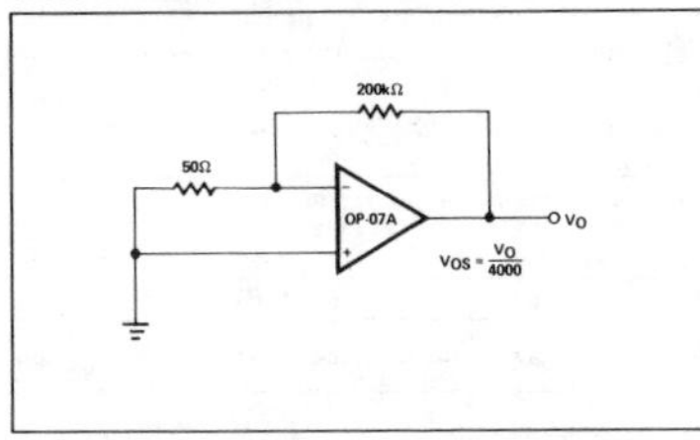


### TYPICAL LOW-FREQUENCY NOISE TEST CIRCUIT

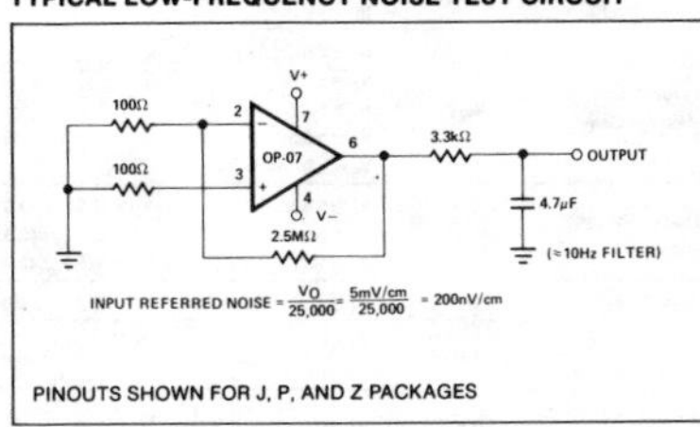


### OPTIONAL OFFSET NULLING CIRCUIT

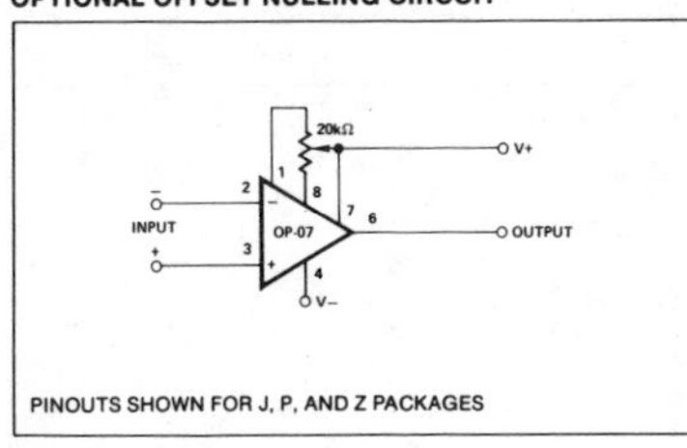


### BURN-IN CIRCUIT

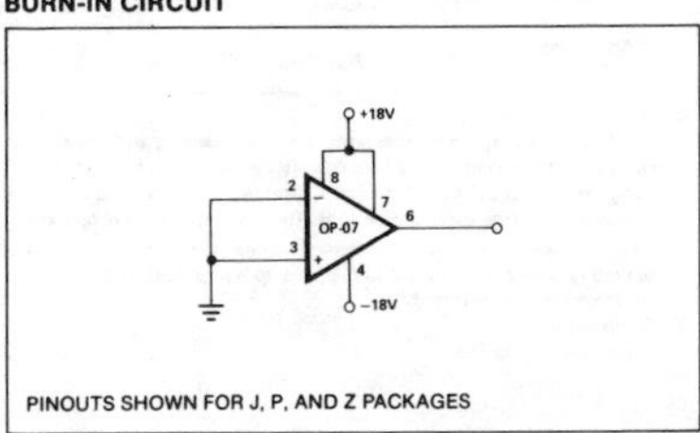


*Printed with permission from Precision Monolithics Incorporated.*

**ELECTRICAL CHARACTERISTICS** at $V_S = \pm 15V$, $T_A = 25°C$, unless otherwise noted.

| | | | OP-07E | | | OP-07C | | | OP-07D | | | |
|---|---|---|---|---|---|---|---|---|---|---|---|---|
| **PARAMETER** | **SYMBOL** | **CONDITIONS** | **MIN** | **TYP** | **MAX** | **MIN** | **TYP** | **MAX** | **MIN** | **TYP** | **MAX** | **UNITS** |
| Input Offset Voltage | $V_{OS}$ | (Note 1) | — | 30 | 75 | — | 60 | 150 | — | 60 | 150 | $\mu V$ |
| Long-Term $V_{OS}$ Stability | $V_{OS}$/Time | (Note 2) | — | 0.3 | 1.5 | — | 0.4 | 2.0 | — | 0.5 | 3.0 | $\mu V$/Mo |
| Input Offset Current | $I_{OS}$ | | — | 0.5 | 3.8 | — | 0.8 | 6.0 | — | 0.8 | 6.0 | nA |
| Input Bias Current | $I_B$ | | — | ±1.2 | ±4.0 | — | ±1.8 | ±7.0 | — | ±2.0 | ±12 | nA |
| Input Noise Voltage | $e_{np\text{-}p}$ | 0.1Hz to 10Hz (Note 3) | — | 0.35 | 0.6 | — | 0.38 | 0.65 | — | 0.38 | 0.65 | $\mu V_{p\text{-}p}$ |
| Input Noise Voltage Density | $e_n$ | $f_O = 10Hz$<br>$f_O = 100Hz$ (Note 3)<br>$f_O = 1000Hz$ | —<br>—<br>— | 10.3<br>10.0<br>9.6 | 18.0<br>13.0<br>11.0 | —<br>—<br>— | 10.5<br>10.2<br>9.8 | 20.0<br>13.5<br>11.5 | —<br>—<br>— | 10.5<br>10.3<br>9.8 | 20.0<br>13.5<br>11.5 | $nV/\sqrt{Hz}$ |
| Input Noise Current | $i_{np\text{-}p}$ | 0.1Hz to 10Hz (Note 3) | — | 14 | 30 | — | 15 | 35 | — | 15 | 35 | $pA_{p\text{-}p}$ |
| Input Noise Current Density | $i_n$ | $f_O = 10Hz$<br>$f_O = 100Hz$ (Note 3)<br>$f_O = 1000Hz$ | —<br>—<br>— | 0.32<br>0.14<br>0.12 | 0.80<br>0.23<br>0.17 | —<br>—<br>— | 0.35<br>0.15<br>0.13 | 0.90<br>0.27<br>0.18 | —<br>—<br>— | 0.35<br>0.15<br>0.13 | 0.90<br>0.27<br>0.18 | $pA/\sqrt{Hz}$ |
| Input Resistance — Differential-Mode | $R_{IN}$ | (Note 4) | 15 | 50 | — | 8 | 33 | — | 7 | 31 | — | MΩ |
| Input Resistance — Common-Mode | $R_{INCM}$ | | — | 160 | — | — | 120 | — | — | 120 | — | GΩ |
| Input Voltage Range | IVR | | ±13 | ±14 | — | ±13 | ±14 | — | ±13 | ±14 | — | V |
| Common-Mode Rejection Ratio | CMRR | $V_{CM} = \pm 13V$ | 106 | 123 | — | 100 | 120 | — | 94 | 110 | — | dB |
| Power Supply Rejection Ratio | PSRR | $V_S = \pm 3V$ to ±18V | — | 5 | 20 | — | 7 | 32 | — | 7 | 32 | $\mu V/V$ |
| Large-Signal Voltage Gain | $A_{VO}$ | $R_L \geq 2k\Omega$, $V_O = \pm 10V$<br>$R_L \geq 500\Omega$ $V_O = \pm 0.5V$ $V_S = \pm 3V$ (Note 4) | 200<br>150 | 500<br>400 | —<br>— | 120<br>100 | 400<br>400 | —<br>— | 120<br>— | 400<br>400 | —<br>— | V/mV |
| Output Voltage Swing | $V_O$ | $R_L \geq 10k\Omega$<br>$R_L \geq 2k\Omega$<br>$R_L \geq 1k\Omega$ | ±12.5<br>±12.0<br>±10.5 | ±13.0<br>±12.8<br>±12.0 | —<br>—<br>— | ±12.0<br>±11.5<br>— | ±13.0<br>±12.8<br>±12.0 | —<br>—<br>— | ±12.0<br>±11.5<br>— | ±13.0<br>±12.8<br>±12.0 | —<br>—<br>— | V |
| Slew Rate | SR | $R_L \geq 2k\Omega$ (Note 3) | 0.1 | 0.3 | — | 0.1 | 0.3 | — | 0.1 | 0.3 | — | $V/\mu s$ |
| Closed-Loop Bandwidth | BW | $A_{VCL} = +1$ (Note 3) | 0.4 | 0.6 | — | 0.4 | 0.6 | — | 0.4 | 0.6 | — | MHz |
| Open-Loop Output Resistance | $R_O$ | $V_O = 0$, $I_O = 0$ | — | 60 | — | — | 60 | — | — | 60 | — | Ω |
| Power Consumption | $P_d$ | $V_S = \pm 15V$, No Load<br>$V_S = \pm 3V$, No Load | —<br>— | 75<br>4 | 120<br>6 | —<br>— | 80<br>4 | 150<br>8 | —<br>— | 80<br>4 | 150<br>8 | mW |
| Offset Adjustment Range | | $R_P = 20k\Omega$ | — | ±4 | — | — | ±4 | — | — | ±4 | — | mV |

**NOTES:**

1. Input Offset Voltage measurements are performed by automated test equipment approximately 0.5 seconds after application of power.
2. Long-Term Input Offset Voltage Stability refers to the averaged trend line of $V_{OS}$ vs. Time over extended periods after the first 30 days of operation. Excluding the initial hour of operation, changes in $V_{OS}$ during the first 30 operating days are typically 2.5μV — refer to typical performance curves. Parameter is sample tested.
3. Sample tested.
4. Guaranteed by design.

*Printed with permission from Precision Monolithics Incorporated.*

**ELECTRICAL CHARACTERISTICS** at $V_S = \pm 15V$, $0°C \le T_A \le +70°C$, unless otherwise noted.

| PARAMETER | SYMBOL | CONDITIONS | OP-07E MIN | OP-07E TYP | OP-07E MAX | OP-07C MIN | OP-07C TYP | OP-07C MAX | OP-07D MIN | OP-07D TYP | OP-07D MAX | UNITS |
|---|---|---|---|---|---|---|---|---|---|---|---|---|
| Input Offset Voltage | $V_{OS}$ | (Note 1) | — | 45 | 130 | — | 85 | 250 | — | 85 | 250 | μV |
| Average Input Offset Voltage Drift Without External Trim | $TCV_{OS}$ | (Note 3) | — | 0.3 | 1.3 | — | 0.5 | 1.8 | — | 0.7 | 2.5 | μV/°C |
| With External Trim | $TCV_{OSn}$ | $R_P = 20k\Omega$ (Note 3) | — | 0.3 | 1.3 | — | 0.4 | 1.6 | — | 0.7 | 2.5 | μV/°C |
| Input Offset Current | $I_{OS}$ | | — | 0.9 | 5.3 | — | 1.6 | 8.0 | — | 1.6 | 8.0 | nA |
| Average Input Offset Current Drift | $TCI_{OS}$ | (Note 2) | — | 8 | 35 | — | 12 | 50 | — | 12 | 50 | pA/°C |
| Input Bias Current | $I_B$ | | — | ±1.5 | ±5.5 | — | ±2.2 | ±9.0 | — | ±3.0 | ±14 | nA |
| Average Input Bias Current Drift | $TCI_B$ | (Note 2) | — | 13 | 35 | — | 18 | 50 | — | 18 | 50 | pA/°C |
| Input Voltage Range | IVR | | ±13.0 | ±13.5 | — | ±13.0 | ±13.5 | — | ±13.0 | ±13.5 | — | V |
| Common-Mode Rejection Ratio | CMRR | $V_{CM} = \pm 13V$ | 103 | 123 | — | 97 | 120 | — | 94 | 106 | — | dB |
| Power Supply Rejection Ratio | PSRR | $V_S = \pm 3V$ to $\pm 18V$ | — | 7 | 32 | — | 10 | 51 | — | 10 | 51 | μV/V |
| Large-Signal Voltage Gain | $A_{VO}$ | $R_L \ge 2k\Omega$<br>$V_O = \pm 10V$ | 180 | 450 | — | 100 | 400 | — | 100 | 400 | — | V/mV |
| Output Voltage Swing | $V_O$ | $R_L \ge 2k\Omega$ | ±12 | ±12.6 | — | ±11 | ±12.6 | — | ±11 | ±12.6 | — | V |

**NOTES:**
1. Input offset voltage measurements are performed by automated test equipment approximately 0.5 seconds after application of power.
2. Sample tested.
3. Guaranteed by design.

## DICE CHARACTERISTICS (125° C TESTED DICE AVAILABLE)

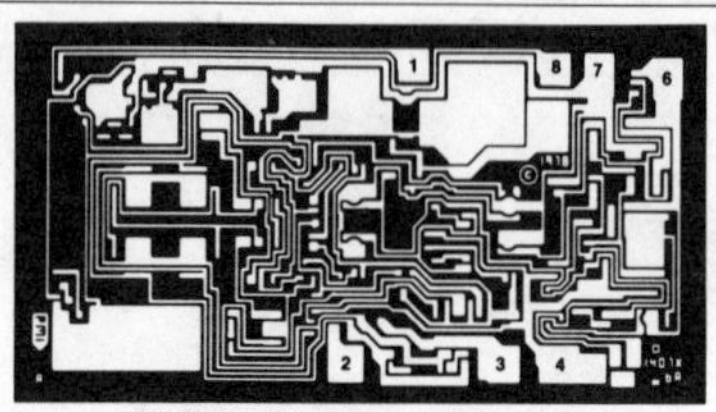

DIE SIZE 0.100 × 0.053 inch, 5300 sq. mils
(2.54 × 1.35 mm, 3.42 sq. mm)

1. BALANCE
2. INVERTING INPUT
3. NONINVERTING INPUT
4. V−
6. OUTPUT
7. V+
8. BALANCE

For additional DICE information refer to 1984 Data Book, Section 2.

**WAFER TEST LIMITS** at $V_S = \pm 15V$, $T_A = 25°C$ for OP-07N, OP-07G and OP-07GR devices; $T_A = 125°C$ for OP-07NT and OP-07GT devices, unless otherwise noted.

| PARAMETER | SYMBOL | CONDITIONS | OP-07NT LIMIT | OP-07N LIMIT | OP-07GT LIMIT | OP-07G LIMIT | OP-07GR LIMIT | UNITS |
|---|---|---|---|---|---|---|---|---|
| Input Offset Voltage | $V_{OS}$ | | 140 | 40 | 210 | 80 | 150 | μV MAX |
| Input Offset Current | $I_{OS}$ | | 4.0 | 2.0 | 5.6 | 2.8 | 6.0 | nA MAX |
| Input Bias Current | $I_B$ | | ±4 | ±2 | ±6 | ±3 | ±7 | nA MAX |
| Input Resistance Differential-Mode | $R_{IN}$ | (Note 2) | — | 20 | — | 20 | 8 | MΩ MIN |
| Input Voltage Range | IVR | | ±13 | ±13 | ±13 | ±13 | ±13 | V MIN |
| Common-Mode Rejection Ratio | CMRR | $V_{CM} = \pm 13V$ | 100 | 110 | 100 | 110 | 100 | dB MIN |
| Power Supply Rejection Ratio | PSRR | $V_S = \pm 3V$ to $\pm 18V$ | 20 | 10 | 20 | 10 | 30 | μV/V MAX |
| Output Voltage Swing | $V_O$ | $R_L = 10k\Omega$<br>$R_L = 2k\Omega$<br>$R_L = 1k\Omega$ | —<br>±12.0<br>— | ±12.5<br>±12.0<br>±10.5 | —<br>±12.0<br>— | ±12.0<br>±11.5<br>±10.5 | ±12.0<br>±11.5<br>— | V MIN |
| Large-Signal Voltage Gain | $A_{VO}$ | $R_L = 2k\Omega$<br>$V_O = \pm 10V$ | 200 | 200 | 150 | 120 | 120 | V/mV MIN |
| Differential Input Voltage | | | ±30 | ±30 | ±30 | ±30 | ±30 | V MAX |
| Power Consumption | $P_d$ | $V_{OUT} = 0V$ | — | 120 | — | 120 | 150 | mW MAX |

**NOTES:**
1. For 25° C characteristics of OP-07NT and OP-07GT, see OP-07N and OP-07G characteristics, respectively.
2. Guaranteed by design.

Electrical tests are performed at wafer probe to the limits shown. Due to variations in assembly methods and normal yield loss, yield after packaging is not guaranteed for standard product dice. Consult factory to negotiate specifications based on dice lot qualification through sample lot assembly and testing.

**TYPICAL ELECTRICAL CHARACTERISTICS** at $V_S = \pm 15V$, $T_A = +25°C$, unless otherwise noted.

| PARAMETER | SYMBOL | CONDITIONS | OP-07NT TYPICAL | OP-07N TYPICAL | OP-07GT TYPICAL | OP-07G TYPICAL | OP-07GR TYPICAL | UNITS |
|---|---|---|---|---|---|---|---|---|
| Average Input Offset Voltage Drift | $TCV_{OS}$ | $R_S = 50\Omega$ | 0.2 | 0.2 | 0.3 | 0.3 | 0.7 | μV/°C |
| Nulled Input Offset Voltage Drift | $TCV_{OSn}$ | $R_S = 50\Omega$, $R_P = 20k\Omega$ | 0.2 | 0.2 | 0.3 | 0.3 | 0.7 | μV/°C |
| Average Input Offset Current Drift | $TCI_{OS}$ | | 5 | 5 | 8 | 8 | 12 | pA/°C |
| Slew Rate | SR | $R_L \geq 2k\Omega$ | 0.3 | 0.3 | 0.3 | 0.3 | 0.3 | V/μs |
| Closed-Loop Bandwidth | BW | $A_{VCL} = +1$ | 0.6 | 0.6 | 0.6 | 0.6 | 0.6 | MHz |

*Printed with permission from Precision Monolithics Incorporated.*

## TYPICAL PERFORMANCE CHARACTERISTICS

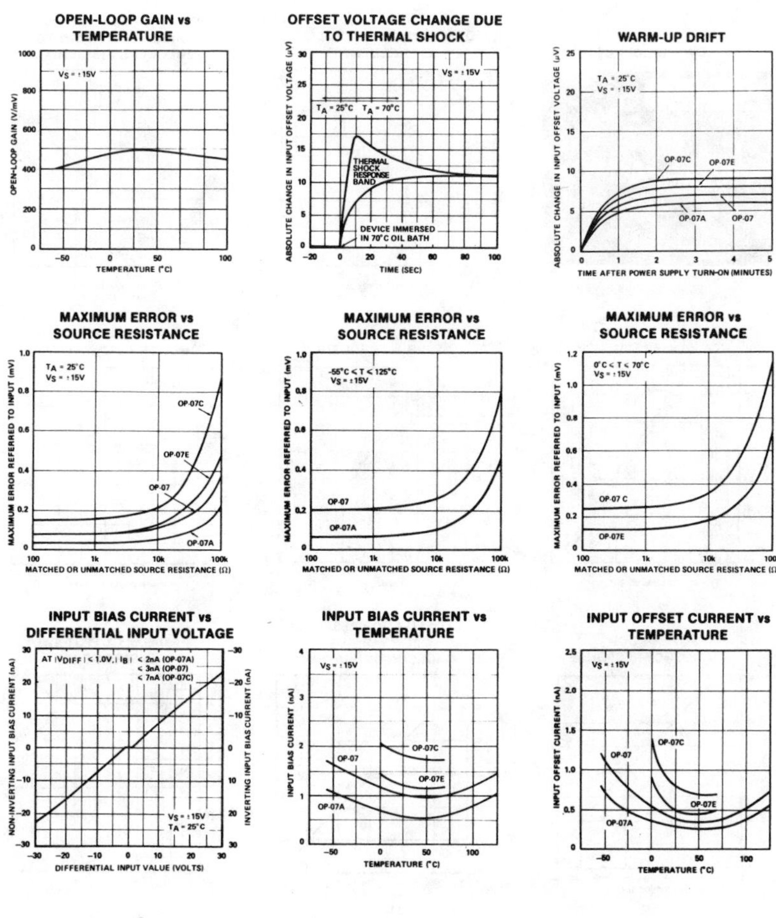


*Printed with permission from Precision Monolithics Incorporated.*

## TYPICAL PERFORMANCE CHARACTERISTICS

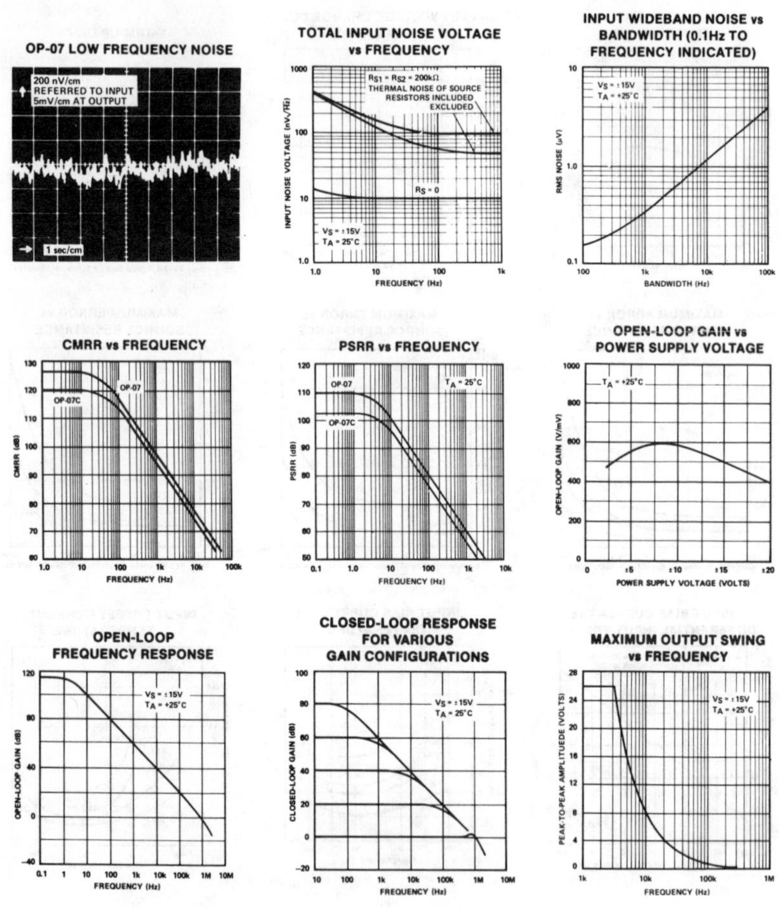


*Printed with permission from Precision Monolithics Incorporated.*

## TYPICAL PERFORMANCE CHARACTERISTICS

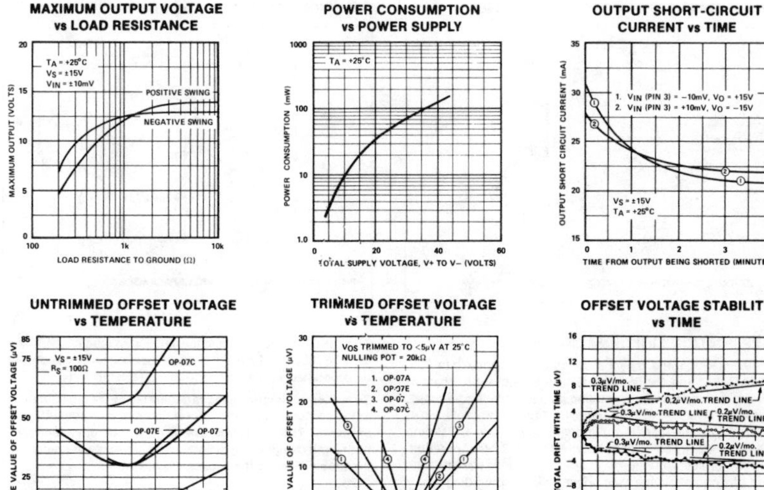


## TYPICAL APPLICATIONS

### HIGH SPEED, LOW $V_{OS}$, COMPOSITE AMPLIFIER

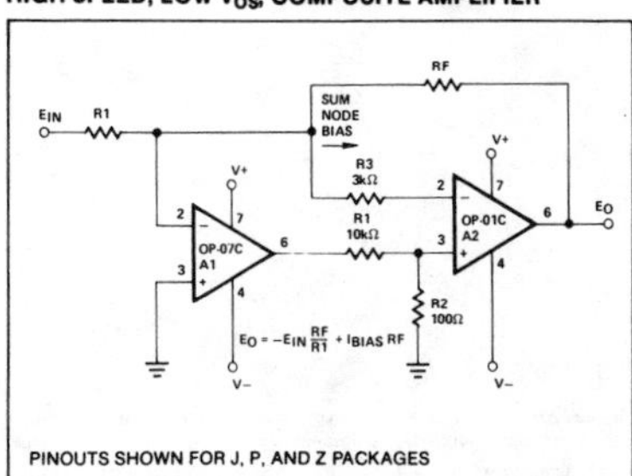


### ADJUSTMENT-FREE PRECISION SUMMING AMPLIFIER

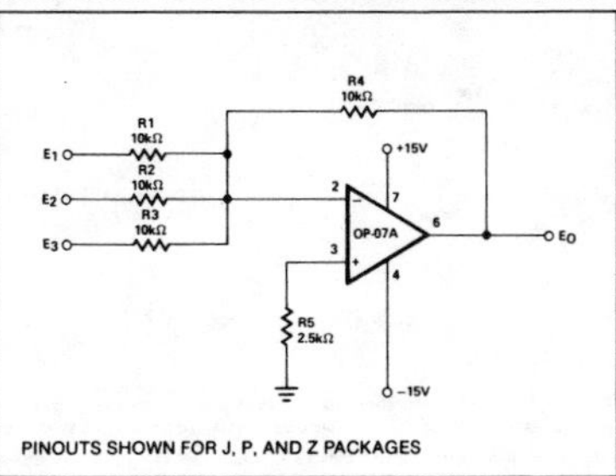


*Printed with permission from Precision Monolithics Incorporated.*

## TYPICAL APPLICATIONS

### HIGH-STABILITY THERMOCOUPLE AMPLIFIER

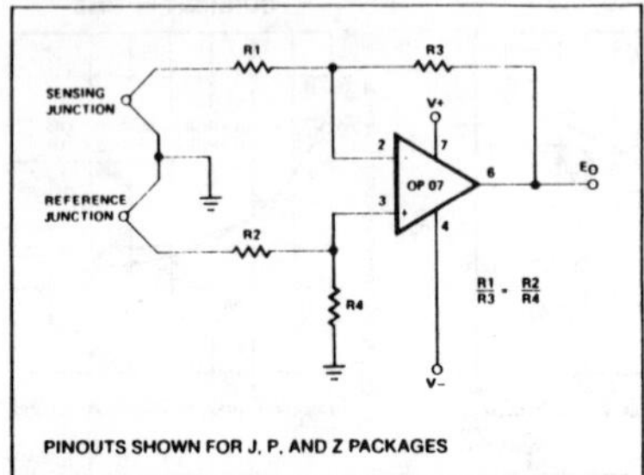


PINOUTS SHOWN FOR J, P, AND Z PACKAGES

### PRECISION ABSOLUTE-VALUE CIRCUIT

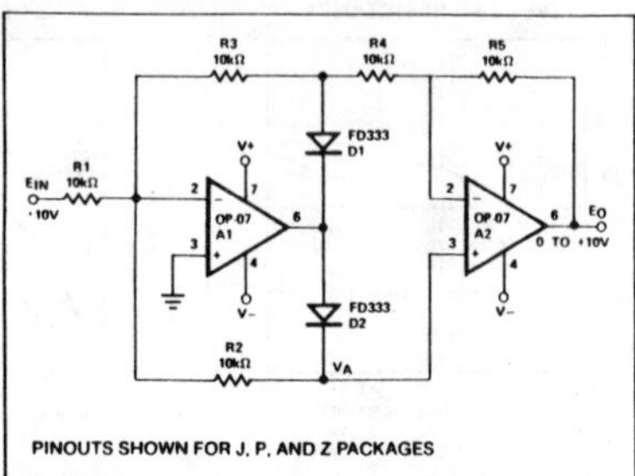


PINOUTS SHOWN FOR J, P, AND Z PACKAGES

## APPLICATIONS INFORMATION

OP-07 series units may be substituted directly into 725, 108A/308A* and OP-05 sockets with or without removal of external compensation or nulling components. Additionally, the OP-07 may be used in unnulled 741-type sockets. However, if conventional 741 nulling circuitry is in use, it should be modified or removed to enable proper OP-07 operation. OP-07 offset voltage may be nulled to zero through use of a potentiometer (see offset nulling circuit diagram).

The OP-07 provides stable operation with load capacitance of up to 500pF and ±10V swings; larger capacitances should be decoupled with a 50Ω decoupling resistor.

Stray thermoelectric voltages generated by dissimilar metals at the contacts to the input terminals can degrade drift performance. Therefore, best operation will be obtained when both input contacts are maintained at the same temperature, preferably close to the package temperature.

*TO-99 Package only



PMI® Precision Monolithics Incorporated A Bourns Subsidiary

1500 SPACE PARK DR., SANTA CLARA, CA 95050 • TEL (408) 727-9222 • TWX 910-338-0218 • TLX 172 070

05840110G5M PRINTED IN USA

*Printed with permission from Precision Monolithics Incorporated.*

# LOW-NOISE PRECISION OPERATIONAL AMPLIFIER

# OP-27

**PMI**

## FEATURES

- **Low Noise** ............... 80nV$_{p-p}$ (0.1Hz to 10Hz) / 3nV/$\sqrt{Hz}$
- **Low Drift** ................................ 0.2μV/°C
- **High Speed** .............. 2.8V/μs Slew Rate / 8MHz Gain Bandwidth
- **Low $V_{OS}$** ........................................ 10μV
- **Excellent CMRR** ............... 126dB at $V_{CM}$ of ±11V
- **High Open-Loop Gain** ...................... 1.8 Million
- **Fits 725, OP-07, OP-05, AD510, AD517, 5534A sockets**

## ORDERING INFORMATION†

| $T_A$ = 25°C $V_{OS}$ MAX (μV) | PACKAGE HERMETIC TO-99 8-PIN | PACKAGE HERMETIC DIP 8-PIN | PACKAGE PLASTIC DIP 8-PIN | OPERATING TEMPERATURE RANGE |
|---|---|---|---|---|
| 25 | OP27AJ* | OP27AZ* | | MIL |
| 25 | OP27EJ | OP27EZ | OP27EP | IND/COM |
| 60 | OP27BJ* | OP27BZ* | | MIL |
| 60 | OP27FJ | OP27FZ | OP27FP | IND/COM |
| 100 | OP27CJ* | OP27CZ* | | MIL |
| 100 | OP27GJ | OP27GZ | OP27GP | IND/COM |

*Also available with MIL-STD-883B processing. To order add /883 as a suffix to the part number. Screening Procedure: 1984 Data Book, Section 3.

†All commercial and industrial temperature range parts are available with burn-in per MIL-STD-883. Ordering Information: 1984 Data Book, Section 2.

## GENERAL DESCRIPTION

The OP-27 precision operational amplifier combines the low offset and drift of the OP-07 with both high-speed and low-noise. Offsets down to 25μV and drift of 0.6μV/°C maximum make the OP-27 ideal for precision instrumentation applications. Exceptionally low noise, $e_n = 3.5nV/\sqrt{Hz}$, at 10Hz, a low 1/f noise corner frequency of 2.7Hz, and high gain (1.8 million), allow accurate high-gain amplification of low-level signals. A gain-bandwidth product of 8MHz and a 2.8V/μsec slew rate provides excellent dynamic accuracy in high-speed data-acquisition systems.

A low input bias current of ±10nA is achieved by use of a bias-current-cancellation circuit. Over the military temperature range, this circuit typically holds $I_B$ and $I_{OS}$ to ±20nA and 15nA respectively.

The output stage has good load driving capability. A guaranteed swing of ±10V into 600Ω and low output distortion make the OP-27 an excellent choice for professional audio applications.

PSRR and CMRR exceed 120dB. These characteristics, coupled with long-term drift of 0.2μV/month, allow the circuit designer to achieve performance levels previously attained only by discrete designs.

Low cost, high-volume production of OP-27 is achieved by using an on-chip zener-zap trimming network. This reliable and stable offset trimming scheme has proved its effectiveness over many years of production history.

The OP-27 provides excellent performance in low-noise high-accuracy amplification of low-level signals. Applications include stable integrators, precision summing amplifiers, precision voltage-threshold detectors, comparators, and professional audio circuits such as tape-head and microphone preamplifiers.

The OP-27 is a direct replacement for 725, OP-06, OP-07 and OP-05 amplifiers; 741 types may be directly replaced by removing the 741's nulling potentiometer.

## PIN CONNECTIONS

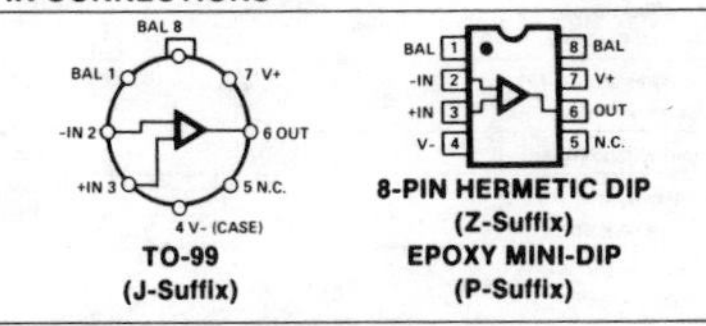


## SIMPLIFIED SCHEMATIC

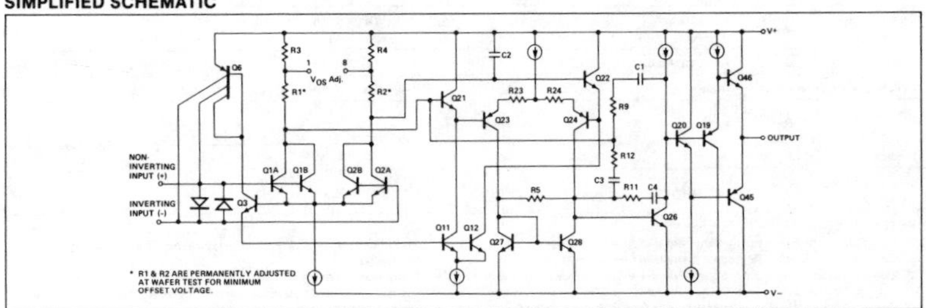


Precision Monolithics Incorporated

*Printed with permission from Precision Monolithics Incorporated.*

## ABSOLUTE MAXIMUM RATINGS (Note 4)

Supply Voltage .................................. ±22V
Internal Power Dissipation (Note 1) ............. 500mW
Input Voltage (Note 3) ........................... ±22V
Output Short-Circuit Duration ................. Indefinite
Differential Input Voltage (Note 2) ................ ±0.7V
Differential Input Current (Note 2) .............. ±25mA
Storage Temperature Range .......... −65° C to +150° C
Operating Temperature Range
 OP-27A, OP-27B, OP-27C (J, Z) ..... −55° C to +125° C
 OP-27E, OP-27F, OP-27G (J, Z) ...... −25° C to +85° C
 OP-27E, OP-27F, OP-27G (P) ........... 0° C to +70° C
Lead Temperature Range (Soldering, 60 sec) ...... 300° C
DICE Junction Temperature .......... −65° C to +150° C

**NOTES:**

1. See table for maximum ambient temperature rating and derating factor.

| PACKAGE TYPE | MAXIMUM AMBIENT TEMPERATURE FOR RATING | DERATE ABOVE MAXIMUM AMBIENT TEMPERATURE |
|---|---|---|
| TO-99 (J) | 80° C | 7.1mW/° C |
| 8-Pin Hermetic DIP (Z) | 75° C | 6.7mW/° C |
| 8-Pin Plastic DIP (P) | 62° C | 5.6mW/° C |

2. The OP-27's inputs are protected by back-to-back diodes. Current limiting resistors are not used in order to achieve low noise. If differential input voltage exceeds ±0.7V, the input current should be limited to 25mA.
3. For supply voltages less than ±22V, the absolute maximum input voltage is equal to the supply voltage.
4. Absolute maximum ratings apply to both DICE and packaged parts, unless otherwise noted.

## ELECTRICAL CHARACTERISTICS at $V_S = \pm15V$, $T_A = 25°C$, unless otherwise noted.

| PARAMETER | SYMBOL | CONDITIONS | OP-27A/E | | | OP-27B/F | | | OP-27C/G | | | UNITS |
|---|---|---|---|---|---|---|---|---|---|---|---|---|
| | | | MIN | TYP | MAX | MIN | TYP | MAX | MIN | TYP | MAX | |
| Input Offset Voltage | $V_{OS}$ | (Note 1) | — | 10 | 25 | — | 20 | 60 | — | 30 | 100 | μV |
| Long-Term $V_{OS}$ Stability | $V_{OS}$/Time | (Note 2) | — | 0.2 | 1.0 | — | 0.3 | 1.5 | — | 0.4 | 2.0 | μV/Mo |
| Input Offset Current | $I_{OS}$ | | — | 7 | 35 | — | 9 | 50 | — | 12 | 75 | nA |
| Input Bias Current | $I_B$ | | — | ±10 | ±40 | — | ±12 | ±55 | — | ±15 | ±80 | nA |
| Input Noise Voltage | $e_{np-p}$ | 0.1Hz to 10Hz (Notes 3, 5) | — | 0.08 | 0.18 | — | 0.08 | 0.18 | — | 0.09 | 0.25 | μVp-p |
| Input Noise Voltage Density | $e_n$ | $f_O$ = 10Hz (Note 3) | — | 3.5 | 5.5 | — | 3.5 | 5.5 | — | 3.8 | 8.0 | $nV/\sqrt{Hz}$ |
| | | $f_O$ = 30Hz (Note 3) | — | 3.1 | 4.5 | — | 3.1 | 4.5 | — | 3.3 | 5.6 | |
| | | $f_O$ = 1000Hz (Note 3) | — | 3.0 | 3.8 | — | 3.0 | 3.8 | — | 3.2 | 4.5 | |
| Input Noise Current Density | $i_n$ | $f_O$ = 10Hz (Notes 3,6) | — | 1.7 | 4.0 | — | 1.7 | 4.0 | — | 1.7 | — | $pA/\sqrt{Hz}$ |
| | | $f_O$ = 30Hz (Notes 3,6) | — | 1.0 | 2.3 | — | 1.0 | 2.3 | — | 1.0 | — | |
| | | $f_O$ = 1000Hz (Notes 3, 6) | — | 0.4 | 0.6 | — | 0.4 | 0.6 | — | 0.4 | 0.6 | |
| Input Resistance — Differential-Mode | $R_{IN}$ | (Note 4) | 1.5 | 6 | — | 1.2 | 5 | — | 0.8 | 4 | — | MΩ |
| Input Resistance — Common-Mode | $R_{INCM}$ | | — | 3 | — | — | 2.5 | — | — | 2 | — | GΩ |
| Input Voltage Range | IVR | | ±11.0 | ±12.3 | — | ±11.0 | ±12.3 | — | ±11.0 | ±12.3 | — | V |
| Common-Mode Rejection Ratio | CMRR | $V_{CM} = \pm11V$ | 114 | 126 | — | 106 | 123 | — | 100 | 120 | — | dB |
| Power Supply Rejection Ratio | PSRR | $V_S = \pm4V$ to $\pm18V$ | — | 1 | 10 | — | 1 | 10 | — | 2 | 20 | μV/V |
| Large-Signal Voltage Gain | $A_{VO}$ | $R_L \geq 2k\Omega$, $V_O = \pm10V$ | 1000 | 1800 | — | 1000 | 1800 | — | 700 | 1500 | — | V/mV |
| | | $R_L \geq 600\Omega$, $V_O = \pm10V$ | 800 | 1500 | — | 800 | 1500 | — | 600 | 1500 | — | |
| Output Voltage Swing | $V_O$ | $R_L \geq 2k\Omega$ | ±12.0 | ±13.8 | — | ±12.0 | ±13.8 | — | ±11.5 | ±13.5 | — | V |
| | | $R_L \geq 600\Omega$ | ±10.0 | ±11.5 | — | ±10.0 | ±11.5 | — | ±10.0 | ±11.5 | — | |
| Slew Rate | SR | $R_L \geq 2k\Omega$ (Note 4) | 1.7 | 2.8 | — | 1.7 | 2.8 | — | 1.7 | 2.8 | — | V/μs |
| Gain Bandwidth Prod. | GBW | (Note 4) | 5.0 | 8.0 | — | 5.0 | 8.0 | — | 5.0 | 8.0 | — | MHz |
| Open-Loop Output Resistance | $R_O$ | $V_O = 0$, $I_O = 0$ | — | 70 | — | — | 70 | — | — | 70 | — | Ω |
| Power Consumption | $P_d$ | $V_O$ | — | 90 | 140 | — | 90 | 140 | — | 100 | 170 | mW |
| Offset Adjustment Range | | $R_P = 10k\Omega$ | — | ±4.0 | — | — | ±4.0 | — | — | ±4.0 | — | mV |

**NOTES:**

1. Input offset voltage measurements are performed ~ 0.5 seconds after application of power. A/E grades guaranteed fully warmed-up.
2. Long-term input offset voltage stability refers to the average trend line of $V_{OS}$ vs. Time over extended periods after the first 30 days of operation. Excluding the initial hour of operation, changes in $V_{OS}$ during the first 30 days are typically 2.5μV — refer to typical performance curve.
3. Sample tested.
4. Guaranteed by design.
5. See test circuit and frequency response curve for 0.1Hz to 10Hz tester.
6. See test circuit for current noise measurement.

*Printed with permission from Precision Monolithics Incorporated.*

**ELECTRICAL CHARACTERISTICS** for $V_S = \pm 15V$, $-55°C \le T_A \le +125°C$, unless otherwise noted.

| PARAMETER | SYMBOL | CONDITIONS | OP-27A MIN | OP-27A TYP | OP-27A MAX | OP-27B MIN | OP-27B TYP | OP-27B MAX | OP-27C MIN | OP-27C TYP | OP-27C MAX | UNITS |
|---|---|---|---|---|---|---|---|---|---|---|---|---|
| Input Offset Voltage | $V_{OS}$ | (Note 1) | — | 30 | 60 | — | 50 | 200 | — | 70 | 300 | μV |
| Average Input Offset Drift | $TCV_{OS}$ $TCV_{OSn}$ | (Note 2) | — | 0.2 | 0.6 | — | 0.3 | 1.3 | — | 0.4 | 1.8 | μV/°C |
| Input Offset Current | $I_{OS}$ | | — | 15 | 50 | — | 22 | 85 | — | 30 | 135 | nA |
| Input Bias Current | $I_B$ | | — | ±20 | ±60 | — | ±28 | ±95 | — | ±35 | ±150 | nA |
| Input Voltage Range | IVR | | ±10.3 | ±11.5 | — | ±10.3 | ±11.5 | — | ±10.2 | ±11.5 | — | V |
| Common-Mode Rejection Ratio | CMRR | $V_{CM} = \pm 10V$ | 108 | 122 | — | 100 | 119 | — | 94 | 116 | — | dB |
| Power Supply Rejection Ratio | PSRR | $V_S = \pm 4.5V$ to $\pm 18V$ | — | 2 | 16 | — | 2 | 20 | — | 4 | 51 | μV/V |
| Large-Signal Voltage Gain | $A_{VO}$ | $R_L \ge 2k\Omega$, $V_O = \pm 10V$ | 600 | 1200 | — | 500 | 1000 | — | 300 | 800 | — | V/mV |
| Output Voltage Swing | $V_O$ | $R_L \ge 2k\Omega$ | ±11.5 | ±13.5 | — | ±11.0 | ±13.2 | — | ±10.5 | ±13.0 | — | V |

**ELECTRICAL CHARACTERISTICS** for $V_S = \pm 15V$, $-25°C \le T_A \le +85°C$ for OP-27J and OP-27Z, $0°C \le T_A \le +70°C$ for OP-27P, unless otherwise noted.

| PARAMETER | SYMBOL | CONDITIONS | OP-27E MIN | OP-27E TYP | OP-27E MAX | OP-27F MIN | OP-27F TYP | OP-27F MAX | OP-27G MIN | OP-27G TYP | OP-27G MAX | UNITS |
|---|---|---|---|---|---|---|---|---|---|---|---|---|
| Input Offset Voltage | $V_{OS}$ | | — | 20 | 50 | — | 40 | 140 | — | 55 | 220 | μV |
| Average Input Offset Drift | $TCV_{OS}$ $TCV_{OSn}$ | (Note 2) | — | 0.2 | 0.6 | — | 0.3 | 1.3 | — | 0.4 | 1.8 | μV/°C |
| Input Offset Current | $I_{OS}$ | | — | 10 | 50 | — | 14 | 85 | — | 20 | 135 | nA |
| Input Bias Current | $I_B$ | | — | ±14 | ±60 | — | ±18 | ±95 | — | ±25 | ±150 | nA |
| Input Voltage Range | IVR | | ±10.5 | ±11.8 | — | ±10.5 | ±11.8 | — | ±10.5 | ±11.8 | — | V |
| Common-Mode Rejection Ratio | CMRR | $V_{CM} = \pm 10V$ | 110 | 124 | — | 102 | 121 | — | 96 | 118 | — | dB |
| Power Supply Rejection Ratio | PSRR | $V_S = \pm 4.5V$ to $\pm 18V$ | — | 2 | 15 | — | 2 | 16 | — | 2 | 32 | μV/V |
| Large-Signal Voltage Gain | $A_{VO}$ | $R_L \ge 2k\Omega$, $V_O = \pm 10V$ | 750 | 1500 | — | 700 | 1300 | — | 450 | 1000 | — | V/mV |
| Output Voltage Swing | $V_O$ | $R_L \ge 2k\Omega$ | ±11.7 | ±13.6 | — | ±11.4 | ±13.5 | — | ±11.0 | ±13.3 | — | V |

**NOTES:**

1. Input offset voltage measurements are performed by automated test equipment approximately 0.5 seconds after application of power. A/E grades guaranteed fully warmed-up.
2. The $TCV_{OS}$ performance is within the specifications unnulled or when nulled with $R_P = 8k\Omega$ to $20k\Omega$.

*Printed with permission from Precision Monolithics Incorporated.*

## DICE CHARACTERISTICS

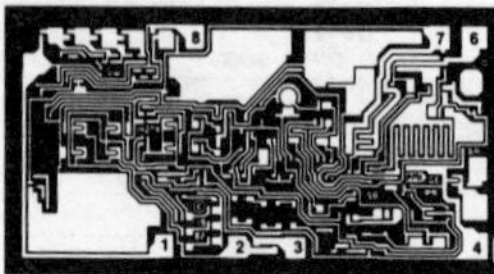

DIE SIZE 0.054 × 0.108 inch, 5832 sq. mils (1.37 × 2.74mm, 3.76 sq. mm)

1. NULL
2. (−) INPUT
3. (+) INPUT
4. V−
6. OUTPUT
7. V+
8. NULL

For additional DICE information refer to 1984 Data Book, Section 2.

**WAFER TEST LIMITS** at $V_S = \pm 15V$, $T_A = 25°C$ for OP-27N, OP-27G, and OP-27GR devices; $T_A = 125°C$ for OP-27NT and OP-27GT devices, unless otherwise noted.

| PARAMETER | SYMBOL | CONDITIONS | OP-27NT LIMIT | OP-27N LIMIT | OP-27GT LIMIT | OP-27G LIMIT | OP-27GR LIMIT | UNITS |
|---|---|---|---|---|---|---|---|---|
| Input Offset Voltage | $V_{OS}$ | (Note 1) | 60 | 35 | 200 | 60 | 100 | μV MAX |
| Input Offset Current | $I_{OS}$ | | 50 | 35 | 85 | 50 | 75 | nA MAX |
| Input Bias Current | $I_B$ | | ±60 | ±40 | ±95 | ±55 | ±80 | nA MAX |
| Input Voltage Range | IVR | | ±10.3 | ±11 | ±10.3 | ±11 | ±11 | V MIN |
| Common-Mode Rejection Ratio | CMRR | $V_{CM}$ = IVR | 108 | 114 | 100 | 106 | 100 | dB MIN |
| Power Supply Rejection Ratio | PSRR | $V_S = \pm 4V$ to $\pm 18V$ | — | 10 | — | 10 | 20 | μV/V MAX |
| Large-Signal Voltage Gain | $A_{VO}$ | $R_L \geq 2k\Omega$, $V_O = \pm 10V$<br>$R_L \geq 600\Omega$, $V_O = \pm 10V$ | 600<br>— | 1000<br>800 | 500<br>— | 1000<br>800 | 700<br>600 | V/mV MIN |
| Output Voltage Swing | $V_O$ | $R_L \geq 2k\Omega$<br>$R_L \geq 600\Omega$ | ±11.5<br>— | ±12.0<br>±10.0 | ±11.0<br>— | ±12.0<br>±10.0 | ±11.5<br>±10.0 | V MIN |
| Power Consumption | $P_d$ | $V_O = 0$ | — | 140 | — | 140 | 170 | mW MAX |

**NOTE:**
Electrical tests are performed at wafer probe to the limits shown. Due to variations in assembly methods and normal yield loss, yield after packaging is not guaranteed for standard product dice. Consult factory to negotiate specifications based on dice lot qualification through sample lot assembly and testing.

**TYPICAL ELECTRICAL CHARACTERISTICS** at $V_S = \pm 15V$, $T_A = +25°C$, unless otherwise noted.

| PARAMETER | SYMBOL | CONDITIONS | OP-27N TYPICAL | OP-27G TYPICAL | OP-27GR TYPICAL | UNITS |
|---|---|---|---|---|---|---|
| Average Input Offset Voltage Drift | $TCV_{OS}$ or $TCV_{OSn}$ | Nulled or Unnulled $R_P = 8k\Omega$ to $20k\Omega$ | 0.2 | 0.3 | 0.4 | μV/°C |
| Average Input Offset Current Drift | $TCI_{OS}$ | | 80 | 130 | 180 | pA/°C |
| Average Input Bias Current Drift | $TCI_B$ | | 100 | 160 | 200 | pA/°C |
| Input Noise Voltage Density | $e_n$ | $f_O = 10Hz$<br>$f_O = 30Hz$<br>$f_O = 1000Hz$ | 3.5<br>3.1<br>3.0 | 3.5<br>3.1<br>3.0 | 3.8<br>3.3<br>3.2 | $nV/\sqrt{Hz}$ |
| Input Noise Current Density | $i_n$ | $f_O = 10Hz$<br>$f_O = 30Hz$<br>$f_O = 1000Hz$ | 1.7<br>1.0<br>0.4 | 1.7<br>1.0<br>0.4 | 1.7<br>1.0<br>0.4 | $pA/\sqrt{Hz}$ |
| Input Noise Voltage | $e_{np-p}$ | 0.1Hz to 10Hz | 0.08 | 0.08 | 0.09 | μVp-p |
| Slew Rate | SR | $R_L \geq 2k\Omega$ | 2.8 | 2.8 | 2.8 | V/μs |
| Gain Bandwidth Product | GBW | | 8 | 8 | 8 | MHz |

**NOTE:**
1. Input offset voltage measurements are performed by automated test equipment approximately 0.5 seconds after application of power.

*Printed with permission from Precision Monolithics Incorporated.*

## TYPICAL PERFORMANCE CHARACTERISTICS

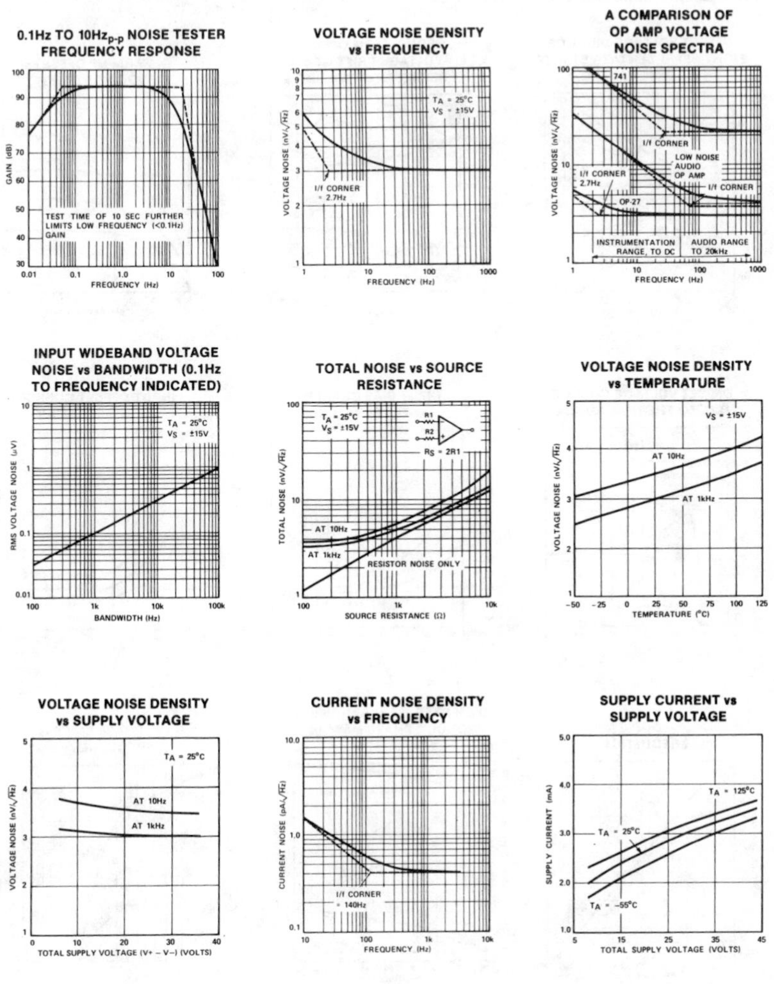


*Printed with permission from Precision Monolithics Incorporated.*

## TYPICAL PERFORMANCE CHARACTERISTICS

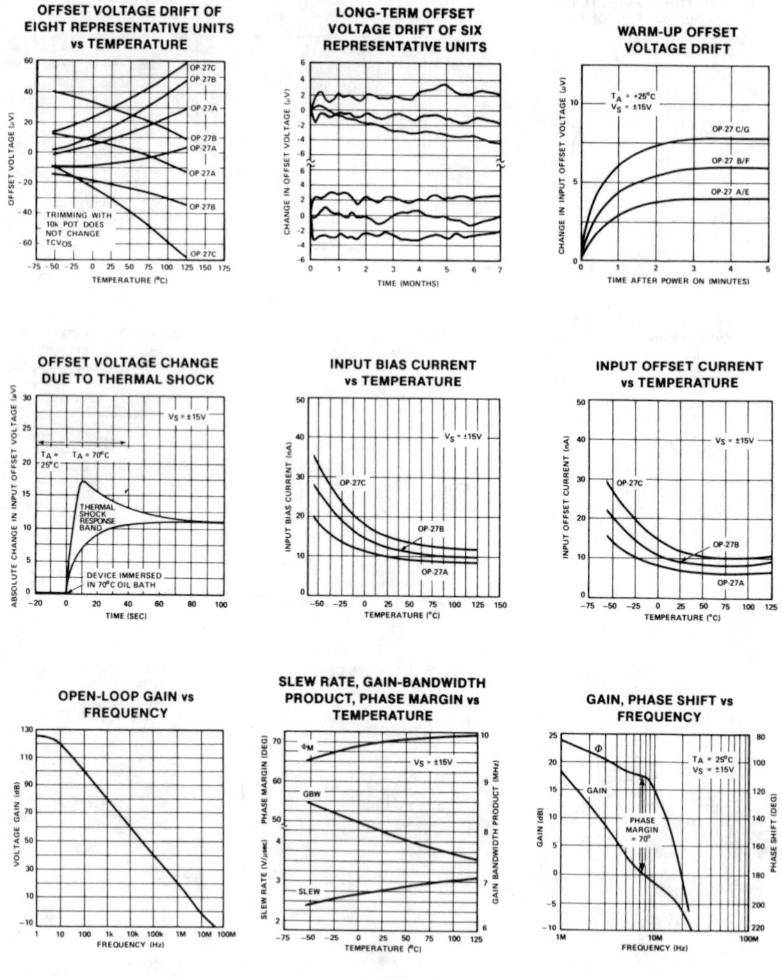


*Printed with permission from Precision Monolithics Incorporated.*

## TYPICAL PERFORMANCE CHARACTERISTICS

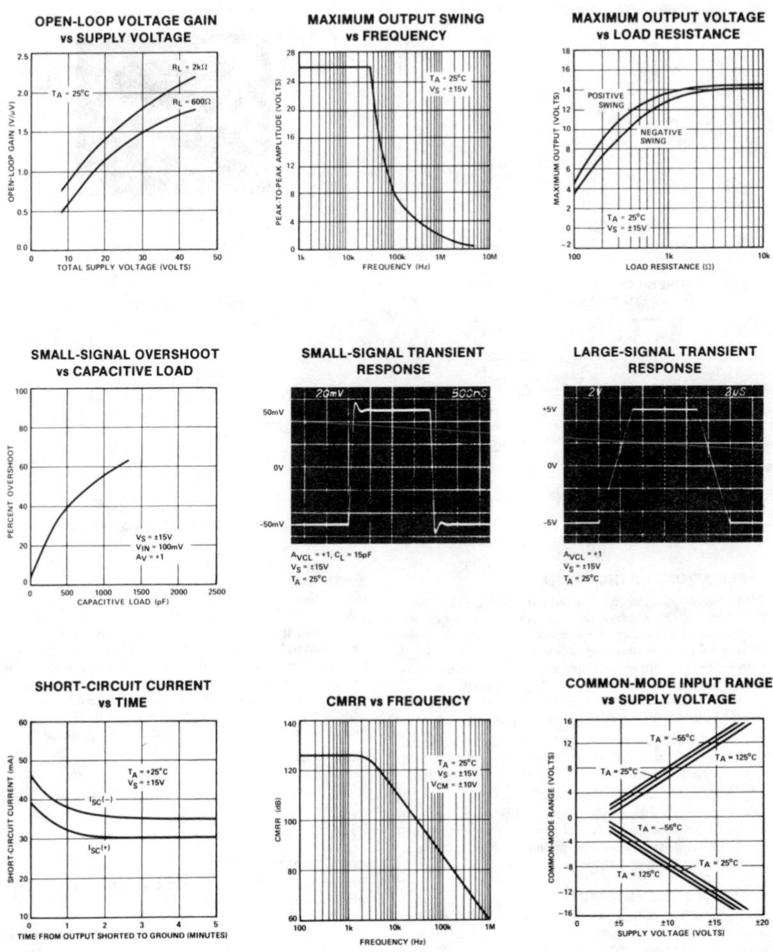


*Printed with permission from Precision Monolithics Incorporated.*

## TYPICAL PERFORMANCE CHARACTERISTICS

### VOLTAGE NOISE TEST CIRCUIT (0.1Hz-TO-10Hz)

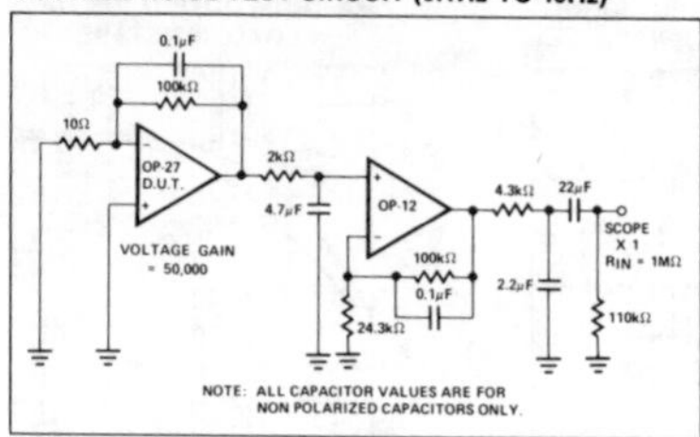


### LOW-FREQUENCY NOISE

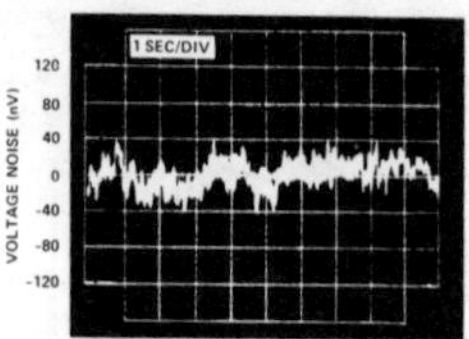


**NOTE:**
Observation time limited to 10 seconds.

### OPEN-LOOP VOLTAGE GAIN vs LOAD RESISTANCE

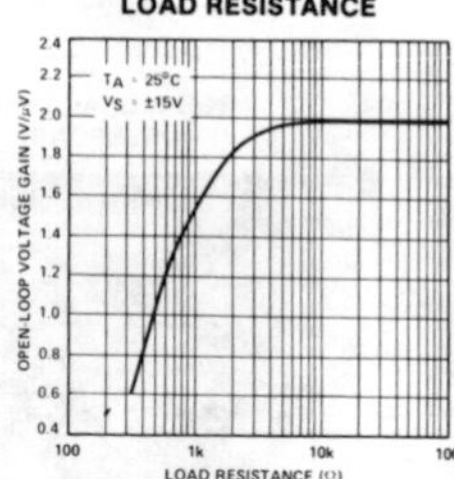


### PSRR vs FREQUENCY

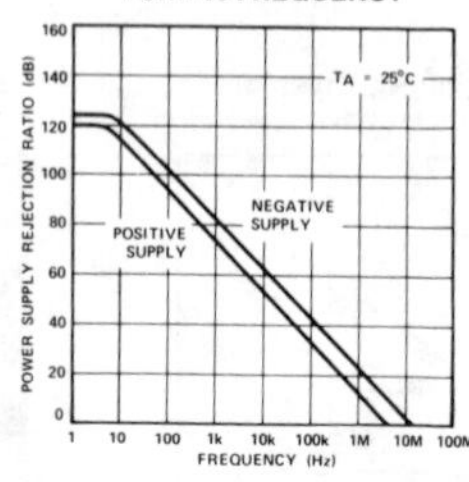


## APPLICATIONS INFORMATION

OP-27 Series units may be inserted directly into 725, OP-06, OP-07 and OP-05 sockets with or without removal of external compensation or nulling components. Additionally, the OP-27 may be fitted to unnulled 741-type sockets; however, if conventional 741 nulling circuitry is in use, it should be modified or removed to ensure correct OP-27 operation. OP-27 offset voltage may be nulled to zero (or other desired setting) using a potentiometer (see Offset Nulling Circuit).

The OP-27 provides stable operation with load capacitances of up to 2000pF and ±10V swings; larger capacitances should be decoupled with a 50Ω resistor inside the feedback loop. The OP-27 is unity-gain stable.

Thermoelectric voltages generated by dissimilar metals at the input terminal contacts can degrade the drift performance. Best operation will be obtained when both input contacts are maintained at the same temperature.

### OFFSET VOLTAGE ADJUSTMENT

The input offset voltage of the OP-27 is trimmed at wafer level. However, if further adjustment of $V_{OS}$ is necessary, a 10kΩ trim potentiometer may be used. $TCV_{OS}$ is not degraded (see Offset Nulling Circuit). Other potentiometer values from 1kΩ to 1MΩ can be used with a slight degradation (0.1 to 0.2μV/°C) of $TCV_{OS}$. Trimming to a value other than zero creates a drift of approximately ($V_{OS}$/300) μV/°C. For example, the change in $TCV_{OS}$ will be 0.33μV/°C if $V_{OS}$ is adjusted to 100μV. The offset-voltage adjustment range with a 10kΩ potentiometer is ±4mV. If smaller adjustment range is required, the nulling sensitivity can be reduced by using a smaller pot in conjuction with fixed resistors. For example, the network below will have a ±280μV adjustment range.

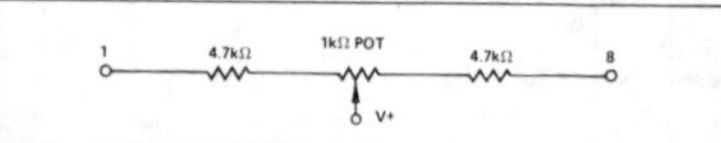


### NOISE MEASUREMENTS

To measure the 80nV peak-to-peak noise specification of the OP-27 in the 0.1Hz to 10Hz range, the following precautions must be observed:

(1) The device has to be warmed-up for at least five minutes. As shown in the warm-up drift curve, the offset voltage

*Printed with permission from Precision Monolithics Incorporated.*

typically changes 4μV due to increasing chip temperature after power-up. In the 10-second measurement interval, these temperature-induced effects can exceed tens-of-nanovolts.

(2) For similar reasons, the device has to be well-shielded from air currents. Shielding minimizes thermocouple effects.

(3) Sudden motion in the vicinity of the device can also "feed-through" to increase the observed noise.

(4) The test time to measure 0.1Hz-to-10Hz noise should not exceed 10 seconds. As shown in the noise-tester frequency-response curve, the 0.1Hz corner is defined by only one zero. The test time of 10 seconds acts as an additional zero to eliminate noise contributions from the frequency band below 0.1Hz.

(5) A noise-voltage-density test is recommended when measuring noise on a large number of units. A 10Hz noise-voltage-density measurement will correlate well with a 0.1Hz-to-10Hz peak-to-peak noise reading, since both results are determined by the white noise and the location of the 1/f corner frequency.

### UNITY-GAIN BUFFER APPLICATIONS

When $R_f \leq 100\Omega$ and the input is driven with a fast, large signal pulse (>1V), the output waveform will look as shown in the pulsed operation diagram below.

During the fast feedthrough-like portion of the output, the input protection diodes effectively short the output to the input and a current, limited only by the output short-circuit protection, will be drawn by the signal generator. With $R_f \geq 500\Omega$, the output is capable of handling the current requirements ($I_L \leq 20mA$ at 10V); the amplifier will stay in its active mode and a smooth transition will occur.

When $R_f > 2k\Omega$, a pole will be created with $R_f$ and the amplifier's input capacitance (8pF) that creates additional phase shift and reduces phase margin. A small capacitor (20 to 50pF) in parallel with $R_f$ will eliminate this problem.

### PULSED OPERATION

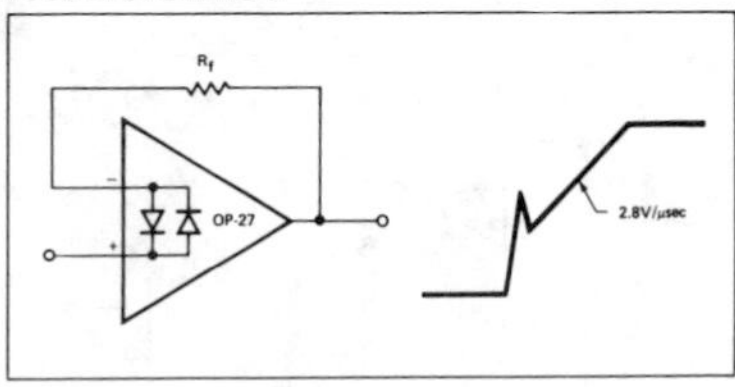


### COMMENTS ON NOISE

The OP-27 is a very low-noise monolithic op amp. The outstanding input voltage noise characteristics of the OP-27 are achieved mainly by operating the input stage at a high quiescent current. The input bias and offset currents, which would normally increase, are held to reasonable values by the input-bias-current cancellation circuit. The OP-27A/E has $I_B$ and $I_{OS}$ of only ±40nA and 35nA respectively at 25°C. This is particularly important when the input has a high source-resistance. In addition, many audio amplifier designers prefer to use direct coupling. The high $I_B$, $V_{OS}$, $TCV_{OS}$ of previous designs have made direct coupling difficult, if not impossible, to use.

Voltage noise is inversely proportional to the square-root of bias current, but current noise is proportional to the square-root of bias current. The OP-27's noise advantage disappears when high source-resistors are used. Figures 1, 2, and 3 compare OP-27 observed total noise with the noise performance of other devices in different circuit applications.

Total noise $= [(\text{Voltage noise})^2 + (\text{current noise} \times R_S)^2 + (\text{resistor noise})^2]^{1/2}$

Figure 1 shows noise-versus-source-resistance at 1000Hz. The same plot applies to wideband noise. To use this plot, just multiply the vertical scale by the square-root of the bandwidth.

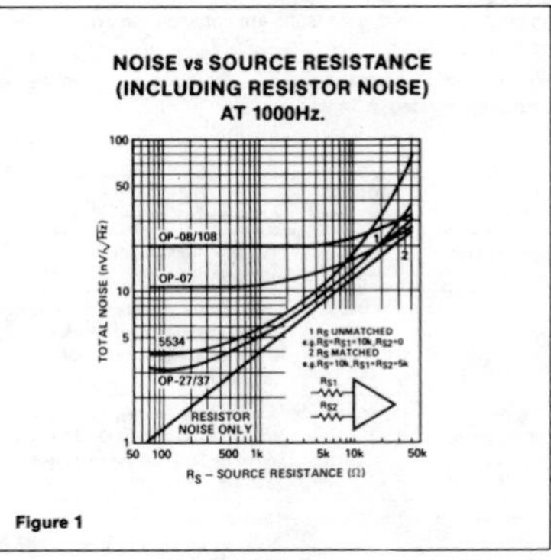


Figure 1

At $R_S < 1k\Omega$, the OP-27's low voltage noise is maintained. With $R_S > 1k\Omega$, total noise increases, but is dominated by the resistor noise rather than current or voltage noise. It is only beyond $R_S$ of 20kΩ that current noise starts to dominate. The argument can be made that current noise is not important for applications with low-to-moderate source resistances. The crossover between the OP-27 and OP-07 and OP-08 noise occurs in the 15-to-40kΩ region.

Figure 2 shows the 0.1Hz-to-10Hz peak-to-peak noise. Here the picture is less favorable; resistor noise is negligible, current noise becomes important because it is inversely proportional to the square-root of frequency. The crossover with the OP-07 occurs in the 3-to-5kΩ range depending on whether balanced or unbalanced source resistors are used (at 3kΩ the $I_B$, $I_{OS}$ error also can be three times the $V_{OS}$ spec.).

*Printed with permission from Precision Monolithics Incorporated.*

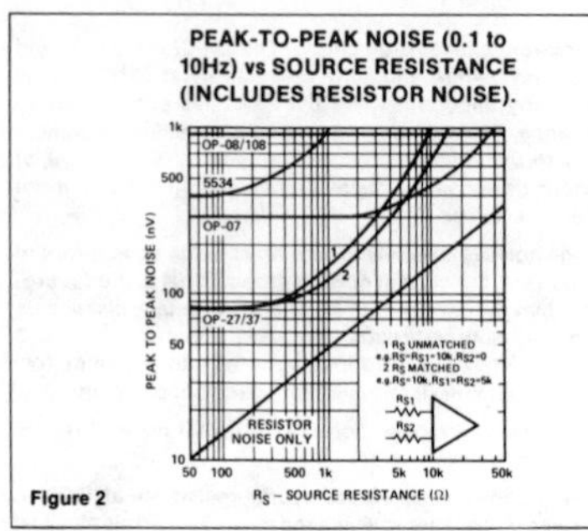


Figure 2

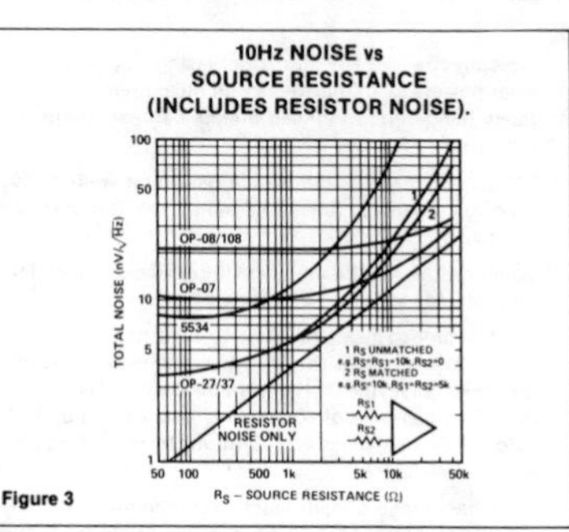


Figure 3

Therefore, for low-frequency applications, the OP-07 is better than the OP-27/37 when $R_S > 3k\Omega$. The only exception is when gain error is important. Figure 3 illustrates the 10Hz noise. As expected, the results are between the previous two figures.

For reference, typical source resistances of some signal sources are listed in Table 1.

**Table 1**

| DEVICE | SOURCE IMPEDANCE | COMMENTS |
|---|---|---|
| Strain gauge | <500Ω | Typically used in low-frequency applications. |
| Magnetic tapehead | <1500Ω | Low $I_B$ very important to reduce self-magnetization problems when direct coupling is used. OP-27 $I_B$ can be neglected. |
| Magnetic phonograph cartridges | <1500Ω | Similar need for low $I_B$ in direct coupled applications. OP-27 will not introduce any self-magnetization problem. |
| Linear variable differential transformer | <1500Ω | Used in rugged servo-feedback applications. Bandwidth of interest is 400Hz to 5kHz. |

**OPEN-LOOP GAIN**

| FREQUENCY AT: | OP-07 | OP-27 | OP-37 |
|---|---|---|---|
| 3Hz | 100dB | 124dB | 125dB |
| 10Hz | 100dB | 120dB | 125dB |
| 30Hz | 90dB | 110dB | 124dB |

For further information regarding noise calculations, see "Minimization of Noise in Op-Amp Applications", Application Note AN-15.

## AUDIO APPLICATIONS

The following applications information has been abstracted from a PMI article in the 12/20/80 issue of Electronic Design magazine and updated.

Figure 4 is an example of a phono pre-amplifier circuit using the OP-27 for $A_1$; $R_1$-$R_2$-$C_1$-$C_2$ form a very accurate RIAA network with standard component values. The popular method to accomplish RIAA phono equalization is to employ frequency-dependent feedback around a high-quality gain block. Properly chosen, an RC network can provide the three necessary time constants of 3180, 318, and 75μs.[1]

For initial equalization accuracy and stability, precision metal-film resistors and film capacitors of polystyrene or polypropylene are recommended since they have low voltage coefficients, dissipation factors, and dielectric absorption.[4] (High-K ceramic capacitors should be avoided here, though low-K ceramics—such as NPO types, which have excellent dissipation factors, and somewhat lower dielectric absorption—can be considered for small values.)

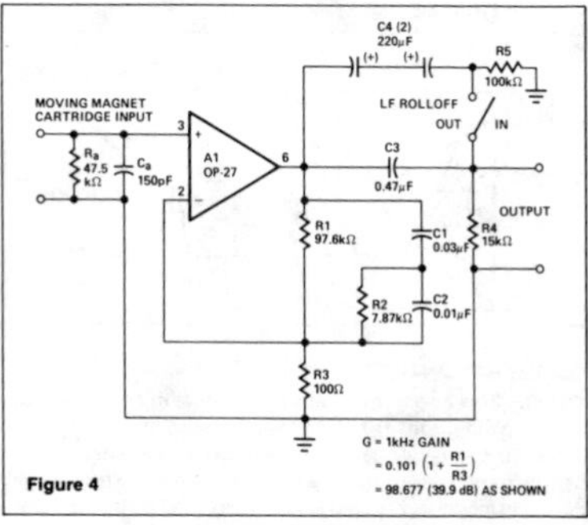


Figure 4

*Printed with permission from Precision Monolithics Incorporated.*

The OP-27 brings a $3.2nV/\sqrt{Hz}$ voltage noise and 0.45 $pA/\sqrt{Hz}$ current noise to this circuit. To minimize noise from other sources, $R_3$ is set to a value of 100Ω, which generates a voltage noise of $1.3nV/\sqrt{Hz}$. The noise increases the $3.2nV/\sqrt{Hz}$ of the amplifier by only 0.7dB. With a 1kΩ source, the circuit noise measures 63dB below a 1mV reference level, unweighted, in a 20kHz noise bandwidth.

Gain (G) of the circuit at 1kHz can be calculated by the expression:

$$G = 0.101 \left(1 + \frac{R_1}{R_3}\right)$$

For the values shown, the gain is just under 100 (or 40dB). Lower gains can be accommodated by increasing $R_3$, but gains higher than 40dB will show more equalization errors because of the 8MHz gain-bandwidth of the OP-27.

This circuit is capable of very low distortion over its entire range, generally below 0.01% at levels up to 7V rms. At 3V output levels, it will produce less than 0.03% total harmonic distortion at frequencies up to 20kHz.

Capacitor $C_3$ and resistor $R_4$ form a simple −6dB-per-octave rumble filter, with a corner at 22Hz. As an option, the switch-selected shunt capacitor $C_4$, a nonpolarized electrolytic, bypasses the low-frequency rolloff. Placing the rumble filter's high-pass action after the preamp has the desirable result of discriminating against the RIAA-amplified low-frequency noise components and pickup-produced low-frequency disturbances.

A preamplifier for NAB tape playback is similar to an RIAA phono preamp, though more gain is typically demanded, along with equalization requiring a heavy low-frequency boost. The circuit in Fig. 4 can be readily modified for tape use, as shown by Fig. 5.

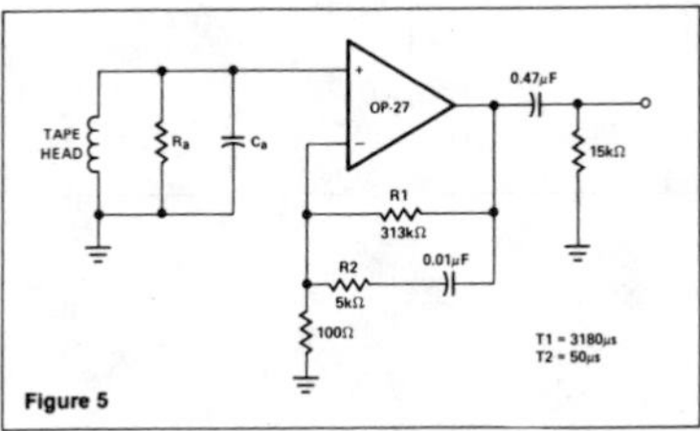


Figure 5

While the tape-equalization requirement has a flat high-frequency gain above 3kHz ($T_2 = 50\mu s$), the amplifier need not be stabilized for unity gain. The decompensated OP-37 provides a greater bandwidth and slew rate. For many applications, the idealized time constants shown may require trimming of $R_1$ and $R_2$ to optimize frequency response for nonideal tape-head performance and other factors.[5]

The network values of the configuration yield a 50dB gain at 1kHz, and the dc gain is greater than 70dB. Thus, the worst-case output offset is just over 500mV. A single 0.47μF output capacitor can block this level without affecting the dynamic range.

The tape head can be coupled directly to the amplifier input, since the worst-case bias current of 80nA with a 400mH, 100 μin. head (such as the PRB2H7K) will not be troublesome.

One potential tape-head problem is presented by amplifier bias-current transients which can magnetize a head. The OP-27 and OP-37 are free of bias-current transients upon power up or power down. However, it is always advantageous to control the speed of power supply rise and fall, to eliminate transients.

In addition, the dc resistance of the head should be carefully controlled, and preferably below 1kΩ. For this configuration, the bias-current-induced offset voltage can be greater than the 100μV maximum offset if the head resistance is not sufficiently controlled.

A simple, but effective, fixed-gain transformerless microphone preamp (Fig. 6) amplifies differential signals from low-impedance microphones by 50dB, and has an input impedance of 2kΩ. Because of the high working gain of the circuit, an OP-37 helps to preserve bandwidth, which will be 110kHz. As the OP-37 is a decompensated device (minimum stable gain of 5), a dummy resistor, $R_p$, may be necessary, if the microphone is to be unplugged. Otherwise the 100% feedback from the open input may cause the amplifier to oscillate.

Common-mode input-noise rejection will depend upon the match of the bridge-resistor ratios. Either close-tolerance (0.1%) types should be used, or $R_4$ should be trimmed for best CMRR. All resistors should be metal-film types for best stability and low noise.

Noise performance of this circuit is limited more by the input resistors $R_1$ and $R_2$ than by the op amp, as $R_1$ and $R_2$ each generate a $4nV/\sqrt{Hz}$ noise, while the op amp generates a $3.2nV/\sqrt{Hz}$ noise. The rms sum of these predominant noise sources will be about $6nV/\sqrt{Hz}$, equivalent to 0.9μV in a 20kHz noise bandwidth, or nearly 61dB below a 1mV input signal. Measurements confirm this predicted performance.

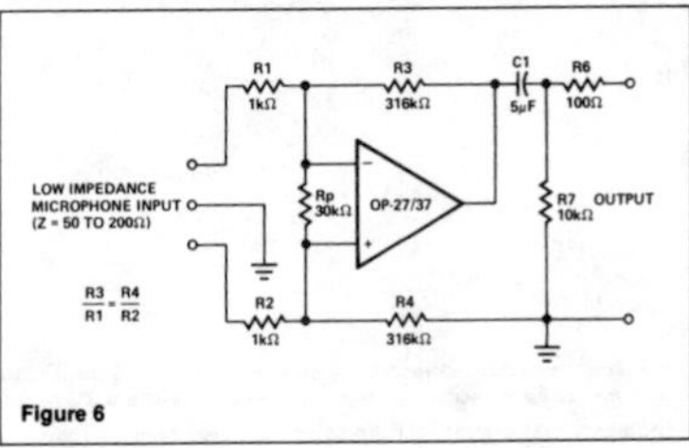


Figure 6

*Printed with permission from Precision Monolithics Incorporated.*

For applications demanding appreciably lower noise, a high-quality microphone-transformer-coupled preamp (Fig. 7) incorporates the internally-compensated OP-27. $T_1$ is a JE-115K-E 150Ω/15kΩ transformer which provides an optimum source resistance for the OP-27 device. The circuit has an overall gain of 40dB, the product of the transformer's voltage setup and the op amp's voltage gain.

*T1 = JENSEN JE-115K-E JENSEN TRANSFORMERS 10735 Burbank Blvd. N. Hollywood, CA 91601

**Figure 7**

Gain may be trimmed to other levels, if desired, by adjusting $R_2$ or $R_1$. Because of the low offset voltage of the OP-27, the output offset of this circuit will be very low, 1.7mV or less, for a 40dB gain. The typical output blocking capacitor can be eliminated in such cases, but is desirable for higher gains to eliminate switching transients.

Capacitor $C_2$ and resistor $R_2$ form a 2μs time constant in this circuit, as recommended for optimum transient response by the transformer manufacturer. With $C_2$ in use, $A_1$ must have unity-gain stability. For situations where the 2μs time constant is not necessary, $C_2$ can be deleted, allowing the faster OP-37 to be employed.

Some comment on noise is appropriate to understand the capability of this circuit. A 150Ω resistor and $R_1$ and $R_2$ gain resistors connected to a noiseless amplifier will generate 220 nV of noise in a 20kHz bandwidth, or 73dB below a 1mV reference level. Any practical amplifier can only approach this noise level; it can never exceed it. With the OP-27 and $T_1$ specified, the additional noise degradation will be close to 3.6dB (or −69.5 referenced to 1mV).

**References**


1. Lipshitz, S.P., "On RIAA Equalization Networks," *JAES*, Vol. 27, June 1979, p. 458-481.
2. Jung, W.G., *IC Op Amp Cookbook*, 2nd Ed., H.W. Sams and Company, 1980.
3. Jung, W.G., *Audio IC Op Amp Applications*, 2nd Ed., H.W. Sams and Company, 1978.
4. Jung, W.G., and Marsh, R.M., "Picking Capacitors," *Audio*, February & March, 1980.
5. Otala, M., "Feedback-Generated Phase Nonlinearity in Audio Amplifiers," London AES Convention, March 1980, preprint 1976.
6. Stout, D.F., and Kaufman, M., *Handbook of Operational Amplifier Circuit Design*, New York, McGraw Hill, 1976.


**BURN-IN CIRCUIT**

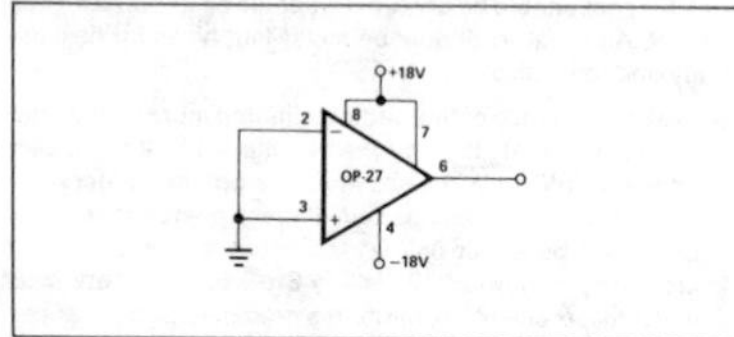


**OFFSET NULLING CIRCUIT**



PMI® Precision Monolithics Incorporated A Bourns Subsidiary
1500 SPACE PARK DR., SANTA CLARA, CA 95050 • TEL (408) 727-9222 • TWX 910-338-0218 • TLX 172 070
06840137G5M PRINTED IN USA

*Printed with permission from Precision Monolithics Incorporated.*

# Index

## D

**M**



**N**

## O

**P**



**R**



**S**

# MORE FROM SAMS

# MORE FROM SAMS

☐ **Printer Connections Bible**
*Kim G. House and Jeff Marble, The Waite Group*
At last, a book that includes extensive diagrams specifying exact wiring, DIP-switch settings and external printer details; a Jump Table of assorted printer/computer combinations; instructions on how to make your own cables; and reviews of various printers and how they function.
ISBN: 0-672-22406-2, $16.95

☐ **Electronic Instruments: Instrumentation Training Course (3rd Edition)** *Dale R. Patrick*
Technicians will learn to understand and use electronic instruments, including the latest digital instruments and computer-based systems used in industry today. Geared toward those in industrial training programs and technical schools, this book can also be used as an introduction for those with no prior electronics or math background. Beginning with a general introduction to electricity, the text moves on to real instruments, covering each piece of equipment's operation, circuitry, and uses. Topics covered include measurement circuits and primary devices, power supplies, amplifiers, digital electronics, transmitters, electronic controllers, converters, and computer-based systems.
ISBN: 0-672-22482-8, $21.95

☐ **First Book of Modern Electronics Fun Projects** *Art Salsberg*
Novice and seasoned electronics buffs will enjoy these 20 fun and practical projects. Electronics hobbyists are introduced to many project building areas including making circuit boards, audio/video projects, telephone electronics projects, security projects, building test instruments, computer projects, and home electronics projects. The necessary tools for each project accompany the step-by-step instructions, illustrations, photos, and circuit drawings.
ISBN: 0-672-22503-4, $12.95

☐ **Second Book of Modern Electronics Fun Projects** *Art Salsberg, Editor*
Have fun, save money, and sharpen your technical skills with this unique and useful collection of electronic projects. Each project has been built and tested by experts who share their electronics adventures with you. Parts lists include where-to-buy information for parts, and when available, for printed-circuit boards and complete part kits.
ISBN: 0-672-22504-2, $12.95

Look for these Sams Books at your local bookstore.

To order direct, call 800-428-SAMS or fill out the form below.

**Please send me the books whose titles and numbers I have listed below.**

________________________
________________________
________________________
________________________
________________________

Enclosed is a check or money order for $ ________
Include $2.50 postage and handling.

All states add local sales tax.

Charge my: ☐ VISA ☐ MC ☐ AE
Account No. Expiration Date ________

☐☐☐☐ ☐☐☐☐ ☐☐☐☐ ☐☐☐☐

Name *(please print)* ________________
Address ________________
City ________________
State/Zip ________________
Signature ________________
*(required for credit card purchases)*

Mail to: Howard W. Sams & Co.
Dept. DM
4300 West 62nd Street
Indianapolis, IN 46268

DC060